Smart and Sustainable Cities and Buildings

Rob Roggema • Anouk Roggema

Editors

Smart and Sustainable Cities and Buildings

 Springer

Editors
Rob Roggema
Research Centre for the Built
Environment NoorderRuimte
Hanze University of Applied
Sciences, Groningen, The Netherlands
and CITTA IDEALE, Office for
Adaptive Planning
Wageningen, The Netherlands

Anouk Roggema
AF-Wordsmith
Amsterdam, The Netherlands

ISBN 978-3-030-37637-6   ISBN 978-3-030-37635-2   (eBook)
https://doi.org/10.1007/978-3-030-37635-2

This Springer imprint is published by the registered company Springer Nature Switzerland AG.
The registered company address is: Gewerbestrasse 11, 6330 Cham, Switzerland

Steering Committee

Rob Roggema, Chair, Hanze University, Groningen, the Netherlands
Andy van den Dobbelsteen, Delft University of Technology, the Netherlands
Jeremy Gibberd, CSIR, South Africa
Wim Bakens, Secretary General of CIB, Rotterdam, the Netherlands

Scientific Board

Rob Roggema, Chair, Hanze University Groningen, the Netherlands
Andy van den Dobbelsteen, Delft University of Technology, the Netherlands
Jeremy Gibberd, Regional Coordinator, CSIR, South Africa, CIB W116
Frank Schultmann, Regional Coordinator, Karlsruhe Institute of Technology, Germany CIB W116
Vanessa Gomes, Regional Coordinator, University of Campinas, Brazil CIB W116

Organising Committee

Kiran Kashyap
Sumita Ghosh
Nimish Biloria
Rob Roggema
Stewart Monti

International Scientific Committee

Prof. Rob Roggema | Hanze University Groningen, the Netherlands (Scientific Coordinator)
Prof. Andy van den Dobbelsteen | Delft University of Technology, the Netherlands
Prof. Chris Geurts | Technical University of Eindhoven, the Netherlands
Dr. Chrisna du Plessis | University of Pretoria, South Africa
Dr. Dale Clifford | Carnegie Mellon University, United States
Dr. Dirk Conradie | CSIR, South Africa
Prof. Doris C.C.K. Kowaltowski | University of Campinas, Brazil
Prof. Greg Keeffe | Queen's University Belfast, United Kingdom
Prof. Guomin (Kevin) Zhang BEng | RMIT University, Australia
Dr. Hielkje Zijlstra | Delft University of Technology, the Netherlands
Dr. Hilde Remoy | Delft University of Technology, the Netherlands
Dr. Jeremy Gibberd | CSIR, South Africa
Dr. Karen Allacker | Katholieke Universiteit Leuven, Belgium
Prof. Dr. Maristela G. da Silva | Federal University of Espirito Santo, Brazil
Peter Teeuw | Delft University of Technology, the Netherlands
Dr. Serge Salat | CSTB, France
Prof. Tom Jefferies | Manchester Metropolitan University, United Kingdom
Dr. Vanessa Gomes | University of Campinas, Brazil
Prof. Wanglin Yan I KEIO University, Japan
Prof. John Bell | Queensland University of Technology, Australia
Dr. Nimish Biloria | University of Technology Sydney, Australia
Prof. Janis Birkeland | University of Auckland, New Zealand
Dr. Regina Bokel | Delft University of Technology, the Netherlands
Prof. Dr. Ellen van Bueren | Delft University of Technology, the Netherlands
Prof. Laurie Buys | Queensland University of Technology, Australia
Ass. Prof. Nicholas Chileshe | University of South Australia, Australia
Dr. Fidelis Emuze | Central University of Technology, South Africa Ass.
Prof. Usha Inger-Raniga | RMIT, Australia
Dr. Geci Karuri-Sebina | South African Cities Network, South Africa

Acknowledgements

We acknowledge the traditional custodians of the land on which the institution is located, the Gadigal people of the Eora Nation.

We honour and celebrate their elders and the elders of all aboriginal and Torres Strait Islander nations past and present.

Contents

The original version of this book was revised: This book was inadvertently published with an incorrect version of Chapter 9, and which was incorrectly named as Chapter 11. The book has been repaginated due to these changes. The correction to this book is available at https://doi.org/10.1007/978-3-030-37635-2_46

Part I
Design and Plan for Smart and Sustainable Cities

Chapter 1
Introduction

Rob Roggema

The Smart and Sustainable Built Environment conference 2018 is lucky to have had a large response of high-quality papers which have been presented at the conference in December 2018 in Sydney. Out of the presented papers and discussions it was clear that the traditional group of academics, interested in technologies, buildings and modelling indoor climates and energy performances is now balanced with a growing group interested in the sustainability and smartness of how to plan and design our cities and neighbourhoods. Though it has always been the ambition of SASBE it is good to see this development continuing and leading to a real broad academic community.

SASBE 2018 has paid tribute to the traditional custodians of the land. This is important to acknowledge, and not only in the way it often is practiced in Australia: with an aunty or elder that welcomes the delegates on the first morning on their traditional land. After which he or she can leave and the conference, or meeting, can really start. Especially when we speak about sustainable development it is a very Western attitude to neglect history, in particular when this history is over 40,000 years. Did you know that aboriginal people built settlements, practiced agriculture and developed smart and sensitive relationships with nature. They even made a deal with killer whales to jointly hunt for fish. After catching the fish, they were so smart to share the catch with the killer whales, who then would drive the fish into the aboriginal settlement next time. This mutually beneficial model ended the moment one of the English first settlers shot a killer whale. They never returned again. This and many other stories, exemplifying the relationship of Aboriginals with land, and nature is captured in the book 'Dark Emu' by Bruce Pascoe. A real recommended read!

R. Roggema (✉)
Research Centre for the Built Environment NoorderRuimte, Hanze University of Applied Sciences, Groningen, The Netherlands and CITTA IDEALE, Office for Adaptive Planning, Wageningen, The Netherlands
e-mail: r.e.roggema@pl.hanze.nl

© Springer Nature Switzerland AG 2020
R. Roggema, A. Roggema (eds.), *Smart and Sustainable Cities and Buildings*,
https://doi.org/10.1007/978-3-030-37635-2_1

This also gave reason to develop the conference as a real mutual experience. The welcome is more than superficial and will give all delegates an impression of traditional Aboriginal thinking. We are extremely happy that Chels Marshall will induct us in some of the basic rituals and thought leadership. Throughout the conference there was Aboriginal food and reminders of the traditional values and sustainability of treating the land and our built environment. This has been an experience all who were part of the conference will always remember. Not only because it has been an impressive experience, but also because it will influence everyone's daily working life from then on.

In this book the papers that were presented during the conference appear, categorized along basic themes: the resilient city, urbanity, smart cities, urban ecology, space and place, and inclusivity in (this) part One, and energy, educational buildings, confort, building design, construction and performance in the sister part Two. Together they respresent an overview over the most recent research into Smart and Sustainable Built Environments, at the scale of individual buildings and cities alike.

Chapter 2
Towards Integration of Smart and Sustainable Cities

Rob Roggema

Abstract In the current academic discourse, there seems to be a dichotomy between smart and sustainable in the built environment. It is treated as separate research fields incidentally connected and the question is how to these worlds could be brought together. In this introductory chapter the proposition is to link smart and sustainable through design, people and data. After reviewing current literature, a Smart Urban Model is presented in which the four components of a smart and sustainable city are an equal part: smart, sustainable, spatial and human. In six examples, one from Sydney, Australia and five from the province of Groningen, the Netherlands, the new model is illustrated. This chapter must be seen as a first start of the discussion only and does not pretend to present the final version of the magical trick to integrate smart and sustainable. It requires further conversations, exploratory research and user-led design processes to experiment with real projects and cities in order to make school and identify what successful smart and sustainable cities can be.

Keywords Smart city · Smart urbanism · Sustainable urban development · Resilient city · Participative planning · Data

2.1 Introduction

By the year 2050 an estimated 2.6 billion people will have moved to or have been born in urban environments. Of these billions of residents, two-thirds will live in Asia or Africa. Many of these cities, should we not act, will emerge out of or swallow-up squatter settlements (New York Times and Shell Oil 2014). Amenities,

R. Roggema (✉)
Research Centre for the Built Environment NoorderRuimte, Hanze University of Applied Sciences, Groningen, The Netherlands and CITTA IDEALE, Office for Adaptive Planning, Wageningen, The Netherlands
e-mail: r.e.roggema@pl.hanze.nl; rob@cittaideale.eu

© Springer Nature Switzerland AG 2020
R. Roggema, A. Roggema (eds.), *Smart and Sustainable Cities and Buildings*,
https://doi.org/10.1007/978-3-030-37635-2_2

such as water, sewer, transportation, electricity, telecommunications, housing, health care or education, have not yet been integrated during the growth of these cities hence will have to be built from the ground up. The discourse around Smart Cities, in contrast, largely focuses on the immediate future and seems to focus its interests on places that are already known and up and functioning. Therefore, to balance the attention between existing, known urban areas and the unknown, novel ones, the focus of Smart City research should shift towards the processes of urban transformation on the longer term in novel urban places, because current smart city explorations will only marginally inform understanding of the real cities of tomorrow. The pace of change is forecast to be so swift that research needs to go beyond current and locally 'sold' technologies. Instead, an intelligent discussion starts with the question which cities we want in the future and whether and how smart urban technologies, and urban design and urbanism, are likely to provide them (Glasmeier and Christopherson 2015). The premise is that Smart + Sustainable + User led design will lead to a more resilient city.

In this chapter after a review of the current smart urbanism discourse, a 'Smart Urban Model' is presented, followed by several prospective examples of this model. The chapter ends with recommendations and conclusions.

2.2   Smart Urbanism

The abundance of distributed sensors, chips with which the infrastructure, streets and buildings are fitted, as well as in the numerous electronic devices of its inhabitants make of the city an intelligent being. The smart city is activated at millions of points, thanks to information and communications technology. This intelligence is profoundly spatial, since it follows the topography of the networks of streets and buildings and the movement of vehicles and its inhabitants, hence is able to produce a map of urban activity in real time (Picot 2015). However, the critique about the inadequacy of the mechanical technological approach to smart cities is heard from different sides (Greenfield 2013; Sennett 2012; Koolhaas 2014). The idea that you put sensors out, measure everything, and on that basis make decisions, is biased because all data is crafted (Van Timmeren et al. 2015). The tech-driven adagio 'give us your data and we'll give you a techno utopia', is impossible to make true. Besides the technology-driven paradigm, a human-driven approach (Kummitha and Crutzen 2017) is equally important. Next to Cyborg City, in which everything is managed, the spontaneous collaborative city, in which 'nothing' is managed exists. Both require design, creativity and spontaneity as well as coordinative power (Picot 2015). Sim City (Terzano and Morckel 2016) should not only be seen as a managed calculation, but even so as a creative process of city design.

The Smart City should emerge as an integrated, sustainable and efficient city with a high 'Quality of Life' for its residents that aims to address urban challenges by (Mosannenzadeh and Vettorato 2014):

1. Application of ICT in its infrastructure and services,
2. Collaboration between its key stakeholders, and
3. Integration of its main domains and investment in social capital.

The underlying promise is that more information will improve the experience of urban social life and lead to the creation of many useful and efficient services (Rabari and Storper 2015). Urban systems in themselves have been complex in terms of their operation, management, assessment, and to plan for in line with the vision of sustainability. Here comes the role of ICT into play, given its foundation on the application of complexity sciences to urban systems and problems (Batty et al. 2012a; Bibri and Krogstie 2016). The development of the Smart City with its various faces has come to the fore in recent years as a promising response to the same challenge of linking smart and sustainable (e.g. Al Nuaimi et al. 2015; Batty et al. 2012b; Neirotti et al. 2014). Smart solutions have been developed for sustainability, optimizing efficiency in urban systems, and enhancing the quality of life of citizens. The fundamental question is, whether that promise is one that is made to everyone. Is the conception of the 'smart city' inclusive or excludes it important groups in society, by the very nature of the data it relies upon?

The interlinked development of sustainability awareness, urban growth, and technological development have recently converged under what is labelled 'smart sustainable cities' (Höjer and Wangel 2015). A 'smart sustainable city', although not always explicitly discussed, is used to denote a city that is supported by a pervasive presence and massive use of advanced ICT, which, in connection with intricately interrelated urban domains and systems, enables cities to become more sustainable. In more detail, it can be described as 'a social fabric made of a complex set of networks of relations between various synergistic clusters of urban entities that, in taking a holistic and systemic approach converge on a common approach into using and applying smart technologies that enable to create, disseminate, and to mainstream solutions and methods that help provide a fertile environment conducive to improving the contribution to the goals of sustainable development' (Bibri and Krogstie 2017). Smart sustainable cities entail thinking about and conceiving of urban environments as constellations across spatial and temporal scales that are networked in multiple ways to provide continuous data coming from urban domains, employing pervasive sensing, processing, and networking technologies, in order to monitor, understand, and analyse how cities function and can be managed so as to guide and direct their development towards sustainability (Bibri and Krogstie 2017).

However, smart urbanism introducing the spatial design perspective, goes beyond the mechanical. Urbanism aims to deliver a city which provides all its basic functions (shelter, welfare, prosperity, social exchange) and shape (i.e. design) it in a way its citizens are serviced and enjoy or consume a convenient life in a sustainable way. It is 'a powerful integrative and action-oriented body of thought on cities that emphasises their particular histories, the social composition of cities, analyses the resources it takes to 'run' a city, provides insights into the intricate ways in which design, politics and business interrelate, and helps to think of the institutional formats and practices that can help deliver on the transition needed. The future calls for smart urbanism rather than smart cities.' (Hajer and Dassen 2014).

The convenient city (Roggema 2019) provides good houses, a clean, healthy and safe environment, access to resources of clean water, renewable energy and healthy food, social interaction, healthy interactive environments, resiliency/low vulnerability for climate impacts, (intelligent) mobility that guides traffic, mode shifts to new tech and innovative transport (autonomous and air vehicles) and arranges collaboration between the constituencies that shape the city. Smart Urbanism integrates technology, knowledge, governance, citizens, and business hence represent a multidisciplinary field, constantly shaped by advancements in technology and urban development (Angelidou 2015). The key smart applications enabled by big data analytics and context–aware computing include smart transport, smart energy, smart environment, smart planning, smart design, smart grid, smart traffic, smart education, smart healthcare, and smart safety (Bibri and Krogstie 2016). Big data analytics and context–aware computing and what these entail in terms of digital sensing technologies, cloud computing infrastructures, middleware architectures, and wireless communication networks, will be the dominant mode of monitoring, understanding, analysing, assessing, operating, organizing, and planning smart (and) sustainable cities to improve their contribution to the goals of sustainable development (Bibri and Krogstie 2017).

The big difference for urban planners is they suddenly have access to real time data, which may alternate and differ over time. City makers and urbanists suddenly have to deal with the option of emerging events (Dosse 2010) and spontaneous developments rather than a determined program or future. This requires in strategic design of temporary uses being it events, temporary urbanism (Bishop and Williams 2012) and including voids and redundancy in the urban fabric (Roggema 2018). The way energy generation and storage can be balanced with real time demand and usage in Smart Grids (Obinna et al. 2017), or how the Living PlanIT in Portugal (Carvalho et al. 2014) and ReGen Villages in the Netherlands (Ehrlich et al. 2015) are monitoring, adapting and closing environmental flow cycles are early examples of these urban design applications of combinations of the virtual and physical city.

No matter what digital input cities undergo, in its essence the city remains the same. The design of the city may be inspired by the fluctuating insights data deliver, and new gadgets and shared bike systems may flock the city, meanwhile the urban form has basically not changed, its physical components and purpose remain. New cities such as Songdo, Masdar or existing cities such as Rio de Janeiro or Barcelona, all dubbed smart cities, do not look any different than the cities before the digital revolution. One may speak about the rise of a new planning paradigm of the intelligent city, other than virtual spaces (Ishida and Isbister 2000) and digital ecosystems enhancing innovation (Komninos 2015) are not distinguished, leaving current physical urban structures intact. The breadth of street widths, the suite of urban block sizes, these have not changed because of digitisation. The city still consists of a street and a façade, no matter whether these are private, public or when a virtual space over cities is created.

There is one fundamental new opportunity for urban life. The way the city can be 3D-mapped (Picot 2015) is new and allows us to access data and info in real time about mobility services, entertainment, food/restaurants, the environmental quality and leisure of places that are near us but not physically visible. This provides us with a convenience tool that makes our urban life potentially of a better quality. We can be informed and make better decisions on where what is available in real time, but even this doesn't fundamentally change the physical appearance of the city.

Therefore, smart urbanism could re-emphasise urban planning principles (Hajer and Dassen 2014):

1. To not limit growth of wealth yet at the same time minimise the use of resources in a socially just and safe way;
2. To present a strong persuasive story, 'beyond the smart gadget', for the use of smart technologies, supporting positive social reform and bringing about urban resilience;
3. To use urban metabolism as the central urban planning framework on which to base urban performance decisions on, and focus on the potential transformations, by increasing urban redundancy by including spatial voids in the city to use whenever suitable (Roggema 2018);
4. To develop urban infrastructure that provides shelter, water, energy and contact in a way that is (hyper)localised, small-scale and off grid;
5. To establish a symbiosis of smart technologies, social innovation, and business models through design;
6. To create the technological environment enhancing open politics of continuous learning, using the intelligence of the 'energetic citizen' (Hajer 2011) as part of the search for solutions. Establish a digital democracy and participatory urban planning using urban living labs (Steen and van Bueren 2017);
7. To practice the urbanism of transplantation, searching for the suitable conditionalities to adapt, correct, adopt and create add-ons to the city and transplant solutions in matching contexts.

The urban consumer turns into a smart citizen, 'prosuming' in an intelligent, agonistic and creative way (McLean and Roggema 2019) while making use of interoperable and open data sources (Van Timmeren et al. 2015). Smart urbanism in practice could work in a quadruple helix model in which:

- innovative companies, investing in developing new concepts and products,
- academia, participating with the brightest minds,
- the government, allowing the novelty to emerge and be tested in the city, and
- the urban prosumer, being the primary user and tests the prototypes come together in an urban ecosystem of exchange, creativily finding new ways of co-design and co-development.

2.3 Smart Urban Model

The traditional model of the city, which is founded on the idea of the city as being a stable or constant structure, is rapidly changing, so too are the associated planning approaches in response to the emerging shifts brought by computing and ICT. These are under-pinned by their foundation on complexity- and data-sciences: from focusing on physical and spatial development to including broader principles (e.g. sustainability) and relying on big data analytics, context information processing, intelligence functions, and simulation models, and what these entail in terms of sensing, computing, data processing, and wireless networking technologies (Bibri and Krogstie 2017).

An integrated smart urban model works toward the following assets (Angelidou 2015):

- Advancement of human capital: citizen empowerment (informed, educated, and participatory citizens), intellectual capital and knowledge creation (Aurigi 2006; Komninos 2009; Liugailaité -Radzvickiené and Jucevic̆ius 2012; Neves 2009; Ratti and Townsend 2011).
- Advancement of social capital: social sustainability and digital inclusion (Batty et al. 2012b; Caragliu et al. 2009; Hodgkinson 2011; Liugailaité -Radzvickiené and Jucevic̆ius 2012).
- Behavioural change – sense of agency and meaning (i.e. the feeling that we are all owners and equally responsible for our city) (Frenchman et al. 2011; Townsend et al. 2010).
- Humane approach: Technology responsive to needs, skills and interests of users, respecting their diversity and individuality (Bria 2012; Lind 2012; Roche et al. 2012; Streitz 2011).

In the smart cities discourse, the conviction that collecting historical data sets will provide the insights for planning and design is often believed in (see e.g. Rathore et al. 2016). However, this is risky as even when big data is collected, for instance through abundant placement of sensors and IoT practice, all relevant data can never be collected and is always subject to (biased) interpretation and choice-making. Especially subjective data such as emotions, values and moods of people are difficult to collect and may change considerably over time. Also, the data of the past does not give any certainty about the future. Especially factors such as climate change, migration and economic change will influence the sort of city that is required in the future. Moreover, the use of data is generally sectoral which might make it useful, but urbanism is integrative for reasons that in the city all different factors are present in conjunction with each other at every given time meanwhile influenced by human beings. For instance, a sectoral approach could solve the traffic problem, while as a result, other problems, such as the loss of biodiversity or water quantities might increase. Finally, design is science with creativity build into its approaches. Creativity implies emergence of unexpected combinations, integration of problems, and the employment of novel propositions.

Paraphrasing Jane Jacobs (Jacobs 1961) social ideas (and laws) shape private investments, which shapes cities, today's planning and procurement practices do not explicitly recognize the value of the smart city vision, and therefore are not shaping the financial instruments to deliver it (Robinson undated). Urban life should come first, then urban place, before thinking technology.

As is often the case with technological change, the producers of the technologies cannot dream users into existence, but instead uptake requires learning by doing through collaboration and risk sharing. The degree of know-how and collateral resources required to use smart city interventions, assuming that everyone owns a smart phone and knows how to operate it at maximum performance, is often taken for granted, but technology audits are necessary to reveal just how flexible, usable and accessible these technology designs are (Offenhuber 2015). Beyond making cities more liveable because their inner political workings are more accessible, local organisations are building tools to make 'sensibility' real, using devices such as Carlo Ratti's City Lab's algorithm, which integrates crowd sourced data from cell phone users who, for instance, are seeking to track night life hot spots. But how much of the smart city research is being directed toward questions of groups in society unlikely to be consulted or enabled to use the sophisticated facets of a cell phone? What of the elderly, the disabled, the economically and socially isolated (Glasmeier and Christopherson 2015)? If the goal is to produce morally balanced and socially aware smart city strategies, then stakeholder engagement is crucial. Stakeholder engagement, or better: user-led design processes, can provide valuable insights about sustainability assets and needs of the city, increase public acceptance of the smart city venture and elevate the 'smartness' of the city to a whole new level, leveraging human capital and collective intelligence (Angelidou 2014). Therefore, the role of technologies in smart cities should lie in enabling sustainable development of cities (Bifulco et al. 2016), not in the new technology as an end in itself (Marsal-Llacuna and Segal 2016). Ultimately, a city that is not sustainable is not really "smart" (Ahvenniemi et al. 2017). Although smart city technology investments are mainly comprised of upgrades rather than true innovations, they potentially offer access to information on local conditions. They can afford communities and interest groups the opportunity to identify negative conditions and the potential to improve the urban experience (Glasmeier and Christopherson 2015). Citizen movements have demonstrated the ability to successfully adopt and adapt the core of smart city technologies to engage in public debate and to advocate for urban improvements (Glasmeier and Christopherson 2015).

The Smart Urban Model comprises of four perspectives that need to be all in the mix and in balance with each other: smart, sustainable, spatial and human (Fig. 2.1). Only then an inclusive city can be developed that is sustainable and supported by technology, is evident. The urban planning process, from abstract-larger scale, to implementation-smaller scale, should therefore be linked with information attributers of smart cities: prosumers (both providers of data and products, even so being end-users), services, infrastructure and data (Anthopoulos and Vakali 2012).

Fig. 2.1 The smart urban model

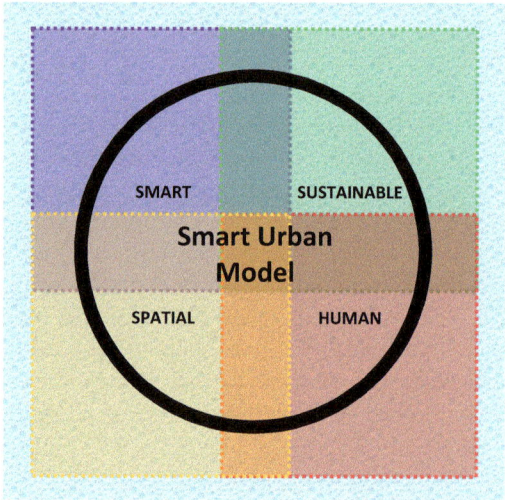

2.4 Imagine

In this paragraph several examples will be presented that have been developed or are being developed that illustrate working according the principles of the New Smart Urban Model. These examples all show, though in different settings and configuration the four principles of the model. They are:

1. Smart. These plans make use of advanced ways of monitoring, data-collection and feedback, using real time sensing, which makes instant feedback and assessment possible and technologies are supporting the systems in the plan;
2. Sustainable. In all these examples optimized solutions for the Food-Energy-Water Nexus are presented, in which no waste is produced, rather new resources are being delivered, minimal resource is used, or the plans show regenerative principles (Cole 2012; Cole et al. 2012; Du Plessis 2012; Gabel undated; Girardet 2013; Mang and Reed 2012; Robinson and Cole 2015) in which they produce and deliver more clean and usable output to the environment than is used and processed, both in a quantitative and qualitative way;
3. Engage. In these projects, engagement is not seen as a necessary box to tick, but as a meaningful process to increase sustainability and resident's satisfaction. Hereto the local citizens are seen as co-producers of information, visions, plans, products and expertise hence are partnering from the early stages of a project as directors of their own future. End-users are one quarter of the quadruple helix together with industries and business, government and academia. The way this type of process can be ideally facilitated is through design-led process;
4. Design. In all these projects design is used as an enabling process in which the strengths of design can be fully flashed out. It is able to illustrate and visualise ideas, unrealised worlds and bring these alive, it is also a tool to facilitate

communication about how different groups, within the community think about the future and, last but not least, as a mean to create an environment beautiful.

2.4.1 M-NEX Western Sydney

The desire to create a green healthy city that is sustainable, even under sometimes harsh climate conditions is at the basis of the Western City Parkland development. Planned around and in the vicinity of the new airport of Badgerys Creek, in the aerotropolis (Kasarda and Lindsey 2012), approximately one million new inhabitants will live, an abundance of jobs is expected, and an extensive agricultural cluster is foreseen (Commonwealth of Australia 2018). In the midst of this, the Sydney Science Park will be developed, a sustainable new precinct for residential, education and science, commercial and ecology, with a special attention to the infrastructure systems of water, energy and waste. Within the Masterplan a primer test-site is designed to test out the integration of the food, energy and water systems. Together with the local stakeholders, such as academia, research institute, developer, and government (residents do not yet exist), a data driven adaptive design is conceived in which the sensored data will give insights in the performance of local recycling, reuse, and regeneration of these infrastructural systems, while continuously operating as an experimental site for education projects, student research and citizen involvement. The idea is to create an autonomous food producing and water cleaning landscape (Fig. 2.2) on a slight slope towards the creek, driven by locally generated renewable energy sources, cleaning waste-, rain-, and stormwater in subsequent productive landscapes: orchard and vineyard, fishponds, greenhouses and crops and herbs will step-by-step clean the water and produce a range of crops, such as fruit, bees and honey in the orchard, fish in the ponds, tomatoes, pepperoni, in the greenhouses and lettuce, legumes, lettuce, cale and herbs on the field. Out of the monitoring the planting and systems can be immediately understood, and if desirable replicated on site or elsewhere in the Science Park, or at scale of the entire Western Sydney Parkland. As a first step the entire system is foreseen to be flipped 90° and scaled up on the N-S slope along the creek.

2.4.2 Foodscape Groningen

The recent analysis published in the Lancet on the implications of our global food system and the ways food need to be produced to stay within the planetary boundaries (Willett et al. 2019) has shown that in different continents different changes to the diets are required. The new diet for the Netherlands has been derived for the regional context (Fig. 2.3). In the project Foodscape Groningen, a design investigation will be undertaken on the crops that need to be grown in the Dutch northern province context to be able to cook this diet, what the nutritional values are and what

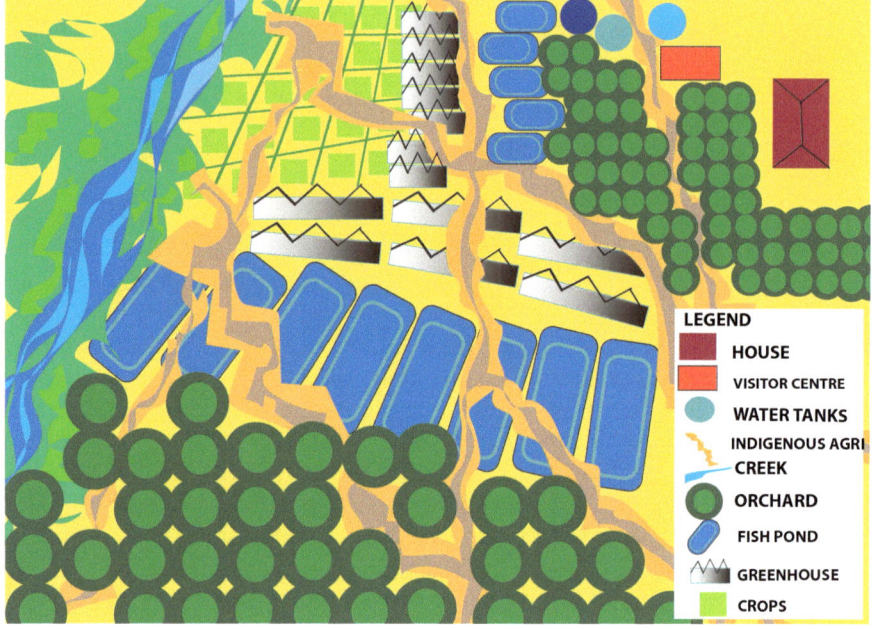

Fig. 2.2 Integrated food-energy-water nexus in western Sydney

kind of dishes can be created. During the growth of the crops the amount and types of crops will be monitored in relation to the soil quality, climate change and weather changes. The plan will be co-designed in a collaboration between local chefs, who will design the new menu and dishes, local growers, who will grow the crops, and academics who will sense the metabolism and a government body which will analyse the nutrient values, qualities and uptake. The crops will be used to cook and eat a dinner with it together.

2.4.3   Aquaponic Wall

One of the technologies to increase productivity of food production in confined urban environments is aquaponics (Somerville et al. 2014; Pollard et al. 2017). Within the Hanze University of Applied Sciences Groningen an Aquaponic Wall (Fig. 2.4) will be built in a co-designed project with the Hanze University-Facility Management, industries, such as design offices and builders/constructors, and the student community. This wall will produce food, clean water, harvest fish, and will be sensored for measuring the growth of the produce in quality and size, the quality of the water in the system and its environment (temperature, humidity, light, air quality). This will give insights about the an aquaponic system in these conditions and the factors determining the output.

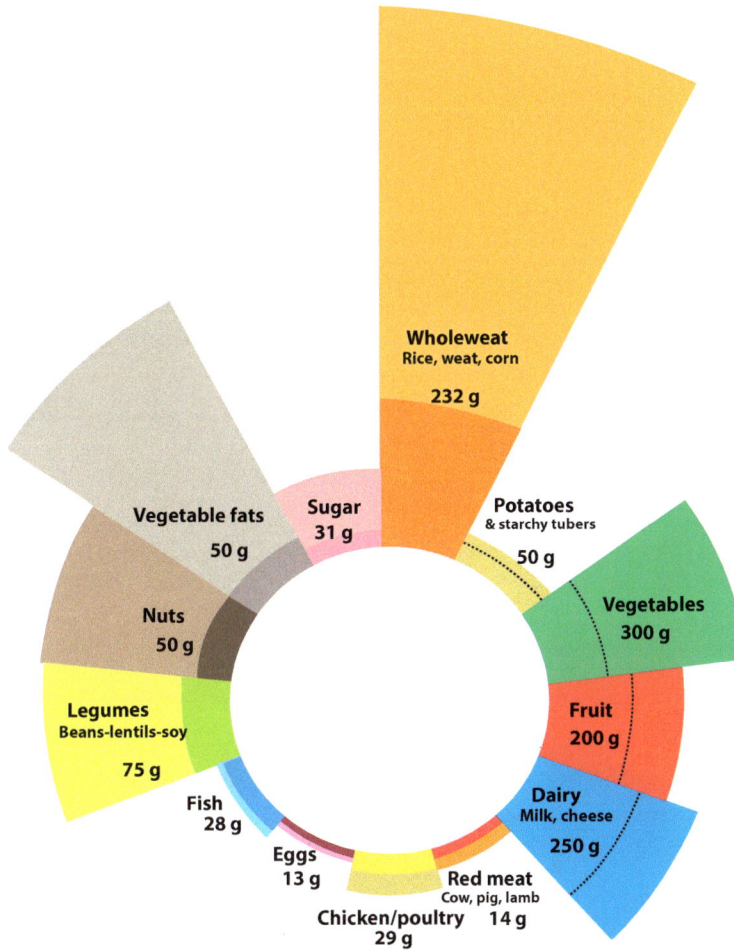

Fig. 2.3 Dutch new diet, based on the global assessment (after: De Volkskrant, 17 January 2019)

2.4.4   Climate Neighbourhood

Many projects related to climate adaptation have been developed in many cities around the globe. Most of these relate to water, have been instigated by councils and were set up with a technological eye. In Europapark Groningen a climate neighbourhood is about to be developed, which is thrived by the local community. The residents started an initiative for greening their urban environment and succeeded in attracting support from the local government and academia. The project is designing a system of green and water throughout the area, mutually connected and operating as a living system. Aspects of the spatial vision were brainstormed

Fig. 2.4 The aquaponic wall in the Hanze University (image: Alex van Spyk)

under guidance of the local residents (Fig. 2.5), and bound together in an overall plan with four pillars:

(a) Ground: breaking concrete reduces street- and garden pavements and redesign these as green spaces and gardens;
(b) Façade: greening, and hanging gardens, such as food-walls, or aquaponics;
(c) Roof: gardens, FoodRoofs (Roggema 2017), eventually with aquaponics;
(d) Climate-gardens: based on the local bureau of meteorology-scenarios (KNMI 2014) four climate gardens represent a scenario each to test out future possible climates. In these gardens the climate is simulated and planted accordingly, fully monitored in their growth, success-plants, and required adjustments. This way the climate neighborhood can prepare for any future climate and is an experimental example which findings can be used throughout the Netherlands.

2.4.5 Positive Energy Districts

In the Making City project so-called Positive Energy Districts will be designed, constructed and monitored. Together with the local community appropriate measures and investments are discussed and implemented, after which the new system

Fig. 2.5 Planning with residents: Climate neighbourhood Europapark Groningen

will be sensored. The amount of electricity will be measured to ensure the district is energy-positive, but also to guarantee the residents they will receive the financial benefits of their in- and around-house generation of energy and how energy can be exchanged between several uses within the neighbourhood to be most efficient. This H2020 project is a collaboration between resident groups, government, academia and industries.

2.4.6 Beyond Circularity Loskade

The old industrial site of the sugar factory in the western part of the city of Groningen, the Netherlands is subject to postponed urban design and development due to the GFC (Meissner 2017). This meant the expectation was that this site would not be developed before 2030. The implications of this decision are large and reach well into the future. The municipality of Groningen decided to create a regulation-low zone, which meant for instance that residential development would be possible, but people could only inhabit houses for periods shorter than 6 months. Developer Van Wijnen jumped in this opportunity to create a small neighbourhood 'De Loskade' (Van Wijnen 2017) where could be experimented with circularity of technical solutions in the houses and apartments as well as circular principles in the public space (Fig. 2.6). Because inhabitants will move in and out quite

Fig. 2.6 The loading loskade just after realisation in August 2019

frequently, additions and adjustments of the houses themselves and the local urban environment can have a fast lead time as well, exquisitely suitable to investigate how circular the area really is and increase circularity over the 10 years of operation. A network of sensors will provide the intelligence to constantly adjust the built form in relation to customer satisfaction and daily use and sustainability against the background of the flux of weather change and long-term climate change.

2.5 Conclusion

A smart city is nothing new. Civilisations have always used the appropriate and available technologies to shape their cities. However, due to the IoT and sensing technology the possibility of an integrative collaboration and continuous testing and improving the quality of the urban environments is novel. The density of data allows to think laterally and use information in real time, and hence influencing, managing and directing daily uses in the city, such as crowd control, traffic management or water systems manipulation has come within reach. As illustrated in the exemplary projects, this can make the city both more convenient to use, but also healthier and more resilient.

A smart city is first and foremost a city, while smartness, gained by cyber-physical intelligence and services, is 'just' another urban asset, which either improves/automates typical functions (transportation, waste management, etc.) or generates jobs and increases citizen satisfaction (from traffic awareness, energy efficiency, etc.) (Anthopoulos 2017). On the one hand side the role of smart technologies is to make our lives more convenient, while on the other hand it provides the tools to become more resilient.

There are several uncertainties whether this will be effectuated and successful. It is most likely and easy to believe the Homo Ludens (Huizinga 1938) will take up novel smart applications that will make his life easier and more joyful. Most probably, the market for smart gadgets will continue to emerge. More uncertain is whether the smart city movement will be able to enforce the implementation of the urban infrastructure needed for the new urban population for a fraction of the costs of current infrastructure. Secondly, the question is if the smart city will it be able to deliver on the promise to create the smart technology for the eco-efficiency needed for cities to really become resilient?

Apart from these uncertainties but directly related to the development and the promise of smart cities, some big questions should also be addressed. Not the least because these questions are an inevitable part of good urbanism, so smart urbanism should, as a self-evident given, contribute to finding solutions for these big questions:

- Will smart cities help to control climate change and keep the earth below a reasonable rise in temperature?
- Can smart city technologies play a role in moderating rapid population growth at global level?

- Is smart urbanism capable of preventing large scale migration, of which a large amount is caused by climate impacts?
- Will the smart city promise contribute also to social justice and equity at world scale?
- Could the smart city provide sufficient healthy food for everyone?
- Could it bring corruption-free democracy everywhere?

Urbanism, not even smart urbanism will instantly pull the switch and aim to solve these and other big issues. But at the same time, it would be a matter of negligence if smart thinking, with the availability of all algorithms, big data and the Internet of Things would not try to make big changes in these fields. This way a smart city should be a humanitarian effort, bringing a better quality of life for the all its citizens, rich, poor, displaced or newly arrived.

References

Ahvenniemi H, Huovila A, Pinto-Seppä I, Airaksinen M (2017) What are the differences between sustainable and smart cities? Cities 60:234–245

Al Nuaimi E, Al Neyadi H, Nader M, Al-Jaroodi J (2015) Applications of bigdata to smart cities. J Internet Serv Appl 6(25):1–15

Angelidou M (2014) Smart city policies: a spatial approach. Cities 41:S3–S11

Angelidou M (2015) Smart cities: a conjuncture of four forces. Cities 47:95–106

Anthopoulos L (2017) Smart utopia VS smart reality: learning by experience from 10 smart city cases. Cities 63:128–148

Anthopoulos LG, Vakali A (2012) Urban planning and smart cities: interrelations and reciprocities. In: Álvarez F et al (eds) FIA 2012, LNCS 7281, pp 178–189

Aurigi A (2006) New technologies, same dilemmas: Policy and design issues for the augmented city. J Urban Technol 13(3):5–28. https://doi.org/10.1080/10630730601145989

Batty M, Axhausen K, Fosca G, Pozdnoukhov A, Bazzani A, Wachowicz M et al (2012a) Smart cities of the future (paper 188). UCL CASA working paper series

Batty M, Axhausen KW, Giannotti F, Pozdnoukhov A, Bazzani A, Wachowicz M et al (2012b) Smart cities of the future. Eur Phys J 214:481–518

Bibri SE, Krogstie J (2016) Big data analytics and context–aware computing for smart sustainable cities of the future. Norwegian Big Data Symposium (NOBIDS). Trondheim, 15 November 2016

Bibri SE, Krogstie J (2017) Smart sustainable cities of the future: an extensive interdisciplinary literature review. Sustain Cities Soc 31:183–212

Bifulco F, Tregua M, Amitrano C, d'Auria A (2016) ICT and sustainability in smart cities management. Int J Public Sect Manag 29(2):132–147. https://doi.org/10.1108/IJPSM-07-2015-0132

Bishop P, Williams L (2012) The temporary city. Taylor and Francis, Abingdon

Bria F (2012) New governance models towards an open Internet ecosystem for smart connected European cities and regions. Open Innovation, Directorate-General for the Information Society and Media, European Commission, pp 62–71

Caragliu A, Del Bo C, Nijkamp P (2009) Smart cities in Europe. Serie Research Memoranda 0048. VU University Amsterdam, Faculty of Economics, Business Administration and Econometrics

Carvalho L, Santos IP, van Winden W (2014) Knowledge spaces and places: from the perspective of a "born-global" start-up in the field of urban technology. Expert Syst Appl 41 (12):5647–5655. https://doi.org/10.1016/j.eswa.2014.02.015

Cole RJ (2012) Regenerative design and development: current theory and practice. Build Res Inf 40 (1):1–6

Cole RJ, Busby P, Guenther R, Briney L, Blaviesciunaite A, Alencar T (2012) A regenerative design framework: setting new aspirations and initiating new discussions. Build Res Inf 40 (1):95–111

Commonwealth of Australia (2018) Western Sydney city deal. Vision. Partnership. Delivery. Commonwealth of Australia, Department of Infrastructure, Regional Development and Cities, and NSW Department of Premier & Cabinet, Western Sydney City Deal, Canberra

Dosse F (2010) Renaissance de l'événement: Un défi pour l'historien – entre sphinx et phénix. PUF, Paris

Du Plessis C (2012) Towards a regenerative paradigm for the built environment. Build Res Inf 40 (1):7–22

Ehrlich J, Leifer L, Ford C (2015) RegenVillages – integrated village designs for thriving regenerative communities. Sustainable Development Knowledge Platform, Department of Economic and Social Affairs, United Nations. URL: https://sustainabledevelopment.un.org/content/docuuments/622766_Ehrlich_Integrated%20village%20designs%20for%20thriving%20regenerative%20communities.pdf. Accessed 15 July 2018

Frenchman D, Joroff M, Albericci A (2011) Smart cities as engines of sustainable growth. Massachusetts Institute of Technology, prepared for the World Bank Institute

Gabel M (undated) Regenerative development; Going beyond sustainability. http://www.designsciencelab.com/resources/Regenerative-Development.pdf. Accessed 21 June 2018

Girardet H (2013) Sustainability is unhelpful: we need to think about regeneration. http://www.theguardian.com/sustainable-business/blog/sustainabilityunhelpful-think-regeneration

Glasmeier A, Christopherson S (2015) Thinking about smart cities. Camb J Reg Econ Soc 8:3–12. https://doi.org/10.1093/cjres/rsu034

Greenfield A (2013) Against the smart city: a pamphlet. Verso, New York

Hajer MA (2011) The energetic society. PBL Publishers, Den Haag

Hajer M, Dassen T (2014) Smart about cities. Visualising the challenge for the 21st century urbanism. nai010 publishers/PBL publishers, Rotterdam/Den Haag

Hodgkinson S (2011) Is your city smart enough? Digitally enabled cities and societies will enhance economic, social, and environmental sustainability in the urban century. OVUM report

Höjer M, Wangel S (2015) Smart sustainable cities: definition and challenges. In: Hilty L, Aebischer B (eds) ICT innovations for sustainability. Springer, Berlin, pp 333–349

Huizinga J (1938) Homo Ludens. Proeve eener bepaling van het spel-element der cultuur. Wolters-Noordhoff, Groningen

Ishida T, Isbister K (eds) (2000) Digital cities: technologies, experiences and future perspective. Springer Verlag, Berlin

Jacobs J (1961) The death and life of great American cities. Random House, New York

Kasarda JD, Lindsey G (2012) Aerotropolis: the way we'll live next. Penguin Books, London

KNMI (2014) KNMI'14: climate change scenarios for the 21st century – a Netherlands perspective. In: Van den Hurk B, Siegmund P, Klein Tank A (eds) Attema J, Bakker A, Beersma J, Bessembinder J, Boers R, Brandsma T, Van den Brink H, Drijfhout S, Eskes H, Haarsma H, Hazeleger W, Jilderda R, Katsman C, Lenderink G, Loriaux J, Van Meijgaard E, Van Noije T, Van Oldenborgh GJ, Selten F, Siebesma P, Sterl A, De Vries H, Van Weele M, De Winter R and Van Zadelhoff G-J. Scientific report WR2014–01. De Bilt: KNMI. www.climatescenarios.nl

Komninos N (2009) Intelligent cities: towards interactive and global innovation environments. Int J Innov Reg Dev, Special Issue: Intelligent Clusters, Communities and Cities: Enhancing Innovation with Virtual Environments and Embedded Systems 1(4):337–355

Komninos N (2015) The age of intelligent cities; Smart environments and innovation-for-all strategies. Routlegde, Abingdon, New York

Koolhaas R (2014) My thoughts on the smart city. Brussels. URL: https://ec.europa.eu/commission_2010-2014/kroes/en/content/my-thoughts-smart-city-rem-koolhaas. Accessed 16 July 2018

Kummitha RKR, Crutzen N (2017) How do we understand smart cities? An evolutionary perspective. Cities 67:43–52

Lind D (2012) Information and communications technologies creating livable, equitable, sustainable cities. In: Starke L (ed) State of the world 2012: moving toward sustainable prosperity. Island Press/Center for Resource Economics, Washington, pp 66–76

Liugailaité -Radzvickiené L, Jucevicˇius R (2012) An intelligence approach to city development. 7th international scientific conference "business and management 2012" May 10–11, 2012, Vilnius, Lithuania

Mang P, Reed B (2012) Designing from place: a regenerative framework and methodology. Build Res Inf 40(1):23–38

Marsal-Llacuna M-L, Segal ME (2016) The intelligenter method (I) for making "smarter" city projects and plans. Cities 55:127–138. https://doi.org/10.1016/j.cities.2016.02.006

McLean L, Roggema R (2019) Planning for a prosumer future: central park case study. Urban Plan 4(1):172–186.. Special issue 'The City of Flows'. https://doi.org/10.17645/up.v4i1.1746

Meissner M (2017) Narrating the global financial crisis. Urban imaginaries and the politics of myth. Palgrave Macmillan, London. https://doi.org/10.1007/978-3-319-45411-5

Mosannenzadeh F, Vettorato D (2014) Defining smart city: a conceptual framework based on keyword analysis. TEMA J Land Use Mobil Environ. Special issue: INPUT 2014 Eighth International Conference INPUT – Naples, 4–6 June 2014

Neirotti P, de Marco A, Cagliano AC, Mangano G, Scorrano F (2014) Current trends in smart city initiatives – some stylized facts. Cities 38:25–36

Neves BB (2009) Are digital cities intelligent? The Portuguese case. Int J Innov Reg Dev 1(4):443

New York Times, Shell Oil (2014) Cities energized: the urban transition. New York Times, 20 November. URL: http://paidpost.nytimes.com/shell/cities-energized.html?_r=1. Accessed 4 July 2018

Obinna U, Joore P, Wauben L, Reinders A (2017) Comparison of two residential smart grid pilots in the Netherlands and in the USA, focusing on energy performance and user experiences. Appl Energy 191:264–275. https://doi.org/10.1016/j.apenergy.2017.01.086

Offenhuber D (2015) Infrastructure legibility – a comparative analysis of open 311-based citizen feedback systems. Camb J Reg Econ Soc 8:93–112

Picot A (2015) Smart cities; A spatialized intelligence. Wiley, Chichester

Pollard G, Ward JD, Koth B (2017) Aquaponics in urban agriculture: social acceptance and urban food planning. Horticulture 3:39. https://doi.org/10.3390/horticulturae3020039

Rabari C, Storper M (2015) The digital skin of cities: urban theory and research in the age of the sensored and metered city, ubiquitous computing and big data. Camb J Reg Econ Soc 8:27–42

Rathore MM, Ahmad A, Paul A, Rho S (2016) Urban planning and building smart cities based on the Internet of things using big data analytics. Comput Netw 101:63–80. https://doi.org/10.1016/j.comnet.2015.12.023

Ratti C, Townsend A (2011) Harnessing residents' electronic devices will yield truly smart cities. Scientific American, 16 August 2011. http://www.scientificamerican.com/article.cfm?id=the-social-nexus Accessed 9 June 2018

Robinson R (undated) Smart city design principles. URL: https://theurbantechnologist.com/smarter-city-design-principles/. Accessed 18 June 2018

Robinson J, Cole RJ (2015) Theoretical underpinnings of regenerative sustainability. Build Res Inf 43(2):133–142

Roche S, Nabian N, Kloeckl K, Ratti C (2012) Are 'smart cities' smart enough? Global geospatial conference 2012, spatially enabling government, industry and citizens, 14–17 May 2012, Québec City, Canada. http://www.gsdi.org/gsdiconf/gsdi13/papers/182.pdf

Roggema R (ed) (2017) The Foodroof of Cantagalo – Rio de Janeiro. Springer, Dordrecht/Heidelberg/London. 201pp

Roggema R (2018) Design with voids. How inverted urbanism increases urban resilience. Archit Sci Rev 61:349–357

Roggema R (2019) The Convenient City. In: Biloria N (ed) Data-driven multivalence in the built environment. Springer, Cham, pp 37–55

Sennett R (2012) No one likes a city that's too smart. The Guardian, 4 December 2012. URL: http://www.theguardian.com/commentisfree/2012/dec/04/smart-city-rio-songdo-masdar. Accessed 16 July 2018

Somerville C, Cohen M, Pantanella E, Stankus A, Lovatelli A (2014) Small-scale aquaponic food production. Integrated fish and plant farming, FAO fisheries and aquaculture technical paper no. 589. FAO, Rome. 262 pp

Steen KYG, van Bueren EM (2017) Urban living labs: a living lab way of working. AMS Institute, Amsterdam

Streitz N (2011) Smart cities, ambient intelligence and universal access. In: Stephanidis C (ed) Universal access in HCI, Part III, HCII 2011, LNCS 6767. Springer, Berlin/Heidelberg, pp 425–432

Terzano K, Morckel V (2016) SimCity in the community planning classroom. Effects on student knowledge, interests, and perceptions of the discipline of planning. J Plan Educ Res 37 (1):95–105. https://doi.org/10.1177/0739456X16628959

Townsend A, Maguire R, Liebhold M, Crawford M (2010) The future of cities, information, and inclusion: A planet of civic laboratories. Institute for the Future, Palo Alto

Van Timmeren A, Henriquez L, Reynolds A (2015) Ubikquity & the illuminated city. Delft University of Technology, Delft

Wijnen V (2017) Loading Loskade, De Loskade op de Suiker, Ruimte voor Innovatie. Brochure. Gorredijk, Van Wijnen BV

Willett W, Rockström J, Loken B, Springmann M, Lang T, Vermeulen S, Garnett T, Tilman D, DeClerck F, Wood A, Jonell M, Clark M, Gorden LJ, Fanzo J, Hawkes C, Zurayk R, Rivera JA, De Vries W, Sibanda LM, Afshin A, Chaudhary A, Herrero M, Augustina R, Branca F, Lartey A, Fan S, Crona B, Fox E, Bignet V, Troell M, Lindahl T, Singh S, Cornell SE, Srinath Reddy K, Naraïn S, Nishtar S, Murray CJL (2019) Food in the Anthropocene: the EAT–Lancet Commission on healthy diets from sustainable food systems. Lancet Comm 393 (10170):447–492. https://doi.org/10.1016/S0140-6736(18)31788-4

Part II
The Resilient City

How can we design a city that is prepared for an uncertain future, no matter by which changes. Most of these papers will discuss different climate impacts and how the city can anticipate and respond to these developments.

Chapter 3
Resilient Spatial Planning for Drought-Flood Coexistence ('DFC'): Outlook Towards Smart Cities

Nguyen Quoc Vinh and Tran Thi Van

Abstract The challenges of booming urbanisation are multi-faceted. Many different concerns have been raised about the urban puzzle during the past few decades. The most recent among them relates to 'smart cities', in which cities embrace available technology in order to become "smart". Many studies show that advanced technologies have been used extensively in managing cities (Verma R, Kumari K, Tiwary R. 2009) (Stefanov et al. 2001) (Akanbi and Fidelis 2013). However, from an environmental point of view, these studies have not been able to explain how a city may become smart while also being environmentally resilient and sustainable. As the concept of 'smart cities' has evolved after being introduced at the beginning of twenty-first century, the notions of resilience – urban resilience and resilient cities – have attracted great attention and interest, in both academia and urban governance. They have since then gradually been adopted in planning practice over recent years. With the help of technologies such as remote sensing and GIS, architects and planning authorities have been able to effectively govern policies which can make cities more and more resilient and sustainable. The more available data GIS can offer, the more effective the task of planning will be.

The scope of this paper is to explore how a city with drought-flood coexistence (DFC) can become resilient through spatial planning. This will be presented in three sections. The first section examines the relationship between smart and resilient cities. The second focuses on research which attempts to turn smart cities into resilient cities by applying (or utilising) remote sensing techniques and GIS to manage natural hazards in urban planning. Last but not least, the case of the Ninh Thuan province located in the South-Central part of the Vietnam Coastal region will

N. Q. Vinh (✉)
Faculty of Civil Engineering, Ho Chi Minh City University of Technology, Ho Chi Minh, Vietnam
e-mail: vinh.bmkt@hcmut.edu.vn

T. T. Van
Faculty of Environment and Resources, Ho Chi Minh City University of Technology, Ho Chi Minh, Vietnam
e-mail: tranthivankt@hcmut.edu.vn

© Springer Nature Switzerland AG 2020
R. Roggema, A. Roggema (eds.), *Smart and Sustainable Cities and Buildings*,
https://doi.org/10.1007/978-3-030-37635-2_3

be discussed in the last section, together with several key findings from its own context of natural conditions resulting in the current DFC. A number of focal principles will be proposed, based on local context, and the new technology of remote sensing and GIS as tools for resilient spatial planning in the case of DFC. The proposed principles not only regulate the water resource management, but also mitigate drought severity for the province, and could become drivers for both future planning implementation and policy.

The study will be concluded with discussions on the critical role of spatial planning in making cities smart and resilient, and doing this with respect to their environment, particularly when remote sensing and GIS are considered.

Keywords Smart city · Resilient city · Resilient spatial planning · Remote sensing and GIS · Sustainability · Extreme weather events · Drought-flood coexistence

3.1 Introduction

The rapid industrialisation and population growth since the industrial stage in urban areas have become a globally problematic challenge. The increased urbanisation requires smart and innovative approaches in order to manage the complexity of urban ecosystems: overpopulation, energy consumption, resource management and environmental protection, etc. This matter is a real puzzle to not only authorities, but also to the professionals, particularly architects and urban planners.

According to the results from the 2015 Revision by the United Nations, the world population reached 7.2 billion by mid-2015, implying that the world has grown by approximately one billion people in the span of 12 years. More than 50% of the global population currently lives in the urban areas (Debnath et al. 2014). Additionally, between 2015 and 2050, the world population will increase by 32%, i.e. from 7.2 to 9.7 billion inhabitants, while the urban population will increase by 63% (from 3.9 to 6.3 billion inhabitants). The current estimations suggest that by 2030, over 60% of the world population will live in the cities, and the most significant growth will be in developing countries in Africa, Asia and Latin America (Eremia et al. 2017). The most important point is that although the cities occupy only 2% of the planet's surface, they accommodate about 50% of the world population, consume 75% of the total generated energy, and are responsible for 80% of the greenhouse effect (Eremia et al. 2017).

With such increasing environmental pressures on urban areas, ways of administrating a city sustainably have been considered and studied for a long time, from the first ideas in ancient times, to later in 1898 with "Garden City of To-morrow" proposed by the British urban planner Ebenezer Howard, with the purpose of transforming slums into neighbourhoods capable of providing humans life opportunities and comfort. The trend afterwards was about transforming the cities into green, sustainable, resilient, intelligent ones, etc., and the currently emerging smart cities.

However, the latest term mainly focuses on managing cities to be technologically smart and ignores environmental sustainability and resilience.

Resilience and resilient cities are the ecology-based concepts to develop a smart city, where a city is basically considered a complexity of Social-Ecological Systems (SES) (Pickett and Cadenasso 2013). Coincidentally, resilient spatial planning needs to be studied to meet the natural services' challenging demand (Lu and Stead 2013). In addition to the current complexity of urban development, urban data sets have become increasingly complicated, with more data being added to the system. Remote sensing and the Geographical Information System (GIS) are the most effective tools for approaching cities information at the moment (Jhawar et al. 2012; Verma et al. 2009).

This paper will explore in depth how these theoretical concepts and contemporary techniques could help establish a system of resilient spatial planning principles for DFC in three sections: (i) the connection between smart and resilient cities, where the focus will be on resilient spatial planning for DFC, (ii) remote sensing and GIS in urban planning and methodology, and (iii) the case of Ninh Thuan generalities and key principles of resilient spatial planning for its DFC.

3.2 The Relationship Between Smart Cities and Resilient Cities

3.2.1 Smart Cities

Sustainable development of urban areas has been a major concern for both architects and authorities since the early nineteenth century. One of the first modern concepts is discussed in the book "Garden Cities of To-morrow". This is further elaborated on by Eremia et al. (2017): "...The cities of Tomorrow will be more readily susceptible to transformation and adornment than the Cities of Yesterday".

The characteristics of future cities have been adopted over many years. Since the interwar period, the key purpose has become alleviating shortcomings caused by industrialisation and creating green cities. Among these concepts, the ideas developed in "The City of Tomorrow and its Planning" (1929) by Le Corbusier, and later in "The City: its growth, its decay, its future" (1943) by Gottlieb Eliel Saarien, have influenced city features in Europe and Northern America. Professional urban planners later focused on how to make a city smarter in the future.

Eremia et al. (2017) reviewed various definitions from specialists from different domains and concluded that a city is 'smart' when it:

- Uses information and communications technology (ICT) to enhance livability, walkability, and sustainability (Smart cities council 2014).
- Monitors and integrates conditions of all critical infrastructures – including roads, bridges, tunnels, rails, subways, airports, seaports, communications, water, power, major buildings – that can optimize its resources, reduce maintenance, and monitor security aspects while maximizing services to its citizens.

- Is a geographical space able to manage all resources (natural, human, building and infrastructure), and waste generated is not environmentally harmful.

Irrespective of the differentiating names and definitions, future cities must adapt to all changes of societies and the natural environment.

3.2.2 Resilient Cities

3.2.2.1 Resilience

The notion of resilience has experienced a long history. It was originally used to refer to systems and their ability to cope with external shocks and disturbance (Lance et al. 2002). A simpler understanding of resilience is the ability to absorb outside disturbance while maintaining former structures and functions (Holling 1973, 2001; Eraydin et al. 2013). Later, various scholars have considered resilience in terms of the adaptive cycle (Folke 2006; Holling and Gunderson 2002; Walker and Salt 2006), which focuses on the dynamics of systems (e.g., ecosystems, societies, and economies) that do not have a stable condition or equilibrium, but repeatedly pass through four characteristic phases: (1) growth and exploitation, (2) conservation, (3) collapse or release, and (4) renewal and reorganisation.

3.2.2.2 Urban Resilience and Resilient Cities

Although originally, the concept of resilience was discussed for the first time in 1973 (Holling 1973), it only got the attention of urban planners in the 1990s (Mileti and Henry 1999). Resilience in planning has often put the emphasis on preparation and mitigation actions, especially at a local scale.

A city is an ecosystem, technically termed a Socio-Ecological System ("SES"), which is supported by services, including: (i) provisional services such as food and water, (ii) regulating services such as regulation of floods, drought, and disease, (iii) supporting services such as soil formation and nutrient cycling, and (iv) cultural services such as recreational, spiritual, religious and other nonmaterial benefits (Pickett and Cadenasso 2013). Therefore, an SES (Pickett 2012; Ahern et al. 2014) is a complex adaptive system which consists of four basic environments: (1) biological, (2) social, (3) physical, and (4) built components (Grove 2009). They cannot be separated from one another (Pickett and Grove 2009).

S. Meerow reviewed and defined urban resilience as such: "Urban resilience refers to the ability of an urban system and all its constituent socio-ecological and socio-technical networks across temporal and spatial scales to maintain or rapidly return to desired functions in the face of a disturbance, to adapt to change and to quickly transform systems that limit current or future adaptive capacity." (Meerow et al. 2016).

Based on literature on resilience and urban resilience, several definitions of resilient cities have been further developed. Among them, the most adapted definition is in Sendai Framework (Cities 2018; Framework and Reduction 2015): "The ability of a city exposed to hazards to resist, absorb, accommodate to and recover from the effects of a hazard in a timely and efficient manner, including through the preservation and restoration of its essential basic structures and functions" ("2009 UNISDR Terminology on Disaster Risk Reduction", 2009).

A number of recent studies (Davoudi and Strange 2009; Fleischhauer and Fleischhauer 2008) acknowledge that spatial planning plays an important role in promoting urban resilience to cope with climate change through the spatial configuration of cities (by way of land-use management). Particularly, Mark Fleischhauer mentioned in his research that urban structures in spatial planning can mitigate multi-hazards (Fleischhauer and Fleischhauer 2008), introducing several indicators used to assess the resilience of spatial planning, such as: keeping areas free of development (under the threat of climate change; differentiated decisions on land use), accepting land use development according to the intensity and frequency of hazards, recommendations in legally binding land use or zoning plans, and contributing on reducing the potential hazards.

With regards to spatial planning for DFC, nature's regulating service (particularly water cycle) becomes the most critical one, whilst others become supporting factors. The natural land such as forest, grassland, wetland, etc. being destroyed for agriculture and urban development could make the region vulnerable to the potential hazardous events of both droughts and floods. To prepare for urban resilience, these land stocks must be spatially planned, and be strictly protected by authorities and urban planners.

Smart and resilient cities are quite similar in terms of promoting city development and sustainability in the future. One is smart, while the other is resilient. Both are moving forward to sustainability.

3.3 Remote Sensing and GIS in Urban Planning

3.3.1 Remote Sensing and GIS for Urban Planning and Natural Hazard Management

The applications of remote sensing and GIS have become an integrated, well developed and successful tool in urban and environmental studies (Bahaa and Al n.d.; Verma et al. 2009; Vân 2011). Indeed, "Urban greenbelt (or open space) mapping, urban encroachment, growth of slums on vacant lands, urban housing, urban utilities and infrastructure, solid waste management, urban transportation and traffic planning, urban hydrology, urban cadastral and real estate, urban ecological hazards, and urban census data can all be mapped, monitored, and analysed using remote sensing" (Rahman 2007). These technologies are also useful for disaster risk management (Westen 2012) (Vorovenci 2011; Thenkabail et al. 2009; Vinh 2014).

The modern technology of remote sensing allows us to collect a lot of physical data rather easily, quickly and on repetitive basis. Together with GIS it can help us analyse the data spatially, offering possibilities of generating various options, thereby optimising the whole planning process. These information systems also offer interpretations of physical (spatial) data combined with other socio-economic data and can thereby provide an important linkage in the total planning process, making it more effective and meaningful.

3.3.2 Applications of Remote Sensing and GIS for Studies of Urban, Drought and Flood

Remote sensing offers cost-effective solutions to city planners' data needs, for both macro and micro level analysis of the land use planning leading to urban environment management. GIS is best utilized for integration of various data sets which helps in identifying the problem areas and suggesting conservation measures. Remote sensing technology together with GIS is an ideal tool to identify, locate and map various types of land associated with different landform units. This is quite helpful for an immediate perspective on the planning of cities.

Currently, drought monitoring management exists in most countries based on site-based information on drought related parameters such as rainfall, weather, crop condition, water availability, etc. However, a limitation in some regions is the availability of data sets: often related to inaccurate and out-of-date information. Remote sensing and GIS technology significantly contribute to all three phases of drought disaster management: preparedness, monitoring and relief. Similarly, all sorts of information related to flood prone/risk zone identification data could be extracted from remote sensing images with quite accurate information.

3.4 Case of Ninh Thuan Province

3.4.1 Location and Natural Conditions

Ninh Thuan is the coastal province of central Vietnam, located at the co-ordinates: $11°18'14''$ to $12°09'15''$ Northern latitude, $108°09'08''$ to $109°14'25''$ Eastern longitude. It is adjacent to the provinces: Khanh Hoa in the North, Binh Thuan in the South, Lam Dong in the West, and the ocean in the East (Fig. 3.1).

Ninh Thuan adheres to common characteristics of the central part of Vietnam, which is divided into 04 ecosystems, from the mountains to the sea: (i) high mountainous area, (ii) hilly area, (iii) plain area, and (iv) coastal sandy area (Sâm et al. 2008).

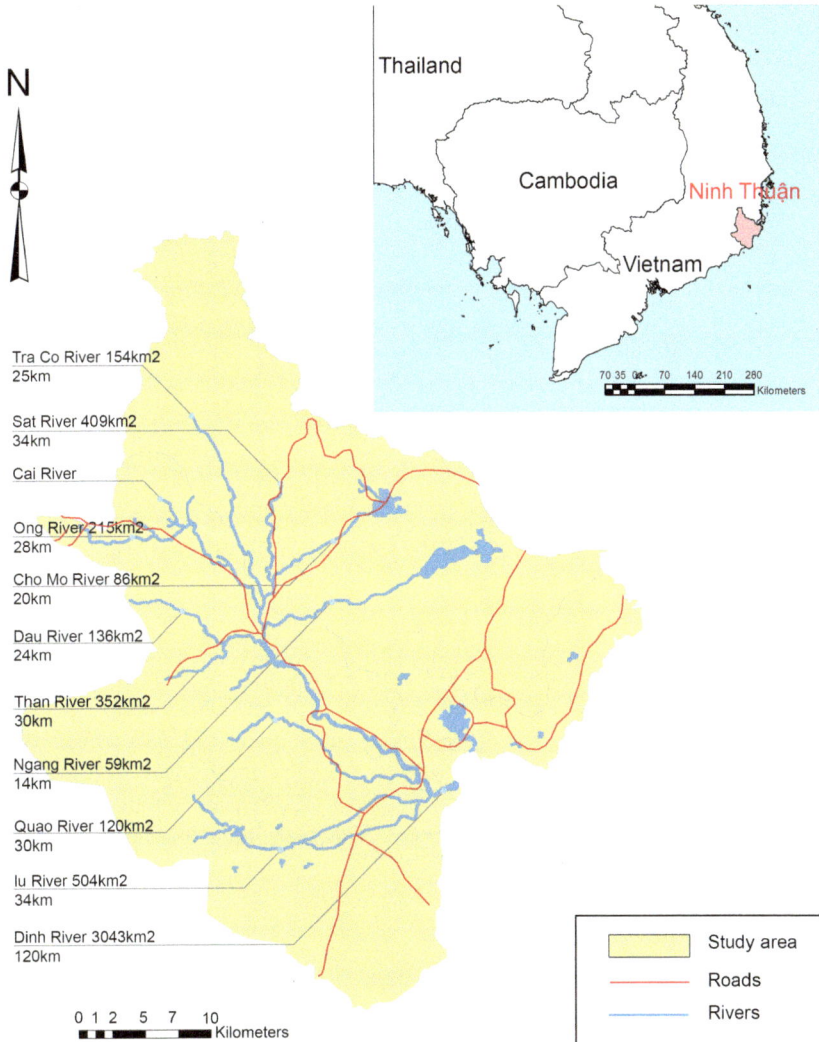

Fig. 3.1 The location of Ninh Thuan province in Vietnam. (Source: Authors)

The geological location of the province has created its natural conditions. They are as follows:

- Hydrology: the total area of rivers basin is 3.092 km^2, including 46 rivers, main streams and four main aquifers, of which the Cai river basin makes up the majority of the area, with 3.000 km^2, and the others make up the remaining 92 km^2. The Cai river is 120 km long, including branches: the Dinh, Me Lam, Sat, and Ong rivers. Water flow is 10 l/s/km^2 during rainy season, and 0.5 l/s/km^2

during dry season. The main geographical characteristic is the non-existence of surface water, there only being mid-low underground water (Sâm et al. 2008).

- Meteorology: dry (75–77%) and hot (26–27 °C), strong wind (2.3–5 m/s, the strongest is of 25 m/s) and high vaporisation. The average amount of rainfall is 700–800 mm in Phan Rang city, which increases to 1.100 mm in mountainous areas.
- Geology: bedrock topography types in the Ninh Thuan are divided into 2 types: "partial down warping" and "constant slope to the sea". The first type has better water retention ability; however, water is lacking in the dry season, therefore, due to the first type's capacity, it is not enough for use (Khuyến 2011).
- Extreme events: A flash flood often follows immediately after a storm upstream, and typically affects areas with harsh topography, meteorology and hydrology. Flash floods and landslides happen during rainy season in riverine and mountainous areas and have significant impacts on the population.

3.4.2 Extreme Events in Ninh Thuan: Droughts, Floods

3.4.2.1 Droughts

Ninh Thuan is the hottest province in Vietnam with an average temperature of 27 °C. The highest temperature ever recorded was 40.5 °C at Nha Ho station in 1937.

The province is divided into three climate sub-regions (Fig. 3.2). The coastal area (region III) has the worst droughts, with an average rainfall of around 500–700 mm per year. This area receives the strong wind from the East Sea, resulting in a hot temperature almost throughout the year. The plain area (region II) also suffers drought with the rainfall allocation from 750–1200 mm per year, while the mountainous area (region I) has a rainfall allocation from 1000–1700 mm per year (Sam and Vuong 2008).

3.4.2.2 Floods and Flash Floods

Weather conditions are alternating from droughts to floods in the Ninh Thuan province. There have been a number of severe floods since the 1960s that have caused negative impacts on the population: in 1964, 1998, 1999, 2000, 2003, 2005, 2009, 2010, and 2011. However, since 1978, flash floods happen regularly on most of the provincial catchments, and seem to be increasing (Viện Hàn Lâm Khoa học và Công Nghệ Việt nam 2016; Sâm et al. 2008).

Fig. 3.2 The three regions of sensitive drought and relevant rainfall. (Source: Authors)

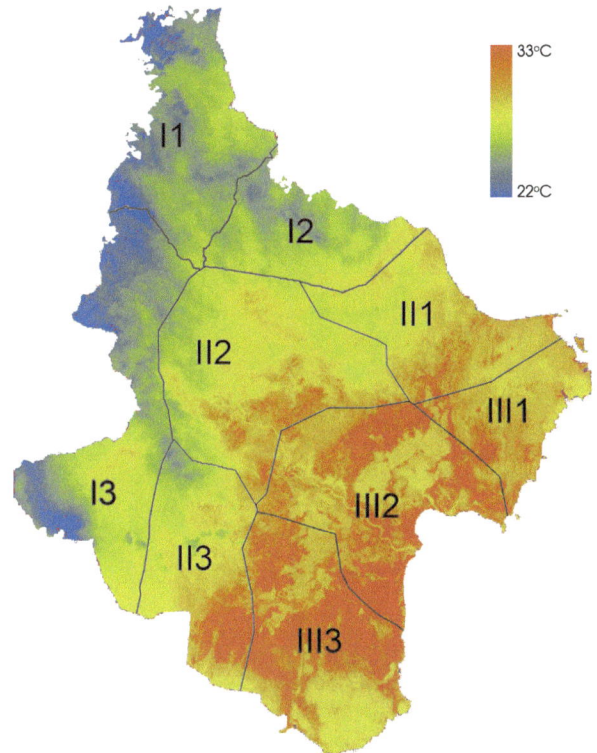

3.5   Resilient Spatial Planning for DFC: Study Methodology and Methods

Methodology: floods and flash floods could be regulated through resilient spatial planning based on local natural conditions. Drought will be mitigated accordingly, based on regulated flood and storm water during the rainy season across scales.

Methods: remote sensing and GIS. Images related to the three studied objects at different periodic times for analyzing their transformations and inter-relationships will be created: (i) urban settlements, based on urbanization statistics, (ii) topography, NDVI, soil types, and (iii) peak times of drought, based on drought and aridity history.

The process of mapping flash flood potential acquires raster (gridded) datasets representing the type of physiographic characteristics influencing the hydrologic response and flash flood potential (Fig. 3.3). A relative flash flood potential index will be assigned to each data layer based on the layer attributes associated with the hydrologic response.

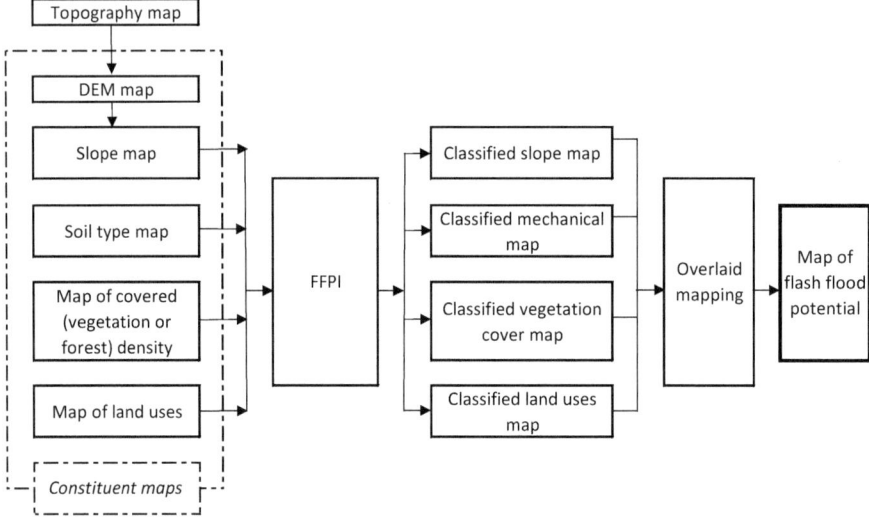

Fig. 3.3 Procedure of mapping flash flood potentials. (Source: Authors)

Mapping sensitive drought will be carried out in 3 steps (Fig. 3.4):

1. Identifying inputs to evaluate sensitive drought. Based on drought impacts and integration with meteorology conditions of the study area, 09 indexes are selected as input criteria.
2. Standardizing the indexes.
3. Overlaying these maps in GIS to set up sensitive drought map.

These indexes are created using GIS to overlay four physical elements, as indicated in Fig. 3.3, that relate to precipitation runoff. These elements are each scaled to create an overall indexed value that forecasters could use to locate basins that may respond more quickly than expected. A small index value indicates a minimum flash flood threat and a large value indicates a maximum flash flood threat.

To identify appropriate spaces in the study area for resilient spatial planning for DFC, besides maps of sensitive drought and flash flood potential, remote sensing images of timing urban transformation will be collected for tracking and evaluating (Fig. 3.5). The three periods to be selected are:

1. Before 1996, a starting point of first urbanisation in Ninh Thuan.
2. After 2000, when the City master plan was approved.
3. At present, 2019.

All these images of urban settlements will be investigated, supported by ENVI software, in terms of drought sensitiveness, especially land surface temperature, and flood potentials respectively. The results will be overlaid and integrated into GIS, to analyse in different models with different assumptions. This will finally help identify causes and effects and define appropriate areas for spatial planning.

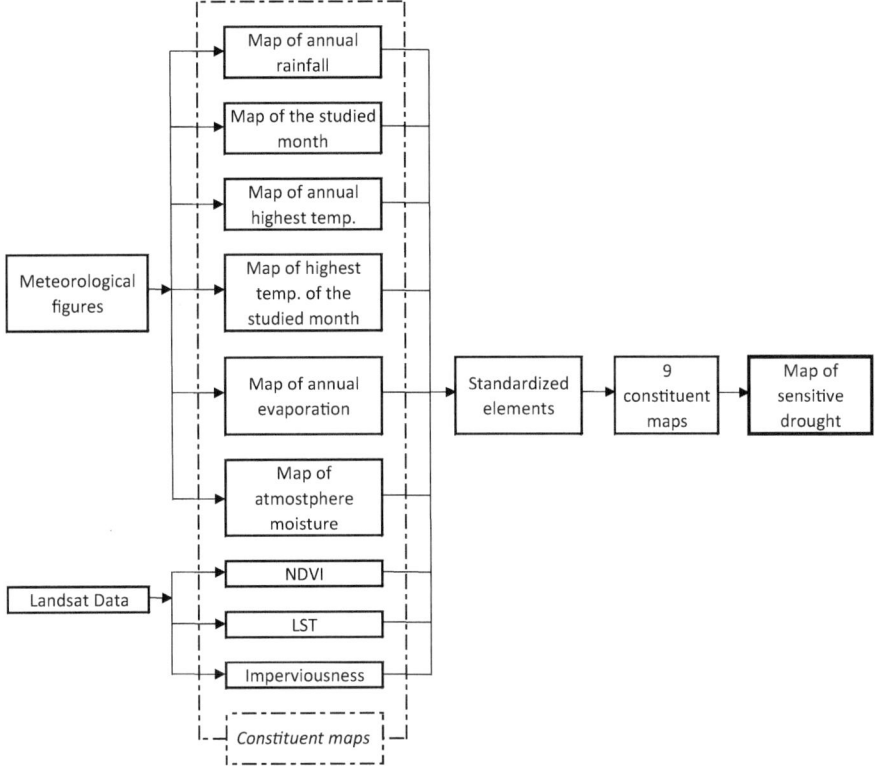

Fig. 3.4   Procedure of mapping sensitive drought. (Source: Authors)

3.6   General Principles of Resilient Spatial Planning for DFC

Based on the new technology of remote sensing and GIS, the proposed general principles of resilient spatial planning for DFC should follow across scales:

1. Regional scale: (i) To respect the natural environments, based on ecological concepts, by fully understanding nature's services, particularly the service of regulating droughts and floods in accordance with the water cycle, (ii) To adapt to the local natural conditions to mitigate DFC by strictly savingand protecting spaces for natural conservation.
2. City scale: (i) To be basically determined (on land use) by both social and environmental considerations, not by commercial land value (or potential land value) alone, (ii) To combine the natural water bodies into city's networked systems of green and blue infrastructures.
3. Site scale: (i) To be reasonably and effectively planned urban functions, especially green and open spaces, and (ii) To technically design buildings that could maximizing recycled water.

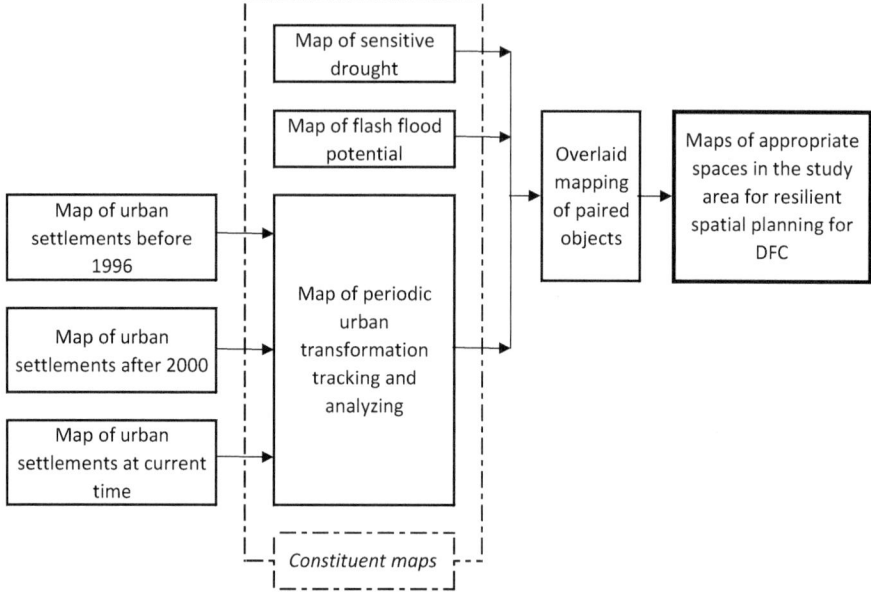

Fig. 3.5 Procedure of mapping spaces in resilient planning. (Source: Authors)

These expected key principles will be further studied in details on the next steps of the research, including concrete indicators of resilient spatial planning for DFC.

3.7 Conclusion

Urban transformation is ubiquitous over the last decades. However, this extraordinary alteration has no set endpoint. Cities, as complex adaptive systems, are not only dynamic but also self- organizing and actively adjusting to cope with all predictable and unforeseen disturbances. This concept, in contrast with the traditional equilibrium paradigm, is heterogeneous, non-linear, hierarchal, and has multiple stable states. This is emphasized in the theory of ecological resilience, which is a very important base to effectively manage SES, through spatial planning and urban design. Spatial planning plays a critical role in progression of cities, particularly on the direction of sustainability.

Besides the traditional methods, this paper introduces the remote sensing and GIS as an integrated, proven to be the most effective method for DFC spatial planning of Ninh Thuan province, to make it smart and resilient. Basing on this, data from remote sensing images of the three main objects and their inter-relationships, such as data on urban settlements, droughts and floods are extracted and analysed. The initial results are later integrated into GIS to identify appropriate spaces for resilient spatial planning. Key ecology-based principles are initially proposed and furtherly studied in the coming research. These principles can become drivers and a useful tool for both policy and implementation of resilient spatial planning, not only for Ninh Thuan, but also for regions facing similar challenges. It is an outlook towards smart cities.

Acknowledgments The authors wish to acknowledge the assistance and encouragement from the colleagues at the Department of Architecture, Faculty of Civil Engineering, Faculty of Environment and Resources, Hochiminh City University of Technology.

References

Ahern J, Cilliers S, Niemelä J (2014) The concept of ecosystem services in adaptive urban planning and design: a framework for supporting innovation. Landsc Urban Plan.. Elsevier B.V. 125:254–259. https://doi.org/10.1016/j.landurbplan.2014.01.020

Akanbi AK, Fidelis U (2013) Application of remote sensing , GIS and GPS for efficient urban management plan - a case study of part of Hyderabad city. In: Novus international journal of Engineering & Technology, (December)

Bahaa E, Al EI (n.d.) The application of remote sensing and Gis in urban planning

Davoudi S, Strange I (2009) Space and place in the twentieth century planning: an analytical framework and an historical review. Conceptions of space and place in strategic spatial planning, pp 7–42

Debnath AK et al (2014) A methodological framework for benchmarking smart transport cities. Cities 37:47–56. https://doi.org/10.1016/j.cities.2013.11.004

Eraydin A et al (2013) Resilience thinking in urban planning. In: Resilience thinking in urban planning, vol 106. Springer, Dordrecht; New York, pp 39–51. https://doi.org/10.1007/978-94-007-5476-8

Eremia M, Toma L, Sanduleac M (2017) The smart city concept in the 21st century. Procedia Eng 181:12–19. https://doi.org/10.1016/j.proeng.2017.02.357

Fleischhauer and Fleischhauer M (2008) The role of spatial planning in strengthening urban resilience

Folke C (2006) Resilience: the emergence of a perspective for social-ecological systems analyses. Glob Environ Chang 16(3):253–267. https://doi.org/10.1016/j.gloenvcha.2006.04.002

Framework S, Reduction DR (2015) Sendai framework for disaster risk reduction 2015–2030, pp 1–25. https://doi.org/A/CONF.224/CRP.1

Grove STAPJM (2009) Urban ecosystems: what would Tansley do?, pp 1–8

Holling CS (1973) Resilience and stability of ecological systems. Annu Rev Ecol Syst 4:1–23

Holling CS (2001) Understanding the complexity of economic, ecological, and social systems. Ecosystems 4:390–405. https://doi.org/10.1007/s10021-001-0101-5

Holling CS, Gunderson LH (2002) Resilience and adaptive cycles. In: Panarchy: understanding transformations in human and natural systems. Island Press, Washington, pp 25–62. https://doi.org/10.1016/j.ecolecon.2004.01.010

Jhawar M, Tyagi N, Dasgupta V (2012) Urban planning using remote sensing. Int J Innov Res Sci Eng Technol 1(1):42–57

Khuyến NM (2011) 'Kết quả nghiên cứu ảnh hưởng của địa hình mặt đá gốc đến khả năng trữ nước dưới đất trong các tầng chứa nước bở rời trầm tích đệ tứ vùng sông Cái Phan Rang, tỉnh Ninh Thuận', Tạp chí Khoa học và Công nghệ thủy lợi số 14 – *2013*, pp 69–76

Lance H, Gunderson LH, Holling CS (2002) Panarchy: understanding transformations in human and natural systems. Panarchy: understanding transformations in human and natural systems, pp 3–24. https://doi.org/10.1016/j.ecolecon.2004.01.010

Lu P, Stead D (2013) Understanding the notion of resilience in spatial planning: a case study of Rotterdam, The Netherlands. Cities 35:200–212. https://doi.org/10.1016/j.cities.2013.06.001

Meerow S, Newell JP, Stults M (2016) Defining urban resilience: a review. Landsc Urban Plan.. Elsevier B.V. 147:38–49. https://doi.org/10.1016/j.landurbplan.2015.11.011

Mileti D, Henry AJ (1999) Disasters by design: a reassessment of natural hazards in the United States, Natural Hazards. The Center, Boulder, pp 1–372. ISBN: 0-309-51849-0

Pickett ST (2012) Ecology of the city: a perspective from science. Urban Des Ecol:160–168

Pickett STA, Cadenasso ML (2013) Ecology of the city as a bridge to urban design. In: Resilience in ecology and urban design. Springer, Dordrecht, pp 7–28. https://doi.org/10.1007/978-94-007-5341-9

Pickett STA, Grove JM (2009) Urban ecosystems: what would Tansley do? Urban Ecosyst 12 (1):1–8. https://doi.org/10.1007/s11252-008-0079-2

Rahman A (2007) Application of remote sensing and GIS technique for urban environmental management and sustainable development of Delhi, India. In: Applied remote sensing for urban planning, . . ., pp 165–197. https://doi.org/10.1007/978-3-540-68009-3_8

Resilient Cities (2018) The new ISO standard for resilient cities indicators: opportunities for city and expert input

Sam L, Vuong ND (2008) Results of monthly isoldrought map building in Ninh Thuan province. Available at: http://www.siwrr.org.vn/tv3_files/i.17-dangkhohanninhthuan.pdf. Accessed 28 July 2017

Sâm L, Lân NV, Vượng NĐ (2008) Ecological zoning, science basis for studying the ecological reservoir systems in the centre of Vietnam. *Viện Khoa học Thủy Lợi Miền Nam*, (3), pp 70–72

Thenkabail PS, Gamage MSDN, Smakhtin VU (2009) The use of remote-sensing data for drought assessment and monitoring in Southwest Asia

Stefanov, W. L., Ramsey, M. S. and Christensen, P. R. (2001) 'Monitoring urban land cover change: An expert system approach to land cover classification of semiarid to arid urban centers', 77, pp. 173–185

Vân TT (2011) 'Nghiên cứu biến đổi nhiệt độ đô thị dưới tác động của quá trình đô thị hóa bằng phương pháp viễn thám và GIS, trường hợp khu vực thành phố Hồ Chí Minh'

Westen C Van (2012) Remote sensing and GIS for natural hazards assessment and disaster risk management. . . . of Space Technology for Disaster Risk Reduction: . . ., pp 1–61. https://doi.org/10.1016/B978-0-12-374739-6.00051-8

Verma R, Kumari K, Tiwary R (2009) Application of remote sensing and GIS technique for efficient urban planning in India. Geomatrix Conference Proceedings, (October 2016), pp 1–23

Viện Hàn Lâm Khoa học và Công Nghệ Việt nam (2016) Nghiên cứu Đánh giá Tài nguyên Nước vùng hoang mạc Ninh Thuận xét đến Biến đổi Khí hậu, Đề xuất giải pháp thích ứng

Vinh PQ (2014) Hạn hán và vấn đề Biến đổi Khí hậu tinh Ninh Thuận

Vorovenci I (2011) Satellite remote sensing in environmental impact assessment

Walker B, Salt D (2006) Resilience thinking: sustaining ecosystems and people in a changing world. Coral Reefs. https://doi.org/10.2174/1874282300802010217

Chapter 4
Globalization and Transformations of the City of Sydney

Shahadat Hossain

Abstract This paper aims to explain the transformations of the city of Sydney, highlighting economic, spatial and cultural changes. It explains the new social polarization, gentrification, and reshaping of urban space and cultural consumption, and transformation of leisure space. The paper is based on research conducted in the context of Sydney in recent decades. It reveals that the emergence of Sydney as a global city is linked to migration, tourism and business, enhanced by globalization. The employment structure has been reorganized due to new technology and the rise of the service economy, which has produced huge inequality and social polarization. Gentrification and migration of new communities have reshaped social and cultural space. The pattern of cultural consumption has been significantly changed and the leisure space has been transformed. Therefore, this paper argues that the new social formations of Sydney are the product of global migration, tourism and privatization of urban space and culture.

Keywords Globalization · Social polarization · Gentrification · Culture and consumption

4.1 Introduction

There is growing global interest for the city of Sydney, which has long been considered synonymous with 'antipodes'. Less than a quarter of a century ago, Sydney was deemed a peripheral node for the Asia-Pacific and little more than a significant national gateway. Today, most urbanists would argue in favour of a much more balanced analysis. In the 'Globalization and World Cities' 2010 roster, Sydney occupied an 'alpha+' position, second only to the ever-dominant NYLON

S. Hossain (✉)
School of Humanities and Communication Arts, Western Sydney University, Sydney, NSW, Australia
e-mail: shahadat.hossain@westernsydney.edu.au

© Springer Nature Switzerland AG 2020   41
R. Roggema, A. Roggema (eds.), *Smart and Sustainable Cities and Buildings*,
https://doi.org/10.1007/978-3-030-37635-2_4

(New York and London). Over the decades, with the growing importance of its CBD as a financial hub, Sydney has been developed around a twofold imagineering underpinning the policies of various state and local governments. The harbor city has centered its globalization on the twin themes of tourism and business, linked by the strategic branding of Sydneysiders' lifestyle rooted in cosmopolitan, modern, and relatively inexpensive features. In this view, Sydney's characteristics are supposed to appeal to both the transient visitor in search of the Australian experience, and the corporate class looking for a livable and styled settlement (Acuto 2011).

Inner-city Sydney has significantly changed due to economic globalization. The inner-city areas have become locations of new finance, banking, education, tourism and hospitality services. The service economy has become the dominant feature and new business and financial institutions have developed. This gentrification process is clearly observed in inner-city areas where high-rise buildings were built for residential purposes. To meet the demands of housing for white collar workers, local developers and real estate are focusing on more high-rise development (Elenius 2013). Sydney has the largest night time economy in Australia, and the largest number of visitors. The night-time activity Sydney offers is central to this. The activities of young people represent a definite dynamic in the city, particularly at night. Dance parties, drinking, live music, hanging out, graffiti and skateboarding occupy an important place within youth culture in Sydney (Rowe and Lynch 2012). The balance of daytime/night-time leisure spaces, which have both social and material affordances, is a key discrimination across neighbourhoods, both within and between cities. Daytime consumer landscapes are more often framed and sociable and inclusive within media, while night-time landscapes are perceived as divisive (Gorman-Murray and Nash 2017).

This paper explains the emergence of Sydney as a global city in recent decades. It addresses transformations of the city highlighting employment, restructuration of urban space and the changing pattern of cultural consumption and leisure spaces.

4.2 Transforming Sydney and the New Social Formations

4.2.1 Globalization and Urban Transformations

In the post-war decades, Sydney's urban planning was characterized by a laissez faire tradition which exempted any attempt to plan and control the city. From the early 1980s, Sydney began to experience economic and social changes, which led to both expansion and concentration in the city centre. The revitalization of central Sydney was driven by new dynamics of the international, national, and state economies. From the late 1970s to the early 1980s, major Western economies which subscribed to Keynesian economic theory began to dismantle trade protectionism and financial regulation, to embrace the neoliberal ideas of unfettered market competition and minimal government intervention. In Australia, the 'rationalist' macroeconomic restructuring under the Hawke-Keating Labour government deregulated the financial and banking systems and floated the Australian dollars, aiming at making Australia an attractive place for international capital and improving Australian economy's competitiveness. These effects were intensified by a 'pro-development state ideology' in New South Wales, seeking to restructure the manufacturing base and relocate warehousing to give space to the commercial facilities needed to serve the growing financial and tourist industries. The state government turned to a neoliberal planning paradigm of market liberalization and place making (Searle and Cardew 2000; Hu 2012, 2013).

The awareness of the need to shape a global Sydney was nurtured in the late 1980s. In the 1990s, a global Sydney became a vision of the state government and the city council. The proposal to bid for the 2000 Olympic Games was one of the initiatives to celebrate and market Sydney in a global arena. Reflecting a greater integration with the global economy and accelerated by the Olympics, the Sydney CBD in the mid and late 1990s was an extraordinary construction zone at the mercy of a patchwork of ambitious private, state government and council development. For the Sydney City Council, Frank Sartor was elected Lord Mayor as the city's first independent mayor in 1991, who continued until 2003. Sartor had passion for the vision of a global Sydney, and attached great importance to the utility of good urban design and open space in realizing such a vision. He was the city's first champion of urban design. His emphasis on better quality design was first recognized by the state government in the Darling Harbour redevelopment and other bicentenary projects. In the pre-Olympics years, both state government and city council initiated a series of urban planning agendas with a focus on the city's livability, accessibility, public space and design excellence, which significantly differentiated from the city's prior planning efforts (Hu 2012).

From the early 2000s, the push for a 'Global Sydney' has been promoted across public and private sectors as a means to facilitate economic growth and address declining levels of productivity. This push was propelled by a number of additional economic challenges facing the city, including resource booms and associated economic prosperity in other cities and states, and intensifying inter-urban competition affecting portions of Sydney's economic base, such as banking and corporate governance. In the wake of the 2008 global financial crisis, the global status of Sydney continues to mobilize as a tool to facilitate economic growth and encourage business confidence. As such, Sydney's planning has been heavily influenced by the quest to secure global city status, which remains evident in ongoing policy ambitions at all levels of government (Baker and Ruming 2014). At a state level, positioning Sydney as Australia's only global city is a long-standing foundation of metropolitan planning, economic policy, and politics.

However, the analysis reveals the contextualization of planning central Sydney, providing a typical example of local planning responses to the forces of globalization. The accelerating process of globalization has been restructuring the geography of the integrated world economy, which is a direct cause of the emergence of a group of strategic sites of global importance, or global cities as 'strategic' sites for the management of the global economy and the production of the most advanced services and financial operations, that have become key inputs for the work of

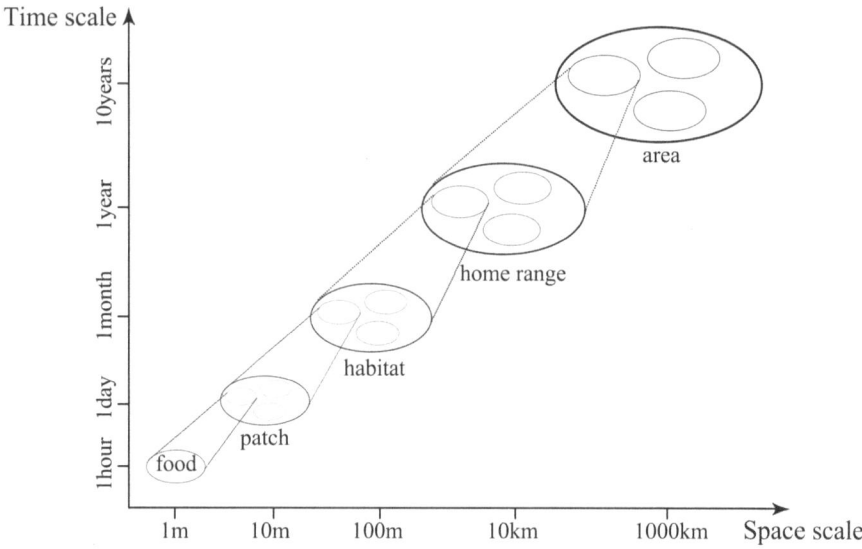

Fig. 4.1 Ecological concepts connecting time and spatail scales

managing global economic operations. Globalization's impacts on Sydney's economic and political landscape gained pace, forging the imperative for planning an entrepreneurial city in a competitive world (Fig. 4.1).

4.2.2   The New Occupational Structure and Social Polarization

The restructuring of Sydney's economy has fundamentally reshaped its class structure. The traditional blue collar working class is now a small proportion of the workforce. Economic rationalization, deregulation and increasingly international rationalism and competition has resulted in a loss of manufacturing jobs in Sydney. A large proportion of jobs associated with the new economy are located in the CBD and adjacent inner-city neighbourhoods, like Pyrmont Ultimo (Fagan 2000; Bounds 2006). The occupational structure in Sydney appears to be polarizing. Whilst in absolute terms, occupations such as trades persons and clerical workers still account for large numbers in a dynamic sense, the major growth occupations are located at the top and bottom of the status-skill continuum. In Sydney, there is significant growth in the occupation category of professionals, and when combined with managers, administrators, and para-professionals, these make up a large proportion of the employed labor force. There is also evidence of polarizing trends – that is, there is growth in professionals and managerial occupations, yet at the same time there is growth in lower-order occupations which lead to a polarizing occupational structure.

There is a wider view of polarization in the labor market in Sydney. It appears that changing polarization may be viewed in terms of three groups. At one end, there is a group of high-skill, high-status occupations which are firmly tied to the global economic structure. At the other end, there may be two groups. One is the group of less-skilled, lower-status occupations only slightly associated with the global economy. The other group comprises the unemployed who have not benefited at all from global changes. As with the changes in the labor market, income polarization in Sydney may reflect the growth of three groups. At one end of the scale, there exists a growing group of high-income, high-status individuals who are strongly attached to the global economy, and have benefited from global integration. At the other end, there is both a growing group of workers who have only weak attachment to the global economy, and a group who are outside the employed labor force, are dependent on welfare, and have benefited very little from global processes.

The analysis of occupational change illustrated that four occupational groups – managers and administrators, professionals, para-professionals, and service and personal service workers – were characterized by growth over decades. The growth of occupations of managers and administrations, professionals and para-professionals tended to cluster towards the high end of the income distribution. In Sydney, huge numbers of members of the managerial workforce with university degrees become the new middle class. Due to the changing nature of employment, most of the people involved in low-paid jobs also consider themselves as the middle class. A proportion of residents living in Pyrmont Ultimo fit the attributes of what is often called the new middle class. This new middle class is comprised of the professional-managerial cohort and professionals in the arts and applied arts, the media, teaching and social services such as social work, and in other public- and non-profit-sector positions. The new middle class in Pyrmont Ultimo has moved to the area to take advantage of its leisure facilities such as cafés, restaurants, clubs, pubs and shopping, and proximity to employment in the CBD. They have chosen an alternative urbanism: suburbanism (Bounds 2001; Bounds and Morris 2005). In Pyrmont Ultimo, many professionals can be found who are involved in managerial jobs in the city.

Manufacturing jobs have decreased due to an increasing number of jobs in service sectors. The new working-class people are mostly involved in hospitality and catering, cleaning, security, and constructions. They are working hard, even in the weekends, in order to improve their economic conditions. They often live far from their workplace, often moving to Western Sydney due to affordable housing, while jobs are being created in CBD and Eastern Sydney. Most of their earnings are spent on housing. Their lives are confined to work and sleep, despite living in a global city.

The global Sydney resident is described as a cosmopolite, a figure who, equipped with various class and taste capacities, performs 'worldliness' through consumption.

Sydney's global arc is an internationally competitive and cosmopolitan area with all the lifestyle attractions and expenses of a world city (Latham 2003). This area is populated by a global urbanite, more inclined to be cosmopolitan, progressively tolerant and confident, and generally embracing the rhetoric and horizons of the global village. Latham (2003) spatializes Sydney as having a 'global' core, a 'middle area' of suburbs which have borne heavy impacts from deindustrialization, experience above-average levels of unemployment and welfare dependency, and which tend to be dominated by the newly arrived immigrants and an 'outer arc' of 'middle Australian' upwardly mobile suburbanites seeking new land release areas on the urban fringe.

4.2.3 Restructuring, Gentrification and the Community

The decline of manufacturing and the growing prominence of a post-fordist economy in Sydney was the basis for new development in the built form and urban design. In Sydney, the demise of the fordist economy is perhaps most dramatically illustrated by the fact that economic rationalism, deregulation, and increasing international competition saw close to 178,000 manufacturing jobs being lost in Sydney in the 1980s (Fagan 2000). In contrast, by the mid-1990s, the proportion of employees in the information and business services sector was approaching the proportions employed in this sector in other global cities (Searle 1996). A large proportion of the jobs associated with the new economy are located in the CBD and adjacent inner-city neighbourhoods like Pyrmont Ultimo (Bounds and Morris 2005).

By the end of the 1980s, the economic base and class structure of Sydney had fundamentally altered. The manufacturing sector had declined considerably and the retail, service, recreation, finance and information economy had become the driving forces of the city's economy (Fagan 2000; Murphy and Watson 1994; Bounds 2001). By the 1990s, Sydney had established itself as a major financial and corporate centre, and in 1997 and 1998 alone, 61 multinationals established regional headquarters in Sydney. Sydney has also established itself as one of the prime tourist spots in the world. In the 1990s, earnings from tourism almost doubled, reaching $16 billion in 1998 (Daly and Pritchard 2001; Bounds and Morris 2006).

The gentrification started in Pyrmont for residential purposes, while the new economy was restructuring. This gentrification process can be clearly observed in Pyrmont Road, where a huge number of high-rise buildings was built for residential purposes. The process of gentrification and urban restructuring are still continuing due to increasing pressure of population in this suburb. In recent decades, huge numbers of migrants from various Asian countries moved to this suburb, being able to afford the costs. Moreover, the second generation of Asian migrants are also buying property here. Due to its location, Pyrmont has become the most densely populated suburb in Sydney. The increasing population has huge pressure on land and housing in this area. To meet the demands of housing for white collar workers, local developers and real estate are focusing on more high-rise development (Elenius 2013).

The new economy has created new social spaces in Pyrmont Ultimo. The middle class, involved in managerial and administrative services in the city, are filling up the suburb. Many of them are second generation Asian migrants who are skilled and involved in professional jobs, and moved here as it is closer to the city. In fact, this inner-city area is dominated by professionals and white collar workers involved in the new service economy. It is becoming very multicultural due to an increasing number of migrants from predominantly Asian countries. Migrants living in this area are socially connected as many of them come from a similar social and cultural background.

The community members have formed the 'Friends of Pyrmont Community Centre' which they have identified as the focal point of the local community. Whilst the building is totally unfit for its purpose and much of it is leased as a childcare centre, they have managed to initiate a range of activities run by volunteers, including monthly community dinners, and a recently held Open Weekend, complete with photographic exhibition and local history display, performances by the local choir and theatre group, and a Sunday children's day to celebrate their success in getting the Centre open on a Sunday (Elenius 2013).

4.2.4   Culture, Consumption and Leisure Spaces

The pattern of cultural consumption significantly changed with the revitalization of the declining, manufacturing-oriented day-time economy of inner-city spaces, caused by de-industrialization through the provision of services in the evening and night-time periods. The accelerating movement of women into the full and part-time workforce, large-scale migration from a wider range of source countries, the flow of population back into the inner city through the growth of densely packed, multiple-occupancy dwellings, and the gentrification of some inner-city neighbourhoods play significant role in the development of the night-time economy. The nature of work is also beginning to change, especially with the rise of the service and information sectors that are much more task- than time-oriented, and forms of post-Fordist continuous production operating on 'just in time' schedules, accompanying a de-regulation of work and leisure hours (Rojek 2010).

The city of Sydney wants to create a vibrant nightlife that offers something for everyone. It has the largest night time economy, and is one of the greater residential and commercial destinations, in Australia. The city has the largest number of visitors, and the offering in Sydney at night is central to this. The fact is, truly great and memorable global cities are those that have many options for all members of the community, irrespective of age. They feel safe, are easy to get around, and are easy to leave at the end of a night. They offer something across the span of the evening, and are designing for both day and night.

The activities of young people represent a definite dynamic in the city particularly at night. The activities of dance parties, drinking, live music, hanging out, graffiti and skateboarding occupy an important place within youth culture in Sydney. For young people the city is a place of electronically monitored public spaces but at the same time contains spaces largely beyond suburban surveillance. It is a place where people and public transport are passing in, through and out, of conviviality, opportunity and options, of public housing, and fast food. The city is a place of concrete shapes and skate-able structures that can be imaginatively accessed using the urban movement practice of 'parkour'. It contains hidden but accessible spaces, with people moving nearby, walking past, standing, gathering, providing. It reveals the constitutes of the dynamic impact of a part of youth culture on the wider city. Its rhythms and spaces articulate with 'made over' city sites as non-or semi-regulated performance spaces with the less regulated structures of living involving casual and part-time work. With post-school education often stretching well into the twenties and some young people experiencing intermittent or long-term unemployment, orderly career progression exerts less of a disciplining force. The live music scene in Sydney, although not only the province of the young, is also instructive in this context. The movement of younger populations into the city has resulted in some conflict with residents over noise, not least that involving amplified music. The displacement of live music by the provision of more lucrative gaming facilities, especially poker machines, has also been a source of conflict over leisure practice. The estimated loss of live music work for 67% of the musicians reveals the precariousness of labor in that sector, alongside the culture and leisure that it produces in specific urban environments (Rowe and Lynch 2012).

Transformations are taking place in consumer landscapes and leisure spaces in inner-city Sydney. The formation and transformation of inner-city spaces associated with sexual and gender minorities became important in the 1980s. Work over the last decade has begun to analyse social and economic changes within gay villages, including critique of neoliberal imperatives and implications of social and political acceptance. The analysis focuses on the use, meaning and social significance of leisure-based consumption sites – clubs, bars, cafes, restaurants. The balance of daytime/night-time leisure spaces, which have both social and material affordances, is a key discrimination across neighborhoods, both within and between cities. Daytime consumer landscapes are more often framed, sociable, and inclusive within the media, while night-time landscapes are perceived as divisive (Gorman-Murray and Nash 2017).

The new political-economic regimes responding to intensified globalization and neoliberalism initiated significant change in the enrolment of gay villages within the socioeconomic urban fabric. This decade sees a shift from the Fordist governance model of municipal administration to a post-industrial model of entrepreneurial urban development and championed market-led revitalization of the inner city as a landscape of consumption and cosmopolitanism, tied to the rise of white-collar labor and the creative class. Given the centrality of consumption sites for public encounters in the inner city, these commercial hospitality venues are closely linked to neighborhood 'feel', ambience, identity, and even politics, as evoked in nomenclature like 'café culture' 'eat street' and 'urban village'. Consumption spaces are being re-conceptualized as sites of hospitality, sociability and conviviality, which can encourage dialogue and foster an appreciation of differences. Commensality is not always a disguise for competitions over taste and status: it can also be about social identification, the sharing not only of food and drink, but of world-views and patterns of living. Touristification marks the commodification and mainstream consumption of gay villages, incorporated in city marketing as a signal to the cities' cosmopolitan diversity (Gorman-Murray and Nash 2017).

4.3 Conclusion

The contextualization of planning central Sydney provides a typical example of local planning responses to the forces of globalization. Globalization's impacts on Sydney's economic and political landscape gained pace, forging the imperative for planning an entrepreneurial city in a competitive world. The cultural practice of an information-rich, techno-elite, global Sydneysider becomes part of a re-mapping exercise. The restructuring of Sydney's economy has fundamentally reshaped its class structure, and the traditional blue collar working class is now a small proportion of the workforce. Economic rationalization, deregulation, and increasingly international rationalism and competition resulted in a loss of manufacturing jobs in Sydney. The inner-city area became the location of new hotels, restaurants and cafés. The number of tourists in Sydney has significantly increased in recent decades. The new economy has created new social and cultural spaces. The middle class, involved in managerial and administrative services in the city, are moving to inner-city areas. The migrants, who are skilled and involved in professional jobs, also moved here. The global Sydney resident is described as a cosmopolite, a figure who, equipped with various class and taste capacities, performs 'worldliness' through consumption. Given the locations of significant sections of the entertainment, hospitality, alcohol and gambling industries, it is evident that the CBD and environs is a complex fabric of work, and leisure flows and activities. There are clusters in inner-city areas which are dominated by hospitality as well as cultural facilities dedicated to exhibition, performance and entertainment. The new political-economic regimes, responding to intensified globalization and neoliberalism, initiated significant change in leisure spaces within the socioeconomic urban fabric. In conclusion, the new social and cultural landscape of Sydney has been produced as a result of global city formation and neoliberal transformations.

References

Acuto M (2011) Sydney, the global harbor city. Diplomatic Courier, December 2011
Baker T, Ruming K (2014) Making 'global Sydney': spatial imaginaries, worlding and strategic plans. Int J Urban Reg Res 39(1):62–78
Bounds M (2001) Economic restructuring and gentrification in the inner city: a case study of Pyrmont Ultimo. Aust Planner 38(3–4):128–132
Bounds M (2006) Urban social theory: city, self and society. Oxford University Press, Melbourne

Bounds M, Morris A (2005) High-rise gentrification: the redevelopment of Pyrmont Ultimo. Urban
    Des Int 10:179–188
Bounds M, Morris A (2006) Second wave gentrification in inner-city Sydney. Cities 23(2):99–108
Daly M, Pritchard B (2001) Sydney: Australia's financial and corporate capital. In: Connell J (ed) In
    Sydney: the emergence of a world city. Oxford University Press, Melbourne, pp 167–188
Elenius E (2013) The Pyrmont and Ultimo experience. Inner Sydney Voice, December 2013
Fagan R (2000) Industrial change in the global city: Sydney's new spaces of production. In: Connell
    J (ed) Sydney: the emergence of a world city. Oxford University Press, Melbourne, pp 144–166
Gorman-Murray A, Nash C (2017) Transformations in LGBT consumer landscapes and leisure
    spaces in the Neoliberal City. Urban Stud 54(3):786–805
Hu R (2012) Shaping a global Sydney: the city of Sydney's planning transformations in the 1980s
    and 1990s. Plan Perspect 27(3):347–368
Hu R (2013) Measuring the changing faces of global Sydney. Paper presented at State of Australian
    Cities Conference Sydney 26–29 November 2013
Latham M (2003) From the suburban: building a nation from our neighbourhoods. Pluto Press,
    Sydney
Murphy P, Watson S (1994) Social polarization and Australian cities. Int J Urban Reg Res
    18:573–590
Rojek C (2010) The labour of leisure. Sage, London
Rowe D, Lynch R (2012) Work and play in the city: some reflections on the night-time leisure
    economy of Sydney. Ann Leis Res 15(2):132–147
Searle G (1996) Sydney as a global city. Department of Urban Affairs and Planning, Sydney
Searle G, Cardew R (2000) Planning, economic development and the spatial outcomes of market
    liberalization. Urban Policy Res 18(3):355–376

Chapter 5
Post-earthquake Recovery in Nepal

Rupesh Shrestha, Alexander Fekete, and Simone Sandholz

Abstract In 2015, a massive earthquake of 7.8 and 7.4 magnitude struck Nepal. This resulted in severe economic and infrastructural damage, not to mention many human casualties. The government of Nepal has identified 625,000 houses as fully destroyed and 180,000 houses as being partially damaged (PDNA Vol B, Post disaster needs assessment – sector reports. Government of Nepal, Kathmandu, 2015; Sector Plans – GoN, Sector plans and financial projections -working documents. National Reconstruction Authority, Government of Nepal, Kathmandu, 2016).

This research is a comparative study of traditional-urban, peri-urban, and remote rural settlements of Nepal which were severely hit by the earthquake. It provides an overview of interests and perceptions of local communities in terms of the recovery process. Furthermore, this research also identifies resilience in terms of basic service recovery (basic shelter, electricity, water supply, telecommunication, groceries/food) and existing challenges in housing recovery programs. Assessing the different settlement types individually also allows for tailored policy recommendations to bridge related gaps. From the survey conducted, it can be seen that earthquake affected people's perception of housing (re)construction has changed considerably and that they are more interested in having earthquake resistant houses after the 2015 events. Analysis also shows that, unlike in urban areas, people in rural areas tend to build stronger houses when they understand the scientific reason behind earthquake-induced damages. Lack of financing is a major hindrance for reconstruction in all study areas, and there is a need for government and financial institutes to engage to create favourable financing schemes.

R. Shrestha (✉)
ITT, TH Köln – University of Applied Sciences, Köln, Germany

A. Fekete
Institute of Rescue Engineering and Civil Protection, TH Köln – University of Applied Sciences, Köln, Germany

S. Sandholz
United Nations University – for Environment and Human Security (UNU-EHS), Bonn, Germany

© Springer Nature Switzerland AG 2020
R. Roggema, A. Roggema (eds.), *Smart and Sustainable Cities and Buildings*,
https://doi.org/10.1007/978-3-030-37635-2_5

Keywords Post-disaster housing reconstruction · Field study · Kathmandu · Sindhupalchok · Bungamati · people's perception · Nepal

5.1 Introduction

On April 25 and May 12 2015, earthquakes of 7.8 and 7.4 magnitude struck Nepal with epicenters at Gorkha district,[1] north-west of the capital Kathmandu (April), and between Sindhupalchok and Dolakha districts north-east of the capital (May), followed by numerous aftershocks. This disaster resulted in major economic and infrastructural damage, as well as 8702 human casualties. The Government of Nepal has identified 625,000 houses as fully destroyed and 180,000 houses as partially damaged in the whole country (PDNA Vol B 2015; Sector Plans – GoN 2016).

In Nepal, the earthquake of 2015 has affected both urban and rural areas. Impacts were comparably higher in undeveloped areas than in developed areas. An estimated 6.695 Billion US$ is required for the recovery process (PDRF 2016), signifying around 32.06% of Nepal's Gross Domestic Product (GDP) of 2015. Housing recovery is estimated to take 3.27 Billion US$. This means 49% of all recovery funds needs to be spent only for the housing, making it the most wanting sector.

People affected with disaster tend to return to 'normality' in post-disaster situations. However, normality in developing countries often means accepting the risk of continuous forms of disaster (Parker et al. 1997; Charles 1995). Disasters are events with multi-dimensional outcomes, ranging from socio-economic to cultural, political, humanitarian, and physical dimensions (El-Masri and Tipple 1997). Hence, recovery and reconstruction should be done appropriately to minimize the different vulnerability dimensions, maximizing resilience rather than just returning to normality, which would mean re-creating the same conditions that have contributed to the previous disaster (UN 2015). Reconstruction can thus bear new opportunities for minimizing risk levels. The reconstruction period offers a chance to strengthen local capacities and to facilitate economic, social, and physical development in the long-term (Berke et al. 1993).

Post-disaster recovery is a complex process which requires multi-sectorial involvement, significant resources, and a wide range of skills. The Sendai Framework for Disaster Risk Reduction (SFDRR) has promulgated the concept of enhancing disaster preparedness for effective response (UN 2015). It further states that "Disasters have demonstrated that the recovery, rehabilitation and reconstruction phase which needs to be prepared ahead of a disaster, is a critical opportunity to 'build back better', through integrating disaster risk reduction into development measures, making nations and communities resilient to disasters." (Sendai

---

[1]Nepal is divided administratively into Federal provinces, Districts, Gaupalika which was previously known as Village development committees (VDC), Metropolitan areas and then Municipalities.

Framework 2015). Hence, current disasters are seen as an opportunity to prevent future disaster and as a means of enhancing resilience. For this research, primary data were collected from 176 households using stratified random sampling in the Kathmandu Valley and the Sindhupalchok district. In addition, three expert interviews with bureaucrats from the Government of Nepal and community leaders were conducted to gather background data. The research compares urban and rural housing recovery. Urban case studies in Kathmandu Valley were Gongabu and Bungamati. For study on rural recovery, Thangpal valley, which lies in Sindhupalchowk district, was chosen.

The purpose of this paper is to:

1. Examine interests and perceptions of local communities towards the recovery approaches being undertaken.
2. Investigate achievements and gaps in the housing recovery process and how it is contributing to resilience of the community.

5.2 Post-disaster Recovery in Literature and in Nepal 2015

Researchers have suggested that shelter reconstruction must be considered a process or a series of actions for fulfilment of certain needs, rather than only as objects or tents (Davis 1978; Jim Kennedy et al. 2008; Turner 1972). Housing plays a central role in both casualties and economic loses. If done properly it can contribute to resilience in long-term (Jones 2006; Jha et al. 2010).

Furthermore, disaster recovery stages are divided into phases, but there is no precise agreement on the number and definitions of these recovery stages (Lindell 2011; Alexander 1993; Haas et al. 1977; Sullivan 2003; UNDRO 1984; Schwab et al. 1998). While confusion and inconsistent definitions still exist, this research uses recovery as a broader term. The point of departure for this study was adapted from Quarantelli (1999) and Jha et al. (2010):

> *the word recovery often seems to imply that attempting to and/or bringing the post disaster situation to some level of acceptability. This may be or may not be the same as the pre-disaster level.* (Quarantelli 1999)

> *Decisions and activities taken after a disaster to rehabilitate or improve the pre-disaster living conditions of the affected communities while encouraging and facilitating the necessary adjustments to reduce disaster risk. It is focused not only on physical reconstruction, but also on the revitalization of the economy, and the restoration of social and cultural life.* (Jha et al. 2010)

Disaster recovery can be sub-divided further into three distinct and interrelated meanings (Lindell 2011):

(a) The goal of recovery means reinstating normal community activities.
(b) The phase of recovery which begins with stabilization of the disaster situation (when the emergency response stops) and ends when the community has returned to its normal state or, as by the definition of Quarantelli (1999), to

some level of acceptability. This research has also adapted the same meaning of recovery phase.

(c) The process of recovery means a series of actions by which a community achieves normality. The process includes activities that were planned before the disaster and activities that were improvised after the disaster.

5.2.1 Challenges and Critiques 'Build Back Better' and SFDRR

After the 2004 Tsunami disaster, Clinton (2006) provided 10 propositions in order to elaborate 'build back better'. As a former US president, Bill Clinton's propositions were prominent in providing a framework for 'build back better'. However, researchers have pointed out the drawbacks in concepts of 'build back better': they do not provide clear goals for post-disaster recovery processes. The term 'better' can have multiple interpretations and does not necessarily go hand in hand with 'safer' construction (Jim Kennedy et al. 2008). This would mean that vulnerabilities are not minimized, and resilience is not maximized. In Aceh which was hit hard by the tsunami, success was measured by the number of houses completed and the speed of execution, rather than focusing on safety, security, and livelihood. Also, 'better' was perceived as 'modern masonry' dwellings rather than structurally safer vernacular architecture (Jim Kennedy et al. 2008). Another risk is recovery efforts that can address only the immediate hazard due to time pressures in the recovery process and not fully mitigate vulnerabilities to other hazards (J. Kennedy et al. 2009; Mannakkara and Wilkinson 2013).

On March 2015, the UN World Conference on Disaster Reduction took place, where Sendai Framework for Disaster Risk Reduction 2015–2030 was signed. SFDRR replaced the Hyogo Framework for Action 2005–2015 (HFA). International frameworks like SFDRR have served as a guideline to carry out the recovery process. SFDRR moves beyond HFA and includes response and recovery timeframes, as a way to reduce risks and build back better (Pearson and Pelling 2015). SFDRR also includes seven global targets, whereas HFA did not include any. SFDRR highly recognises the vital role of civil society, science & respective governments to achieve its global targets.

However, certain conceptual drawbacks have been pointed out in the framework. SFDRR discourse has been termed as top-down. It does not include substantial focus on community participation, while local communities are valued partners in disaster recovery rather than mere "aid recipients" (Tozier de la Poterie and Baudoin 2015). Global targets included in SFDRR lack quantitative guidance on how the targets are to be met. This can lead to ambiguity in assessing how far SFDRR has achieved its goal (Pearson and Pelling 2015).

Jha et al. (2010) states that a disaster which has occurred in both urban and rural areas is challenging and complex to solve. This is a valid assessment of Nepal after the 2015 earthquake, the recovery of which is posing serious urban and rural challenges. To meet these challenges, the Government of Nepal announced the establishment of a National Reconstruction Authority (NRA), a national body that is supposed to report to the Cabinet and which is empowered to set recovery policies, and provide oversight to the recovery efforts of Government as well as the support provided by international and local actors. The underlying visions of the Government of Nepal are well-planned, resilient settlements and a prosperous society through the recovery process, including safe structures, social cohesion, access to services, livelihood support and capacity building (PDRF 2016).

In Nepal, a more systematic disaster response management began only after 1982 when the Natural Calamity Relief Act 1982 or Natural Disaster Relief Act was created (MoHA 2013). In 1999, the Nepalese government promulgated the Local Self Governance Act. These two acts provided legal foundations for disaster response. This was instrumental towards decentralization of authority by focusing on interrelationships between development processes, environment, and disaster. It allows local authorities, namely District Development Committees (DDCs), Municipalities, and Village Development Committees (VDC),[2] to manage environmentally friendly development through their own actions. It is noteworthy that local bodies did not have any elected officials, due to political deadlock for 20 years. However, on May 14 2017, Nepal successfully conducted local elections, and elected officials are taking charge of development matters again (Sharma 2017).

To support people severely affected by the earthquake, Government of Nepal has decided to provide NRs 300,000 (equivalent to US$2792.16, 1US$ is approx. NRs. 107.29 on Oct 2016) financial grant, each increased from the initial NRs. 200,000 (equivalent to US$1864.11) (Post Report Kantipur 2016). The grant amount is the same for urban and rural areas and dispatched on an instalment basis after the completion of each milestone of construction. A loan of 2.5 million NRs. from Banks for Kathmandu Valley and 1.5 million for areas outside Kathmandu valley can also be granted. The government prefers most of the housing reconstruction to be on the original land. Only where it is unavoidable, relocation will be permitted. For relocation, the government intends to prepare settlement plans which include required civic amenities, employment patterns and social networks (PDRF 2016).

Remittance is an important funding source for reconstruction in rural areas of Nepal (Sector Plans – GoN 2016; PDNA 2015). Migrant remittance flow increases in the aftermath of large disasters, which acts as a safety net for households. Such households appear less vulnerable and possess a considerably greater livelihood resilience (Pant 2016). Practical Action (2014) reported that migrant workers from Kathmandu and Jhapa district save up to 37% of their income, 18.1% of this being savings for various construction related practices. A study by Manandhar (2016) found that despite a significant level of remittance money spent on construction work, unsafe construction practices in Nepal increased.

---

[2]VDC were a local governance body before March 2017 which is replaced by Gaupalika that has greater decision-making powers. Development works at local levels were executed through VDC.

5.3   Case Study & Target Population

Gongabu falls into the peri-urban outer fringe of Kathmandu, and is a densely packed urbanized area. Gongabu occupies 2.7 km^2 and has a population of 54,410. The residential neighbourhood is characterized by a mixed land use, with socially mixed classes. It is also considered ecologically important because it serves as a major recharge zone for the valley (Shakya and Shrestha 2013). The same study also found that buildings are used in the most part for housing purposes, and are mostly owner-built with private developers and contractors. The Tokha municipality governs Gongabu but is itself a new municipality, which was formed just 4 months before the Gorkha earthquake of 2015 (Post Report Kantipur 2014). Tokha Municipality (2015) statistics show that 332 houses were fully, and 2446 houses partially, damaged by the earthquake. Altogether, there were 135 deaths in this neighbourhood. A study by Hibino et al. (2015) shows that reinforced concrete and masonry buildings were the most damaged.

Bungamati falls into the sub-urban inner fringe, and is a settlement with traditional value. It occupies 4.03 km^2 area and has 5966 population. It is located 10 km south of Kathmandu's centre. It was settled in the seventh century and its inhabitants are predominantly Newars, the indigenous people of Kathmandu valley. The traditional settlement has a compact built form and is famous for renowned temples, wood-carvings and handicrafts. The Central Bureau of Statistics (2012) states that the people here have mostly Hindu and Buddhist religious values. The town form, street patterns, community infrastructure, and open space hierarchy includes Newari architecture style buildings (Shrestha 2008). 4 months prior to the earthquake of 2015, it was declared a part of the newly established Karyabinayak municipality. The Karyabinayak Municipality damage assessment report (2016) shows that 1273 houses were fully damaged and 222 houses were partially damaged. Total houses recorded to be existent before disaster were 1704 houses. This data also shows that the Bungamati experienced severe damage, as 75% of its housing stock was fully destroyed in the 2015 earthquake.

The Thangpal valley in the rural hinterland of the Sindhupalchowk district is highly inaccessible, a great distance from a market centre, and has little infrastructure. The Thangpal valley has a total population of 8047 and covers is 59.52 km^2. The two villages studied within the valley are Thangpalkot and Gunsa. The Thangpal valley can be categorized as rural hinterland with an agrarian and subsistence lifestyle. The valley lies in the Sindhupalchok district, which has been categorized as severely hit by the Gorkha earthquake (Sector Plans – GoN 2016). The Thangpal valley comprises of 3 VDCs, viz. Thangpalchhap, Thangpalkot, and Gunsa. There are 1924 households in the Thangpal valley (653 households in Thangpalkot and 449 in Gunsa) (CBS 2014). The population density is only 135.19 inhabitants per sq. km. This is 170.5 times less than Gongabu. The area of Thangpal valley is 25 times bigger than Gongabu and lies in an altitude between 5000 ft. to 7000 ft. People in the valley are from different castes, the majority being from "tamang", "lama", and "chhetri" caste. The literacy rate is less than 50%, and further social problems are poverty, lack of education, and child marriage (VDC

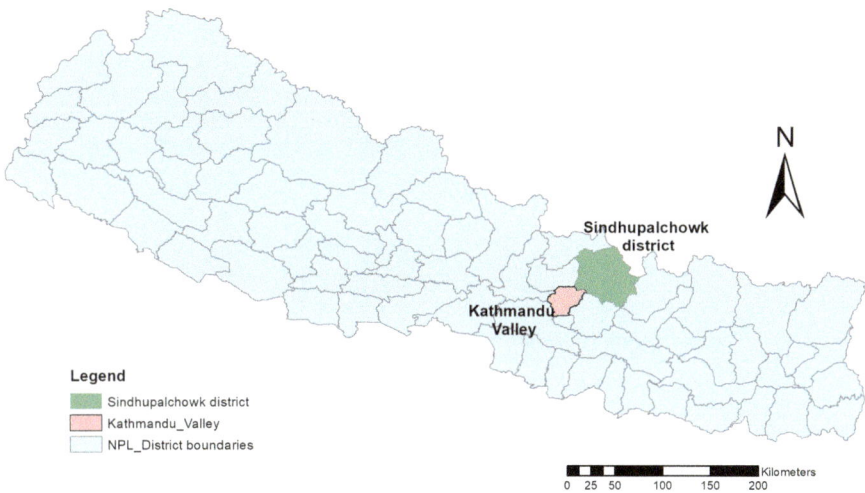

Fig. 5.1 Map of Nepal with location of Kathmandu and Sindhupalchok. (Data source: ArcDiva, KVDA, modified by authors)

office – Gunsa 2011). The moderate climate is considered good for farming; hence, people are mostly engaged in agriculture as a profession, however, a growing number of people from Gunsa also go for foreign employment and send regular remittances.

Damage to 98% of houses was observed in the settlement. 90% of buildings in this area were mud-bonded stone constructions. Data from UN OCHA (2015) indicates that only 7 buildings, from a total of 1102 houses, were left undamaged after the earthquake. Thus, the whole settlement is termed "destroyed". Buildings are planned in a dispersed manner in Thangpalkot, clustered in Gunsa. Buildings were predominantly for housing purposes before the earthquake. The housing construction approach here is mostly owner-built or formal type housing built by a local contractor. The settlements of Gunsa and Thangpalkot each have a Village Development Committee (VDC), which exists as the lower-tier and local body for governance. The second tier consists of Districts, which are governed by the District Development Committee (DDC) (Figs. 5.1 and 5.2).

5.4 Methodology

The World Bank's Handbook on Safer Homes by Jha et al. (2010) and Silva (2010), the Sendai Framework 2015, and the Global Assessment Report on Disaster Risk Reduction 2013 formed the base for conceptualizing this research. Following a multiple-case design, three settlements of Nepal were selected from rural and two different urban contexts. The case study sites of Bungamati, Gongabu, and Thangpal

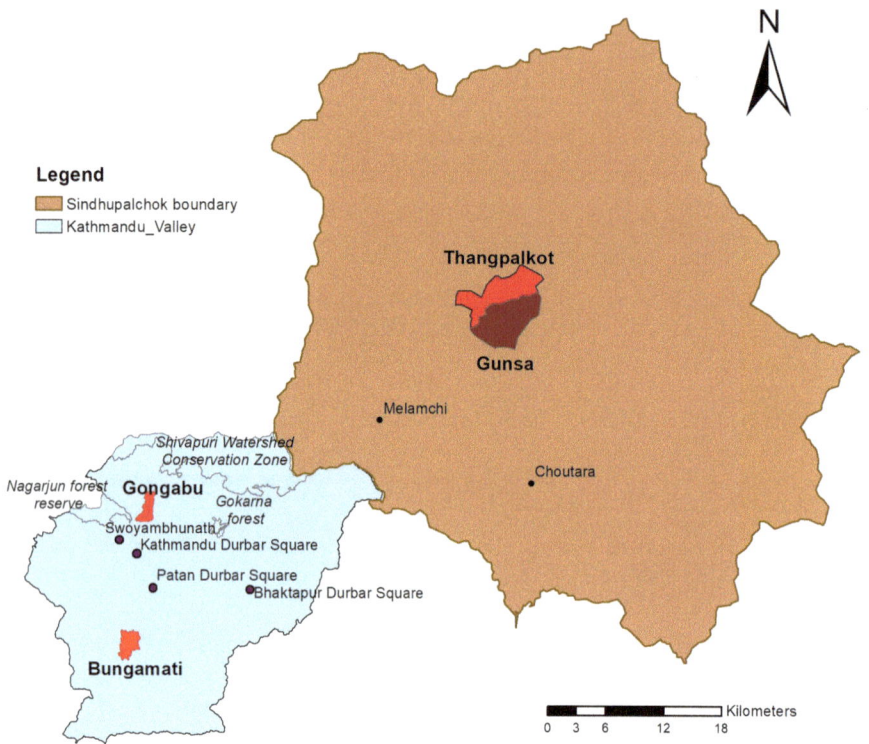

Fig. 5.2 Map of Kathmandu valley and Sindhupalchok. (Data source: ArcDiva, KVDA, modified by authors)

valley were picked based on existing knowledge of the researchers. The reason behind the selection of urban and rural sectors as case study areas, is the distinctive socio-economic patterns, architecture, settlement patterns, governance, damage level, and demographic distribution.

Through a questionnaire, 59 individuals from separate households from Bungamati, 57 from Gongabu, and 60 from Thangpal Valley were surveyed using stratified random sampling. 3 key expert interviews were collected during the fieldwork – this was carried out during June 2016.

Various cases of reconstruction approaches occurring in other countries and its relationship to the whole recovery phase were studied based on literature and secondary data sources. Reconstruction approaches undertaken by government and civil society were compared to people's needs. The needs were categorized in terms of planning, design, and construction of housing recovery (J. D. Barenstein 2006, 2013; Jim Kennedy et al. 2008; Silva 2010; Graf 2012).

The research was carried out following the "Case study method" from Yin (2009). Previous studies in the field of urban and disaster research in Nepal have applied a mix of qualitative and quantitative methods (Marahatta 2013; Bhandari 2010; Sandholz 2015). J. D. Barenstein (2006) uses a quantitative approach in her research, and compares various reconstruction approaches employed after the Gujarat earthquake. Subba (2003) asserts that, to understand the contextual reality and dynamic interplay between human actions and human needs, qualitative method is required.

Studies by Silva (2010) and Jha et al. (2010) were used to prepare a semi-structured questionnaire for interviewing the disaster affected people in the three case study sites. Furthermore, the semi-structured interview technique was also chosen for interviews with key informants. The use of a semi-structured interview led to qualitative data about perceptions, meanings, definitions, and a construction of reality. It also opened new leads to current research (Punch 2005; Chitrakar 2015).

5.5 Design of Survey Questionnaire

The questionnaire sheet was designed with seven sections and 31 questions. It was based on acquiring the data needed to address the purpose of the research.

The sections considered were:

(a) Personal details
(b) Perception of risk, hazard and resilience
(c) Design
(d) Planning
(e) Construction
(f) Participation, beliefs, practices
(g) Needs after earthquake

Most questions were closed-ended with multiple answer choices to allow for a quantitative analysis. For open-ended questions, space was provided to fill the text.

The household survey data was analyzed using SPSS to analyse correlations, and to look for cause and effect relationships. For this purpose, Lambda tests for correlation analysis were conducted. (Creswell 2003).

Analysis and interpretation of qualitative data from interviews with key informants was done by coding in MaxQDA. Interview recordings were transcribed in word processing software and later imported to MaxQDA. The qualitative data from key informant interviews as well as from open-ended questions in the survey was used to further deepen the results gained from the quantitative data. The mixed methodology approach also allowed for comparison of the different actors' perceptions.

5.6 Findings

5.6.1 Hazard That Affects the People Most

Nepal and its inhabitants are at risk from different natural and man-made hazards. To assess the risk perception, the respondents in all three sites were asked about the hazards they feel most at risk to.

Respondents in Gongabu believe that earthquakes are going to affect them more than fire. Few expressed that floods would also affect them. Additionally, nearly 100% of respondents in Bungamati believe that an earthquake is going to affect them the most in terms of natural hazards.

However, in Thangpal, only 43.33% believe that an earthquake is going to affect them while 30% said that thunder and another 20% that landslide would affect them most. It can thus be observed that recovery works have to be tailored to the different sites, to mitigate multiple hazards if necessary and to take people's perception of hazard risk into consideration, in order to come up with sustainable recovery strategies.

5.6.2 Awareness About Building Codes and People's Perception on Rebuilding

Building code compliance is taken as a primary tool for measuring the quality of reconstructed houses. However, research shows that only 52.63% of respondents in Gongabu know that building codes exist at all. The awareness level for Bungamati is similar, where the same number of respondents are aware of building codes. This, however, is lowest in Thangpal valley, where only 47.46% of respondents have heard about building codes implemented by the government. This can be a severe hurdle for post-disaster recovery if end-users are not aware of the quality to be reached. As a result, housing reconstruction might potentially perpetuate pre-disaster risk conditions.

At the same time, the study shows that the majority of respondents from all three sites prefer to use stronger construction materials for rebuilding their houses. In their opinion, the future houses must be of better quality than their previous houses and more resistant to earthquakes. Most respondents believe that concrete buildings are stronger compared to vernacular materials and technology. They also said that they want to receive training for constructing better houses.

Table 5.1 Lambda test for Thangpal valley. Crosstabulation of survey question 'What do you believe about Earthquake?' and 'Which type of house do you prefer to construct after earthquake of 2015?'

Directional measures

			Value	Asymp. std. error[a]	Approx. T[b]	Approx. sig.
Nominal by nominal	Lambda	Symmetric	.147	.064	2.117	.034
		Which type of house do you prefer to construct after earthquake of 2015? Dependent	.051	.035	1.439	.150
		What do you believe about earthquake? Dependent	.276	.124	1.945	.052
	Goodman and Kruskal tau	Which type of house do you prefer to construct after earthquake of 2015? Dependent	.065	.022		.024[c]
		What do you believe about earthquake? Dependent	.244	.065		.013[c]

Area: Thangpal valley, n = 60
[a]Not assuming the null hypothesis
[b]Using the asymptotic standard error assuming the null hypothesis
[c]Based on chi-square approximation

Lambda tests[3] done to understand the correlation between the preferred type of house and the respondents' belief about earthquakes did not show any correlation for Gongabu and Bungamati.

However, for Thangpal valley (Table 5.1), the test shows that the preferred type of house (Reinforced Cement Concrete/RCC structure professionally engineered, stone masonry – old technique, etc.) has a moderate relation with what respondents believe about earthquakes. This shows that people's perception about what is the cause of disaster (either by natural event or God's wrath) changes people's decision on how they would build their house. Also, further data analysis shows that people who believe earthquakes are caused naturally and not due to God's wrath tend to more easily accept buildings with vernacular styles of construction with improved seismic resistant features, indicating a relation between the level of education and lifestyle.

---

[3]Lambda test is a measure of association for nominal variables and its results ranges from 0.00 to 1.00. A lambda of 0.00 reflects no association between variables and a Lambda of 1.00 is a perfect association.

5.6.3   Reconstruction Site Preferences and Restoration of Services

The earthquake also poses questions on whether reconstruction should take place in the same site as before. The household survey shows that in Gongabu, 94.7% of the people desire to reconstruct their houses in the same place. The top four reasons for not relocating to other places are:

(a) 26.3% of respondents said that it is their ancestral property.
(b) 17.5% said that they find this area secure.
(c) 15.8% said that they find enough basic services available.
(d) 17.5% said that they are well-acquainted with the community members there.

In Bungamati, the household survey shows that 67.8% of respondents want to reconstruct their buildings in the same place as before and 18.64% of respondents informed that they want to construct their houses elsewhere. The top two reasons for staying are:

(a) 59% of respondents said that the reason to not move else is because it's their ancestral property.
(b) 13% said that that they find enough basic services available so they are hesitant to leave.

As for the Thangpal valley, the household survey shows that 74% want to reconstruct their house in the same place as before. The top two reasons for reconstructing in the same place are:

(a) 35.5% respondents said that they have no other land.
(b) 35.4% respondents said that it is their ancestral property.

These results again indicate a difference between urban and rural populations, while in all sites ancestral property is among the top two reasons for staying.

Basic services as a reason is given only in the urban case study areas. This is not surprising given that the restoration of basic services took the longest in the Thangpal valley. Restoration of basic services in this case means a supply which is at pre-disaster level. A comparative analysis from Fig. 5.3 shows that the recovery of basic shelter was quicker in Gongabu, compared to Bungamati and Thangpal valley. It took an average of 31.62 days to arrive at a semi-permanent type of shelter for the affected people, whereas in Thangpal valley it took almost 134 days.

In terms of restoring electricity, too, Gongabu performed better than the other areas. The mean value of restoration of the normal supply of electricity was 23 days. For water supply resumption, Bungamati performed better, taking 15 days to resume normal water supply. The Thangpal valley required the most time, taking 42 days to resume a normal water supply. Telecommunication resumed quickly in Bungamati, within 6 days, whereas in Thangpal valley it took 33 days on average. For a normal level of supply of groceries, Bungamati required only 1 day, Gongabu 4 days, and Thangpal valley 16 days. It is noteworthy that Thangpal valley is an agrarian society

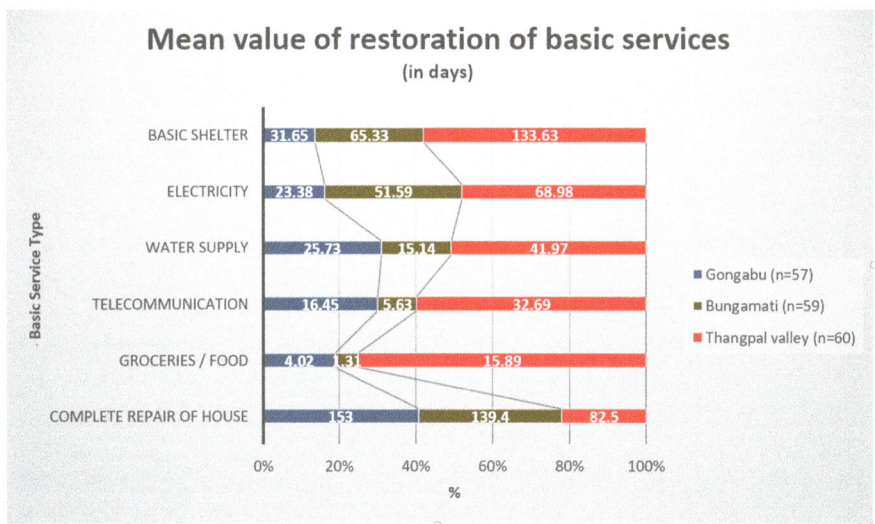

Fig. 5.3 Mean value of restoration of basic services

but it took the longest for a normal supply of grocery items. For those houses which were completely repaired after the earthquake, Thangpal valley took 83 days on average for the repair works, whereas Gongabu took 153 days for complete repair.

5.6.4 People's Perception on Government Capability of Supporting

Research in Gongabu shows that 52.63% of people believe that the government can support post-disaster recovery. Others were sceptical about the government's capacity to support them. Trust in government was better in Bungamati where 62.5% think that the government is capable of supporting them for post-disaster reconstruction. This was also the case in Thangpal valley, where research shows that around 60% of respondents believe that the government can support their community towards post-disaster recovery, and 40% believed that the government is not capable.

5.6.5 People's Demands for Post-disaster Reconstruction

For questions regarding post-disaster recovery needs, the top answer was "financial support" in all case study areas. Figure 5.4 shows that financial needs hindered their reconstruction.

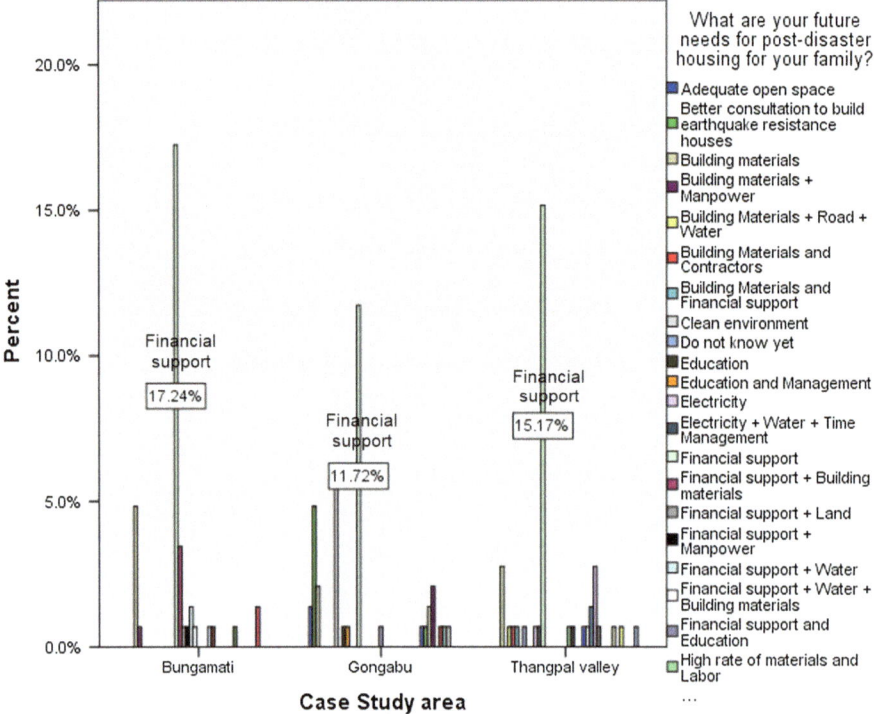

Fig. 5.4 Results for the question "What are your future needs for post-disaster housing for your family?" (n = 176)

In second position for post-disaster housing, respondents of Gongabu said that they need a "clean environment". For people of Bungamati and Thangpal valley this was "building materials".

In third position, people of Gongabu said that they need "better consultation to build earthquake resistant houses". People of Bungamati said both "financial support and building materials" are needed. People of Thangpal valley said that they need "good roads" as it is problematic for transporting building materials.

Financial support in Nepal is closely linked to remittances, which have been a major source of income over the past years. However, the research shows regional differences.

Overall 36,7% of the respondents in Thangpal valley have at least one family member working abroad and 26.67% said that they rely on these remittances for reconstruction. Only 6% of respondents said that farming would provide them with sufficient financial resources for reconstruction.

Thus, remittances are an important source for reconstruction, particularly in the rural case study site. A Lambda test (Table 5.2), however, shows that there is no relationship between the presence of a migrant family member and the type of house an earthquake affected person might choose for reconstruction.

Table 5.2 Lambda test for correlation for Thangpal valley between variables "Are there migrant workers in family?" and "Which type of house do you prefer to construct after earthquake of 2015?", n = 60

Directional measures

			Value	Asymp. std. error[a]	Approx. T[b]	Approx. sig.
Nominal by nominal	Lambda	Symmetric	.066	.030	2.071	.038
		Are there migrant workers in family? dependent	.182	.082	2.071	.038
		Which type of house do you prefer to construct after earthquake of 2015? dependent	.000	.000	.[c]	.[c]
	Goodman and Kruskaltau	Are there migrant workers in family? dependent	.141	.046		.319[d]
		Which type of house do you prefer to construct after earthquake of 2015? dependent	.011	.010		.745[d]

[a]Not assuming the null hypothesis
[b]Using the asymptotic standard error assuming the null hypothesis
[c]Cannot be computed because the asymptotic standard error equals zero
[d]Based on chi-square approximation

5.7   Discussion

There are certain similarities between the three case studies, although they have different social, economic, and cultural contexts. Respondents in all three sites lack awareness of building codes, are reluctant to relocate to new site, believe in the government, and are facing financial constraints for reconstruction.

The government demands compliance to building codes for earthquake resistant buildings, but in all case study areas, more than half of the population did not know such building codes exist. This raises the need for knowledge dissemination campaigns for building codes. However, the study also shows that most people want to construct stronger houses with stronger construction materials. People's awareness towards the need for stronger building construction along with technical input has increased.

The study also shows that a small percentage of earthquake affected people from all the case study areas want to relocate to another site. The government has decided that only in cases when it is unavoidable, relocation can be allowed to discourage scattered settlement and promote larger integrated settlements. In a practical aspect, this is challenging, as many integrated settlements must be designed and land adjustment schemes have to be executed, since not all the affected people might have land in government designated settlement zones.

Results show that most people believe that the government can support their community in post-disaster reconstruction works. However, a proper institutional set-up within government institutions is another hindrance for recovery. Municipalities and VDCs are lacking manpower and elected officials. This creates a lack of clarity for project execution and problems with accountability. On the other hand, the elections of May 2017 have paved the way for newly elected officials to take charge of development and post-disaster recovery matters, which can be a positive change for post-disaster recovery.

The lack of financing is reported to be a major hindrance for affected people to reconstruct in all three sites. Therefore, an adequate financial support scheme is in high demand. The government, along with financial institutes, should engage in creating favorable policies to address this urgent need. For people in Gongabu and Bungamati, a policy review for soft loan is suggested (HRRP/UN-Habitat 2016). A payback period extension to 18 years can be considered. A similar payback period extension can be beneficial for earthquake affected people of Bungamati, where heritage-based tourism can be a good source of income. For the affected people of the Thangpal valley, the best way is to adopt vernacular architecture with earthquake resilient technologies. Other modern construction materials will increase the cost of reconstruction and people there might not be able to pay back their loans on time.

Results also show that remittances are a good source of funds in Nepal, particularly in rural settings, but do not influence the type of construction technology a household might choose. There is a possibility that remittance dependent households are interested in building a new house, thus disregarding the necessary earthquake resistant features. This practice is likely to increase earthquake risk.

There are also contrasting issues in reconstruction for the case study sites – rooted in the differences in pre-earthquake building and settlement patterns as well as differences in recovery actions taken so far.

Bungamati is a heritage zone which has a variety of tangible and intangible values. This would mean that reconstruction will not just be about building materials or earthquake resistive designs, but also about conserving the settlement's heritage value. Careful considerations need to be made during reconstruction, linking it with public space and community infrastructure. The Thangpal valley has not been declared as a heritage zone, but does have its unique vernacular architecture. Contrarily, Gongabu is not a heritage zone. Therefore, the scale of interventions and approach would be different according to each case study sites.

Analysis shows that people in Thangpal valley tend to build stronger houses when they understand the scientific reason behind earthquakes. This can be used for risk communication and sensitizing people to construct better houses. Hence, an approach to help people understand how earthquakes are caused might be one way of sensitizing them to follow earthquake resistive designs. However, no similar correlation is found for Bungamati and Gongabu.

Gongabu has performed better in terms of basic shelter and electricity recovery, while Bungamati has performed better for water supply, telecommunication, and groceries supply recovery. However, Thangpal valley performed poorly on all basic services recovery. This shows that basic service recovery is faster in urban contexts, potentially leaving rural areas, which have been hit hard by the earthquake, behind.

5.8 Conclusions

Housing recovery is crucial in post-disaster Nepal. This study reveals the differences between different sites in urban and rural contexts, posing questions about locally tailored recovery approaches rather than generalized ones on national scale. It also clearly shows that housing reconstruction is going far beyond simple shelter, but is a rather complex issue at the core of overall recovery.

People are keen to construct better earthquake resistive houses after the 2015 earthquakes. However, two main gaps have been identified. Firstly, although people are willing to reconstruct safer houses, they are not aware of existing building codes. As a consequence, they may build non-safe constructions which they believe to be stable as a result of a simple lack of knowledge. Second, belief in government is high, although the proper institutional set-up within government institutions is another hindrance for recovery. This creates a lack of clarity for project execution and problems with accountability. Institutional strengthening of local bodies is thus needed.

The study also shows that a proper risk communication strategy can motivate people from the rural Thangpal valley to build earthquake resistive houses. It further highlights the local importance of remittances, which should be included in recovery strategies.

The high differences in restoration of basic services is a sign of growing inequalities between rural and urban areas. Further study can shed light on the factors that enhanced the resilience of urban areas, and replicability of these factors to rural settlements. Also, pre-disaster settlement features like the heritage value of Bungamati must be considered while reconstructing, to preserve its identity for the local community, improve cohesion, and maintain tourism as a main source of income.

Lack of financing is reported to be another major hindrance for those affected in reconstruction. An adequate financial support scheme is thus in high demand. The government, along with financial institutes, should engage in creating favorable policies which address this need.

Priority 1 of the Sendai Framework for Disaster Risk Reduction articulates that "understanding disaster risk" is necessary to build back better and not repeat the same mistakes while rebuilding. Through understanding interests and perceptions of the affected people in the three case-study sites, this research highlights areas where previously existing risks are not mitigated and at the same time newer risks are (potentially) created during reconstruction. Gaps highlighted in this research can also be useful to understand risks that might increase vulnerability of the affected people in the future.

Sendai Priority 4 has "Build Back Better" at its core. By highlighting some major hindrances, this research can contribute to Nepal's post-disaster recovery efforts.

Acknowledgement The researchers would like to thank the German Academic Exchange Service (DAAD) and the Institute of Rescue Engineering and Civil Protection (TH Köln) for their generous financial support to undertake this research.

The researchers would also like to thank the Kathmandu Valley Preservation Trust (KVPT) & Architecture Sans Frontieres Nepal (ASF Nepal) for support on data gathering and logistics.

References

Alexander D (1993) Natural disasters. Chapman and Hall, New York

Barenstein JD (2006) Housing reconstruction in post-earthquake Gujarat: a comparative analysis, vol 44. ODI, London

Barenstein JD (2013) Communities' perspectives on housing reconstruction in Gujarat following the earthquake of 2001. In: Barenstein JD, Leemann E (eds) Post-disaster reconstruction and change: communities' perspectives. CRC Press, Boca Raton, pp 71–100

Berke PR, Kartez J, Wenger D (1993) Recovery after disaster: achieving sustainable development, mitigation and equity. Disasters 17(2):93–109. https://doi.org/10.1111/j.1467-7717.1993.tb01137.x

Bhandari RB (2010) Analysis of social roles and impacts of urban ritual events with reference to building capacity to cope with disasters: case Studies of Nepal and Japan. Kyoto University, Kyoto

CBS (2014) National population and housing census 2011 (Village Development Committee/Municipality) SINDHUPALCHOWK. CBS, Kathmandu

Central Bureau of Statistics, Government of Nepal (2012) National Population and Housing Census 2011 (National Report) 01 (NPHC 2011), vol 3. Central Bureau of Statistics, Kathmandu, p 205

Charles K (1995) Assessing disaster needs in megacities: perspectives from developing countries. GeoJournal 37(3):381–385

Chitrakar RM (2015) TRansformation of public space in contemporary urban neighbourhoods of Kathmandu Valley, Nepal: an investigation of changing provision, use and meaning. Queensland University of Technology, Brisbane

Clinton WJ (2006) Lessons learned from tsunami recovery: key propositions for building back better. United Nations Secretary-General's Special Envoy for Tsunami Recovery, New York

Creswell JW (2003) Research design: qualitative, quantitative, and mixed methods approaches, 2nd edn. Sage, Thousand Oaks

Davis I (1978) Shelter after disaster. Oxford Polytechnic Press, Oxford

El-Masri S, Tipple G (1997) Urbanisation, poverty and natural disasters: vulnerability of settlements in developing countries. In: Awotona A (ed) Reconstruction after disaster: issues and practices, 1st edn. Ashgate Publishing Limited, Cambridge

Graf A (2012) Unaffordable housing and its consequences: a comparative analysis of two post-mitch reconstruction projects in Nicaragua. In: Post-disaster reconstruction and change. CRC Press, Boca Raton, pp 191–208

Haas J, Eugene RWK, Pijawka D (1977) From rubble to monument: the pace of reconstruction. Reconstruction following disaster. In: Eugene Haas J, Robert W, Kates, Bowden MJ (eds) , 1st edn. MIT Press, Cambridge, pp 1–23

Hibino Y, Onishi N, Nakamura A, Maida Y, Studies E (2015) Field investigation in affected area due to the 2015 Nepal Earthquake by AIJ Reconnaissance Team: damage assessment and seismic capacity evaluation of buildings in gongabu, Kathmandu. AIJ Reconnaissance Team, Kathmandu

HRRP/UN-Habitat (2016) Urban housing recovery policies & strategies for post - earthquake Nepal (Focusing to Kathmandu Valley and market towns). HRRP/UN-Habitat, Kathmandu

Jha AK, Barenstein JD, Phelps PM, Pittet D, Sena S (2010) Safer homes, stronger communities. Construction. Washington, DC, The World Bank. https://doi.org/10.1596/978-0-8213-8045-1

Jones TL (2006) Mind the gap! post-disaster reconstruction and the transition from humanitarian relief. RICS, University of Westminster, London

Karyabinayak Municipality (2016) Damage assessment after 2015 earthquake (Nepali Version). Karyabinayak Municipality, Lalitpur

Kennedy J, Ashmore J, Babister E, Kelman I (2008) The meaning of 'build Back better': evidence from post-tsunami Aceh and Sri Lanka. J Contingen Crisis Manag 16(1):24–36

Kennedy J, Ashmore J, Babister E, Kelman I, Zarins J (2009) Disaster mitigation lessons from 'build back better' following the 26 December 2004 tsunamis. Water Urban Dev Paradig December 2004:297–302

Lindell MK (2011) Recovery and reconstruction after disaster. In: Bobrowsky PT (ed) Encyclopedia of natural hazards. Canada. Springer, Dordrecht. https://doi.org/10.1007/978-1-4020-4399-4

Manandhar B (2016) Remittance and earthquake preparedness. Int J Dis Risk Reduc 15:52–60. https://doi.org/10.1016/j.ijdrr.2015.12.003

Mannakkara S, Wilkinson S (2013) Build back better principles for post-disaster structural improvements. Struct Surv 31(4):314–327. https://doi.org/10.1108/SS-12-2012-0044

Marahatta PS (2013) Community-based earthquake vulnerability reduction in traditional settlements of Kathmandu Valley. Tribhuvan University, Kirtipur

MoHA (2013) National disaster response framework (NDRF), vol 1. Government of Nepal, Kathmandu

Pant B (2016) Role of remittances: reconstruction and recovery. The Himalayan Times, August 24, 2016

Parker D, Islam N, Chan NW (1997) Reducing vulnerability following flood disasters. In: Awotona AA (ed) Reconstruction after Disaster: Issues and Practices, 1st edn. Ashgate Publishing Limited, Cambridge

PDNA (2015) Post disaster needs assessment, vol A. Government of Nepal, Kathmandu

PDNA Vol B (2015) Post disaster needs assessment – sector reports. Government of Nepal, Kathmandu

PDRF (2016) Post disaster recovery framework 2016 – 2020. National Reconstruction Authority, Government of Nepal, Kathmandu

Pearson L, Pelling M (2015) The UN Sendai framework for disaster risk reduction 2015–2030: negotiation process and prospects for science and practice. J Extreme Events 02(01):1571001. https://doi.org/10.1142/S2345737615710013

Post Report Kantipur (2014) "Govt announces 61 municipalities." Kantipur Publication, December 3, 2014

Post Report Kantipur (2016) "Reconstruction of homes: PM's extra assistance promise hard to keep." Kantipur Publication, September 9, 2016

Practical Action (2014) Understanding the role of remittances in reducing risk to earthquakes. Practical Action, Kathmandu

Punch KF (2005) Introduction to social research: quantitative and qualitative approaches. California. Sage Publications, Thousand Oaks

Quarantelli EL (1999) The disaster recovery process: what we know and do not know from research. In: International forum on civil protection

Sandholz S (2015) Our town? heritage and identities in changing urban landscapes of the global south. Leopold-Franzens-Universität Innsbruck, Innsbruck

Schwab J, Topping KC, Eadie CC, Deyle RE, Smith R (1998) Planning for post-disaster recovery and reconstruction, PAS Report 483/484. American Planning Association, Chicago

Sector Plans – GoN (2016) Sector plans and financial projections -working documents. National Reconstruction Authority, Government of Nepal, Kathmandu

Sendai Framework (2015) Sendai framework for disaster risk reduction 2015–2030. United Nations, Sendai. https://doi.org/A/CONF.224/CRP.1

Shakya S, Shrestha MK (2013) Urban growth in the northern fringe of Kathmandu Valley focusing on the residential development: the case of Dhapasi and surrounding VDCs. Pro IOE Grad Conf 1:221–225

Sharma G (2017) Nepal to hold first local elections in 20 years: minister. Thomson Reuters, July 25, 2017

Shrestha, BK (2008) Transformation of machendra bahal at bungamati – conservation and management plan. In alumni papers. Lund university: housing development & management – HDM. http://www.hdm.lth.se/fileadmin/hdm/alumni/papers/CMHB_2008_a/NEPAL_Bijaya_K_Shrestha.pdf

Silva da J (2010) Lessons from Aceh: key considerations in post-disaster reconstruction. Practical Action Publishing, p 98. https://doi.org/10.4324/9781849775137

Subba M (2003) Urban containment policy: does it present a hope to manage an impending urban crisis of the Kathmandu Valley? Norwegian University of Science and Technology, Trondheim

Sullivan M (2003) Integrated emergency management: a new way of looking at a delicate process. Aust J Emerg Manag 18:4–27

Tokha Municipality (2015) Tokha Municipality damage report. Tokha Municipality, Kathmandu

Tozier de la Poterie A, Baudoin M-A (2015) From Yokohama to Sendai: approaches to participation in international disaster risk reduction frameworks. Int J Disaster Risk Sci 6(2):128–139. https://doi.org/10.1007/s13753-015-0053-6

Turner JFC (1972) Housing as a verb. In: Turner JFC, Fichter R (eds) Freedom to build: dweller control of the housing process. Macmillan Company, New York, pp 148–175

UN OCHA (2015) Sindhupalchok coordinated assessment analysis package. Humanitarian Response, Kathmandu

UNDRO (1984) "Disaster prevention and mitigation: a compendium of current knowledge." Preparedness aspects 11. UN, Berkeley

VDC Office – Gunsa (2011) Gunsa VDC Profile.Pdf. VDC Office – Gunsa, Sindhupalchok

Yin RK (2009) Case study Reserach – design and methods, Applied social research methods series, vol 5, 2nd edn. https://doi.org/10.1016/j.jada.2010.09.005

Chapter 6
Analyzing the Potential of Land Use Transformation in the Urban Structuring and Transformation Axes in São Paulo: A Case Study in the Belenzinho Neighbourhood

Rafael Barreto Castelo Da Cruz, Karin Regina De Castro Marins, and Larrisa Santoro Dias Macedo

Abstract Traffic congestion, low-quality public transportation, long distances and long travel time are problems for thousands of Brazilians. The city of São Paulo brings together services, employment, and various activities, attracting thousands of people to the city. The transit-oriented development (TOD) strategies, integrated into São Paulo urban regulations from 2014 to 2016, redirect urban planning strategies to the construction and retrofitting of compact and high density neighbourhoods. These strategies are based on land use, population diversity, and compactness, while they aim to promote the improvement and the increased use of public spaces. The aim of this work is to analyse the requirements for the implementation of TOD strategies in relation to the land use potential of transformation in the selected neighbourhood of Belenzinho, which is the region of influence of the Belém Metro Station. In São Paulo, the Urban Transformation Structuring Axis (UTSA) indicates the areas in the city where it is desired to intensify land use transformation, and that are directly accessed from the high and medium capacity transport network. The case study area of Belenzinho forms a mixed urban fabric among sheds and residential buildings. Possible developing areas were identified and classified into 5 categories: (i) very low transformation potential; (ii) low transformation potential; (iii) high transformation potential; (iv) potential for requalification; and (v) in transformation. Results showed that 78% of the lots registered are likely to be transformed according to TOD strategies. However, this percentage decreases to 47% when it comes to the available area of these lots. There are many small lots with underutilized occupations, but with potential for transformation. Although the region is well located and represents potential for transformation, its location close to public transport should encourage

R. B. C. D. Cruz (✉) · K. R. D. C. Marins · L. S. D. Macedo
Department of Construction Engineering of Polytechnic School, University of São Paulo, São Paulo, Brazil
e-mail: rafaelcastelo@usp.br; karin.marins@usp.br; larissa.santoro.macedo@usp.br

© Springer Nature Switzerland AG 2020
R. Roggema, A. Roggema (eds.), *Smart and Sustainable Cities and Buildings*,
https://doi.org/10.1007/978-3-030-37635-2_6

71

investment and new real estate developments. Though the Master Plan guidelines promote solutions associated with the TOD, it is necessary to evaluate the possibilities of transformation of each urban area, the number of lots, and their available areas for transformation. In addition, it was noted that an improvement in infrastructure is necessary to attract investors aligned with the Masterplan guidelines, and new constructions must attend the diverse social groups to avoid the dispersion of low income people to the periphery. Finally, the challenge of integrating humanised public area to the private constructions remains.

Keywords TOD · Urban planning · Urban transformations · Land use

6.1   Introduction

Gomide (2006) and Gouveia (2017) point out that traffic congestion, low quality public transport, long distances travelled, and travel time are common problems faced by millions of Brazilians living in urban centres. São Paulo boasts services, jobs, universities, entertainment, and cultural attractions. It brings a populational contingent every day to central areas, moving millions of people, further worsening the externalities associated with urban mobility.

According to IPEA (2012), the Urban Mobility National Policy (PNMU) is an important milestone in the management of public policies in Brazilian cities, promoting prioritisation of instruments on non-motorised and public means of transportation rather than individual ones.

In this context, the Transit-Oriented Development (TOD) proposes redirect planning strategies and urban design, through the development of high density compact neighborhoods which offer a diversity of uses, services, and safe and active public spaces, thus favouring social interaction.

The aim of this work is to analyse the requirements for the implementation of TOD strategies correlated to the land use potential of transformation, in the selected neighbourhood of Belenzinho. The question is, what requirements are there when, in fact, implementing development urban strategies aimed at transport in urban areas?

6.2   TOD Urban Parameters

TOD strategies lead to the creation of sustainable urban communities, where the territory, land use, infrastructure and service networks are planned in an integrated manner. Therefore, distances between people, their destinations, and their main activities are shortened, promoting sustainable mobility and reducing daily travel time.

As pointed out by Hidalgo (2018) and Young (2015), transport-oriented planning strategies favour community development, economic improvement, and revitalization of the neighborhood, because of the results of its actions.

Strategies are outlined by the high-density regional growth organisation, in the corridors of the high capacity transport system, and in the main centrality nodes. In stations' surroundings, large sidewalks are added to favour paths by foot and the elaboration of more diverse uses, buildings that provide the first floors to commercial activities and services, and upper floors available to housing as pointed by Hidalgo (2018).

One of the ideas is that buildings that provide the first floors for commercial activities and services, and upper floors available for housing, contribute to a better distribution of activities in the city, and revitalise certain areas.

Almost all principles guiding the design and implementation of TOD zones directly influence land use, circulation, urban form, and overall performance of a place. A creative mix of uses within proximity is vital to attract a general community (Dittmar et al. 2004).

A lively mix of uses strengthens the link between transit and development. Additionally, having a dense mix of uses near transit is important to creating a centre. The process of selecting transit hubs to be developed as TOD zones hinges on the existing, as well as proposed, urban development schemes (Jacobson et al. 2008).

The potential of a successful selection depends to a great extent on the particular characteristics of the transit station's immediate and surrounding area. Experience from numerous cities demonstrates that implementing TOD can result in significant benefits to individuals, communities, and entire regions, by improving the quality of life for people of all ages (Carlton 2009).

6.3 Strategies Adopted in the Revision of the São Paulo Strategic Master Plan

São Paulo Strategic Master's Plan, approved in 2014, orients the rebalancing city growth by shortening the distances between housing and work, facing, consequently, socio-territorial inequalities and diminishing displacement of daily flows. For this purpose, substantial urban transformation would be necessary, balanced and planned by municipality, based on guidelines, goals and referencing principles (Souza et al. 2017).

The Urban Transformation Structuring Axis (UTSA) is made up of fractions of territory destined to promote residential and non-residential uses, with demographic and constructive high densities, and to promote landscape qualification of public spaces, items articulated within the public collective transport system (GESTÃO URBANA 2017). Additionally, according to Fig. 6.1, its purpose is also to avoid dispersed verticalisation, reduce free and fragmented areas, and to develop opportunities to implement social policies.

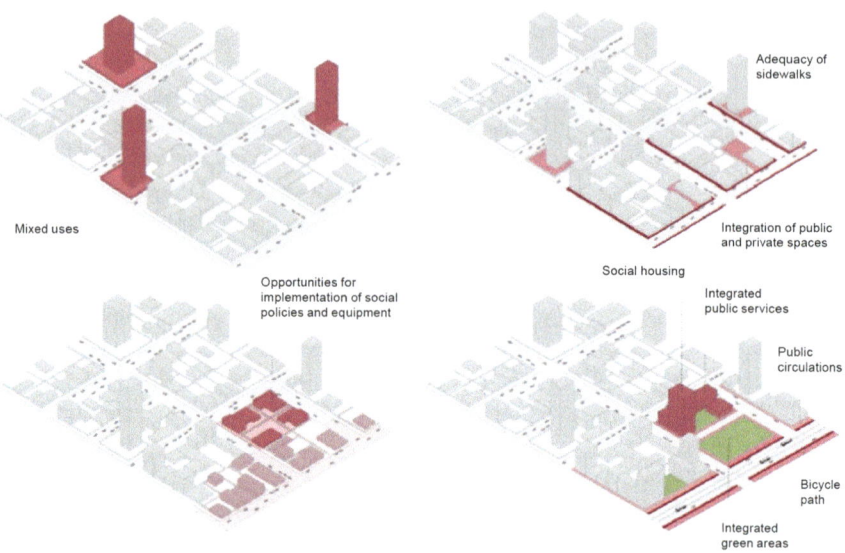

Fig. 6.1 Strategic opportunities adopted in São Paulo

For this transformation, Urban Structuring and Transformation Network was designed. This network connects city areas where the intensification of the soil use process, oriented by the densification of the population and building, is desired for urban public spaces qualification. One important point of these structuring elements is the collective high-capacity public transport.

TOD planning incorporates a multi-objective, which includes not only the economic efficiency aspect of transit ridership, but also the living environment aspect of service facilities, and the social equity aspect of inequalities in land development, between planned areas and other areas.

TOD is a fast-growing development strategy and is becoming more popular among city planners, land developers, and government officials. São Paulo has adopted strategies in this way to advance to the next decades.

One of the premises of the Strategic Master Plan is to guide the growth of the city around the axes of public transportation, such as subway stations and bus corridors. Therefore, it is necessary to transform the use and occupation of the soil near these axes. Thus, this work sought in a preliminary way to evaluate the transformation potential of the region.

6.4 Evaluation of the Opportunities in the Belém Neighbourhood in São Paulo

The work characterised as field research consisted of visits to the study area, with characterisation of the lots and systematisation of the data on a georeferenced basis. The urban area of study is located near Belém Metro station. The region has been marked in the last century by industrial occupations and workers' villages. Recently, it forms a mixed urban fabric, including industrial sheds, horizontal towns, and various real estate ventures.

In addition to the subway station, there is the Radial Leste, an important avenue that connects the most populous regions of the city, and the bus corridor, as can be seen in Fig. 6.2. It is interesting to analyse the process of development of the region around these axes, which behave as physical divisors. On both the north and south sides, the dichotomy between large vertical buildings and horizontal buildings, mixed with an urban mosaic with various nonresidential uses, is striking.

It is necessary to transform the use and occupation of the soil near these axes, therefore, this work sought in a preliminary way to evaluate the transformation potential of the region. The adjacent blocks of the transport axes comprise sectors 27 (blue) and 29 (yellow), respectively, in Figs. 6.3 and 6.4.

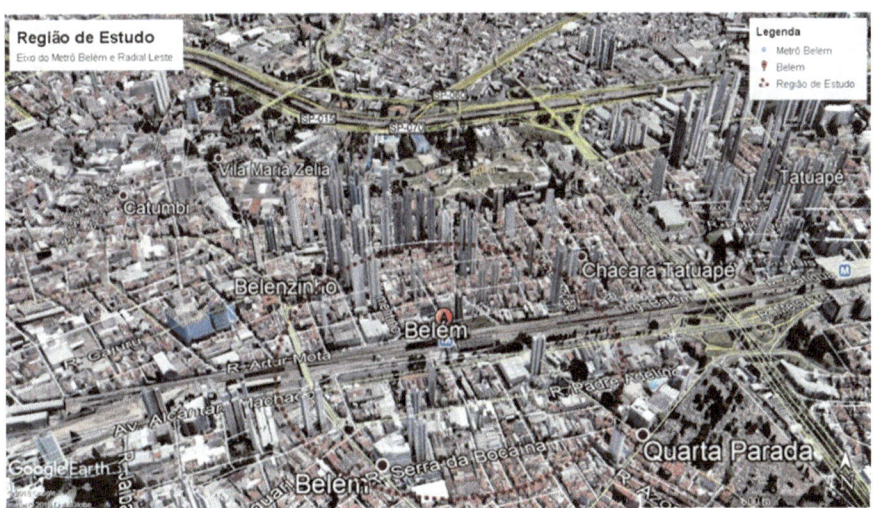

Fig. 6.2 Study area

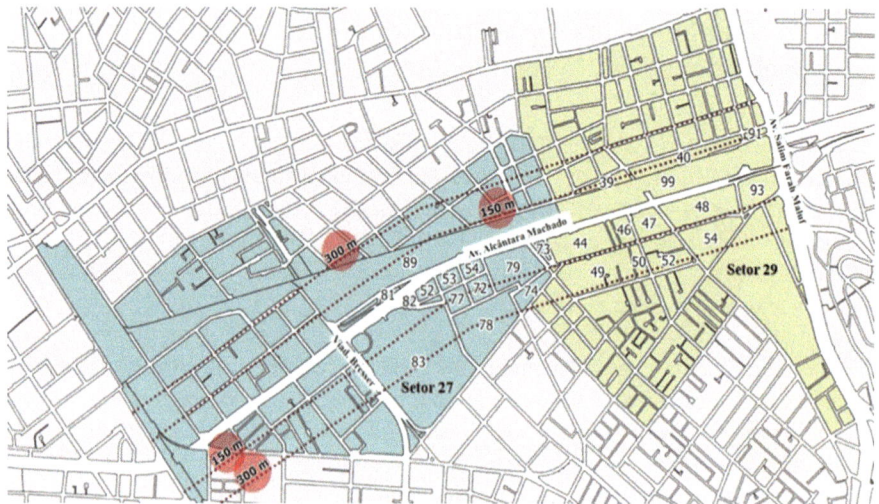

Fig. 6.3 Delimitation of the area of influence of the Radial Leste Avenue axis

Fig. 6.4 Area of influence of the subway station Belém

These sectors were chosen since they are within the area of influence of the axis, comprised by the delimited Radial East, by dashed lines at 150 meters and 300 meters. These sectors were also delimited by the Salim Farah Maluf Avenue

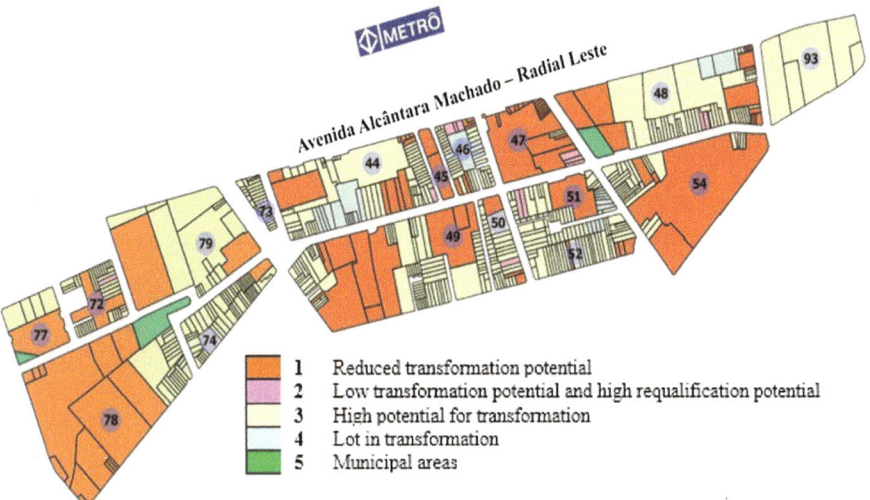

Fig. 6.5 Map of Classification of lots in relation to processing potential

and the Bresser Viaduct, in order to maintain a study area occupied by blocks with similar profiles, and around the block where the subway station is located, at a distance of 400 and 600 meters of the same stroke the range radius of the axis.

The blocks were obtained from the GeoSampa portal of the City of São Paulo. From these can be identified their respective batches, to classify the transformation potential of batches with grades 1–5, in which:

1. Reduced transformation potential: in general, public spaces or consolidated private spaces, in which transformation is difficult due to their characteristics and form of occupation;
2. Low transformation potential and high requalification potential: lots with greater difficulties of expropriation; in general, small and medium-sized buildings, with several lots in the same land, and therefore, more than one owner. In this way, a change in its use is considered, in order to qualify it;
3. High potential for transformation: the lots are subject to expropriation and their constructions are demolished, being replaced by new developments;
4. Lot in transformation: lots that have projects under execution;
5. Municipal areas: in general, public squares.

The region is divided by the subway line into two lanes, one to the north and one to the south. The results of this work concentrate on the analysis of the lots located in the strip to the south. The results of the classification were imported into the digital map of the city of São Paulo and can be seen in Fig. 6.5.

6.5 Results the Potential of Land Use Transformation in in the Belém Neighborhood in São Paulo

Areas with a low processing potential (14.2% in lots and 2.3% in land area) are characterised by residential buildings, hospitals, hotels, shopping centers, educational establishments, churches, gas stations, a municipal deposit, an energy transformation and distribution station, a graphic, and more. Due to its characteristics and the type of use and occupation of the soil, its transformation and requalification is greatly impaired.

The lots already under transformation (4.6% in lots and 6.4% in land area) are mostly residential or commercial buildings, and a new private hospital under construction, which, probably due to the time of drafting the projects, do not follow the proposals given by the updated Master Plan.

Areas with reclassification potential (1.8% in lots and 2.3% in land area) are occupied by larger commercial establishments, and small and medium office buildings. Due to their characteristics, the land purchase process, demolition of the existing building, and construction of a new one is also impaired. Therefore, a re-qualification of its use is recommended, made possible, for example, by the provision of part of the building area for public use, allowing a greater flow of people and activities, and diversifying their occupation.

The results indicate that approximately 78% of the lots and 75% of the area allow for an urban transformation, but when analysed in more depth, the lots in this category representing an opportunity for transformation are small. The largest lots have already gone through a construction process. This can be seen in visual analysis in Fig. 6.5 and Table 6.1.

However, although there is considerable scope for transformation, it is necessary to reconcile the interests of the various segments of society: municipal management, the population that inhabits and uses the region, potential investors in future ventures, the real estate market, developers, and builders.

The classification of lots for each sector was organised according to their quantity and area of land, by number of lots and area in Table 6.1.

In addition, most of the new developments in the region follow the old planning guidelines and do not adopt the measures proposed by the TOD. This table indicates,

Table 6.1 Classification in relation to the quantity of lots and area

Classify potential	Sector 27 e Sector 29		Sector 27 e Sector 29	
	Number of lots	%	Land Area (m²)	%
1	93	14.2%	189,048	15.5%
2	12	1.8%	3,823	2.3%
3	515	78.9%	188,989	75.6%
4	30	4.6%	12,832	6.4%
5	3	0.5%	6,332	0.2%
Total	**653**	**100%**	**400000 m²**	**100%**

therefore, that the new strategies will be developed and executed in the long term. Nevertheless, there will remain great territorial space available for its implantation. Authors such as Ye (2013) point out that these transformations can considerably increase the population density of the affected regions.

Handayeni (2014) points out that this development occurs in an evolutionary way around the centers of the axes, and that a strategy of zoning that favours the progressive occupation of the lots is favourable to the transformation in the case of Indonesian cities.

It is a consensus among authors that plans and standards alone are not sufficient if there is no joint effort. It is stressed that these transformations take time and require support and legitimacy.

6.6 Analysis of the Results from the Literature Perspective

Rufino (2016) and Marques and Bichir (2001) argue that in areas where, traditionally, real estate market production is more intense, there is a transformation of the more attractive lots, that is, the larger lots that are located closer to points of interest. To urban planners, there is the challenge of transforming the territory, and integrating an infrastructure of urban equipment, trades, and services that possibly will be created with the new transformations into the lots that will not be transformed.

The price of the lot can be a factor regulated by the urbanistic instruments. Because lots with differentiated formats (narrower and elongated, or with irregular shapes), have greater difficulty with transformation, they will thus influence the expected result of the built environment resulting from the proposal by the TOD. According to Crane (2000), this behaviour is difficult to predict, because there are several variables linked to the urban form of design neighborhoods.

In this context, urban regulations that encourage transport-oriented development strategies isn't enough. Systematic integration should take place, including the remodeling of the streets system, the occupation and use of the lots, and a distribution of the transformation opportunities in the territory. In this way, it is possible to create a city that combines transportation and cheaper and affordable housing, where residents can access employment opportunities, and where a synergy of public services and economic nuclei takes place, among other transformations (Salat and Ollivier 2017).

6.7 Conclusions

Preliminary results point out that, although the guidelines propose interesting solutions, what is observed in practice is that the need to diagnose the potential areas of transformation, either by the number of plots available or by the existing supply areas, is an important condition in making decisions about the urban transformation strategies that will be adopted for each case.

It is observed that real estate enterprises are in larger distances from the subway station axis and an important avenue which connects the most populous regions of the city. This fact, in a preliminary way, shows the real state sector choice in occupying residential areas not affected by intense traffic, noise and air pollution. These are characteristics of the blocks nearby the axis, which tend to be marked by vulnerability and marginalisation. This interpretation deserves a deeper and better diagnosis to be exploited in future works, since it is an initial researcher's perception.

Therefore, for the study area, it is perceived that the necessary requirements in urban areas to implement transport-oriented urban development strategies are associated with the availability of lots, as well as the potential for transformation.

References

Carlton I (2009) Histories of transit-oriented development: perspectives on the development of the TOD concept real estate and transit, urban and social movements, concept protagonist. Institute of Urban and Regional Development, University of California, Berkeley

Crane R (2000) The influence of urban form on travel: an interpretive review. J Plan Lit 15(1):3–23

Dittmar H, Belzer D, Autler G (2004) An introduction to transit-oriented development. In: Dittmar H, Ohland G (eds) The new transit town: best practices in transit-oriented development. Island Press, Washington, DC, pp 1–18

GESTÃO URBANA. Plataforma Digital de Informações de Participação, Uso e Ocupação de Solo na Prefeitura do Munícipio de São Paulo – Disponível em: http://gestaourbana.prefeitura.sp. gov.br/. Accessed 20 May 2017

Gomide AA (2006) Mobilidade urbana, iniquidade e políticas sociais. Políticas Sociais: Acompanhamento e Análise 12:242–250

Gouveia DS (2017) A formulação de políticas públicas para a mobilidade urbana no Grande ABC. Dissertação de Mestrado. Universidade Municipal de São Caetano do Sul. São Caetano do Sula. São Paulo, 186 p

Handayeni KDME (2014) TOD best practice: lesson learned for GHG mitigation on transportation sector in Surabaya City, Indonesia. Proc – Soc Behav Sci 135:152–158

Hidalgo PA (2018) Desenvolvimento Orientado pelo Transporte – DOT A valorização do transporte público, pedestres e ciclistas. Minha Cidade, [S.L], v. 183, n. 02, out. 2015. Disponível em: http://www.vitruvius.com.br/revistas/read/minhacidade/16.183/5749. Accessed 10 Jan 2018

IPEA – Instituto de Pesquisa Econômica (2012) Aplicada. A Nova Lei de Diretrizes da Política Nacional de Mobilidade Urbana. Disponível em: http://www.ipea.gov.br/portal/images/stories/ PDFs/comunicado/120106_comunicadoipea128.pdf. Accessed 5 Mar 2018

Jacobson, Justin, Forsyth A (2008) Seven American TODs: good practices for urban design in Transit-Oriented Development projects. J Transport Land Use 1:2(fall):51–88

Marques E, Bichir R (2001) Investimentos públicos, infraestrutura urbana e produção da periferia em São Paulo. Revista Espaço e Debates. São Paulo, Neru, v. XVII, n. 42

Rufino MBC (2016) Transformação da periferia e novas formas de desigualdades nas metrópoles brasileiras: um olhar sobre as mudanças na produção habitacional. Cad Metrop São Paulo 18 (35):217–236

Salat S, Ollivier G (2017) Transforming the urban space through transit-oriented development: the 3V approach. World Bank, Washington, DC. https://openknowledge.worldbank.org/handle/10986/26405. License: CC BY 3.0 IGO

Souza AP, Seo HNK, Yamaguti R (2017) Structuring axes of urban transformation: possibilities and gaps. In: XV meeting of the National association of graduate studies and research in urban and regional planning. Procedures, V. 17 no. 1, May, 2017. Sao Paulo

Ye L (2013) Urban transformation and institutional policies: case study of mega-region development in China's Pearl River Delta. J Urban Plan Dev 139(6):292–300

Young G (2015) The emerging role of a landscape bases strategy in the South Africa built environment: a case study of Johannesburg Wemmer Pan Precinct. In: SASBE 2015 Proceedings, 9–11 December, 2015. Pretoria, South Africa, pp 405–41

Part III
Urbanity

How to design a city in which the quality of life is high and at the same time goals around sustainability, energy provision and circularity can be realized.

Chapter 7
Implementing a New Human Settlement Theory: Strategic Planning for a Network of Circular Economy Innovation Hubs

Steven Liaros

Abstract Whilst the energy transition from fossil fuels to renewables offers significant environmental benefits, the other transition—from a centralised to a distributed energy system—underpins a disruptive model for planning cities, towns, and villages. A local energy micro-grid can power a local water micro-grid, which in turn can irrigate a local food system, offering a community the opportunity to harvest, store, and distribute food, water, and energy within their immediate catchment. Designing the layout of the built environment in the form of a campus or resort—with smaller private spaces and a wide range of accessible shared spaces and facilities—would also minimise energy demand, while simultaneously providing opportunities for social interaction and connection.

Creating places where local residents can collaborate to provide their basic needs is a form of Place-Making as well as an achievable alternative to the Universal Basic Income (UBI). The direct delivery of basic needs—consumed by the producing community—rather than the provision of money to pay for the purchase of these same needs, addresses the issue of wealth distribution but also re-imagines how wealth is created. It requires communities to take responsibility for their local environment, supporting infrastructure and others in their community.

Described by the author as a Circular Economy Innovation Hub, such a planning strategy adopts the principles of the Circular Economy—systems thinking, life-cycle planning and striving for zero waste. By integrating the water, energy, food and built systems, waste can be repurposed and the overall efficiency of all component systems is significantly increased. The more efficient delivery of the identified natural needs then offers residents more free time for innovation and creativity. Finally, such places are not isolated villages but hubs or nodes in a network, connecting and collaborating with others in their bio-region and beyond.

S. Liaros (✉)
Department of Political Economy, University of Sydney, Sydney, Australia

PolisPlan, Sydney, Australia
e-mail: steven@polisplan.com.au

© Springer Nature Switzerland AG 2020
R. Roggema, A. Roggema (eds.), *Smart and Sustainable Cities and Buildings*,
https://doi.org/10.1007/978-3-030-37635-2_7

With the planning theory stage completed, a number of further projects with research partners and local governments are now building a knowledge platform, including infrastructure modelling, planning policies, an architecture brief and financial strategies. These would enable this form of regenerative land development to be incorporated in local government planning policies, providing certainty for investors and enabling delivery by developers.

Keywords Circular economy · Energy transition · Regenerative development · Universal basic income · Healthy built environment

7.1 Introduction

Smart Cities, Livable Cities, Green Cities, Biophilic Cities, EcoCities, and Regenerative Cities. Add to the mix Transition Towns, Eco-Villages, Intentional Communities, and Place-Making, and it seems everyone is talking about cities and what cities of the future could, or should, be like. Yet, in all this conversation, there is little to no discussion about what a city actually is. Before adding an adjective, let's try to understand the noun we are trying to describe and change. What is a City?

Most commonly, the 'city' refers to 'urban settlements', where urban implies non-rural. However, while the United Nations' *World Urbanization Prospects (2014 Revision)* was widely quoted, reporting that 54% of the world's population now lives in urban areas, the same report acknowledged that "there is no common global definition of what constitutes an urban settlement" and indeed "the urban definition employed by national statistical offices varies widely". In this paper, it will be argued that given clean energy technologies as well as the need to reconnect humans with the environment on which we live, the fundamental separation between urban and rural areas is inappropriate and that food systems should be integrated with the built environment.

For the purposes of this chapter, we will use the original Greek and Roman concept of the city as simply 'a community of citizens'. Taking this as a starting point, we ought to ask: what do these citizens need? The natural needs are a good place to begin; water, food, energy, and shelter. Then, given advances in technology and knowledge more generally, as well as awareness of human impacts on ecosystems, how would we design a new city to provide these basic needs? How big would it be and how many citizens would it support? Recognising various options for telecommuting and the future of electric vehicles and car sharing, how would this city connect with other cities? How might we design an energy system using renewable energy to generate, store, and distribute energy in a micro-grid? Significant research in water sensitive urban design (WSUD) and cities as water supply catchments, suggest that it is also possible to reimagine our systems for harvesting, storing, and distributing water. Then, of course, water can generate energy and also store both potential energy and heat energy. There are substantial opportunities to increase efficiency by integrating the energy and water systems. What if we then include a food system—an expanded form of diverse urban agriculture—also using it to clean water, while saving the energy used for packaging and transport?

To implement a new human settlement theory, the question to ask is: if we were able to build a city from scratch, knowing what we now know and having access to the technology that is available to us today, how should we go about it? If we were to imagine ourselves as explorers, looking for new lands and wishing to build a new settlement, what characteristics of geography and climate would we look for? How would we determine the appropriateness of a locality so as to complement existing settlement patterns and restore degraded land? It is considered that this approach offers significant opportunities for rural and regional Councils who are seeking to attract people and investment to their localities. Land is less expensive, large parcels are available and farming communities are looking for options not only to attract investment and people but also to regenerate the land that has been so degraded by the chemical-based, large scale monoculture practices that have been imposed on them over the past 50 or so years.

The purpose of this chapter is to describe what is considered possible if we were to reimagine our approach to strategic planning in the manner described above and then, to briefly outline how this model could be implemented through the NSW planning framework.

Section 7.2 provides a general overview of an evolving development model described as a Circular Economy Innovation Hub. This model adopts the principles of the Circular Economy—systems thinking, life-cycle planning and striving for zero waste. The concept of the Circular Economy allows us to think of cities as systems that efficiently provide residents with their basic needs. Rather than connecting detached houses to the energy and water grids, how do we integrate food, water, and energy into the built system? Can a local energy micro-grid power a local water micro-grid to, in turn, irrigate a local food system, offering a community the opportunity to harvest, store, and distribute food, water and energy within their immediate catchment? This section also explains why the term 'Circular Economy Innovation Hub' is used.

Section 7.3 offers a discussion about how the proposed innovation hub addresses a number of significant public debates. These include:

- the future of work due to technological change, including how the internet is creating opportunities for remote work, sometimes referred to as the e-change.
- the future of work and the consequent suggested need for a universal basic income.
- affordability of housing and cost of living pressures.
- one planet living and striving to live within the limits of our ecological systems.
- the characteristics of healthy urban design as developed by public health experts exploring the impacts of the built environment on health outcomes.
- the concept of regenerative development and its intersection with regenerative agriculture.

Section 7.4 outlines how the Circular Economy Innovation Hub can be implemented through the NSW planning system. This identifies how specific strategies, policies and development plans would need to be created or amended to enable this form of development. This is important to provide certainty for investors and future residents.

7.2 Overview of the Development Model

7.2.1 What's in a Name?

The development model is referred to as a 'Circular Economy Innovation Hub'. The term has been carefully chosen to reflect its inherent characteristics and to differentiate it from other 'alternative' approaches to land development.

The first characteristic implicit in the name is that the settlement is referred to as a 'Hub' rather than a village. The aim is to emphasise that the development should be imagined as a node within a broader network and not an isolated entity. The hub should not be imagined as a gated community but would remain connected with, and a part of, the broader society, complying with its laws and participating in the local, regional, and global economy in an open and transparent manner. This model should essentially be viewed as an alternative to greenfield subdivisions. Rather than developers simply providing housing, they would provide housing integrated with shared spaces as well as food, water and energy systems. Additionally, the aim is not just to build a single hub but to create a replicable process that can be implemented by other land developers to eventually create a network of similar hubs.

Secondly, the hub will be designed according to the principles of the Circular Economy. The proposed transition from a linear to a circular economy has gained significant momentum in recent years, receiving the support of the European Parliament (2016), European Investment Bank (2015), and numerous major banks and corporations. A useful definition of the Circular Economy is provided by the Ellen MacArthur Foundation (2018), a leading European advocate of the Circular Economy:

> Looking beyond the current "take, make and dispose" extractive industrial model, the circular economy is restorative and regenerative by design. Relying on system-wide innovation, it aims to redefine products and services to design waste out, while minimising negative impacts. Underpinned by a transition to renewable energy sources, the circular model builds economic, natural and social capital.

From the above definition, we can extract some of the key design principles of our proposed land development model, such as life-cycle planning, systems thinking and striving for zero waste. The Circular Economy is underpinned by renewable energy, therefore, taking a systems approach, an energy micro-grid will generate, store, monitor and distribute renewable energy on site. The energy system will power a water system that will be cycled through the site, providing for residents, irrigating crops and providing water for animals. The living and work spaces will be passively designed to minimise energy demand and more generally, the energy, water, food and built systems will be integrated to maximise efficiency.

The development is referred to as an Innovation Hub to emphasise that it is not a dormitory suburb but will integrate living spaces with work spaces, incorporating a work hub that supports the transition from the work commute to tele-commuting. There will also be a significant amount of work available in the management and maintenance of the water, food and energy systems and the shared spaces and facilities.

It is also proposed to incorporate innovative product development and business models to advance the notion of the Circular Economy. This primarily involves the inclusion of waste to resource micro-factories as currently being developed by the Centre for Sustainable Materials Research and Technology (SMaRT) at the University of New South Wales (UNSW Newsroom, 2018). The SMaRT Centre refers to green materials as those that "are made entirely, or primarily, from the rubbish we throw away".

The reference to innovation also includes the innovative systems approach to the design of the infrastructure, so as to efficiently provide the basic needs of all residents. The efficient provision of necessities maximises the free time available for innovation, art, relaxation and social connection.

The idea of developing a network of hubs that provide residents and visitors with water, food, energy and shelter aligns with the indigenous view of the landscape as a network of waterholes connected by song-lines. Watson (2015), in describing indigenous law, refers to a distributed system, with obligations for people in each place to renew the land. The relationship with the natural world involves a way of being that is cyclical—aligning with natural cycles—rather than our current linear world-view.

7.3 Life Cycle Planning

Designing at a village scale also enables more efficient delivery of living and work spaces, allowing shared spaces to be used for multiple purposes and enabling residents to move to different parts of the hub as their housing needs change through different life stages. Current failure to plan for different household sizes, coupled with significant transaction costs involved in moving houses, have resulted in a significant misalignment between housing structures and household occupancy rates as illustrated in Figs. 7.1 and 7.2. This has been exacerbated by demographic changes in the Australian population over the last half century—an aging population, increasing accounts of late marriages, reduced housing affordability and increasing divorce rates. These have all resulted in a falling number of occupants per household. Currently in NSW, 56% of households have only one or two occupants, while 67% of houses are detached dwellings, essentially designed for families.

The proposed Circular Economy Innovation Hub would design for a microcosm of the NSW age demographic and household size demographic, providing smaller private spaces, with access to a wide range of shared spaces. These ideas are already being developed through flexible house designs and in co-housing developments.

Fig. 7.1 Occupants per occupied dwellings in NSW. (ABS Census 2016, table G31)

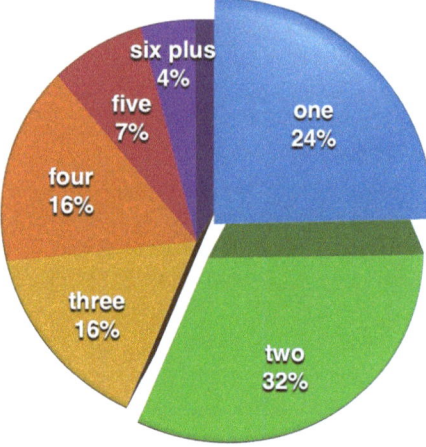

Fig. 7.2 Dwelling Structure in NSW of occupied dwellings. (ABS Census 2016, table G32)

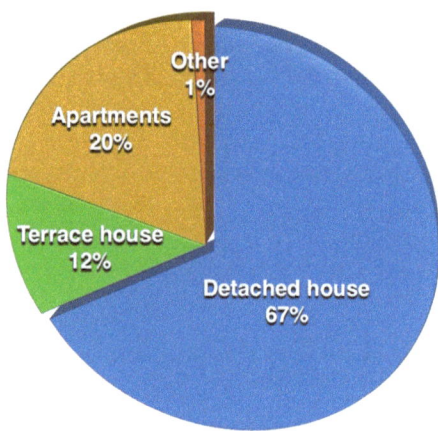

7.4 Responding to Public Debates

7.4.1 The Future of Work and the E-Change

Modelling included in a 2015 report by the Committee for the Economic Development of Australia (CEDA) suggested that *"around 40 per cent of the [Australian] workforce face the high probability of being replaced by computers in the next 10 to 15 years"*. It is difficult to comprehend the transformative effect that the Information Revolution is likely to have on our society. The only comparison we have is the massive changes caused by the Industrial Revolution when technological advancement resulted in the restructuring of society and the consequent reorganisation of cities. According to Grigg (1987,93), *"In the early eighteenth century farmers and farm workers made up three-quarters or more of the labour force in nearly every country."* Yet, by the late twentieth century, this had fallen to as low as 2% of the workforce in Sweden, Switzerland, the UK and the US, while in Europe it employs about 8% (Grigg 1987:95).

From a widely distributed population mostly involved in agriculture, the Industrial Revolution created a centralising force, resulting in high population concentrations in urban centres. From the late nineteenth century, various solutions have been proposed to address the issues arising from congestion in large cities. Ebenezer Howard (1896) expressed concern that *"the people should continue to stream into the already over-crowded cities, and should thus further deplete the country districts"*. Howard argued for the need to draw people out of large cities and into rural areas, by designing places that provide a better balance between town and country life. He referred to these as 'Garden Cities'. Despite the significant role of the Garden Cities Movement in town planning in the UK, plans and policies in Australia to encourage people to leave the cities have been largely non-existent or ineffective. The proposed innovation hub can be imagined as a type of garden city in which the gardens are diverse, integrated food systems, rather than simply pleasant open spaces or rural landscapes.

Over recent decades, much of the migration from the cities to rural and regional areas has been attributed to individuals seeking a more relaxed lifestyle, that is, a sea-change (to coastal towns) or a tree change (to rural or farming areas).

Demographer and Business Analyst Bernard Salt argued in a report for NBN Co. (2016) that access to the internet is adding another dimension to this shift:

> We are witnessing a quiet lifestyle revolution in suburban Australia. The fusion of a relaxed lifestyle in tree-change and sea-change locations combined with super connectivity provided by the NBN network, is giving people even greater scope to take greater control of where they live and how they work.
>
> I predict a cultural shift or 'e-change movement' which could see the rise of new silicon suburbs or beaches in regional hubs as universal access to fast broadband drives a culture of entrepreneurialism and innovation outside our capital cities.

Embracing this e-change represents an important economic development strategy for rural and regional councils. The Circular Economy Innovation Hub development model, with internet co- working spaces and waste to resource micro-factories, can be imagined as a futuristic 'garden city' that provides the co-working spaces as well as a low-cost platform (see 3.3 Housing Affordability) for innovators and entrepreneurs willing to relocate to a rural or regional area as part of an e-change movement.

7.5   The Future of Work and Universal Basic Income

The study of city planning over the last century or so, has evolved to address the problems of congestion and pollution caused by the agglomeration of manufacturing and factory workers in cities. Whilst these urban problems are important, the urban environment as we know it was created by changes in the agricultural system. The production of the food needed to sustain urban populations is entirely missing from the planning of cities. The Circular Economy Innovation Hub is a system that provides for the needs of the citizens by incorporating food as an essential part of that system.

It is no longer appropriate for the planning of cities and towns to be based on the consequences of agglomeration and centralisation, ignoring food production and the possibilities for decentralisation offered by the internet. Planning should extrapolate the future from the world as it is today, rather than from the world of the Industrial Revolution. As technology continues to advance, making many traditional jobs obsolete, it is important to start creating resilient places where people can work to directly satisfy their basic needs, relying less on jobs that provide an income to satisfy these same basic needs.

There is growing interest in the concept of a Universal Basic Income (UBI) as a means of addressing the likelihood of future job losses as well as a means of addressing inequality in wealth distribution. Rather than debating how to fund a basic income in monetary terms, a far more effective and efficient strategy would be to create places that provide peoples' basic needs directly.

This also addresses a significant gap in the UBI debate, which aims to address the inequality in the distribution of wealth but does not address how that wealth is created. Regenerative land development complements the UBI debate as it aims to increase our natural capital through restoration and maintenance of land and water, and also plant and animal life, while minimising waste and other negative impacts.

7.6   Housing Affordability

The housing affordability crisis is a symptom of much broader structural issues. Housing affordability is not just about house prices, it is also about access to and availability of work, transport costs and other costs of living. These issues can no longer be addressed in isolation but need to be tackled holistically and systematically. This has been our approach in the design of the Circular Economy Innovation Hub.

With regards to the cost of living, the hub would be designed to provide food, water and energy for a discrete population. Having a known and fixed population allows the design process to provide for an abundance of these basic necessities. An over-supply of food, water and energy—the demand for which doesn't vary significantly with price—drives their price towards zero. Whilst work is still necessary and so a fair system for allocating responsibility for this work will be required, food, water and energy—having zero marginal cost—would not be market exchange commodities. The passive architectural design of the built environment also reduces energy demand and therefore cost.

The design of a village as a live and work hub also substantially reduces transport costs by having work opportunities within walking distance of living environments. A compact design with up to 200 people makes vehicle sharing more feasible as

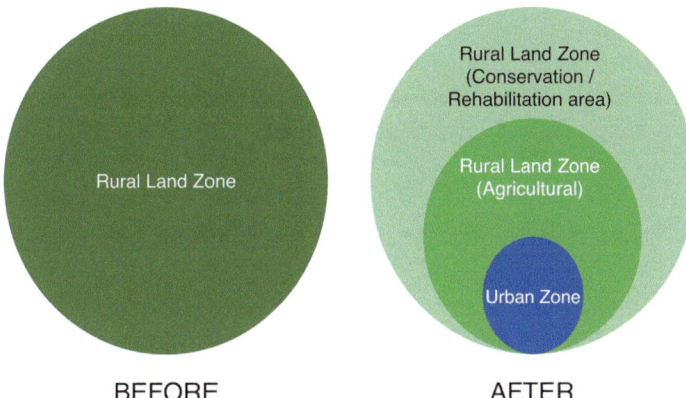

BEFORE   AFTER

Fig. 7.3 Rezoning rural land into three precincts

access to the shared vehicles will be within walking distance. Quality internet connection at the co-working spaces enhances the option of tele-commuting. Meanwhile, the local energy micro-grid would be designed to incorporate an electric vehicle (EV) charging station—a shift to EVs drives fuel costs towards zero.

In addition to these cost-of-living and transport factors, tackling housing affordability requires that the various components of house prices be addressed. The first component is land value, which can be minimised by purchasing rural land and capturing the land value uplift when it is rezoned, as illustrated in Fig. 7.3.

The principal objective of this approach is to internalise the land value uplift. It is well understood that the process of rezoning land increases the value of that land. This increased land value—the land value uplift—currently benefits the land owner at the time the land is identified for rezoning. They may then sell the land and retain all the financial benefit of this value uplift. As the process of rezoning land is a public function, the increased value should provide a public benefit, either in the form of public infrastructure, lower housing costs or a combination of the two. This project intends to make housing more affordable while also funding the local water, food and energy infrastructure by capturing this land value uplift. This would be done by requiring certain activities and infrastructure works to be carried out at certain stages prior to the rezoning of land.

Housing construction costs would also be reduced by creating smaller private spaces. Unlike a tiny house village, the reduction in house sizes is compensated by access to a wide range of shared community spaces, such as for work, cooking and eating areas, entertainment facilities, swimming pools and the like.

Infrastructure costs are kept to a minimum by the compact design, which is more like a resort or campus than like current residential subdivisions. Also, rather than subdividing the land and setting up a complex regulatory planning framework for the construction of each lot, the Innovation Hub would be built in a coherent manner by a single entity, thus further reducing construction and financing costs.

Further work is now being undertaken with respect to land tenure and investment structures to minimise property transfer costs and land speculation. Given that residents would have access to various shared spaces and assets provided in different parts of the site, a collective ownership arrangement would be preferred. The Community Land Trust (CLT) model developed for the Australian context by Crabtree et al. (2013) would be an appropriate structure. CLTs provide for collective ownership of the land, which is held in trust in perpetuity to avoid land speculation. We are also proposing to integrate other emerging models in the development industry such as 'build-to-rent' and 'baugruppen', or development collectives.

7.7   One Planet Living

Kate Raworth (2017) expands on the concept of One Planet Living by arguing that "wellbeing depends on enabling every person to lead a life of dignity and opportunity, while safeguarding the integrity of Earth's life-supporting systems". As well as an ecological ceiling, an economic system should also provide a social platform. Raworth calls this living within the doughnut.

It is a well-known principle in economics that, for basic necessities, people buy about the same amount whether the price rises or falls. That is, the demand for basic necessities is broadly 'inelastic', which also means that the total amount of food, water and energy needed by households is relatively stable. It is therefore possible to design a place for the needs of a discrete population by estimating their demand for food, water, energy and living and work spaces. A key design approach for the development is match the population to the capacity of the land and its infrastructure, starting with an assumed population size of no greater than 200 persons. The final design size may be reduced and would be determined by the capacity of the chosen site and its supporting infrastructure to sustain that population. The land area required to feed 200 people will significantly influence total land requirements, so a research project is currently being developed to determine land and water requirements for different nutrition plans.

7.8   Healthy Urban Design

According to research by public health professionals, the built environment has an important role to play in supporting human health. In a review of the literature in this field of Healthy Urban Design by Kent et al. (2011), three key interventions were identified that could support human health. These are; getting people active, connecting and strengthening communities and providing healthy food options.

The design of the Circular Economy Innovation Hub integrates a food system of significant scale into the built environment, providing not just healthy food options but the opportunity to collaborate with others in the community to provide that food. A walkable environment connecting a wide range of daily activities also allows people to get more active. This development model therefore has the potential to significantly improve health outcomes for the resident community.

7.9 Regenerative Development

The concept of regenerative development is emerging as a new approach to land development. Proponents argue that we need to move beyond sustainability—sustaining ourselves and the environment—to regenerative development where we have a positive impact. This is best illustrated in the diagram in Fig. 7.4 by Bill Reed from Regenesis. Meanwhile, forward thinking farmers like Charles Massy in his book Call of the Reed Warbler are advocating for a revolution in farming practices, calling the new approach 'regenerative agriculture'.

The Circular Economy Innovation Hub integrates regenerative development with regenerative agriculture.

7.10 Implementation Through the NSW Planning System

7.10.1 Strategic Planning

In order for Circular Economy Innovation Hubs to be financed, developed and replicated, the development model must be clearly articulated in local government

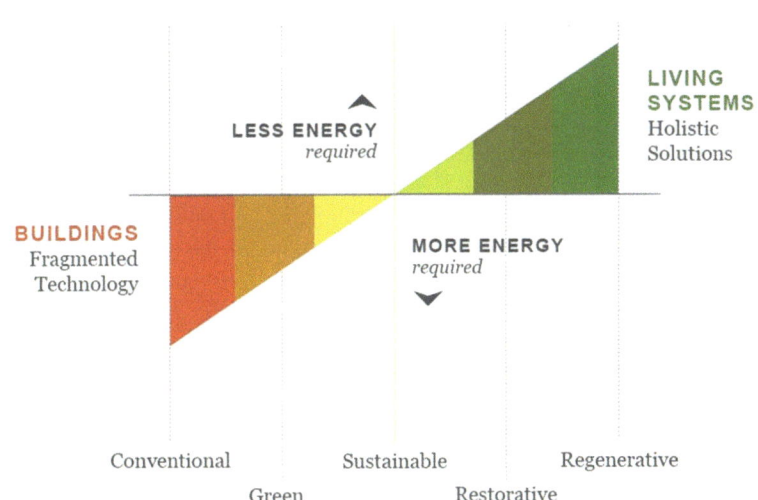

Fig. 7.4 Trajectory of Ecological Design (image used with permission) © All rights reserved. Regenesis – Contact Bill Reed, bill@regenesisgroup.com for permission to use

strategies as a desired form of development. In accordance with section 3.9 of the Environmental Planning and Assessment Act 1979 (EP & A Act, NSW), all Councils must prepare a Local Strategic Planning Statement (LSPS). This will set out a 20-year vision for land use in the local government area. This is a recent amendment to the EP & A Act and regional councils must have their first statement in place by 1 July 2020.

The LSPS provides an ideal opportunity for Councils—in consultation with their community—to introduce this development model into their local planning framework. The LSPS should explicitly refer to Circular Economy Innovation Hubs (or equivalent terminology) and should include objectives such as for future land development to be regenerative in character, that development integrates regenerative agriculture and that it provides for all the energy and water needs plus a specified proportion of food for a discrete population. The LSPS should also:

- Identify the general localities (not the specific sites) where this development would be permitted.
- Refer to a policy document or chapter in the Development Control Plan (DCP) for more information.

7.10.2 Policy Document or Chapter in DCP

The policy or plan should clearly describe the development form and the process through which this development outcome could be achieved. As a minimum, the following should be included:

- Requirements to prepare Concept Development Application in accordance with Division 4.4 of the EP & A Act together with a Planning Instrument Amendment in accordance with Division 3.5 of the EP & A Act.
- Requirements at different stages prior to the rezoning of land (eg. what must be done on land that is not to be rezoned, what must be done before Council resolves to refer to the Department of Planning, prior to advertisement, prior to final resolution, prior to referral to Minister for signing and publication).
- The minimum total land area and the proportions of the site area for the three precincts (i.e. (a) conservation/rehabilitation area, (b) agriculture and (c) live/work hub).
- Minimum requirements for harvesting, management, storage and distribution of water, food and energy.
- Design principles for buildings.
- Preparation of a transport plan for the site as well as impacts on the surrounding road network.

A useful approach would be to first identify one site for a pilot project, the development of which would assist in refining the development controls and wording in the strategy.

7.10.3 *Voluntary Planning Agreement Policy*

The proposed development will include various facilities, assets, and open spaces to service the population within the development site. Some of these will also be available to the proximate township and other communities in the broader area. It would therefore be appropriate to prepare a Voluntary Planning Agreement (VPA) policy, including a standard template VPA that provides a framework for the delivery and management of infrastructure, both on the subject site and the surrounding area. This should address:

- Effect on any development contributions required pursuant to sections 7.11 or 7.12 of the EP&A Act.
- Effect on any charges for water supply, sewerage and storm-water drainage facilities under s64 of the Local Government Act 1993.
- Requirement for the VPA to run with the land pursuant to section 7.6 of the EP&A Act.
- Effect on waste levies.
- Effect on ordinary rates or requirement for any special rates.

7.11 Conclusion

With the research and planning theory stage completed, a number of further projects with research partners and local governments are now helping to create a knowledge platform, including infrastructure modelling, planning policies, an architecture brief and financial strategies. These would enable this form of regenerative land development to be incorporated in Local Government planning policies, providing certainty for investors and enabling delivery by developers.

References

Committee for Economic Development of Australia (2015) Australia's future workforce? At: https://www.ceda.com.au/CEDA/media/ResearchCatalogueDocuments/Research%20and%20Policy/PDF/26792-Futureworkforce_June2015.pdf. Accessed 5 Sept 2018

Crabtree L, Blunden H, Phibbs P, Sappideen C, Mortimer D, Shahib-Smith A, Chung L (2013) The Australian community land trust manual. The University of Western Sydney, Sydney

Ellen MacArthur Foundation. What is a circular economy? At: https://www.ellenmacarthurfoundation.org/circular-economy. Accessed on 4 Sept 2018

European Investment Bank (EIB) (2015) Financing the circular economy. At: http://www.eib.org/en/infocentre/events/all/financing-the-circular-economy.html. Accessed on 4 Sept 2018

European Parliament briefing (2016) Closing the loop – new circular economy package. At: http://www.europarl.europa.eu/RegData/etudes/BRIE/2016/573899/EPRS_BRI(2016)573899_EN.pdf. Accessed on 4 Sept 2018

Grigg D (1987) The industrial revolution and land transformation. In: Wolman, Fourier (eds) Land transformation in agriculture. SCOPE, John Wiley & Sons Ltd, Chichester and New York

Howard E (1896) Garden cities of tomorrow. Available at: https://ebooks.adelaide.edu.au/h/how ard/ebenezer/garden_cities_of_to-morrow. Accessed on 4 Sept 2018

Kent J, Thompson SM, Jalaludin B (2011) Healthy built environments: a review of the literature. Healthy Built Environments Program, City Futures research Centre, UNSW, Sydney. ISBN: 978-0-7334-3046-6

Raworth K (2017) A Doughnut for the Anthropocene: humanity's compass in the 21st century. Lancet: Planet Health 1:48–49

Salt B (2016) Superconnected lifestyle locations: the rise of the e-change movement. NBN Co. Ltd

United Nations Department of Social and Economic Affairs (2014) World urbanization prospects: the 2014 revision, highlights. United Nations, New York

UNSW Newsroom (2018) World-first e-waste microfactory launched at UNSW. At: https://news room.unsw.edu.au/news/science-tech/world-first-e-waste-microfactory-launched-unsw. Accessed on 4 Sept 2018

Watson I (2015) Aboriginal peoples, colonialism and international law: raw law. Routlege, Oxford & New York

Chapter 8
Density and Quality of Life in Mashhad, Iran

Fereshteh Moradi and Rob Roggema

Abstract Quality of Life (QoL) is a concept which can be evaluated in urban environments through consideration of social, physical, environmental, and economic indicators. The strategy of a high- density building typology, including the need to expand vertical public space as a by-product of this urban planning, also implying rise in the number of urban residents, has gained popularity in Iran. Housing and the way it is shaped is influencing the Quality of Life. The present study depicts an analytical and comparative assessment of the Quality of Life as an attribute of sustainability in the Sheshsad Dastgah residential complex, Firooze residential complex, and the Eastern part of the Goharshad neighborhood in Mashhad, Iran. Quality of Life is seen as an integrated way to describe the experienced quality of the resident's direct environment. The study adopts a descriptive-analytical methodology. Correlations among variables and step-wise regression analysis are the statistical methods used. The overall population consists of 7033 people, of whom 370 are selected as agents in the investigation. As indicators of the Quality of Life in this research, physical, social, and environmental aspects were used and are measured using sub-indicators pertaining to environmental quality, hygiene, access to recreational spaces, access to educational facilities, security, physical belonging, social solidarity, collaboration, access to daily facilities, and housing and infrastructure. The Quality of Life scores revealed that the average score of QoL in Sheshsad Dastgah (SD) is 4.17, while it is 3.46 in Firooze (F) and 2.88 in

F. Moradi
School of Architecture, University of Technology Sydney, Ultimo, Australia
e-mail: fereshteh.moradi@student.uts.edu.au

R. Roggema (✉)
Research Centre for the Built Environment NoorderRuimte, Hanze University of Applied Sciences, Groningen, The Netherlands and CITTA IDEALE, Office for Adaptive Planning, Wageningen, The Netherlands
e-mail: r.e.roggema@pl.hanze.nl; rob@cittaideale.eu

© Springer Nature Switzerland AG 2020
R. Roggema, A. Roggema (eds.), *Smart and Sustainable Cities and Buildings*,
https://doi.org/10.1007/978-3-030-37635-2_8

Goharshad East (GE), where 5 is the highest possible and 1 is the lowest possible score. Hence, the overall quality of life in Sheshsad Dastgah is closer to the ideal situation. The conclusion drawn from the three cases studied in this research, is that medium high-density building with a rich amount of green spaces (Sheshsad Dastgah) has a positive effect on improving the residents' Quality of Life rather than a villa pattern of residence (Goharshad East) or a more concentrated high-rise urban fabric without abundant green spaces (Firooze).

Keywords Quality of life · Urban fabric · Density · High-rise building · Stepwise regression analysis

8.1   Introduction

In recent decades, identifying, measuring, and improving the Quality of Life has been a major goal for researchers, planners, and government officials alike. The Quality of Life is highly influenced by time, place, and its constituent factors, and elements vary from one period, place, and geographical location to another. The World Health Organization defines Quality of Life as "people's subjective evaluation of their place in life regarding values, cultural systems, and their living environment, in line with their objectives, expectations, benchmarks, and concerns" (Noll 2004). Although considerable attention has been given to the topic, there is still room for further development of comprehensive and transparent frameworks for the assessment of Quality of Life in urban environments (Faria et al. 2018). A wide range of studies has used QoL-indicators to show the level of sustainability in different regions of the world (Senlier et al. 2009; Lepage 2009; Turkoglu 2015). The indicators that measure the Quality of *Urban* Life normally contain the concept of sustainability so that they can be used to balance and monitor the need among stakeholders to improve sustainable development (Senlier et al. 2009).

The options to physically expand cities in a horizontal direction is limited, due to cities' natural localities and the accessibility of urban services. Therefore, on one hand, further outward expansion meets its limitations, and on the other hand, the increasing and ongoing need for housing as result of population growth, necessitates the construction of high(er) density neighborhoods. The present study is undertaken in line with the previously mentioned necessities and aims to investigate the influence of density and urban patterns on the urban QoL. Regarding the relationship between QoL and urban spaces, the main objective is to improve and enhance the quality and accessibility of these spaces, so not only are the negative effects in urban spaces prevented, but their residents' QoL is also improving by creating a high-quality public space. This study aims to undertake an analytical investigation of urban density and patterns, whilst comparing QoL in the Sheshsad Dastgah (SD) residential complex representing medium density residential complexes, and Firooze (F) representing a high-rise residential gated complex, while in Eastern

Goharshad (GE) two-three story houses are common. Comparing these three cases will make it possible to correlate the indicators of experienced QoL with urban densities and patterns.

Following this, the research objective of this study is to investigate and measure the indicators of QoL in the three densities of urban fabric and to present an appropriate pattern of residence, based on the mentioned juxtaposition in Mashhad. Currently, empirical research that studies the implications of higher-density neighborhoods for the QoL is lacking. While planners increase the density of housing, they do not have data to which they can refer in order to assess the impact of their plans on the QoL of individuals (Mitrany 2005). This study aims to address the following questions: Which criteria influence QoL? and how are urban densities correlated to QoL? In essence, the study emphasizes the relationship between densities in urban precincts and the experienced QoL, as reflected in a range of indicators.

8.2   Background

QoL has been studied in a variety of academic fields, ranging from medicine to the real estate market (Faria et al. 2018). In recent years, many studies regarding urban QoL have been conducted, which denotes the importance of the subject. Most papers review both the objective conditions and subjective satisfaction. As an example, the so-called Q1999 study, entitled Economic Health, investigated QoL indicators, such as employment, remuneration, proper housing, and environmental features like supporting family, children, the poor, and elderly, together with transportation infrastructure, was conducted in 1999, questioning 1888 persons over 18 years old in five neighborhoods of Grand Trawas, USA (Rahnama 2010).

Sani Roychansyah et al. conducted research on the effects of urban compactness on the distributions of quality houses. They aimed to find the influence of urban compactness on the quality of houses in the city of Yogyakarta, Indonesia. In the end, they concluded that the region with the highest level of compactness had higher quality houses (33%) than the regions where compactness was medium (22%) and low (18%) (Saniroychansyah et al. 2016).

To date, research into the subjective aspects of high density focuses mainly on the negative consequences of overcrowding. On the contrary, a study by Michal Mitrany outlines positive aspects of higher density in neighborhoods, exploring the physical–spatial environment of two neighborhoods in the city of Haifa, Israel. Where physical planning enabled the potential advantages of high density to be realized, this was positively perceived and evaluated by residents. Such advantages mainly comprise accessibility to a variety of services, more frequent public transportation, and access to open spaces within walking distance. Particularly advantageous were the increased opportunities for social gathering. At the same time, however, high density did not foster increase of social relationships on a neighborhood level (Mitrany 2005).

8.3 Methodology

Both descriptive and inferential statistical methods were used in this study. With regards to the inferential statistics, the correlation between variables was measured. To test the significance of the differences among several groups with ordinal variables, step-wise regression was used. The data derived from the questionnaires, concerning the views of the residents, was analyzed using SPSS. The method consisted of semi-structured interviews with closed questionnaires, designed to obtain residents' evaluations of various aspects of the residential context anonymously.

8.4 Research Indicators

Considering both objective and subjective indicators of QoL, this study aims to investigate the effects of densities and the urban pattern on the residents' QoL. Environmental, social, and physical indicators were taken as the main indicators, with variables to environmental quality, hygiene, access to recreational spaces, access to educational facilities, security, physical belonging, social solidarity, collaboration, access to daily facilities, and housing and infrastructure as the sub-components of these main indicators. At the end of the questionnaires, respondents were asked to state their level of satisfaction with living in their neighborhood. There are economic indicators like income level (Soleimani et al. 2014; Forouhar and Hasankhani 2018), costs of living, costs of rent, costs of utilities (Forouhar and Hasankhani 2018), that were looked at. We consciously excluded these economic factors because they are very similar in each neighborhood.

8.5 Research Population and Sampling

The population of this research consists of all the residents of the Firooze district, the Eastern part of the Goharshad district, and the Sheshsad Dastgah residential complex in the city of Mashhad, Iran (Fig. 8.1). According to the 2016 census, the total population of these areas adds up to 7033 individuals. Using Cochrane's formula, with a maximum acceptable error ($d = 0.05$) and confidence level ($z = 1.96$) and (p and $q = 0.5$) ((Khaef and Zebardast 2016), the sample size of the study was determined to be 370 people.

Fig. 8.1 Location of the study areas in Mashhad. (Based on Google Maps)

At the first stage, pilot pre-tests by application of Cronbach's Alpha as a tool to evaluate the reliability of planned questions were directed with 30 households. Cronbach's Alpha rate ranges from 0 to 1. Cronbach's Alpha for this study is 0.88. According to Nunnally and Bernstein (1978), values of 0.7 and above are measured as acceptable reliability coefficients. Hence, the test and planned questions could be considered reliable.

8.6 Theoretical Framework

At the onset of every discussion, it is essential to define the keywords and terms that are used. The first step in assessing urban QoL, based on density and urban patterns, is to provide a comprehensive definition of such terms as the urban patterns, residential complexes, high-rise building, density, and QoL.

8.7 Urban Patterns

According to the Statistical Center of Iran, housing or a residential unit is a place, space or area which one or more families inhabit and which has passageways to one or more entrances. Urban patterns can be categorized into various types, depending on residential density, number of floors, and the number of families or people who live there. Considering all of these indicators, urban patterns can be divided into three major categories: Villas: single-family houses, Condos: multiple-family houses, and residential complexes (Tash 2012).

8.8 Residential Complexes

Such dwellings include several residential units which are built on a single piece of land, with shared entrances and common spaces. The distribution of residential blocks in such properties is a function of each block's form and geometry, its surface area, lighting, density, and other construction rules, and often results in disconnected units (Naqsh-e-Mohit 2011).

8.9 High-Rise Building

Various points of view have offered different definitions of high-rise buildings, each viewing it from their own perspective. From the civil engineers' point of view, a building is called high-rise if its height causes the lateral forces of wind and earthquakes to have a considerable influence on its design (Jordan 1994; Lundstrom and Rex 1990; Gane and Haymaker 2009). Accordingly, in terms of height, buildings with more than 10 stories are considered high-rises. From a fire-safety point of view, a building whose height (the distance between its lowest level of the highest floor and its lowest level within the reach of fire trucks) exceeds 23 meters is considered a high-rise. From a geometrical perspective, a building is considered a high-rise if its height-to-diameter ratio is equal to or greater than 3.14. Although such definitions exist for high-rise buildings, no specific criterion has been presented for defining such structures. In structures, height is a relative concept and depends on such conditions as social situations, individual's perception of the environment and, to a large degree, is defined by the norms of the location (Craighead and Inc 2009; Kwak et al. 2015; Mirrahimi et al. 2016; Samuelson et al. 2016).

Different countries, or even various pieces of legislation within one country, do not use a fixed height or a specific number of floors to define a high-rise building. According to the latest Urban Development Strategic-Structural Plan of Mashhad, buildings that have at least 12 stories are considered high-rises (Farnahad 2016).

8.10 Quality of Life

The term quality of life was first introduced by A.C. Pigou (Pigou and Co 1932) in "The Economics of Welfare" in 1920. However, its widespread use dates to the 1950's, when it was first used to emphasize materialistic indicators, to the extent that a country's Gross Domestic Product (GDP) was considered its single major criterion. Following a swathe of criticism of this materialistic interpretation of QoL in the late fifties, J.K. Galbraith redefined the concept and added immaterial values within environmental, political, and social domains (Galbraith 1950). Many thinkers and experts within this field believe that education, hygiene, the status of women, defense expenditure, economy, population, environment, social issues, cultural differences, welfare, etc. are all among the factors influencing QoL (Shultz et al. 2006).

Britannica describes QoL as the degree to which an individual is healthy, comfortable, and able to participate in, or enjoy, life events (Jenkinson 2016). Urban QoL is synonymous with attention to social, economic, cultural, environmental and mental indicators, in both objective (quantitative) and subjective (qualitative) terms, throughout the planning process. Criteria for QoL include the accessibility of facilities, quality of housing, quality of leisure spaces, the creation of opportunities for social interaction, social occasions, employment, welfare, and social collaboration (Kokabi 2005).

The concept of QoL touches upon two aspects of human life: Objective conditions of the society and the urban environment, and subjective-cognitive perception of life experiences by social groups and individuals (Fukuyama 1995).

8.11 Density

Gross density is measured as the ratio of the total urban population and the total residential area. In contrast to the net or urban density, the values of residential densities are usually larger as the same population size is referred to a smaller reference area (residential instead of the total built-up area) (Wolff et al. 2018).

Past research on the effects of high density on people has focused either on urban areas (population density) or on the home environment (indoor density) (Mitrany 2005). In this paper, we measure gross density and net density of the three precincts which are analyzed. FAR (Floor Area Ratio) is calculated by dividing the total floor space to the total size of the site/plot (Figs. 8.2, 8.3, and 8.4).

Fig. 8.2 Google Earth map of Firooze precinct

Fig. 8.3 Google Earth map of Sheshsad Dastgah precinct

8.12 Findings

As mentioned above, the average score, in all sections of the research, falls within
the range of 1–5, with 5 indicating a high degree of satisfaction with the item and
1 indicating a dissatisfaction with or paucity of the mentioned factor in the neighborhood in question.

Fig. 8.4 Google Earth map of Goharshad precinct

Table 8.1 shows the residents' satisfaction in each of the case study areas for the different criteria, based on the results of the questionnaire. In the Sheshsad Dastgah residential complex, the most appreciated aspects are those of recreational spaces, environmental quality, physical belonging, collaboration, security, and access to educational and daily facilities. Indicators such as hygiene, housing and infrastructure, and social solidarity are less appreciated but still score relatively well.

In the Firooze precinct, which can be characterized as a high-density urban fabric, the results show that environmental quality, security and housing, and infrastructure are very well appreciated. Indicators such as recreational spaces, environmental quality, hygiene, and access to daily facilities scored an average level, while social solidarity and collaboration are underperforming.

In the Goharshad East (GE) district, the residents' average scores for indicators of environmental quality, hygiene, access to educational services, and housing and infrastructure are highest, but, compared to the other case studies, do not score very high. The scores for indicators of security, social solidarity, physical belonging, and access to recreational spaces are all close to the mean and close to the two lowest scores amongst all criteria in the other two neighborhoods. The average scores for access to daily facilities and collaboration are the lowest in the entire investigation, and relatively close to 1, the lowest possible level.

To obtain the overall chart for each individual's quality of life, the measured indicators were normalized and then added. Descriptive statistics for QoL show that

Table 8.1 Average score for each indicator in three case studies

Precinct	Recreational spaces	Hygiene	Environmental quality	Physical belonging	Security	Social solidarity	Collaboration	Education	Access to daily facilities	Housing and infrastructure	Total QoL
Sheshsad Dastgah	3.78	3.97	4.59	4.37	4.33	3.99	4.3	4.34	4.49	3.57	4.17
Firooze	3.42	3.42	4.08	3.61	4.31	2.41	2.44	3.22	3.39	4.27	3.46
Goharshad	2.76	3.3	4.25	2.91	2.76	2.16	1.56	3.68	1.81	3.58	2.88

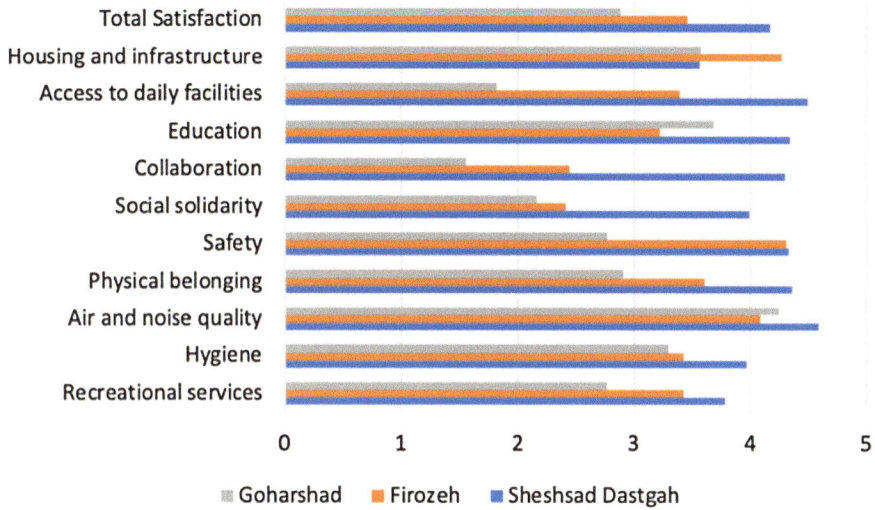

Fig. 8.5 Average score for each indicator in three case study areas

the average score for residents of Sheshsad Dastgah (SD) is 4.17, 3.46 for residents of Firooze (F) district, and 2.88 for Goharshad East (GE) dwellers (Fig. 8.5).

8.13   Quantitative Analysis of the Survey Findings

As mentioned in the introduction of this article, one of the purposes of this research was to examine the existence or non-existence of a correlation between density and QoL in the case study areas. To answer this question, firstly the density of buildings of three different urban fabric was calculated (see Table 8.2). Secondly, the residents' responses to the 30 questions on the questionnaire were computed into 10 main indicators. Finally, the correlation between density and every single indicator of QoL was analyzed. Table 8.3 shows the results of this correlation analysis.

As can be derived from Table 8.3, the density (number of floors) and indicators of security, hygiene, physical belonging, social solidarity, collaboration, access to daily facilities, access to recreational spaces, and housing and infrastructure are positively correlated, because their calculated sig. is less than 0.05. Correlating every single indicator with the density revealed that the two indicators of education and environmental quality are not correlated (Their sig. is more than 0.05; 0.99 and 0.977 respectively). This means that the availability of education and environment qualities are not related to the density of residential areas as much as other indicators.

Table 8.2 Physical attributes and densities in case studies

Precinct	Population	Total units	Total gross floor area (m^2)	Total site area (m^2); incl. plots, roads, open spaces	Gross density (#dwellings/ ha)	Gross density (#people/ ha)	Area of site (m^2)	Net density (#dwellings/ ha)	Net density (#people/ ha)	FAR
Firooze	1560	1306	141,048	99,265	132	157	84,291	155	185	1.67
Goharshad E	1386	360	27,278	57,282	63	242	41,293	87	336	0.66
Sheshsad Dastgah	2787	899	82,141	222,731	40	125	198,426	45	140	0.41

Table 8.3 Correlation between density and indicators of quality of life

Indicators		Education (I1)	Security (I2)	Physical belonging (I3)	Social solidarity (I4)	Collaboration (I5)	Hygiene (I6)	Environment quality (I7)	Access to recreational spaces (I8)	Access to daily facilities (I9)	Housing and infrastructure (I10)
Density	Pearson correlation	.086	.134**	.459**	−.304**	−.167**	−.272**	.002	.119*	−.460**	.559**
	Sig. (2-tailed)	.099	.010	.000	.000	.001	.000	.977	.022	.000	.000
	N	370	370	370	370	370	370	370	370	370	370

**Correlation is significant at the 0.01 level (2-tailed)
*Correlation is significant at the 0.05 level (2-tailed)

On the other hand, there is a direct correlation between density and eight of the 10 main indicators of QoL which are applied in this research. Hence, it can be concluded that the density of a residential area influences the quality of dwellers' life (objective indicators) and their sense of satisfaction (subjective indicators).

8.14 Regression Analysis

In regression, the presence of multiple correlations among dependent and independent research variables is investigated. As the results of a regression analysis on an interval scale are more robust, all the variables are analyzed using an interval scale. To achieve this goal, all questions on different dimensions of QoL were computed and transformed from the Likert spectrum (ordinal) into interval (scale) data. After computing of the QoL indicators, in order to see the extent to which each indicator affects respondents' life satisfaction and the indicator's level of importance, stepwise regression analysis was conducted. Total life satisfaction is considered a dependent variable and the computed indicators of QoL are considered predictors: Education (I 1), Security (I 2), Physical Belonging (I 3), Social Solidarity (I 4), Collaboration (I 5), Hygiene (I 6), Environment Quality (I 7), Access to Recreational Spaces (I 8), Access to Daily Facilities (I 9), Housing and Infrastructure (I 10). The result of the stepwise regression summary is shown in Table 8.4.

Table 8.4 shows that the coefficient of determination is 0.560. The coefficient of determination denotes that 56% of the variation in life satisfaction in the study area can be explained by the independent variables of the study (i.e. environmental, social and physical indicators). According to the level of significance of 0.000 (the level of significance should be less than 0.05), it can be said, with a probability of 99%, that the correlation between the dependent variable and the independent variables is 0.756 and that the independent variables of the study explain 56% of variation in the dependent variable. In other words, more than half of the differences in quality of life in these neighborhoods are related to their level of the examined indicators.

According to the results depicted in Table 8.5, the regression equation is reflected as below:

$$
\begin{aligned}
\text{Life satisfaction} &= 0.259\,(I4) - 0.522\,(I3) + 0.393\,(I2) + 0.264(I5) - 0.148(I8) \\
&\quad + 0.151\,(I7) + 0.183\,(I6)\,R2 \\
&= 0.572
\end{aligned}
$$

Table 8.4 Stepwise regression model summary

Model	R	R square	Adjusted R square	Std. the error of the estimate
7	.756a	.572	.560	.543

Table 8.5 Stepwise regression analysis results

Model	Unstandardized coefficients		Standardized coefficients	t	Sig.	95% confidence interval for B		Collinearity statistics	
	B	Std. error	Beta			Lower bound	Upper bound	Tolerance	VIF
(Constant)	0.738	.248		7.005	.000	1.250	2.226		
(I4) Social solidarity	.259	.051	.313	5.063	.000	.158	.360	.313	3.198
(I3) Physical belonging	−.522	.057	−.521	−9.209	.000	−.633	−.410	.372	2.691
(I2) Security	.393	.062	.348	6.383	.000	.272	.515	.400	2.502
(I5) Collaboration	.264	.040	.399	6.559	.000	.185	.343	.322	3.107
(I8) Access to recreational spaces	−.148	.045	−.213	−3.265	.001	−.238	−.059	.279	3.587
(I7) Environment quality	.151	.055	.140	2.754	.006	.043	.259	.461	2.169
(I6) Hygiene	.183	.070	.131	2.609	.009	.045	.321	.471	2.121

a. Dependent variable: total satisfaction

Table 8.6 Excluded variables

Model	Beta in	t	Sig.	Partial Correlation	Collinearity statistics		Minimum tolerance
					Tolerance	VIF	
(I1) Education	.002	.040	.968	.002	.852	1.174	.278
(I9) Access to daily facilities	.029	.443	.658	.023	.280	3.570	.251
(I 10) Housing and infrastructure	.079	1.449	.148	.076	.402	2.490	.191

The above equation displays that there is no statistically significant relationship between life satisfaction and education (I 1), access to daily facilities (I 9) and housing and infrastructure (I 10). In the regression analysis, the beta coefficient denotes the level of importance of each indicator. The beta coefficient in Table 8.5 shows that social solidarity (I 4), security (I 2), and collaboration (I 5) are the three most important aspects of QoL that significantly affect total life satisfaction in the study areas. The negative beta score for physical belonging (I 3) and access to recreational spaces (I 8) could be anticipated since the answers to these questions are very different in each precinct. Especially the access to recreational spaces has widespread responses, because this criterion consists of three variables, indicating access to green, walking/cycling and exercise spaces, while two neighborhoods of the three case studies hardly have any of these facilities. Therefore, their life satisfaction is adversely affected by this indicator.

In fact, many indicators are included in research that may affect people's subjective QoL. Therefore, to select indicators among all applied indicators that best describe QoL, step-wise regression assists us in excluding indicators which statistically are not involved in determining QoL in our case studies. Based on Table 8.6, three indicators (education, access to daily facilities, and housing and infrastructure) can be excluded from the regression model because statistically, the sig amount which is calculated for them is above 0.05.

8.15   Conclusion and Suggestions

To answer the first question of the study which concerned the influence of density on increasing or decreasing urban QoL, considering the urban fabric of the three case studies, one with a low-density pattern (East Goharshad), the other with a medium density fabric (Sheshsad Dastgah residential complex) and the latter one as a high-rise building complex (Firooze), a comparison of QoL scores in the three neighborhoods reveal the influence of density on the QoL. Based on the findings of the present research, the correlation analysis shows that there is a direct correlation between density and eight of 10 indicators of QoL. Therefore, it can be said that the density of a residential area influences the residents' QoL (objective indicators) and their sense of satisfaction (subjective indicators). The chosen neighborhoods have a similar economic situation. Hence, the economic indicators have been excluded from this research consciously, because they do not significantly affect the results.

Descriptive statistics show that the average score for QoL among the residents of the Sheshsad Dastgah residential complex equals 4.17, whereas it is 3.46 for the Firooze district and 2.88 for East Goharshad. In this study, the highest and the lowest possible scores for quality of life are 5 and 1 respectively. Thus, the overall score of QoL in the Sheshsad Dastgah residential complex is closer to the ideal situation (score 5) compared with that of the Firooze and Goharshad East precincts.

Accordingly, with regards to the cases studied in this research, and drawing from the results of the step-wise regression, medium-rise buildings with green space amidst them has a positive influence on the residents' QoL, rather than high-rise complexes or detached houses.

To answer the second research question, which concerned the criteria and factors influencing the QoL, we should refer to the results of the regression analysis, where the influence of various indicators on residents' QoL was measured. Based on those results, all in all, the independent variables of the study explain 56% of the variation in the dependent variable. In addition, the beta coefficient shows that social solidarity (I 4), security (I 2), and collaboration (I 5) are the three most important aspects of QoL that significantly affect total life satisfaction in the study areas. On the other hand, the equation displays that there is no statistically significant relationship between life satisfaction and access to educational facilities (I 1), access to daily facilities (I 9), and housing and infrastructure quality (I 10). This can be concluded from various levels of these indicators in the three surveyed neighborhoods, which affected the results of regression analysis.

Thus, it can be said that planning schemes for every precinct should consider the preferred density and urban fabric for enhancing the Quality of Life, and as such, they should not only consider enhancing physical arrangements (I 1, I 9 and I 10), but more importantly, they should try to improve the districts' social (I 4, I 2 and I 5) and environmental (I 7 and I 6) foundations. More emphasis should be placed on achieving higher levels of health and access to walking/cycling infrastructure. These precincts should enhance their social solidarity, create and reinforce local communities, and empower neighborhood-based institutions. Taking these measures, along with the physical and spatial improvement of living spaces, can increase the quality of life across neighborhoods.

Our investigation concludes that the best quality of life can be developed in the medium dense urban fabric where abundant attention is given to social and environmental quality.

Appendices

Correlation Between All Indicators

Correlations

		Density	Education	Security	Hygiene	Environment quality	Physical belonging	Social solidarity	Collaboration	Access to daily facilities	Access to recreational spaces	Housing and infrastructural
Density	Pearson correlation	1	.086	-.134**	-.272**	.002	.459**	-.304**	-.167**	-.460**	.119*	.559**
	Sig. (2-tailed)		.099	.010	.000	.977	.000	.000	.001	.000	.022	.000
	N	370	370	370	370	370	370	370	370	370	370	370
Education	Pearson correlation	.086	1	.176**		.174**	.310**	.217**	.239**	.344**	.146**	.100
	Sig. (2-tailed)	.099		.001		.001	.000	.000	.000	.000	.005	.055
	N	370	370	370		370	370	370	370	370	370	370
Security	Pearson correlation	-.134**	.176**	1		.711**	.431**	.564**	.437**	.303**	.617**	-.057
	Sig. (2-tailed)	.010	.001			.000	.000	.000	.000	.000	.000	.278
	N	370	370	370		370	370	370	370	370	370	370

Hygiene	Pearson correlation	−.272**	.174**	.711**	1	.321**	.437**	.394**	.251**	.751**	.480**	.058
	Sig. (2-tailed)	.000	.001	.000		.000	.000	.000	.000	.000	.000	.262
	N	370	370	370	370	370	370	370	370	370	370	370
Environment quality	Pearson correlation	.002	.310**	.431**	.321**	1	.591**	.578**	.625**	.421**	.634**	.013
	Sig. (2-tailed)	.977	.000	.000	.000		.000	.000	.000	.000	.000	.810
	N	370	370	370	370	370	370	370	370	370	370	370
Physical belonging	Pearson correlation	.459**	.217**	.564**	.437**	.591**	1	.410**	.413**	.358**	.733**	.469**
	Sig. (2-tailed)	.000	.000	.000	.000	.000		.000	.000	.000	.000	.000
	N	370	370	370	370	370	370	370	370	370	370	370
Social solidarity	Pearson correlation	−.304**	.239**	.437**	.394**	.578**	.410**	1	.779**	.605**	.667**	−.236**
	Sig. (2-tailed)	.000	.000	.000	.000	.000	.000		.000	.000	.000	.000
	N	370	370	370	370	370	370	370	370	370	370	370

Collaboration	Pearson correlation	-.167**	.344**	.303**	.251**	.625**	.413**	.779**	1	.421**	.639**	-.235**
	Sig. (2-tailed)	.001	.000	.000	.000	.000	.000	.000		.000	.000	.000
	N	370	370	370	370	370	370	370	370	370	370	370
Access to daily facilities	Pearson correlation	-.460**	.146**	.617**	.751**	.421**	.358**	.605**	.421**	1	.599**	-.123*
	Sig. (2-tailed)	.000	.005	.000	.000	.000	.000	.000	.000		.000	.018
	N	370	370	370	370	370	370	370	370	370	370	370
Access to recreational Spaces	Pearson correlation	.119*	.239**	.533**	.480**	.634**	.733**	.667**	.639**	.599**	1	.072
	Sig. (2-tailed)	.022	.000	.000	.000	.000	.000	.000	.000	.000		.168
	N	370	370	370	370	370	370	370	370	370	370	370
Housing and infrastructural	Pearson correlation	.559**	.100	-.057	.058	.013	.469**	-.236**	-.235**	-.123*	.072	1
	Sig. (2-tailed)	.000	.055	.278	.262	.810	.000	.000	.000	.018	.168	
	N	370	370	370	370	370	370	370	370	370	370	370

**Correlation is significant at the 0.01 level (2-tailed)

*Correlation is significant at the 0.05 level (2-tailed)

Correlation Between Density and Other Indicators

	Density	Education	Security	Hygiene	Environment quality	Physical belonging	Social solidarity	Collaboration	Access to daily facilities	Access to recreational spaces	Housing and infrastructural
Pearson correlation	1	.086	−.134**	−.272**	.002	.459**	−.304**	−.167**	−.460**	.119*	.559**
Sig. (2-tailed)		.099	.010	.000	.977	.000	.000	.001	.000	.022	.000
N	370	370	370	370	370	370	370	370	370	370	370

**Correlation is significant at the 0.01 level (2-tailed)
*Correlation is significant at the 0.05 level (2-tailed)

Cronbach's Alpha Result (Test of reliability)

Case processing summary		N	%
Cases	Valid	370	100.0
	Excluded[a]	0	0.0
	Total	370	100.0

[a]Listwise deletion based on all variables in the procedure

F. Moradi and R. Roggema

Reliability statistics

Cronbach's alpha	N of items
.886	27

Regression

ANNOVA

Model		Sum of squares	df	Mean square	F	Sig.
1	Regression	75.399	1	75.399	161.5	.000b
	Residual	171.791	368	.467		
	Total	247.190	369			

2	Regression	97.792	2	48.896	120.1	.000c
	Residual	149.398	36	.407		
	Total	247.190	369			

3						
	Regression	119.831	3	39.944	114.8	.000d
	Residual	127.359	366	.348		
	Total	247.190	369			

a. Dependent variable: total satisfaction
b. Predictors: (constant), social solidarity
c. Predictors: (constant), social solidarity, physical belonging
d. Predictors: (constant), social solidarity, physical belonging, security
e. Predictors: (constant), social solidarity, physical belonging, security, collaboration
f. Predictors: (constant), social solidarity, physical belonging, security, collaboration, access to recreational spaces
g. Predictors: (constant), social solidarity, physical belonging, security, collaboration, access to recreational spaces, environment quality
h. Predictors: (constant), social solidarity, physical belonging, security, collaboration, access to recreational spaces, environment quality, hygiene

ANOVA

Model		Sum of squares	df	Mean square	F	Sig.
4	Regression	134.373	4	33.593	108.7	.000e
	Residual	112.816	365	.309		
	Total	247.190	369			
5	Regression	136.569	5	27.314	89.9	.000f
	Residual	110.621	364	.304		
	Total	247.190	369			
6	Regression	138.585	6	23.097	77.2	.000g
	Residual	108.605	363	.299		
	Total	247.190	369			
7	Regression	140.588	7	20.084	68.2	.000h
	Residual	106.601	362	.294		
	Total	247.190	369			

a. Dependent variable: total satisfaction
b. Predictors: (constant), social solidarity
c. Predictors: (constant), social solidarity, physical belonging
d. Predictors: (constant), social solidarity, physical belonging, security
e. Predictors: (constant), social solidarity, physical belonging, security, collaboration
f. Predictors: (constant), social solidarity, physical belonging, security, collaboration, access to recreational spaces
g. Predictors: (constant), social solidarity, physical belonging, security, collaboration, access to recreational spaces, environment quality
h. Predictors: (constant), social solidarity, physical belonging, security, collaboration, access to recreational spaces, environment quality, hygiene

Model Summary

Model	R	R square	Adjusted R square	Std. error of the estimate	Change statistics				
					Square change	F change	df1	df2	Sig. F change
1	.552a	.305	.303	.683	.305	161.515	1	368	.000
2	.629b	.396	.392	.638	.091	55.009	1	367	.000
3	.696c	.485	.481	.590	.089	63.335	1	366	.000
4	.737d	.544	.539	.556	.059	47.051	1	365	.000
5	.743e	.552	.546	.551	.009	7.223	1	364	.008
6	.749f	.561	.553	.547	.008	6.738	1	363	.010
7	.754g	.569	.560	.543	.008	6.805	1	362	.000

Predictors: (constant), social solidarity

Predictors: (constant), social solidarity, physical belonging predictors: (constant), social solidarity, physical belonging, security

Predictors: (constant), social solidarity, physical belonging, security, collaboration predictors: (constant), social solidarity, physical belonging, security, collaboration, access to recreational spaces

Predictors: (constant), social solidarity, physical belonging, security, collaboration, access to recreational spaces, environment quality

Predictors: (constant), social solidarity, physical belonging, security, collaboration, access to recreational spaces, environment quality, hygiene

Dependent variable: total satisfaction

Coefficients[a]

Model		Unstandardized coefficients		Standardized coefficients	t	Sig.	95.0% confidence interval for B		Collinearity statistics	
		B	Std. error	Beta			Lower bound	Upper bound	Tolerance	VIF
1	(Constant)	1.703	.114		18.480	.000	1.879	2.327	1.000	1.000
	Social solidarity	.458	.036	.552	12.709	.000	.387	.528		
2	(Constant)	1.095	.171		18.122	.000	2.759	3.430		
	Social solidarity	.570	.037	.688	15.454	.000	.497	.642	.832	1.202
	Physical belonging	-.330	.045	-.330	-7.417	.000	-.418	-.243	.832	1.202
3	(Constant)	1.691	.175		14.222	.000	2.147	2.835		
	Social solidarity	.493	.035	.594	13.902	.000	.423	.562	.770	1.299
	Physical belonging	-.504	.047	-.504	-10.820	.000	-.596	-.413	.649	1.542
	Safety	.424	.053	.376	7.958	.000	.320	.529	.630	1.586
4	(Constant)	1.518	.165		15.249	.000	2.193	2.843		
	Social solidarity	.240	.050	.290	4.843	.000	.143	.338	.348	2.872
	Physical belonging	-.573	.045	-.573	-12.718	.000	-.662	-.484	.616	1.622
	Safety	.483	.051	.427	9.466	.000	.382	.583	.613	1.631
	Collaboration	.263	.038	.398	6.859	.000	.188	.339	.372	2.689
5	(Constant)	1.357	.174		13.514	.000	2.014	2.700		
	Social solidarity	.282	.052	.340	5.462	.000	.180	.383	.317	3.151
	Physical belonging	-.485	.055	-.485	-8.770	.000	-.594	-.376	.402	2.488
	Safety	.496	.051	.440	9.770	.000	.397	.596	.607	1.648
	Collaboration	.286	.039	.432	7.333	.000	.209	.362	.355	2.818
	Access to recreational spaces	-.123	.046	-.176	2.688	.008	-.212	-.033	.286	3.491
	(Constant)	1.164	.188		11.492	.000	1.794	2.534		
	Social solidarity	.272	.051	.328	5.297	.000	.171	.373	.316	3.168

(continued)

Coefficients[a]

Model		Unstandardized coefficients		standardized coefficients	t	Sig.	95.0% confidence interval for B		Collinearity statistics	
		B	Std. error	Beta			Lower bound	Upper bound	Tolerance	VIF
	Physical belonging	−.526	.057	−.526	9.212	.000	−.638	−.413	.372	2.689
	Safety	.487	.051	.431	9.630	.000	.387	.586	.603	1.657
6	Collaboration	.255	.040	.385	6.304	.000	.175	.334	.324	3.083
	Access to recreational spaces	−.131	.045	−.188	2.892	.004	−.220	−.042	.285	3.510
	Environment quality	.144	.055	.133	2.596	.010	.035	.252	.462	2.162
	(Constant)	0.738	.248		7.005	.000	1.250	2.226		
	Social solidarity	.259	.051	.313	5.063	.000	.158	.360	.313	3.198
	Physical belonging	−.522	.057	−.521	9.209	.000	−.633	−.410	.372	2.691
	Safety	.393	.062	.348	6.383	.000	.272	.515	.400	2.502
	Collaboration	.264	.040	.399	6.559	.000	.185	.343	.322	3.107
7	Access to recreational spaces	−.148	.045	−.213	−3.265	.001	−.238	−.059	.279	3.587
	Environment quality	.151	.055	.140	2.754	.006	.043	.259	.461	2.169
	Hygiene	.183	.070	.131	2.609	.009	.045	.321	.471	2.121

a. Dependent variable: total satisfaction

Coefficients[a]

	Model	Unstandardized coefficients		Standardized coefficients	t	Sig.	95.0% confidence interval for B		Collinearity statistics	
		B	Std. error	Beta			Lower bound	Upper bound	Tolerance	VIF
	(Constant)	0.738	.248		7.005	.000	1.250	2.226		
	Social solidarity	.259	.051	.313	5.063	.000	.158	.360	.313	3.198
	Physical belonging	-.522	.057	-.521	9.209	.000	-.633	-.410	.372	2.691
	Safety	.393	.062	.348	6.383	.000	.272	.515	.400	2.502
7	Collaboration	.264	.040	.399	6.559	.000	.185	.343	.322	3.107
	Access to recreational spaces	-.148	.045	-.213	-3.265	.001	-.238	-.059	.279	3.587
	Environment quality	.151	.055	.140	2.754	.006	.043	.259	.461	2.169
	Hygiene	.183	.070	.131	2.609	.009	.045	.321	.471	2.121

a. Dependent variable: total satisfaction

Excluded variables[a]

Model		Beta in	t	Sig.	Partial correlation	Collinearity statistics		
						Tolerance	VIF	m tolerance
	Education	.047b	1.050	.294	.055	.943	1.061	.943
	Safety	.136b	2.840	.005	.147	.809	1.236	.809
	Hygiene	.118b	2.513	.012	.130	.845	1.184	.845
	Environment quality	.000b	−.004	.997	.000	.666	1.501	.666
	Physical belonging	−.330b	−7.417	.000	−.361	.832	1.202	.832
1	Collaboration	.220b	3.208	.001	.165	.393	2.545	.393
	Access to daily facilities	.140b	2.595	.010	.134	.634	1.577	.634
	Access to recreational spaces	−.293b	−5.206	.000	−.262	.555	1.801	.555
	Housing and infrastructure	−.292b	−6.928	.000	−.340	.944	1.059	.944
	Education	.090c	2.151	.032	.112	.926	1.080	.808
	Safety	.376c	7.958	.000	.384	.630	1.586	.630
2	Hygiene	.253c	5.626	.000	.282	.754	1.326	.742
	Environment quality	.226c	4.087	.000	.209	.516	1.939	.516
	Collaboration		4.811	.000	.244	.382	2.615	.382
	Access to daily facilities	.203c	4.008	.000	.205	.620	1.614	.591
	Access to recreational spaces	−.038c	−.507	.612	−.027	.302	3.317	.302
	Housing and infrastructure	−.159c	−2.957	.003	−.153	.560	1.787	.493

a. Dependent variable: total satisfaction
b. Predictors in the model: (constant), social solidarity
c. Predictors in the model: (constant), social solidarity, physical belonging
d. Predictors in the model: (constant), social solidarity, physical belonging, safety
e. Predictors in the model: (constant), social solidarity, physical belonging, safety, collaboration
f. Predictors in the model: (constant), social solidarity, physical belonging, safety, collaboration, access to recreational spaces
g. Predictors in the model: (constant), social solidarity, physical belonging, safety, collaboration, access to recreational spaces, environment quality
h. Predictors in the model: (constant), social solidarity, physical belonging, safety, collaboration, access to recreational spaces, environment quality, hygiene

Collinearity statistics

Model		Beta in	t	Sig.	Partial correlation	Tolerance	VIF	m tolerance
	Education	.084d	2.159	.032	.112	.925	1.081	.630
	Hygiene	.074d	1.371	.171	.072	.486	2.059	.406
	Environment quality	.216d	4.220	.000	.216	.515	1.941	.515
	Collaboration	.398d	6.859	.000	.338	.372	2.689	.348
3	Access to daily facilities	.025d	.467	.641	.024	.475	2.104	.475
	Access to recreational spaces	−.073d	−1.070	.285	−.056	.300	3.331	.300
	Housing and infrastructure	−.016d	−.300	.764	−.016	.486	2.057	.335
4	Education	.022e	.588	.557	.031	.868	1.152	.346
	Hygiene	.102e	2.008	.045	.105	.483	2.072	.341
	Environment quality	.122e	2.367	.018	.123	.465	2.151	.336
	Access to daily facilities	.045e	.877	.381	.046	.474	2.110	.298
	Access to recreational spaces	−.176e	−2.688	.008	−.139	.286	3.491	.286
	Housing and infrastructure	.113e	2.107	.036	.110	.432	2.316	.284
5	Education	.019f	.514	.607	.027	.867	1.153	.286
	Hygiene	.124f	2.441	.015	.127	.473	2.115	.281
	Environment quality	.133f	2.596	.010	.135	.462	2.162	.285
	Access to daily facilities	.104f	1.925	.055	.101	.419	2.389	.253
	Housing and infrastructure	.096f	1.784	.075	.093	.424	2.357	.210
6	Education	.009g	.241	.810	.013	.857	1.167	.285
	Hygiene	.131g	2.609	.009	.136	.471	2.121	.279
	Access to daily facilities	.099g	1.854	.065	.097	.418	2.391	.252
	Housing and infrastructure	.106g	1.981	.048	.104	.422	2.368	.197
7	Education	.002h	.040	.968	.002	.852	1.174	.278
	Access to daily facilities	.029h	.443	.658	.023	.280	3.570	.251

(continued)

Collinearity statistics

Model	Beta in	t	Sig.	Partial correlation	Tolerance	VIF	m tolerance
Housing and infrastructure	.079h	1.449	.148	.076	.402	2.490	.191

a. Dependent variable: total satisfaction
b. Predictors in the model: (constant), social solidarity
c. Predictors in the model: (constant), social solidarity, physical belonging
d. Predictors in the model: (constant), social solidarity, physical belonging, safety
e. Predictors in the model: (constant), social solidarity, physical belonging, safety, collaboration
f. Predictors in the model: (constant), social solidarity, physical belonging, safety, collaboration, access to recreational spaces
g. Predictors in the model: (constant), social solidarity, physical belonging, safety, collaboration, access to recreational spaces, environment quality
h. Predictors in the model: (constant), social solidarity, physical belonging, safety, collaboration, access to recreational spaces, environment quality, hygiene

Excluded variables[a]

Model		Beta in	t	Sig.	Partial correlation	Collinearity statistics		
						Tolerance	VIF	m Tolerance
7	Education	.002h	.040	.968	.002	.852	1.174	.278
	Access To Daily Facilities	.029h	.443	.658	.023	.280	3.570	.251
	Housing And Infrastructure	.079h	1.449	.148	.076	.402	2.490	.191

a. Dependent variable: total satisfaction
h. Predictors in the model: (constant), social solidarity, physical belonging, safety, collaboration

References

Craighead G (2009) High-rise security and fire life safety. Butterworth-Heinemann, Oxford

Faria PA, Ferreira FA, Jalali MS, Bento P, António NJ (2018) Combining cognitive mapping and MCDA for improving quality of life in urban areas. Cities 78:116–127

Farnahad CE (2016) Urban development strategic-structural planning. Ministry of Housing Urban Development, Tehran

Forouhar A, Hasankhani M (2018) Urban renewal mega projects and residents' quality of life: evidence from historical religious center of Mashhad metropolis. J Urban Health 95:232–244

Fukuyama F (1995) Trust: the social virtues and the creation of prosperity. Free Press Paperbacks, New York

Galbraith JK (1950) America and Western Europe (no.159). Public affairs committee, New York

Gane V, Haymaker J (2009) Benchmarking current conceptual high-rise design processes. J Archit Eng 16:100–111

Jenkinson C (2016) Quality of life. Encyclopaedia Britanica, New York

Jordan P (1994) High-rise building. Google Patents

Khaef S, Zebardast E (2016) Assessing quality of life dimensions in deteriorated inner areas: A case from Javadieh neighborhood in Tehran metropolis. Soc Indic Res 127:761–775

Kokabi A (2005) Planning for the improvement of urban life quality in central regions in cities: a case study of Khorramabad. Master, Tarbiat Modares University

Kwak K-H, Baik J-J, Ryu Y-H, Lee S-H (2015) Urban air quality simulation in a high-rise building area using a CFD model coupled with mesoscale meteorological and chemistry-transport models. Atmos Environ 100:167–177

Lepage A (2009) The quality of life as attribute of sustainability. TQM J 21:105–115

Lundstrom B, Rex O (1990) High-rise building. Google patents

Mirrahimi S, Mohamed MF, Haw LC, Ibrahim NLN, Yusoff WFM, Aflaki A (2016) The effect of building envelope on the thermal comfort and energy saving for high-rise buildings in hot–humid climate. Renew Sustain Energy Rev 53:1508–1519

Mitrany M (2005) High density neighborhoods: Who enjoys them? GeoJournal 64:131–140

Naqsh-e-Mohit (2011) A development model and comprehensive plan for the north-west region in Mashhad (Examining the housing status). Naqsh-e-Mohit Consulting Engineers, Tehran

Noll H-H (2004) The European system of social indicators: a tool for welfare measurement and monitoring social change. In ZUMA-Centre for Survey Research and Methodology, Socia Indicators Department, Mannheim, Germany, Workshop on Measurement of Wellbeing in Developing Countries, Hanse Kolleg, Delmenhorst

Nunnally J, Bernstein I (1978) Psychometric theory. McGraw-Hill, New York Google Scholar

Pigou ACJM, Co L (1932) The economics of welfare, 1920. Macmillan, London

Rahnama M (2010) Planning for the improvement of urban life quality in neighborhoods (A case study of Davoodieh, Tehran). Master, Tarbiat Modares University, Tehran, Iran

Samuelson H, Claussnitzer S, Goyal A, Chen Y, Romo-Castillo A (2016) Parametric energy simulation in early design: high-rise residential buildings in urban contexts. Build Environ 101:19–31

Saniroychansyah M, Farmawati A, Anindyah DS, Atianta L (2016) Urban compactness effects on the distributions of healthy houses in Yogyakarta City. Proc – Soc Behav Sci 227:168–173

Senlier N, Yildiz R, Aktaş ED (2009) A perception survey for the evaluation of urban quality of life in Kocaeli and a comparison of the life satisfaction with the European cities. Soc Indic Res 94:213–226

Shultz C J, Westbrook M D, Dinh Tho N (2006) Subjective quality of life and market activity in vietnam: new data and extensions. Macromarketing 2006 seminar proceedings: macromarketing the future of marketing?, Queenstown, New Zealand

Soleimani M, Tavallaei S, Mansuorian H, Barati Z (2014) The assessment of quality of life in transitional neighborhoods. Soc Indic Res 119:1589–1602

Tash (2012) A development model and comprehensive plan for the central region in Mashhad (Examining the housing status). Tash Consulting Architects and Urban Planners, Tehran

Turkoglu H (2015) Sustainable development and quality of urban life. Proc – Soc Behav Sci 202:10–14

Wolff M, Haase D, Haase A (2018) Compact or spread? A quantitative spatial model of urban areas in Europe since 1990. PLos One 13:e0192326

Chapter 9
Deep Renovation in Sustainable Cities: Zero Energy, Zero Urban Sprawl at Zero Costs in the Abracadabra Strategy

Annarita Ferrante, Anastasia Fotopoulou, Lorna Dragonetti, and Giovanni Semprini

Abstract The energy efficiency challenge in buildings mainly concerns the energy efficient refurbishment and investments in existing buildings. Yet, today, only 1,2% of existing buildings is renovated every year in Europe. The actual investment gap in the deep renovation sector is due to the fact that high investments are required up-front, and they are generally characterized by an excessively high degree of risk, long payback times, and the general "invisibility of the energy benefit". ABRACA-DABRA is an H2020 project that aims to activate a market for the deep renovation of existing buildings through a major transformation of the buildings aiming at the increase of the real estate value. This increase is essentially given by a volumetric addition (Add-ons) whose added value, once capitalized in terms of selling or renting, is able to reduce the payback time of the investment. Several pilot case studies have been used to test the efficacy of the strategy. At this stage of the project, a challenging sector like the social housing sector is also being explored to verify if a retrofit strategy including add-ons and densification could help boost the renovation of the public and residential housing stock. The process is based on a cost-effectiveness analysis. In this paper, to demonstrate how the densification action could be an effective solution to promote energy efficiency interventions and new business models to shorten the payback time of renovation investments, five

---

The original version of this chapter was revised: This chapter was inadvertently published with an incorrect version of Chapter 9. The book has been repaginated due to these changes. The correction to this chapter is available at https://doi.org/10.1007/978-3-030-37635-2_46

A. Ferrante (✉) · A. Fotopoulou · L. Dragonetti
Department of Architecture, School of Engineering and Architecture, University of Bologna, Bologna, Italy
e-mail: annarita.ferrante@unibo.it

G. Semprini
Department of Industrial Engineering, School of Engineering and Architecture, University of Bologna, Bologna, Italy
e-mail: giovanni.semprini@unibo.it

different buildings have been studied. The simulation made on these case studies is divided into three steps: a feasibility study, an energy saving analysis, and a payback time calculation. In the last phase of the study, the financial assumptions are fundamental. In the case of social housing, the sale, rental, and social values were considered and combined to find the best opportunity of incomes and the shortest payback time. Moreover, additional issues were taken into account regarding the regulatory aspects and the technical feasibility of this type of approach. Implementing this strategy necessitates adding new units on the rooftop or on the side of an existing building. This might meet obstacles, such as urban regulation restrictions and the consensus among tenants. To overcome these obstacles, the project promotes new policy recommendations that municipalities could approve, and counterbalanced measures to help residents embrace the ABRACADABRA strategy.

Keywords Energy retrofit · Urban densification · Add-ons · Housing · Market attractiveness

9.1   Introduction

It is widely acknowledged that the residential stock in Europe is one of the most energy consuming sectors. The EU is trying to reverse this trend by promoting energy retrofit actions on existing buildings, notably though the implementation of the Energy Efficiency Directive (EU Parliament and the EU Council 2012) and the Energy Performance of Building Directive (European Parliament and the Council of the European Union 2010). Despite these efforts, deep renovation actions cover only about 1% of the construction sector activities (European Commission 2016). There is clearly a lack of investments from the potential investors in deep renovation activities. This is due mostly to the high up-front costs, long payback times and legislative barriers. The European H2020 project ABRACADABRA has identified these key obstacles and aims to overcome them, based on the assumption that an increase of the real estate value of a renovated building could trigger deeper renovation interventions. ABRA strategy is based on volumetric Add-ons and Renewable Energy Sources (named AdoRES), such as side or façade additions, rooftop extensions or even an entire new building construction, that "adopt" the existing buildings (the so-called "Assisted Buildings") to achieve nearly zero energy and to activate a new real estate market, decreasing payback times. In this paper, we will describe how this strategy is applied in two challenging sectors: the private owned buildings, and the social housing, showing the results obtained in different cases: two case studies of residential housing in Reggio Emilia, the Student House in Athens, and two cases of residential private buildings in Bologna; demonstrating how AdoRES can increase the attractiveness of deep renovation market, reducing payback times, increasing their real estate value, and adding new units (Semprini et al. 2017).

9.1.1 Challenges and Barriers of Energy Retrofit in the Residential Sector

Building renovation is challenging, both in the private and social housing sector. Although the social housing sector presents specific character, some of the reasons for the lack of investments in such activities are common to both sectors and are both of social and economic nature. Literature to explain the low renovation rate in the housing sector is abundant. The financial aspects, such as the high upfront costs, long payback times, and the lack, instability or complexity of available funding or fiscal incentives are often considered as the main barriers to renovation. But they are not the only ones. Despite the acknowledged non-energy related benefits of energy efficiency renovation – such as health and comfort, architectural and aesthetic improvements, end-users might not recognize the benefits of an energy efficiency renovation. They might also mistrust new technologies and construction professionals, or simply might not be aware of the possibilities. Hence the importance of awareness-raising campaigns about all the benefits of energy efficiency renovation. On top of that, regulatory factors and administrative procedures further hinder renovation. This includes urban planning rules and construction permit procedures, but also rules linked to property and housing law, such as decision-making rules in multi-apartment block buildings, and contractual obligations towards the tenants (including rent increase limitation and relocation obligations). Such barriers are considered to influence the low renovation rate in the entire housing sector, although at different levels depending on the sector. Overcoming them has become a political priority in order to foster a more energy efficient building stock. Taking these challenges into account, it is necessary to promote a cultural change among owners and tenants, by informing them about which benefits they might gain after the intervention and by promoting good habits regarding energy consumption. A non-informed user could indeed take unsuccessful renovation actions (i.e. energy waste) (Sousa Monteiro et al. 2017). ABRA strategy promotes a user-orientated renovation to overcome all these difficulties, providing counterbalancing measures such as adding extra rooms, balconies or sunspaces (facade addition) to existing units. From a social point of view, it could be an opportunity to reduce social exclusion and a general renovation of the urban area. The specific challenges that energy retrofit has to overcome in public buildings are linked to property regime. It is, in fact, necessary that public bodies start the action and in some cases, is very difficult to have a short payback time because they have a particular business model to capitalize the additions. It also true that, in general, municipalities and public bodies can burden long-term investments. The simulations in this case have to be modeled on each specific case, taking into account technical, social and market limits.

9.2   The Methods and the Tools

The research study was carried out following the same process for both case studies, with the following steps:

1. An architectural feasibility study of the possible Add-ons for the building.
2. The energy consumption analysis before and after the deep renovation using a Simplified Energy Model (SEM).
3. The calculation of the renovation and construction costs.
4. The Payback time calculation for different scenarios.

Figure 9.1 illustrates the different renovation options that are ideally possible in a punctual densification at the scale of the building. Starting from the standard energy renovation of the original building, which is also assumed as a constant in all the incremental scenarios, five other options are displayed. This feasibility study is the starting point of the ABRA strategy, since it is very rare that all the AdoRES can be applied to one single case study due to regulatory or architectural issues. In addition, for a renovation resulting in a successful intervention, it is necessary to know the number of possible surfaces that can be added. The renovation measures include action on the envelope (external coating, windows replacement) and the HVAC system. The necessary actions are identified by targets to maximize the energy savings (i.e. specific U-values for each opaque surface). Subsequently, the energy consumption analysis was conducted using a Simplified Energy Model (SEM)

Fig. 9.1 Renovation scenarios

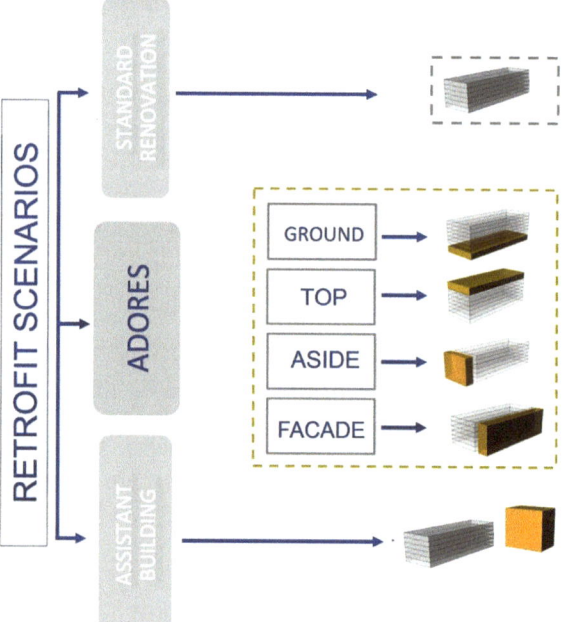

calculator. The calculation is conducted in stationary mode according to EN ISO 13790 (EN ISO 13790:2008) and ISO EN 52016-1 (ISO 52016-1:2017). The main inputs needed for calculation are the principal climate and energetic data (geometric values of the building, heat sources, transmission and ventilation properties, set points etc.). As a result of the calculation model, the SEM gives outputs monthly, and annual energy needs of the building before and after the deep renovation. All energy parameters are calculated as monthly mean values and then used to calculate seasonal values (Fig. 9.2).

These results are fundamental for the economic evaluation of the deep renovation; in fact, since there is a standard to reach, every case study will have different parametric renovation cost (€/m2) depending on the current state of the building. Regarding the construction cost, it is obviously necessary, for conducting the feasibility study, to have an idea of the intervention, and to agree on a standard construction. The Add-ons are built with timber panel for opaque surfaces and aluminum triple glaze windows filled with argon, in order to reach a zero-energy target with the use of PV panels and heat pumps for heating and cooling. Renewable Energy Sources (RES), like photovoltaic panels, are also installed in the existing building to reach the nZeb target. Renovation and construction costs (ELENCO REGIONALE 2011) and the energy consumption are the principal inputs for the assessment and the calculation of the payback time. Energy savings compensate the negative cash flow linked to the renovation and construction costs, and the profit realized from selling or renting the added units. In the case of sale transaction, we simulated that all the new dwellings would be sold in the first two years (this is a hypothesis based on the state of the market). This cost-effectiveness comparison allows for immediate identification of the most relevant scenarios for the investors and stakeholders. Figure 9.3 shows the cost estimation units and input for the deep renovation and construction.

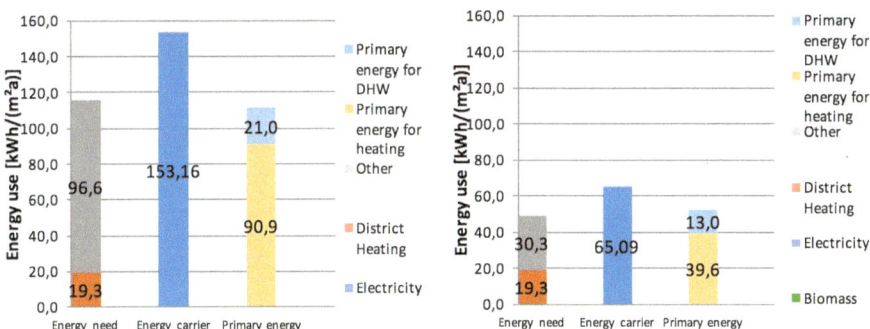

Fig. 9.2 Energy use BEFORE (on the left) and AFTER the deep renovation (right)

COST ESTIMATION

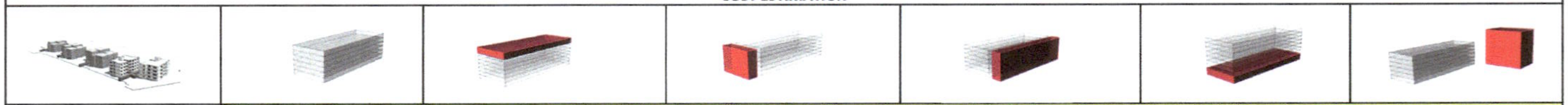

PACKAGES		DEEP RENOVATION COST FOR THE SCENARIOS																	
		DEEP RENOVATION			TOP			ASIDE			FACADE			GROUND FLOOR			ASSISTANT BUILDING		
	€/ UoM	Unit of Measure		$€_{tot}$	Unit of Measure		$€_{tot}$	Unit of Measure		$€_{tot}$	Unit of Measure		$€_{tot}$	Unit of Measure		$€_{tot}$	Unit of Measure		$€_{tot}$
		$SQM_{horiz.}$	SQM_{vertic}		$SQM_{horiz.}$	SQM_{vertic}		$SQM_{horiz.}$	SQM_{vertic}		$SQM_{horiz.}$	SQM_{vertic}		$SQM_{horiz.}$	SQM_{vertic}		$SQM_{horiz.}$	SQM_{vertic}	
1.1 THERMAL ENVELOPE																			
1.1.1 Structure	2,66 €			- €			- €			- €			- €			- €			- €
1.1.2 Facade	150,00 €		685,7	102.847,50 €		685,7	102.847,50 €		562,15	84.322,50 €		405,7	60.847,50 €		685,7	102.847,50 €		685,7	102.847,50 €
1.1.3 Roof	42,00 €	280,0		11.760,00 €	-65,0		- 2.730,00 €	280,0		11.760,00 €	280,0		11.760,00 €	280,0		11.760,00 €	280,0		11.760,00 €
1.1.4 Floors	23,00 €	1442,0		33.166,00 €	1442,0		33.166,00 €	1442,0		33.166,00 €	1442,0		33.166,00 €	0,0		- €	1442,0		33.166,00 €
1.1.5 Openings	220,00 €		121,4	26.697,0 €		121,4	26.697,00 €		121,4	26.697,00 €		54,7	12.028,50 €		121,4	26.697,00 €		121,4	26.697,00 €
1.2 INSTALLATIONS		Unit			Unit			Unit			Unit			Unit			Unit		
1.2.1 RES	284,00 €			- €			- €			- €			- €			- €			- €
1.2.2 HVAC	20.000,00 €	12		240.000,00 €	12		240.000,00 €	12		240.000,00 €	12		240.000,00 €			- €	1		20.000,00 €
1.2.3 Electrical system	50,00 €	12		600,00 €	12		600,00 €	12		600,00 €	12		600,00 €			- €	50		2.500,00 €
		€/ SQM	Cond. Area, m²	$€_{tot}$	€/ SQM	Cond. Area, m²	$€_{tot}$	€/ SQM	Cond. Area, m²	$€_{tot}$	€/ SQM	Cond. Area, m²	$€_{tot}$	€/ SQM	Cond. Area, m²	$€_{tot}$	€/ SQM	Cond. Area, m²	$€_{tot}$
Total		402,98 €	1030	415.070,5 €	388,91 €	1030	400.580,5 €	385,00 €	1030	396.545,5 €	347,96 €	1030	358.402,0 €	137,19 €	1030	141.304,5 €	191,23 €	1030	196.970,5 €
		CONSTRUCTION COST FOR THE SCENARIOS																	
					TOP			ASIDE			FACADE						ASSISTANT BUILDING		
	€/ UoM				Unit of Measure		$€_{tot}$	Unit of Measure		$€_{tot}$	Unit of Measure		$€_{tot}$	Unit of Measure		$€_{tot}$	Unit of Measure		$€_{tot}$
					$SQM_{horiz.}$	SQM_{vertic}		$SQM_{horiz.}$	SQM_{vertic}		$SQM_{horiz.}$	SQM_{vertic}		$SQM_{horiz.}$	SQM_{vertic}		$SQM_{horiz.}$	SQM_{vertic}	
2.1 ADORES STRUCTURE																			
2.1.1 AdoRes Structure	800,00 €				345,0	0,0	276.000,00 €	0,0		- €	200,0	0,0	160.000,00 €	0,0	0,0	- €	0,00	0,00	- €
2.2 ADORES THERMAL ENVELOPE																			
2.2.1 AdoRes Facade	150,00 €					40,7	6.111,00 €		57,5	8.622,00 €		851,2	127.680,00 €		0,0	- €		0,00	- €
2.2.2. AdoRes Roof	42,00 €				336,00		14.112,00 €	44,3		1.859,76 €	84,0		3.528,00 €			- €	0,00		- €
2.2.3 AdoRes Floors	23,00 €						- €	44,3		1.018,44 €	84,0		1.932,00 €	0,0		- €	0,00		- €
2.2.4 AdoRes Openings	220,00 €					25,4	5.588,00 €		84	18.480,00 €		187,2	41.184,00 €		0	- €		0,00	- €
2.3 ADORES INSTALLATIONS					Unit			Unit			Unit			Unit			Unit		
2.3.1 AdoRes RES	284,00 €				336		95.424,00 €	336		95.424,00 €			- €			- €	0		- €
2.3.2 AdoRes HVAC	17.201,97 €				4		68.807,88 €	0		- €	0		- €			- €	2		34.403,94 €
2.3.3 AdoRes Electrical system	50,00 €				4		200,00 €	0		- €	0		- €			- €	20		1.000,00 €
2.4 Yard general costs	15%						69.936,43 €			18.810,63 €			50.148,60 €			- €			5.310,59 €
					€/ SQM	Cond. Area, m²	$€_{tot}$	€/ SQM	Cond. Area, m²	$€_{tot}$	€/ SQM	Cond. Area, m²	$€_{tot}$	€/ SQM	Cond. Area, m²	$€_{tot}$	€/ SQM	Cond. Area, m²	$€_{tot}$
Total					1.554,14 €	345	536.179,31 €	542,16 €	266,0	144.214,83 €	659,47 €	583	384.472,60 €	0,0	0	- €	0,0	0	40.714,53 €
		DEEP RENOVATION AND ADORES CONSTRUCTION COST FOR THE BUILDING																	
					€/ SQM	Cond. Area, m²	$€_{tot}$	€/ SQM	Cond. Area, m²	$€_{tot}$	€/ SQM	Cond. Area, m²	$€_{tot}$	€/ SQM	Cond. Area, m²	$€_{tot}$	€/ SQM	Cond. Area, m²	$€_{tot}$
Total					681,28 €	€ 1.375,0	€ 936.759,8	417,25 €	€ 1.296,0	€ 540.760,3	460,55 €	€ 1.613,0	€ 742.874,6	137,19 €	€ 1.030,0	€ 141.304,5	230,76 €	€ 1.030,0	€ 237.685,0

Fig. 9.3 Cost estimation summary

9.3   Residential Housing Case Studies

Several case studies have been used to test the retrofit action trough Add-ons or ADORES, as named in the ABRACADABRA project. The first two cases shown in this study are both owned by ACER RE (a social housing corporation), in the Reggio Emilia area, Emilia Romagna, Italy. Here, in the case of the construction of a stand-alone assistant building, they could sell or rent at market prices. To prove the technical and architectural feasibility of the Add-ons, the building and the additions have been 3D modeled.

As Fig. 9.4 shows, in this case, it is not possible to add a facade addition or an assistant building. Therefore, the only scenarios that can be taken into account to calculate the payback times are the top addition and the side addition. As the figures demonstrate, in this case, the optimal scenario is the one that maximizes the densification (side addition). Similar results have been performed in other residential buildings, both public and private owned buildings: Bagnolo in Piano, Zografou in Athens, and tower buildings and block buildings in Bologna. The main economic data assumed for the calculations are reported in Table 9.1.

Based on the values provided by the economic data illustrated in Table 9.1, a comparison between the value of the building and the investment can be achieved, as illustrated in Fig. 9.5.

However, the most interesting results are provided by the assessment of the different volumetric options in relation to the payback times in the various market contexts, as displayed in Fig. 9.6.

As the results in Fig. 9.6 demonstrate, each scenario is competitive when compared to the deep renovation option. However, the optimal scenario is the side addition, as it maximizes the densification and the rentable/selling surfaces.

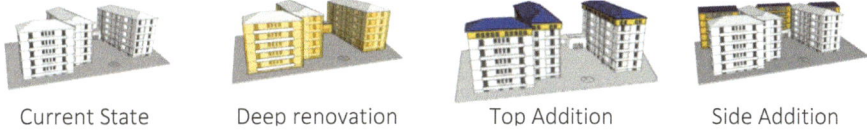

| Current State | Deep renovation | Top Addition | Side Addition |

Fig. 9.4 3D model of the ADORES illustrating the possible options for the Reggio Emilia case of Viale Magenta

Table 9.1 Economic data assumed for the calculations of the payback time and the real estate value		
Real estate value	1.800,00 €/m^2
Real estate value (add-ons)	2.800,00 €/m^2
Monthly rent for new units	8,80 €/m^2
Social monthly rent for new units	5,00 €/m^2

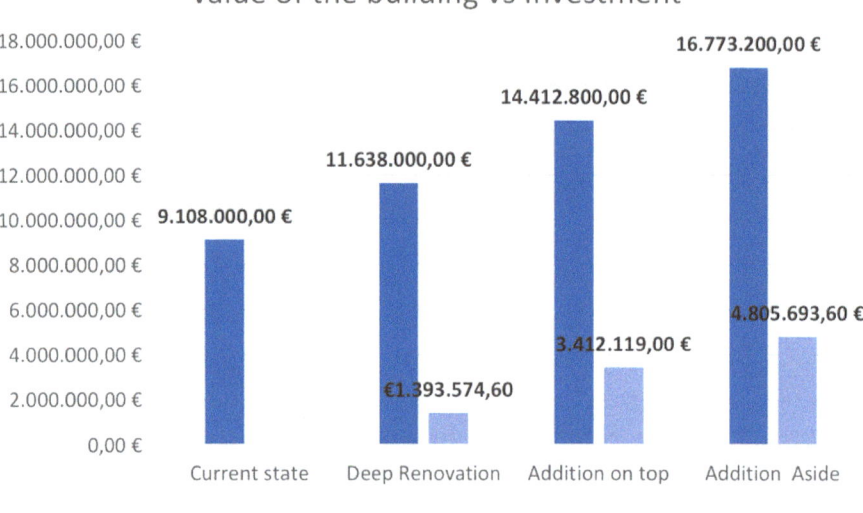

Fig. 9.5 Add-ons feasibility tables in different cases

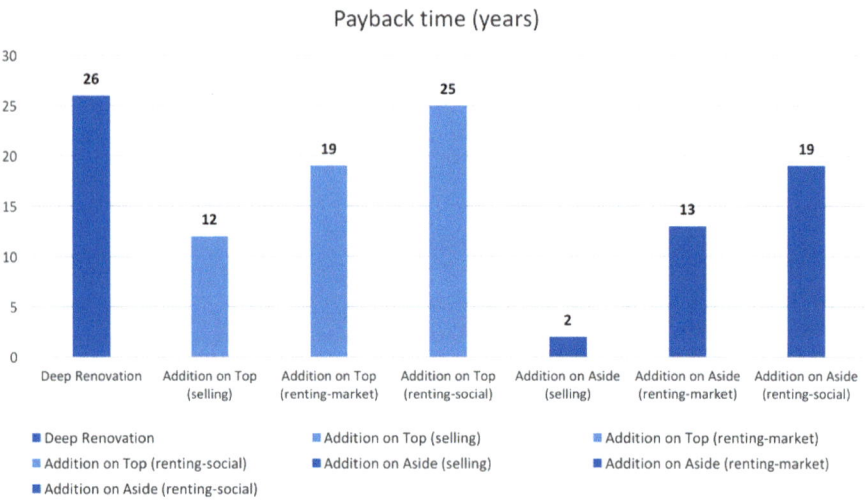

Fig. 9.6 Comparison of the payback time in the different volumetric options in the various market contexts (selling – private market; renting at market prices and renting in the social housing sector)

Similar studies have been performed in other residential buildings, in particular for two other buildings on the social housing market (Bagnolo in Piano, in Reggio Emilia, and Zografou in Athens) as illustrated in Fig. 9.7, and the tower buildings and block buildings in the private market of Bologna, reported in Fig. 9.8.

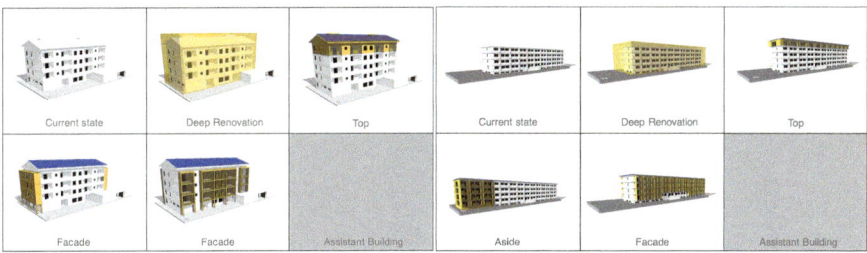

Fig. 9.7 Add-ons feasibility studies for Bagnolo in Piano, Reggio Emilia on the left and the students' house in Zografou, Athens, Greece, on the right

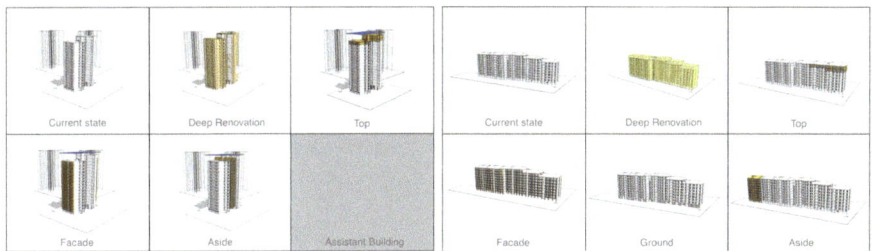

Fig. 9.8 Add-ons Feasibility tables in two different cases of the private market in the city of Bologna, Italy

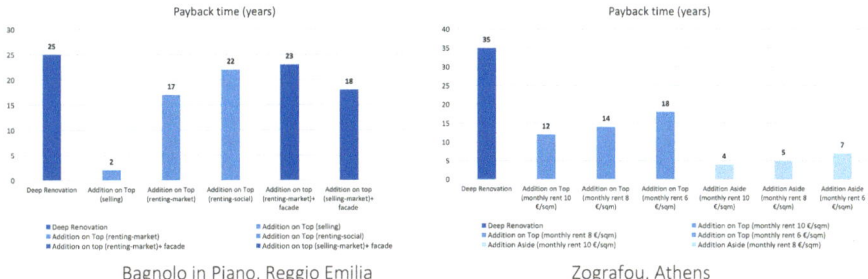

Fig. 9.9 Comparison of the payback time in the different volumetric options in the various market contexts (selling – private market; renting at market prices and renting in the social housing sector) for the second case of Reggio Emilia (on the left) and the Students house in Athens (on the right)

As for the case of Viale Magenta, the most interesting results are provided by the assessment of the different volumetric options in relation to the pay back times in the various market contexts, as displayed in Fig. 9.9.

As for the previous cases, also for the private real estate market, the most interesting results are provided by the assessment of the different volumetric options in relation to the pay back times, as displayed in Figs. 9.10 and 9.11.

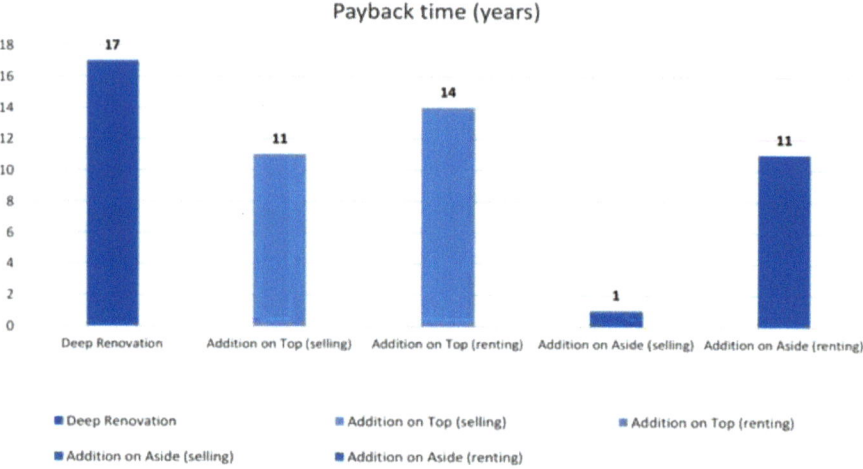

Fig. 9.10 Comparison between the payback times in the two different volumetric options (on Top and Aside) in the private market context (selling – private market; renting at market prices) for the case of the Towers in Bologna

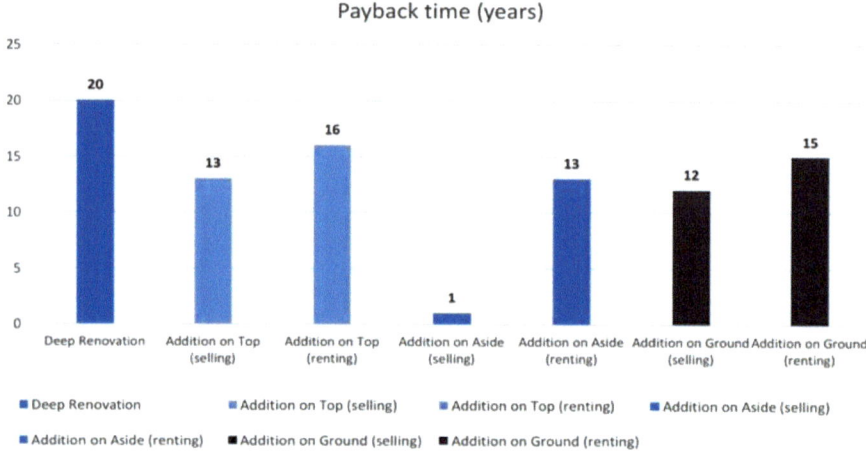

Fig. 9.11 Comparison between the payback times in the three possible volumetric options (on Top, Aside and ground) in the private market context (selling – private market; renting at market prices) for the case of the block buildings in Bologna

As the results in Figs. 9.10 and 9.11 demonstrate, each scenario is competitive when compared to the deep renovation option. However, the optimal scenario is given by the side addition, as it maximizes the amount of rentable/selling surfaces.

9.4   Brief Discussion of the Results and Conclusions

The simulations in the different scenarios have been conducted with specific assumptions. Clearly, using social prices for renting and selling does not have the same effectiveness of using market values. However, the facade addition when combined with the vertical extension (top addition) has various benefits, such as a major increase of the real estate value of the existing building and the space extension in the existing units. This can certainly be a measure to encourage the acceptance of the roof-top addition from the tenants. From the obtained results, it can be observed that the real estate value of the building is always far higher than the value of the deep renovated building in all the cases. Although the results are very different in terms of quantity, which are in turn depending on the possible amount of space addition in the different cases, there are many common aspects. In almost all cases, the side addition is the best option for the building, presenting the major increase of the value of the building with a minor investment. This case is an example of a valid implementation of renting business model and densification actions that can be replicated in other public buildings of the same typology.

Moreover, in all cases, the cost-benefit evaluation has been proved to be a valid method to identify the optimal scenario (Ferrante et al. 2017). Implementing such an approach would allow the addition of new surfaces avoiding soil sealing and could be a strategy for the urban and architectural renovation, including the social housing sector. In this framework though, there are some issues to be solved. Some are specific to social housing sector others might be more general and linked to the split incentive dilemma. A possible solution is to create a new business model, where the social housing associations could act like ESCO (Energy Service Company). In the case of public buildings, it would be a very interesting option to add new spaces to rent in order to shorten the payback times. Furthermore, there is the possibility of implementing this strategy on other typologies of building, not only the ones with residential function. Altogether, the results of ABRACADABRA, very briefly discussed in this paper, demonstrate that Add-ons are a solution that can help to boost deep energy efficiency renovation. The impact of these solutions, however, will vary according to the local market(s). The estimated payback time will moderately or considerably differ if the property renovated and its extension are to be sold or rented, if the rents are subject to market restrictions – as is the case in the social housing sector and in rent-controlled housing regimes.

Acknowledgements The paper presented is part of the project ABRACADABRA, funded by the EU under the program H2020, G. A. n. 696126.

References

ELENCO REGIONALE DEI PREZZI DELLE OPERE PUBBLICHE DELLA REGIONE EMILIA-ROMAGNA, L.R. 10/2011

EN ISO 13790:2008- Energy performance of buildings – calculation of energy use for space heating and cooling. International Organization for Standardization

European Commission (2016) Good practice in energy efficiency. Accompanying the document Proposal for a Directive of the European Parliament and of the Council amending Directive 2012/27/EU on Energy Efficiency. Brussels

European Parliament and the Council of the European Union (2010) 2010/31/EU, Energy Performance of Buildings Directive (EPBD recast). Brussels

EU Parliament and the EU Council (2012) 2012/27/EU, Energy Efficiency Directive. Brussels

Ferrante et al (2017) La strategia del progetto ABRACADABRA per azzerare il consumo energetico e bilanciare i costi nella riqualificazione degli edifici esistenti. In: Demolition or reconstruction? Colloqui.AT.e 2017. A cura di: Bernardini, Gabriele; Di Giuseppe, Elisa. EdicomEdizioni, Monfalcone, pp 482–492

ISO 52016-1:2017- Energy performance of buildings – energy needs for heating and cooling, internal temperatures and sensible and latent heat loads – Part 1: calculation procedures, International Organization for Standardization

Semprini G et al (2017) New strategies towards nearly zero energy in existing buildings: the ABRACADABRA project. Energy Procedia 140:151–158

Sousa Monteiro C et al (2017) Addressing the challenges of public housing retrofits, 9th International Conference on Sustainability in Energy and Buildings, SEB-17, 5–7 July 2017, Chania, Crete, Greece. Energy Procedia 134:442–451

Part IV
Smart Cities

How can data collections, sensing and data analytics contribute to increasing the sustainability of the city.

Chapter 10
Application of Fuzzy Analytic Hierarchy Process (AHP) for Ranking and Selection of Innovation in Infrastructure Project Management

Mohammadali Noktehdan, Mohammad Reza Zare, Johnson Adafin, Suzanne Wilkinson, and Mehdi Shahbazpour

Abstract In the absence of strong market forces driving innovation in infrastructure construction environments, the role of the project owner becomes critical in creating the motivation to innovate. Successful infrastructure projects addressed innovation systematically, by employing incentivizing mechanisms. Whilst this is recognized, this study seeks to answer the question: In what order can innovations be ranked when infrastructure projects are set for developing innovation incentive system? Addressing this, a Fuzzy Analytic Hierarchy Process (FAHP)-based questionnaire was used to run a pairwise comparison matrix for the classified types of innovation. These were ranked, and unique weights were developed for each of the classified innovations. This paper proposes a procedure for determination of the weights of alternative innovations in the FAHP, which is based on expected values of the fuzzy numbers and their products. The results show that classified innovations have various levels of importance in infrastructure projects. A case study of SCIRT's infrastructure rebuild in Christchurch, New Zealand, which reported more than 500 innovations, has been undertaken and analyzed. This enables current research to gain a better insight and understanding of the behaviors of innovation in different types. As a main contribution, the study demonstrates how the classification system could be used by project owners, to put in place mechanisms to influence the development and adoption of various types of construction innovation. In addition, the findings inform industry professionals of how to enable classified innovations in infrastructure projects to maximize productivity performance.

M. Noktehdan (✉) · M. R. Zare
Department of Civil Engineering, University of Isfahan, Esfahan, Iran
e-mail: mnok946@aucklanduni.ac.nz

J. Adafin · S. Wilkinson · M. Shahbazpour
Department of Civil Engineering, University of Auckland, Auckland, New Zealand

© Springer Nature Switzerland AG 2020
R. Roggema, A. Roggema (eds.), *Smart and Sustainable Cities and Buildings*,
https://doi.org/10.1007/978-3-030-37635-2_10

Keywords SCIRT (Stronger Christchurch Infrastructure Rebuild Team) ·
Innovation · Incentivizing

10.1 Introduction

Internationally, the construction industry is identified as a traditional or
low-technology sector with insignificant contribution (3% of total expenditure) to
Research and Development (R&D) activities associated with innovation (Seaden
et al. 2003). Challenges with performance and productivity, quality achievement,
and uptake of innovation are endemic in the industry (Xue et al. 2014). Substantive
research has found that innovation is a requirement for organizations to survive in
any dynamic economy (Gambatese and Hallowell 2011; Smith and Estibals 2011).
In general terms, innovation is an avenue to improve productivity and efficiency
(Maghsoudi et al. 2016), hence, it appears to be a crucial subject of research
(Maghsoudi et al. 2016). Firms have been pushed to stay cost effective instead of
high-quality intensive, due to the low-cost competitive contracts as a predominant
approach in construction (Kale and Arditi 2002). Reichstein et al. (2005) comprehensive survey of UK construction firms indicated that many construction firms do
not motivate to innovate, and hence remain competitive. They can sustain themselves by meeting local needs of their undemanding customers. This increases the
tendencies for clients and companies involved in project delivery to continue with
the conservative approaches to practice, thus resulting in slow innovation uptake
(Tawiah and Russell 2008). As evidenced in a recent study of the Australian
construction industry, Loosemore and Richard (2015) observed that most construction clients are not interested in innovation. Instead, they seem to be driven mainly
by price. To help motivate the global industry, Loosemore and Richard (2015)
suggested that project owners need to have a better understanding of what innovation
is and how it can benefit them.

   In the absence of strong market forces driving innovation in the construction
industry, the role of the client or project owner becomes critical in creating the
motivation to introduce or develop innovative solutions throughout the various
phases of the construction project (Mahpour and Mortaheb 2018). To entice construction companies to be more innovative, some clients (especially local or national
governments) have started to incentivize firms by incorporating innovation related
performance indicators as part of the construction contract. A systematic attention to
innovation was launched by the New Zealand government, by targeting a 20%
performance enhancement (innovation included) by 2020 (Wilkinson et al. 2012).
This industry-level attention to innovation is still being followed by the NZ government under a national program called "National Science Challenge 11: Building
Better Homes, Towns and Cities Strategic Area 6: Transforming the Building
Industry" (Wilkinson et al. 2018). Numerous studies have shown that clients can
use their purchasing power to demand innovation (Egbu 2008; Ozorhon 2012;
Widén et al. 2008). The need for a strategic plan to incentivize innovation has

been identified by various innovative infrastructure projects around the world. 'Crossrail' is an innovative infrastructure project that introduced an innovation strategy as a practical plan intended to improve innovation potential through a project. This plan is considered as uniquely positioned to lead innovation in the construction industry (L. Crossrail 2015b). 'Crossrail', as a partnering infrastructure project, is one which developed a successful innovation strategy (Davies et al. 2014): "Innovation has been defined as one of the visions in different layers of our organization" (Crossrail 2013). The London Olympic Park is another successful mega-project that formally managed innovation by employing a strategic plan throughout the project (Davies et al. 2014). Stronger Christchurch Infrastructure Rebuild Team (SCIRT) as a current research case study is another successful infrastructure project, defining an innovation strategic plan throughout the project's life.

Despite recent successes in innovation awarding systems around the world, there is still room for opportunity for upgrading innovation benchmarking mechanism. Since innovation has not been defined in a comprehensive manner, different types of innovation have been treated in the same way in existing awarding system. This single-view approach has miss-motivated teams to innovate not through an ambitious approach, but just to meet simple expectations. A new approach is needed in order to first identify innovation from a variety of different viewpoints, rank the identified types, and finally award teams based on both the number/month and the correspondence importance of reported innovations. Towards this mechanism, different classified types of possible innovations should first be identified. Each of the identified innovation types should be ranked in order find the correspondence importance in the awarding system. Instead of a sole treatment for every single type of innovation, the multiple-view approach of awarding mechanism provides an opportunity to motivate teams based on both number/months and the innovation types.

The AHP approach has been widely used as a practical method in multi-criteria decision-making (Andrić and Lu 2016; Mahpour and Mortaheb 2018). However, the conventional AHP method is not found capable of dealing with uncertainty and vagueness involved by the criteria such as an 'innovation topic' in construction by Jato-Espino et al. (2014). Taking this into account, they proposed "AHP + FSs" as a "hybrid approach" that can deal with the vagueness of innovation. To the best of the knowledge of the researchers, the application of the concept of Fuzzy AHP and Fuzzy logic theory for innovation assessment has not been studied yet. Andrić and Lu (2016) introduced the main advantage of the Fuzzy approach as "the ability to operate with linguistic variables since some events cannot be described numerically".

Current research is set to address a fundamental question: what are the innovation benchmarks that could take into account both quality and quantity of innovation in order to award teams throughout project lifecycle? To date the effectiveness of a financial-based incentive system to improve innovation has not been tested through empirical research. This is an important question, because if research findings verify variations in innovation awards through various classified types, it strengthens the argument for developing an innovation incentive approach in the construction industry that is sophisticated and contingent on the both 'Quality' and 'Quantity' of monthly reported innovations in projects.

This research aims to empirically verify whether there are variations in the importance level of innovation throughout seven classified types, by analyzing experts' opinions. Using the developed ranking result, this study further aims to upgrade an innovation assessment protocol, currently used in SCIRT as a case study. At first, a fuzzy logic-based innovation assessment framework is proposed in order to rank seven classified types of innovation. Using the result of this ranking, an upgrade is proposed on SCIRT innovation awarding system. The proposed upgrade classifies innovation monthly scores in three identified levels considering both 'Numbers/Month' and 'Seven classified types'. The findings of the review show the result of fuzzy-based framework and how the changes were applied in order to upgrade the SCIRT benchmarking model. The result of this application is then discussed more at the discussion section.

10.2   Research Background

Financial-based reward systems have been widely used (Rose and Manley 2005) in various managerial areas such as waste reduction (Mahpour and Mortaheb 2018), safety management (Hasan and Jha 2013; Hinze 2002), and sustainability (Pitt et al. 2009). Almost all reward systems consist of three main stages; 'well-defining the indicator', 'designing performance benchmarks' and finally 'award teams financially'. Through these three steps, 'identifying key performance indicators' (Lauras et al. 2010; Marques et al. 2011) and the 'pavement system' (Love et al. 2010; Shen et al. 2004) have been well addressed in literature. Meanwhile, researchers technically ignored the 'benchmarking method' as one of the main components of financial-based awarding system.

"One of the fastest ways to reach maturity and excellence in project management is through the use of benchmarking" (Kerzner 2017). Benchmarking is defined as such: "A benchmark is a measurement or standard against which comparisons can be made. Benchmarking is the process of comparing business processes and performance metrics to industry bests or best practices from other industries" (Kerzner 2017). Performance improvement can fail as a result of an incapable benchmarking system. Benchmarking mistakes were already identified as a key to benchmarking failures (Kerzner 2017). He stated this as: "Some of these mistakes include: Failing to have a benchmarking plan and not knowing what to look for". "Without effective metrics, managers will not respond to situations correctly and will end up reinforcing undesirable actions by the project team - keeping the project team headed in the right direction cannot be done easily without effective identification and measurement of metrics". Teams were encouraged through vague benchmarks, which resulted in unwanted performance. Innovation should first be measured appropriately in order to be improved throughout the project lifecycle. This fact is also well-supported by

Kerzner (2017): "You cannot correct or improve something that cannot be effectively identified and measured".

A 'Three-level' benchmarking model has been widely used by recent infrastructure projects. 'Crossrail', 'SCIRT', 'London Olympic park' and 'Heathrow Terminal 5' (Basu et al. 2009; Crossrail 2015a; Wilkinson et al. 2012) are examples of infrastructure projects that identified a 'three-level' benchmark for innovation assessment. Considering the amount of monthly reported innovations by project teams, they were categorized in one of the three levels. In SCIRT, as the case that is studied in this paper, three levels of "Minimum condition of satisfaction", "Stretch" and "Outstanding" were identified. This approach of rewarding which considered only the 'number/month', means that teams in SCIRT were not well-incentivized for harder types of innovation. They just relied on more numbers of monthly reported innovation regardless of the types and the importance. This study found this an inefficient approach, which was caused as a consequence of incomplete benchmarking in this project. T. M. Rose and Manley (2005) introduced this as "Disruptive Justice-Reward Intensity" in the fifth 'Motivational variable' of their framework. He stated that "the award must be significant enough to motivate the agent but should not exceed the value of the benefit to the principals". Addressing this, an upgrade is proposed by this study as a key contribution to knowledge. Using the result of this study, the project practitioners are able to identify a multiple-dimensional innovation assessment method that awards project teams, based on both the quantity (number/months) and quality (seven types) in appropriate timeframe throughout project lifecycle.

10.3   Research Method

A fuzzy-based theory is used by this paper in order to first rank different types of innovation, and then award teams based on the ranking result. A case study approach was also employed by this research in order to further investigate the ability of the proposed assessment framework in a real context in construction sector.

10.4   Fuzzy Logic-Based Innovation Benchmark for Incentivizing Teams in Project

Developing a tool for project managers to assess project team's innovation level can be divided into three main parts. This assessment tool begins with an innovation classification model. This model is used as a comprehensive definition that divides different types of innovation. Then, the identified types of innovation will be ranked based on experts' judgments, using a fuzzy logic-based questionnaire. Finally, an upgrade will be suggested on the innovation benchmarking system, using indexed classified types of innovation. As a result, a three-level innovation benchmark can be recommended as the final outcome of the innovation framework in SCIRT.

10.4.1 First Part: Innovation Types Identification

Innovation should clearly be defined with a comprehensive and multiple-view classification model as the first stage. Different types of innovation with a variety of levels of Novelty and Benefits have been identified through literatures in various classification models (Garcia and Calantone 2002; Lim and Ofori 2007; OECD 2005; Tidd et al. 1998).

One of the most recent classification models, developed by Noktehdan et al. (2015) consolidated a variety of innovation types with relative levels of 'Novelty' and 'Benefits' in single and multiple-view conceptual models. Six different types of innovation were identified by this model: '*Technology*', '*Method*', '*Design*',' *Product*', '*Function*', and '*Tool*'. This study will use this classification model component in order to rank different types of innovation. By adding a seventh category called 'Hi-Technology', this study optimized the model before it was used in this paper. Hi-Technology is defined as a new Technology that is coupled with a new Method. In order to gain an in-depth understanding about the seven categories, the authors strongly recommend to study Noktehdan et al. (2015)'s innovation classification model.

10.4.2 Second Part: Fuzzy AHP-Based Method for Seven Types Ranking of Innovation

A method based on the fuzzy AHP concept is proposed to estimate weights of innovation types and perform innovation ranking. Opinions of decision makers are used to estimate innovation rankings by the proposed procedure. Every decision maker has been asked to express their judgment on innovation through seven types importance compared to other types in the designed questionnaires. The fuzzy knowledge representation technique was used for denoting the importance of each type, to form fuzzy judgement matrices. Furthermore, fuzzy weights have been computed. Afterwards, a single value will be formed using the fuzzy weights from different decision makers that have been aggregated. The aggregation process is carried out by an aggregation operator. At the end, fuzzy weight values are converted into numerical values which will be used in the process of innovation seven types assessment. The process of ranking classified types of innovation is categorized in seven steps, which have been outlined below.

Step 1: *Linguistic scale for evaluation of fuzzy judgment matrix*: Judgement matrices or pairwise comparison matrices (Tables 10.4, 10.5, 10.6 and 10.7) are used in the traditional AHP approach, in order to shape the relationships between goal and factors, and between factors and sub-factors. An element in row i and column j represents the judged value of attribute i over the attribute j. Despite the crisp numerical values in the traditional AHP, in fuzzy AHP, a linguistic scale with linguistic variables is used to compare the importance of sub-factors and the fuzzy judgment matrix.

Step 2: *Data collection for establishment of pair-wise comparison matrices*: The experts' opinion is collected using the designed questionnaire. The experts were asked to compare each pairs of the classified types. Their decisions were collected through a set of questionnaires. The fuzzy judgement matrices were created using the collected data. The linguistic scale (Table 10.2) and the triangular fuzzy number are used through this data collection process.

Step 3: *Consistency test:* The next step is to check the consistency. The triangular fuzzy numbers should first be converted into crisp values in order to run the consistency test. Thus, the fuzzy comparison matrices will be firstly converted into traditional comparison matrices using the defuzzification process. The 'center-of-centroid' method is suggested to be used for this defuzzification process, which represents the central value of triangular fuzzy numbers, because of its symmetry. If the traditional comparison matrix is consistent, then the fuzzy comparison matrix will also be consistent (Table 10.1).

Firstly, the maximum eigenvalue, k_{max} from the comparison matrix X should be estimated. Then, the consistency test of the comparison matrix is performed in two steps:

$$CI = \frac{\lambda max - n}{n - 1},$$

where CI – consistency index, n is the dimension of square matrix.

$$CR = \frac{CI}{RI},$$

where RI – random consistency index and CR is consistency ratio.
Random consistency index RI is the consistency index value from the random generated matrix. Saaty (1980) generated a randomly reciprocal matrix using

Table 10.1 Scale of relative importance used in the pair-wise comparison of fuzzy AHP

Linguistic variable	Fuzzy number	Membership function
Equally important	1	(1,1,3)
More important	3	(1,3,5)
Strongly more important	5	(3,5,7)
Very strongly more important	7	(5,7,9)
Absolute important	9	(7,9,9)

Table 10.2 Random consistency index RI

N	1	2	3	4	5	6	7	8	9	10
RI	0	0	0.58	0.9	1.12	1.24	1.32	1.41	1.45	1.49

scale $1/9$, $1/8$,..., 1,..., 8, 9. The average value for the random consistency index of the randomly 500 generated matrixes for different values of n is specified in Table 10.2. If the value of CR is smaller than 0.1, then the judgment matrix has good consistency. Otherwise, the judgment matrix doesn't have acceptable consistency, hence it should be modified.

Step 4: *Priority weights are developed:* The next step is to calculate the weight of innovation types with respect to an element on a higher level. First, the *eigenvector* is computed by normalization of the geometric mean:

$$vi = \left(\prod_{j=1}^{n} xij \right)^{\frac{1}{n}}$$

Where n is the dimension of fuzzy matrix $\tilde{U}$, and u_{ij} is the element in matrix U belonging to the i-th row and j-th column. $\tilde{v}i$ is the geometric mean of criterion, and is given as a triangular fuzzy number, since the elements of the matrix are fuzzy triangular numbers.

The weights for each innovation types wi, which is a triangular fuzzy number, are developed by calculating as such:

$$\widehat{W}i = \frac{v}{\sum_{i=1}^{n} vi}$$

Step 5: *Opinions of experts are aggregated:* When the weights of each of the seven types are obtained, then the next step is to aggregate the results of different experts into a single combined preference for each innovation. The aggregation value of innovation weights calculated based on the experts' opinions is estimated by the following formula:

$$FWi = \frac{1}{m} \sum_{k=1}^{m} \widehat{W}ik$$

Where FWi represents fuzzy weight of the i_{th} innovation, m is the number of experts, and $\widehat{W}_{ik}$ is the fuzzy weight for the i-th innovation types which result from the opinion of the k-th experts.

Step 6: *Defuzzification of the priority weights:* Defuzzification is the process of transferring a triangular fuzzy number to the crisp number. This should be happening at the end of the ranking process when the innovation types are ranked, and the relative weights are developed. In this case, the representative method, which is applied for defuzzification of fuzzy weights, is based on

$$Wi = \frac{wi + 2w2 + w3}{4.}$$

Where W_i is the crisp value of the i-th innovation weight; and (w1; w2; w3) is the triangular fuzzy value of the i-th innovation weight.

Step 7: *Normalization*: Crisp weights of innovation types are normalized.

10.4.3 Third Part: Designing Innovation Benchmark

The innovation factors developed as the crisp numbers are used, at this stage, in order to classify project teams into three different levels. Covering the main concern of the current paper, the innovation benchmarking system is upgraded using the outcome of the previous seven steps. SCIRT's three-level benchmarking style is focused as a case study in the next section, in order to be able to propose an optimization. '*Minimum Condition of Satisfaction (MCOS)*', '*Stretch*' and '*Outstanding*' are the three levels that indicate the amount of reward.

10.5 Case Study

The above proposed innovation benchmark is being tested through a real infrastructure project that incentivizes innovation through a strategic plan. In fact, the idea of upgrading the innovation benchmark in the previous section was actually triggered after a deep study that was conducted by the researchers on a real innovation benchmarking system. Actually, the entire process of upgrading was shaped based on the three-level style of benchmarking that SCIRT used for incentivizing innovation. The 'three-level' innovation benchmarks in SCIRT project counted just the numbers of monthly reported innovations, regardless of the types (Table 10.3).

This research followed the result of Noktehdan et al. (2015), a paper that assessed the innovation stance of SCIRT by applying the innovation classified on the 500 innovations database. The results clearly showed a weakness of the benchmarking system which cause a biased trend through the developed

Table 10.3 SCIRT innovation benchmark

Innovation benchmark	Scale
Minimum condition of satisfaction (MCOS)	2 innovations a month
Stretch	3 innovations a month
Outstanding	5 innovations a month

innovations. Innovations with a higher importance level were in the minority, and the simpler problem-solving types shaped the majority of the SCIRT innovations. This was identified as a disability of the benchmarking system that counted just the number/month, regardless of the type for rewarding innovation throughout the teams. Addressing this limitation, current research proposes an upgrade of the benchmarking system by ranking seven types of innovation. This is in order to prioritize bigger innovation by rewarding more through the benchmarking system. In this section, FAHP as a well-known method was used to rank the seven types of innovation based on the SCIRT people judgments. After the weights are developed, a new benchmarking system will be proposed as the main contribution to knowledge by this paper.

Stronger Christchurch Infrastructure rebuild Team (SCIRT) is an organization established under an alliance agreement, and is responsible for rebuilding horizontal infrastructure in Christchurch following the earthquakes of 2010 and 2011. Innovation was given a special consideration from the beginning, when the SCIRT alliance was formed. In fact, members of the alliance were encouraged to innovate, and report on their innovations, on a monthly basis as one of their KPIs. These KPIs were linked directly to the pay/reward aspect of the contract. As a result, the alliance members had ample motivation to report their innovations. To date, more than 500 innovations have been reported by SCIRT. This has provided a unique opportunity to analyze and better understand the relationship between construction innovation and productivity improvements.

10.6 Procedure of Innovation Prioritization for the SCIRT

Step 1: Linguistic scale for evaluation of fuzzy judgment matrix: The linguistic scale of relative importance for pair-wise comparison with a triangular fuzzy number adopted from Cheng (1997) is used to measure subjective judgements of experts.

Step 2: *Data collection for establishment of pair-wise comparison matrices*: In this survey, eight professors from the University of Esfahan were asked to give their opinions for ranking the seven classified types of innovation. These experts were all well-informed about the construction projects' condition, since they were all very active in several national construction projects. After the consistency test was applied, four of the eight responses were found inconsistent and were therefore dismissed for future research. Another four questionnaires were very consistent according to the testing results. The pair-wise comparisons were made according to Tables 10.4, 10.5, 10.6 and 10.7. The innovation classification model includes Product (P), Design (D), Tool (T), Function (F), Method (M), Technology (Te) and Hi-Technology (HT). The pair-wise comparison matrices of the factors and sub-factors are summarized in Tables 10.4, 10.5, 10.6 and 10.7.

Step 3: *Consistency test*: The consistency test for every matrix is performed and results of the test are provided at the bottom rows of Tables 10.4, 10.5, 10.6 and 10.7. Every judgment matrix is consistent.

Table 10.4 Computed weights of appraisal factors by DM1

	Product	Design	Tool	Function	Method	Technology	H-Technology	
P	(1,1,3)	(1/5,1/3,1)	(1/3,1/2,1)	(1/5,1/3,1)	(1/7,1/5,1/3)	(1/9,1/7,1/3)	(1/9,1/8,1/7)	(0.013, 0.029, 0.091)
D	(1,3,5)	(1,1,3)	(1/3,1/3,1)	(1/3,1/2,1)	(1/5,1/4,1/3)	(1/7,1/5,1/3)	(1/9,1/7,1/5)	(0.02, 0.052, 0.146)
T	(1,2,3)	(1,1,3)	(1,1,3)	(1/5,1/3,1)	(1/7,1/5,1/3)	(1/7,1/6,1/5)	(1/9,1/8,1/7)	(0.023, 0.042, 0.144)
F	(1,3,5)	(1,2,3)	(1,3,5)	(1,1,3)	(1/3,1/2,1)	(1/7,1/5,1/3)	(1/9,1/8,1/5)	(0.029, 0.078, 0.236)
M	(3,5,7)	(3,4,5)	(3,5,7)	(1,2,3)	(1,1,3)	(1/5,1/4,1/3)	(1/7,1/5,1/3)	(0.074, 0.134,0.346)
Te	(5,7,9)	(3,5,7)	(5,6,7)	(3,5,7)	(3,4,5)	(1,1,3)	(1/3,1/2,1)	(0.133, 0.268, 0.526)
HT	(7,8,9)	(5,7,9)	(7,8,9)	(5,7,9)	(3,5,7)	(1,2,3)	(1,1,3)	(0.189, 0.397,0.662)

Consistency check: $\lambda max = 7.66$; CI $= (7.66 \_ 7)/6 = 0.111$; CR $=$ CI/RI $= 0.111/1.32 = 0.083 < 0.1$

Table 10.5 Computed weights of appraisal factors by DM2

	Product	Design	Tool	Function	Method	Technology	H-Technology	
P	(1,1,3)	(1/5,1/3,1)	(1/5,1/3,1)	(1/5,1/3,1)	(1/7,1/5,1/3)	(1/9,1/7,1/5)	(1/9,1/9,1/7)	(0.0125, 0.027, 0.098)
D	(1,3,5)	(1,1,3)	(1/3,1,1)	(1/3,1,1)	(1/5,1/3,1)	(1/7,1/5,1/3)	(1/9,1/7,1/5)	(0.0199, 0.057, 0.169)
T	(1,3,5)	(1,1,3)	(1,1,3)	(1/3,1,1)	(1/5,1/3,1)	(1/7,1/5,1/3)	(1/9,1/7,1/5)	(0.024, 0.057, 0.199)
F	(1,3,5)	(1,1,3)	(1,1,3)	(1,1,3)	(1/5,1/3,1)	(1/7,1/5,1/3)	(1/7,1/6,1/5)	(0.028, 0.057, 0.228)
M	(3,5,7)	(1,3,5)	(1,3,5)	(1,3,5)	(1,1,3)	(1/5,1/4,1/3)	(1/7,1/6,1/5)	(0.047, 0.124, 0.376)
Te	(5,7,9)	(3,5,7)	(3,5,7)	(3,5,7)	(3,4,5)	(1,1,3)	(1/5,1/3,1)	(0.116, 0.248, 0.574)
HT	(7,9,9)	(5,7,9)	(5,7,9)	(5,7,9)	(5,6,7)	(1,3,5)	(1,1,3)	(0.185, 0.431, 0.751)

Consistency check: $\lambda_{max} = 7.57$; CI $= (7.57 - 7)/6 = 0.096$; CR $=$ CI/RI $= 0.096/1.32 = 0.072 < 0.1$

Table 10.6 Computed weights of appraisal factors by DM3

	Product	Design	Tool	Function	Method	Technology	H-Technology	
P	(1,1,3)	(1/3,1/2,1)	(1/3,1,1)	(1/5,1/3,1)	(1/3,1/4,1/5)	(1/7,1/5,1/3)	(1/7,1/5,1/3)	(0.020, 0.049, 0.171)
D	(1,2,3)	(1,1,3)	(1/3,1,1)	(1/3,1/2,1)	(1/3,1/2,1)	(1/3,1/2,1)	(1/3,1/2,1)	(0.030, 0.091, 0.274)
T	(1,1,3)	(1,1,3)	(1,1,3)	(1/3,1/2,1)	(1/3,1/2,1)	(1/5,1/3,1)	(1/5,1/3,1)	(0.033, 0.074, 0.324)
F	(1,3,5)	(1,2,3)	(1,2,3)	(1,1,3)	(1/3,1/2,1)	(1/3,1/2,1)	(1/3,1/2,1)	(0.041, 0.131, 0.425)
M	(3,4,5)	(1,2,3)	(1,2,3)	(1,2,3)	(1,1,3)	(1/3,1,1)	(1/3,1/2,1)	(0.063, 0.179, 0.625)
Te	(3,5,7)	(1,2,3)	(1,3,5)	(1,2,3)	(1,1,3)	(1,1,3)	(1/5,1/3,1)	(0.067, 0.189, 0.675)
HT	(3,5,7)	(1,2,3)	(1,3,5)	(1,2,3)	(1,2,3)	(1,3,5)	(1,1,3)	(0.074, 0.288, 0.725)

Consistency check: $\lambda max = 7.27$; CI $= (7.27 - 7)/6 = 0.046$; CR $=$ CI/RI $= 0.046/1.32 = 0.035 < 0.1$

Table 10.7 Computed weights of appraisal factors by DM4

	Product	Design	Tool	Function	Method	Technology	H-Technology	
P	(1,1,3)	(1/3,1,1)	(1/5,1/3,1)	(1/3,1,1)	(1/7,1/5,1/3)	(1/7,1/5,1/3)	(1/9,1/9,1/7)	(0.018, 0.036, 0.096)
D	(1,1,3)	(1,1,3)	(1/5,1/3,1)	(1/3,1,1)	(1/7,1/5,1/3)	(1/7,1/5,1/3)	(1/9,1/9,1/7)	(0.021, 0.036, 0.125)
T	(1,3,5)	(1,3,5)	(1,1,3)	(1/3,1,1)	(1/5,1/3,1)	(1/5,1/3,1)	(1/9,1/7,1/5)	(0.028, 0.067, 0.230)
F	(1,1,3)	(1,1,3)	(1,1,3)	(1,1,3)	(1/7,1/5,1/3)	(1/7,1/5,1/3)	(1/9,1/9,1/7)	(0.032, 0.058, 0.182)
M	(3,5,7)	(3,5,7)	(1,3,5)	(3,5,7)	(1,1,3)	(1/3,1,1)	(1/7,1/5,1/3)	(0.091, 0.165, 0.431)
Te	(3,5,7)	(3,5,7)	(1,3,5)	(3,5,7)	(1,1,3)	(1,1,3)	(1/7,1/5,1/3)	(0.091, 0.165, 0.460)
HT	(7,9,9)	(7,9,9)	(5,7,9)	(7,9,9)	(3,5,7)	(3,5,7)	(1,1,3)	(0.247, 0.473, 0.754)

Consistency check: $\lambda\text{max} = 7.78$; CI $= (7.78 - 7)/6 = 0.131$; CR $=$ CI/RI $= 0.131/1.32 = 0.0996 < 0.1$

Table 10.8 Fuzzy, crisp and normalized innovation rank

Innovation	Fuzzy weight of the hazards	Crisp rank	Normalized rank
Tool	(0.015, 0.0352, 0.114)	0.049	0.039
Function	(0.022, 0.059, 0.178)	0.079	0.064
Design	(0.027, 0.06, 0.224)	0.092	0.074
Product	(0.0325, 0.081, 0.26)	0.113	0.091
Method	(0.068, 0.150, 0.444)	0.203	0.165
Technology	(0.10, 0.217, 0.55)	0.271	0.220
HI-technology	(0.173, 0.397, 0.723)	0.422	0.343

Step 4: *Calculation of the priority weights*: The priority weights are expressed as triangular fuzzy numbers, and the results are given in the last Column in Tables 10.4, 10.5, 10.6 and 10.7.

Step 5: *Aggregation of decision group's opinions*: After the aggregation process of decision maker groups is completed, the estimated values of fuzzy weights for each hazard are given in Table 10.8. The decision of each decision maker is weighted equally.

Step 6: *Defuzzification of the priority weights*: The fuzzy weights are converted into crisp values and the result is shown in the 3rd Column in Table 10.8.

Step 7: *Normalization*: When the crisp rank values are assigned to hazards, the next step is the normalization process of the rank values. The results are obtained in the 4th Column in Table 10.8.

10.7   Designing the Benchmark

The project team innovation could be categorized into three levels, depending on the innovation index developed for each classified type and the monthly number of reported innovations. The SCIRT innovation benchmark could be divided, not only based on the monthly reported number, but also based on the seven innovation indexes that developed for the classified types. Seven different crisp innovation values have been computed by the defuzzification process using the 'Center-of-Gravity' method. Corresponding to these crisp values, three possible innovation categories (innovation Category 1–3) are defined, with an index range of:

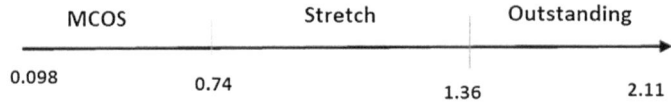

Fig. 10.1 Proposed Three- Level innovation benchmark

Minimum condition of Satisfaction: two monthly innovations classified in Tool; Innovation index; 2×0.049=0.098

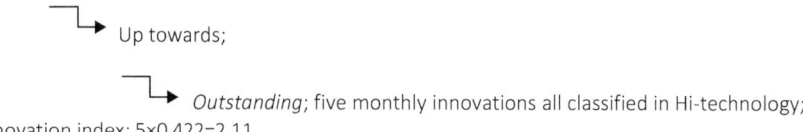

Therefore, the upgrade on innovation benchmarking describes three innovation categories as: Minimum Condition of Satisfaction (MCOS), Stretch (ST) and Outstanding (OU), as is summarized in Fig. 10.1.

Minimum Condition of Satisfaction (MCOS) is used as a level of reward if project teams introduce at least two innovations, both in 'Tool' type. The minimum range of innovation index at this level was identified as: innovation index: $(0.049) * 2 = 0.98$. Teams will be categorized at this level at most up until the innovation index of 0.73. This means the teams will still be awarded this level even with fifteen monthly innovations that all are classified in 'Tool': $(0.049 \times 3 = 0.735)$. Furthermore, even if the team would come up with two innovations that are highly ranked in this paper, such as one Hi-technology and one Technology, the award would be placed in this level as well, as the innovation index will be calculated as: $(0.422 \times 1 + 0.271 \times 1 = 0.693)$.

Stretch is where the team will come up with innovation index range from (0.74 to 1.36). This index range might be started with at least sixteen monthly innovations in 'Tool' type, or three innovations that could be classified into two 'Method' and one 'Technology'.

Outstanding is the highest level of innovation award that could be awarded, for the innovation index range between (1.36 to 2.11). This index range could be covered by a team with least twenty-eight 'Tools' (28×0.049= 1.37) or only four Hi-Technology'. This could continue towards five Hi-technology innovations that would be reported by a team in one month.

10.8   Practical Application

The whole concept of this study has been developed based on an in-depth knowledge about the SCIRT's innovation awarding system. This was an opportunity for the proposed benchmarking system to be well-fitted to a real project. A full access of the researcher to the SCIRT project provided a unique opportunity to study about the awarding system, by analyzing the database of 500 innovations. The result of this study has been presented by (Noktehdan et al. 2015). They showed clear trends toward simpler innovation types due to a single-view approach applied by the SCIRT's innovation reward system. This biased trend triggered a need to further study about any optimization available for upgrading the SCIRT innovation awarding system. Reviewing literature also showed a similar biased approach, through awarding mechanisms recently used by other infrastructure projects. Bearing in mind that any optimization should be based on a practical context, current research focused on the SCIRT awarding system in order to propose a well-practiced innovation benchmarking system. This approach guarantees the practical application of the proposed benchmarking system. a classified and easy to understand format provides an opportunity for the proposed innovation benchmarking to be practically applied in future infrastructure projects.

10.9   Discussion

Owners in infrastructure projects often encourage project teams b y using financial incentives (Rose and Manley 2010). The impact of incentive systems on performance improvement was identified by Meng and Gallagher (2012) as such: "on the whole, the embrace of incentives proves to play a driving role in encouraging best practice and ensuring project success". Despite the fact that the financial reward system is being used widely in construction projects, there is still room for opportunity about the means of optimizing outcomes. T. Rose and Manley (2007) indicated this as: "yet, very little empirical research has been conducted into *how* financial incentive should be applied in the context of particular project types in order to maximize their effectiveness". This means that not all incentive systems should necessarily be considered as a successful mechanism in order to improve performance. Volker and Rose (2012) identified this fact as: "Despite a general belief that incentive mechanisms can improve value for money during procurement and performance during project execution, empirical research on the actual effects is nascent". A significant impact was identified as a result of failure in an incentive system implication by T. Rose and K. Manley (2010b). He stated this as: "Aligning the motivation of contractors and consultants to perform better than 'business-as-usual' on a construction project is a complex undertaking and the costs of failure are high as misalignment can compromise project outcomes". Addressing this predominant concern throughout literature, current studies investigated an optimization on an innovation incentive system by focusing on a case study in the NZ construction industry.

Different types of innovation with a variety of levels of importance were not separated by SCIRT, and hence they were all awarded in a same manner by the incentive system. This study found this an unfair approach that decreases the motivations for more important types of innovation. T. Rose and Manley (2011) defined this inefficiency as "Distributive justice theory". Distributive justice theory suggests that the financial reward amount offered will be judged by its fairness relative to the effort required achieve the reward (Rose and Manley 2011). Avoiding this biased trend, innovation should have been monthly measured, considering both the quantity and quality. This is exactly what SCIRT innovation benchmarking system failed to do. This study recommends more flexibility on the awarding system by developing variety of awarding levels for different types of innovation. The importance of flexibility for developing a sustainable awarding system was also identified by T. Rose and K. Manley (2010a): "Interviewees from Projects A and B felt a lack of flexibility in the incentive goal and measurement process negatively impacted on their perceptions of fairness". Addressing this, a fuzzy theory was employed in order to rank seven types of innovation based on the experts' judgments. Questionnaires were designed based on Fuzzy theory concepts, and sent to the experts. After analyzing the data, each of the seven types of innovation was given an individual importance index, which was further used in upgrading the SCIRT benchmarking system. The SCIRT innovation benchmark was upgraded using a fuzzy based innovation assessment framework. This achievement is based on a theory that takes into consideration the types of monthly reported innovation in assessing monthly innovation scores. The results show that innovation with a variety of different importance levels will be treated differently using the upgraded benchmarking system. This is a very important finding, since teams will be awarded fairly. According to the new concept, teams are not miss-motivating to just improve the number/month, but will also consider the type and importance of reported innovation. The biased-trend of the team towards the simple types of innovation as a result of an inflexible benchmarking system is well-addressed by using this paper result. This means that according to the new benchmarking system, even with almost 15 monthly innovations which all reported in 'Tool' category, the corresponding team will be still awarded in the lowest level (MCOS). On the other hand, with only five monthly innovations which all classified in 'Hi-Technology' category, the corresponding team will be awarded in highest level (Outstanding). Since this upgrade was applied on a real case study, the results of this study were kept away from personal bias. This finding at least can improve the project practitioners' knowledge of different types of innovation definitions. This means that innovation is not a single concept, but a multi-type concept that should be classified in different types. Another contribution of this paper is about the notion that identified individual importance levels for different types of innovation and then awarded each of the seven types respectively.

10.10   Conclusion

Innovation has recently been identified by huge infrastructure projects as one of the project KPI's in construction industry. Financial-based reward models have been used by these projects in order to improve innovation. The innovation level of teams should be measured through the benchmarking system. Afterwards, teams are rewarded based on their monthly innovation scores. This strategic attention requires researchers to develop tools for identifying, measuring and awarding innovation in construction projects. Recent attentions to innovation reward mechanisms by some successful infrastructure projects around the world show the importance of the current study. The current paper addressed this opportunity by focusing on an innovation awarding mechanism that was used by an innovative infrastructure project in New Zealand. The SCIRT innovation incentive system was deeply studied by the researchers in order to address an area of weakness. This weakness was about an unfair awarding mechanism that miss-motivated teams to just improve the monthly numbers of innovation regardless of the importance. The three-level reward mechanism was found inefficient to improve more important types of innovation in SCIRT. Innovation was identified in a single definition by SCIRT, which caused the benchmarking system to award all types of innovation similarly. This was misunderstood by SCIRT team in the way that innovation is a single-view value that would be awarded not based on the importance but just based on the number/month. This research criticized this view as a cause of a biased trend through the +500 innovations. Therefore, a new method was proposed in order to award teams. The fuzzy set theory is suitable to express experts' logic opinion about innovation indicators. Seven classified types of innovation have been identified from literature. Furthermore, the fuzzy AHP procedure is applied to rank and developed innovation weights. Based on the result, an upgrade was proposed on the SCIRT's three-level awarding format. The new benchmarking system considers both the quality (types) and quantity (number/month) of innovation through awarding teams. This means that the overall monthly scores are calculated not only based on the numbers, but also considering the developed seven crisp numbers for seven types of innovation. At the end, according to the calculated innovation scores, the team would be rewarded based on one of the three levels. The authors believe that there is still room for future researchers to keep digging deeper in the incentive systems in the construction sector.

Appendix

Questionnaire

This questionnaire is based on our research into the viability of a flexible incentive system for construction contracts based on innovation. Innovations would be ranked using seven types of measure, and those that score higher would be awarded more. The innovation classification model below aims to rank the seven types, based on expert opinion from construction industry professionals. Your response to this questionnaire is important in order for us to validate our research.

Step 1: Read a description of the 7 different definitions developed for each type of innovation (Table 10.9 details an example from the SCIRT project).

Step 2: Examples of 7 new innovations and their various features are detailed in the diagram below (Fig. 10.2);

Table 10.9 Example of innovation in SCIRT project

Innovation	Example SCIRT Innovations
Hi Technology	"Pressure waste water system" was one of the breakthrough innovations that was used by SCIRT project.
Technology	"Lightweight Localized Storm Water Pump Station"
Product	"Bridge St Cathodic Protection": Whilst working on the repairs to the piers of Bridge St Bridge, we have installed cathode protection to the piers.
Design	"Rationalization of waste water pipe in Hawkesbury Ave": When the drawings for Hawkesbury Avenue were reviewed by the Delivery team they identified that a section of pipe could be removed if one additional manhole was installed at the position of the first lateral.
Method	"Pipe bursting the water main in Buckingham "
Function	"CSS Workshop": These workshops were a great way to communicate the updates to the team at once, also discussed was the best way to communicate the changes to our subcontractors.
Tool	"Hydraulic Aluminum Shoring ": Aluminum hydraulic shores and shields are an excellent lightweight resource for working around existing utilities, supporting trench walls near structures, curbs, or sidewalks.

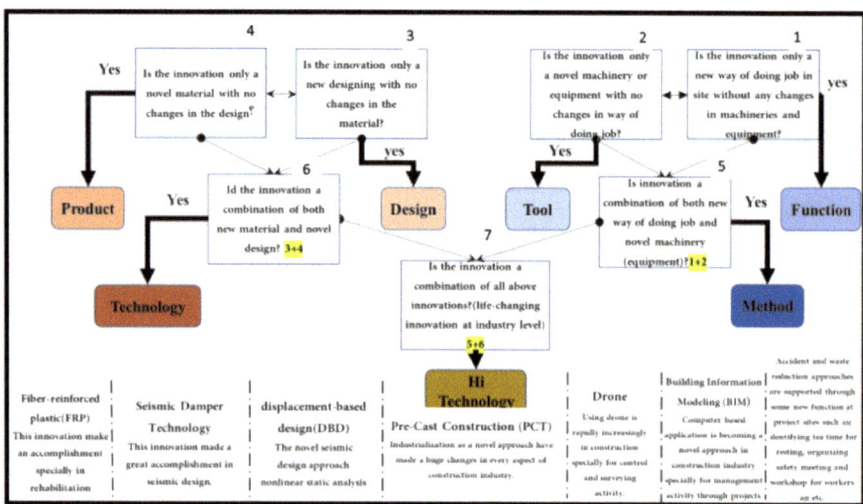

Fig. 10.2 A guideline for innovation classification model

Thinking about this group of new innovations, we would like you to compare the relative importance of their performance in the following categories and complete

Table 10.10 Pairwise comparison

	9	7	5	3	1	3	5	7	9	
Tool	9	7	5	3	1	3	5	7	9	Function
Tool	9	7	5	3	1	3	5	7	9	Product
Tool	9	7	5	3	1	3	5	7	9	Design
Tool	9	7	5	3	1	3	5	7	9	Method
Tool	9	7	5	3	1	3	5	7	9	Technology
Tool	9	7	5	3	1	3	5	7	9	Hi Technology,
Function	9	7	5	3	1	3	5	7	9	Product
Function	9	7	5	3	1	3	5	7	9	Design
Function	9	7	5	3	1	3	5	7	9	Method
Function	9	7	5	3	1	3	5	7	9	Technology
Function	9	7	5	3	1	3	5	7	9	Hi technology
Product	9	7	5	3	1	3	5	7	9	Design
Product	9	7	5	3	1	3	5	7	9	Method
Product	9	7	5	3	1	3	5	7	9	Technology
Product	9	7	5	3	1	3	5	7	9	Hi Technology,
Design	9	7	5	3	1	3	5	7	9	Method
Design	9	7	5	3	1	3	5	7	9	Technology
Design	9	7	5	3	1	3	5	7	9	Hi Technology,
Method	9	7	5	3	1	3	5	7	9	Technology

Function Tool Design	Value of a_{jk}	Interpretation
Product Method	1	j and k are equally important
Improved Technology	3	j is slightly more important than k
Brand New Technology	5	j is more important than k
	7	j is strongly more important than k
	9	j is absolutely more important than k

For example: In line 1, if you think 'Tool' is a more important innovation-type in comparison to 'Function', it is ranked 9 (on the left-hand side of the row). But in case you found 'Function' as much more important, it would be ranked 9 (on the right-hand side). Finally, in case you found that both are equal in importance, the rank would be 1(in the middle row).

References

Andrić JM, Lu D-G (2016) Risk assessment of bridges under multiple hazards in operation period. Saf Sci 83:80–92

Basu R, Little C, Millard C (2009) Case study: a fresh approach of the balanced scorecard in the Heathrow terminal 5 project. Meas Bus Excell 13(4):22–33

Cheng C-H (1997) Evaluating naval tactical missile systems by fuzzy AHP based on the grade value of membership function. Eur J Oper Res 96(2):343–350

Crossrail L (2013) Crossrail innovation strategy Crossrail learning legacy project: Crossrail learning legacy project

Crossrail (2015a) Performance assurance procedure. In: PROGRAMME CONTROL. Crossrail Learning Legacy, p 12 London

Crossrail L (2015b) Innovation Programme Overview. from London's Crossrail learning legacy project   http://learninglegacy.crossrail.co.uk/wp-content/uploads/2016/04/11A_021_Innovation-Programme-OverviewProspectus.pdf

Davies A, MacAulay S, DeBarro T, Thurston M (2014) Making innovation happen in a megaproject: London's crossrail suburban railway system. Proj Manag J 45(6):25–37

Egbu C (2008) Clients' roles and contributions to innovations in the construction industry: when giants learn to dance. In: Clients driving innovation. Wiley-Blackwell, Chichester, pp 69–77

Gambatese JA, Hallowell M (2011) Factors that influence the development and diffusion of technical innovations in the construction industry. Constr Manag Econ 29(5):507–517

Garcia R, Calantone R (2002) A critical look at technological innovation typology and innovativeness terminology: a literature review. J Prod Innov Manag 19(2):110–132

Hasan A, Jha KN (2013) Safety incentive and penalty provisions in Indian construction projects and their impact on safety performance. Int J Inj Control Saf Promot 20(1):3–12

Hinze J (2002) Safety incentives: do they reduce injuries? Pract Period Struct Des Constr 7 (2):81–84

Jato-Espino D, Castillo-Lopez E, Rodriguez-Hernandez J, Canteras-Jordana JC (2014) A review of application of multi-criteria decision-making methods in construction. Autom Constr 45:151–162

Kale S, Arditi D (2002) Competitive positioning in United States construction industry. J Constr Eng Manag 128:238–247

Kerzner H (2017) Project management metrics, KPIs, and dashboards: a guide to measuring and monitoring project performance. Wiley, Hoboken

Lauras M, Marques G, Gourc D (2010) Towards a multi-dimensional project performance measurement system. Decis Support Syst 48(2):342–353

Lim JN, Ofori G (2007) Classification of innovation for strategic decision making in construction businesses. Constr Manag Econ 25(9):963–978

Loosemore M, Richard J (2015) Valuing innovation in construction and infrastructure: getting clients past a lowest price mentality. Eng Constr Archit Manag 22(1):38–53

Love PE, Davis PR, Chevis R, Edwards DJ (2010) Risk/reward compensation model for civil engineering infrastructure alliance projects. J Constr Eng Manag 137(2):127–136

Maghsoudi S, Duffield C, Wilson D (2016) Innovation in infrastructure projects: an Australian perspective. Int J Innov Sci 8(2):113–132

Mahpour A, Mortaheb MM (2018) Financial-based incentive plan to reduce construction waste. J Constr Eng Manag 144(5):04018029

Marques G, Gourc D, Lauras M (2011) Multi-criteria performance analysis for decision making in project management. Int J Proj Manag 29(8):1057–1069

Meng X, Gallagher B (2012) The impact of incentive mechanisms on project performance. Int J Proj Manag 30:352–362

Noktehdan M, Shahbazpour M, Wilkinson S (2015) Driving innovative thinking in the New Zealand construction industry. Buildings 5(2):297–309

OECD (2005) *Oslo manual: guidelines for collecting and interpreting innovation data*. OECD publishing, Paris

Ozorhon B (2012) Analysis of construction innovation process at project level. J Manag Eng 29(4):455–463

Pitt M, Tucker M, Riley M, Longden J (2009) Towards sustainable construction: promotion and best practices. Constr Innov 9(2):201–224

Reichstein T, Salter AJ, Gann DM (2005) Last among equals: a comparison of innovation in construction, services and manufacturing in the UK. Constr Manag Econ 23(6):631–644

Rose TM, Manley K (2005) A conceptual framework to investigate the optimization of financial incentive mechanisms in construction projects

Rose T, Manley K (2007) Effective financial incentive mechanisms: an Australian study

Rose TM, Manley K (2010) Financial incentives and advanced construction procurement systems. Proj Manag J 41(1):40–50

Rose T, Manley K (2010a) Client recommendations for financial incentives on construction projects. Eng Constr Archit Manag 17(3):252–267

Rose T, Manley K (2010b) Motivational misalignment on an iconic infrastructure project. *Build Res Inf* 38:144–156

Rose T, Manley K (2011) Motivation toward financial incentive goals on construction projects *Journal of Business Research*, 64(7), 765–773

Saaty T (1980) The analytical hierarchical process. Wiley, New York

Seaden G, Guolla M, Doutriaux J, Nash J (2003) Strategic decisions and innovation in construction firms. Constr Manag Econ 21(6):603–612

Shen L, Li Q, Drew D, Shen Q (2004) Awarding construction contracts on multi-criteria basis in China. J Constr Eng Manag 130(3):385–393

Smith K, Estibals A (2011) Innovation and research strategy for growth. Stationery Office, London

Tawiah PA, Russell AD (2008) Assessing infrastructure project innovation potential as a function of procurement mode. *J Manag Eng* 24(3):173–186

Tidd J, Bessant J, Pavitt K, Wiley J (1998) Managing innovation: integrating technological, market and organizational change. Wiley and Sons, Chichester

Volker L, Rose TM (2012) Incentive mechanisms in infrastructure projects: a case-based comparison between Australia and the Netherlands. In: Paper presented at the Working Paper Series, Proceedings of the 2012 engineering project organizations conference-global collaboration

Widén K, Atkin B, Hommen L (2008) Setting the game plan: the role of clients in construction innovation and diffusion. In: Clients driving innovation. Wiley, Oxford, pp 78–87

Wilkinson S, Kempton T, Gleeson A (2012) Identifying Canterbury Rebuild Project KPI's (Baseline Report). Retrieved from http://www.mbie.govt.nz/publications-research/research/construction-sector-productivity/canterbury-rebuild-kpis-project-full-report.pdf

Wilkinson S, Tookey J, Regan P, Casimir M, Melanie M, Nariman G, … Karen B (2018) Transforming the building industry, State of nation knowledge report

Xue X, Zhang R, Yang R, Dai J (2014) Innovation in construction: a critical review and future research. Int J Innov Sci 6(2):111–126

Chapter 11
The Role of Smart City Initiatives in Driving Partnerships: A Case Study of the Smart Social Spaces Project, Sydney Australia

Homa Rahmat, Nancy Marshall, Christine Steinmetz, Miles Park, Christian Tietz, Kate Bishop, Susan Thompson, and Linda Corkery

Abstract This chapter explores the potential of smart cities initiatives as a driver of partnership formation. It presents lessons learnt from the collaboration between the Faculty of Built Environment at the University of New South Wales Sydney, Street Furniture Australia, and Georges River Council, New South Wales, as partners in a Commonwealth funded smart cities grant awarded in 2017. The research pilots how environmental sensors can inform the potential to improve the amenity and use of public open spaces and contribute to the asset management system of small-scale street furniture. This project provides a basis from which to explore the opportunities and challenges of collaboration across three domains (academia, industry, government) while conducting a smart cities project. We demonstrate how mutually collaborative efforts can better harness real-time data to identify and address citizens' needs, interests and demands, for public space in parks and plazas, in addition to assisting council with developing an efficient asset management system. These critical insights (concerning processes, outputs and outcomes) can be applied to develop an effective model of research, practice, and local government collaboration that stimulates urban innovations to address complex problems of twenty-first century cities.

Keywords Smart cities partnership · Triple helix model · Interdisciplinary collaboration

---

The original version of this chapter was revised: This chapter was inadvertently published with an incorrect version of Chapter 9, and which was incorrectly named as Chapter 11. The book has been repaginated due to these changes. The correction to this chapter is available at https://doi.org/10.1007/978-3-030-37635-2_46

H. Rahmat (✉) · N. Marshall · C. Steinmetz · M. Park · C. Tietz · K. Bishop · S. Thompson · L. Corkery
Faculty of Built Environment, University of New South Wales, Sydney, NSW, Australia
e-mail: h.rahmat@unsw.edu.au; n.marshall@unsw.edu.au

© Springer Nature Switzerland AG 2020, corrected publication 2020
R. Roggema, A. Roggema (eds.), *Smart and Sustainable Cities and Buildings*,
https://doi.org/10.1007/978-3-030-37635-2_11

11.1   Introduction

As cities are growing and densifying, they are struggling with challenges such as overcrowding, traffic congestion, noise, pollution and infrastructure that needs updating to meet current demands. In response, technology is viewed as a panacea, but this promise should be met with caution. Contemporary urban issues are different from what occurred in the past. Smart cities require transdisciplinary knowledge and collaboration amongst different industries to address 'wicked' (complex and interlinked) problems. Innovative solutions which transcend disciplinary boundaries, are needed.

We are at the tipping point of the smart city movement. Local and state governments, alongside private industry and academia are jockeying for position to be leaders in the space. As a result, disparate initiatives and pilot projects in this space are often less informed by evidence-based research. In turn, documented efforts of conceptualising and testing innovative smart city methodologies go unnoticed and hence have limited impact and/or remain with a select few. Meaningful collaboration to advance the field is essential. This chapter positions and demonstrates a working cross-sectorial collaboration, outlining the strengths and challenges associated with complex smart city projects which seek to benefit the community and its governance structures.

11.2   Smart Cities

Much of the initial research about smart cities attempted to interrogate the concept and develop definitions (Hollands 2008; Allwinkle and Cruickshank 2011; Batty et al. 2012; Albino et al. 2015). Key elements of these definitions include the utilisation of new technologies (mainly Information and Communication Technologies), enabling sustainable economic, social, and urban development, an urban entrepreneurial element (mostly hi-tech), and knowledge-based economy (knowledge/information capital). Smart cities can also focus on human and social capital, social learning and community development (Hollands 2008). Kitchin (2014) signifies the role of data analytics to understand, monitor, regulate and plan the city; this is widely recognised as contributing to the big data industry. This chapter draws on Albino et al. (2015), who propose that smart cities can be applied to 'hard' domains such as buildings, energy grids, natural resources, water and waste management, and mobility, as well as 'soft' domains including education, culture, policy innovations, social inclusion, and government. The discussion and practice surrounding smart cities is often about big data sets, large scale infrastructure projects, reconfiguring urban systems that involve major technological applications and sometimes, altering business culture (e.g. Tel Aviv as a start-up city with one start-up for every 290 residents—the highest per capita figure in the world. Oren 2017, p. 118). What is often overlooked are different scale projects and their impacts which may be at a local scale and directed at specific human-centered problems aimed at enhancing people-place relationships using open, accessible, and user-friendly technologies.

Different urban issues operate at a national to micro local scale (e.g. Australia's National Broadband Network [NBN] initiative down to a local government's place-based smart initiatives. The range of potential problems need different solutions at varying scales. As Cruickshank (2011) notes, the innovative and creative capacities of the Smart Cities partnership is needed to deal with this complex and multi-scale landscape. The smaller scale projects tend to enable urban entrepreneurialism through getting start-ups and tech-based companies engaged in these projects/initiatives and public-private partnerships to support new digital businesses. Citizens also have a role in creating smart cities and need to be involved to ensure the outcomes are beneficial to them at human scale. The Australian Federal Government has the following four priority areas that encapsulate different scales of smart city projects: smart infrastructure, smart precincts, smart services and communities, and smart planning and design (Australian Government *Smart Cities and Suburbs* grant program 2017). The span of content areas and scale of these priority areas indicate interdisciplinary and cross-sectoral collaboration is a necessity.

11.3 Collaboration

To frame collaboration in a smart city project, this chapter draws on the triple helix model. This model is based on genetics and DNA coding. *"The triple helix of university-industry-government interactions is a universal model for the development of the knowledge-based society, through innovation and entrepreneurship"* (Etzkowitz and Zhou 2017, p. 4). *"The university is the generative source of knowledge-based societies . . . Industry . . . a key factor as the locus of production, government as the source contractual relations that guarantee stable interactions and exchange"* (Etzkowitz and Zhou 2017, p. 23). The triple helix model has been applied in many circumstances: for instance, *"University-Industry-Government (U-I-G) interactions and relationships provide an optimum methodology for entrepreneurship and innovation, moving research/knowledge into practice/use."* (Etzkowitz and Zhou 2017, p. 2). For Leydesdorff and Deakin (2011), *"the triple-helix model . . . proposes that the three evolutionary functions shaping the selection environments of a knowledge-based economy are: (i) organized knowledge production, (ii) economic wealth creation, and (iii) reflexive control"* (p. 56).

Cruickshank (2011) has applied the triple helix model to the smart city, arguing that collaboration is needed for the intellectual capital of universities and wealth creation of industry. He focuses on the underlying institutional relations that support the involvement of universities, industry, and government in knowledge production processes. In a more recent study, Dameri et al. (2016, p. 2974) note that *"the smart city success depends on the synergic action by the triple helix key actors: public bodies, universities, and private companies"*. They further argue that there are differences in smart city vision among these actors with respect to three factors: technology, human factor, and institutional factors. They conclude that *"without a central direction, coordinating the interests of all the key actors with the stake-holders' expectations and needs, the smart city will remain an interesting innovative laboratory, but failing in creating public and private value for all in the long term"* (Dameri et al. 2016, p. 2980).

11.4   Design Thinking

Smart city issues are more than technical problems to be solved by computer engineers, data scientists or business administrators. They require creative and collaborative approaches to develop innovative solutions to address urban challenges. Design thinking is an often overused and somewhat misunderstood term (Dorst 2010). However, when applied at the front-end of a design or planning process, design thinking can lead to meaningful insights and greater understanding of opportunities to address problems. At its simplest, design thinking can be information gathering, but more sophisticated approaches involve analytical (benchmarking and data acquisition) and observational (ethnographic or other) techniques, collaborative design strategies, and the testing of iterative prototypes. These methods can heighten empathy and understanding by placing people at the center of the process and thereby propose new insights to solve problems (Brown 2008; Adams et al. 2011). The collaboration between university, industry and local government enables an inter-institutional project team to take a transdisciplinary approach (Brown and Katz 2009). A range of methodologies, from qualitative observational studies to quantitative data capture and analytics, builds a rich and nuanced understanding of people and place. Design thinking also refers to the iterative approach which is applied to the current project. As it is a pilot project, insights were gathered from the information and data collected and used to revise and fine-tune our approach and interpretation of the data based on this feedback loop.

11.5   Equal Contributors in the Triple Helix Model

Employing the triple helix model of collaboration to achieve smart cities objectives and aspirations, each partner plays important roles and the resulting outcomes vary as discussed below.

11.5.1   University

Boyer (1990) suggests that the university can function as the scholarship of integration and application; to explore how knowledge can be applied to consequential problems; and how it can be helpful to individuals and institutions. In doing so, universities serve the interests of the larger community (see Ian Jacobs' talk to the National Press Club in which it is stated that 'for every $1 investment in universities it creates a $10 return in the community!' https://youtu.be/trNPuFp9Doo). Boyer argues that the process should be dynamic rather than one-way in which knowledge is first "discovered" and then "applied": *"New intellectual understandings can arise out of the very act of application. . . theory and practice vitally interact, and one renews the other"* (Boyer 1990, p. 23). He highlights *"the need to move beyond traditional disciplinary boundaries, communicate with colleagues in other fields, and discover patterns that connect"* (p. 20). Charles (2011, p. 282) who discusses the role of universities in building knowledge cities, notes that *"ensuring that the knowledge assets of the university are applied to meet the needs of the local community, in terms of business or indeed in wider social and cultural impacts, requires a set of policies and initiatives, and a form of scholarship, to ensure that local needs inform the development of high-quality research and teaching, that those activities engage with local partners"*. Researchers can scientifically observe, identify, and evaluate innovations and provide directions for further smart city investigations. This is difficult for others who might not have the time, resources or training. The relevant outputs that research-intensive universities require are publications, grant monies and industry networks. This requires researchers to *explore, create, innovate* without immediate commercial gain, which is the prime concern of industry. Researchers can enquire without citizen pressure to deliver goods and services which are normally the focus of government.

11.5.2 Industry

The private sector plays a key role in smart city projects. At times, it is not only a partner in Public-Private Partnerships, it can also be one of the essential driving forces behind projects, alongside city and public-sector initiators. Private sector businesses are interested in widening their business interests and are keen to boost their image by being publicly concerned about the future of cities (Hatzelhoffer 2012). Businesses also want to be perceived as having a 'green' and 'social' conscience, alongside their need to make a profit. Smart city projects, like the one presented here, are an excellent opportunity for industry to identify and test before going to market. There is value in prototyping products *in situ*. It *"allows for the testing of the potential solutions early on in the design process . . . [it] involves turning design concepts into functional solutions. . . In particular, in urban environments, the process of turning a design concept into a functional solution is not straightforward"* (Tomitsch 2018, p. 153). Industry cannot do this without a partner who has a need and a place for their products to be tested in a real urban setting.

11.5.3 Government

Local government can benefit from the living lab concept as it is at the forefront of assisting research and practice to identify and solve real-world problems affecting the local community. Despite local governments competing with each other for businesses and reputations as sustainable and liveable place creators, government administrations are networked with each other and share knowledge and solutions to urban problems. They learn from each other. They also reach out to universities and industry because they need and want cutting-edge knowledge and access to the latest technology, which they do not have in-house, to address their local challenges.

Local governments offer the place to host technology and trial innovative solutions and to collect evidence for new, potentially beneficial outcomes for the community. As the Australia Government noted: *"smart infrastructure to improve efficiency, smart precincts to make communities more liveable, smart services and communities that will deliver community-focused local government services and smart planning and design to build adaptable and resilient cities"* (Australian Government Smart Cities and Suburbs Program 2017: website).

11.6 Case Study: A Smart Cities Partnership

This smart city case study involves a collaboration between the Faculty of Built Environment at the University of New South Wales Sydney [UNSW], Street Furniture Australia [SFA], and Georges River Council [GRC], in the southwest district of the metropolitan area. The study provides an opportunity to design, test and implement new, smart street furniture in the public domain. It also pilots a flexible and smart infrastructure management system. This is significant for smaller local government areas that are typically 'lower-tech' and cannot afford complicated, outsourced, technical infrastructure needed to run a 'smart city management system'. The project overall will give the local council hard evidence to inform open space, urban design and public infrastructure decisions. Both components of this project are designed to improve healthy and connected living.

This project is a case study, which we present in detail in this chapter. Findings from this study are not intended to be generalisable. However, a case study can include 'lessons learnt' from within the project scope (Patton 2002). These details, although specific to this case study, are relatable to other local governments with similar small-scale projects where collaboration is critical for success.

This research project entitled 'Smart Social Spaces: Smart Street Furniture Supporting Social Health' was a recipient of the Australian Government's *Smart Cities and Suburbs* grant program, 2017–2019. There were 49 successful projects in round one of funding worth $27.7 million; round two funding is valued at another $22 million and was distributed in late 2018. The program demonstrates the Australian Federal Government's commitment to the smart cities movement. Our research pilots the use of environmental sensors to determine the extent to which improvements can be made to increase the amenity and use of public open spaces and the asset management system of small-scale street furniture.

Using digital sensors installed on street furniture and existing park facilities, sensors record real time use of urban furnishings in two public spaces in the GRC – a plaza, Memorial Square in Hurstville, and a park, Olds Park in Penshurst. These different types of public space were selected: one an urban setting that is intensively used on a daily basis, and the other, a park less intensively used during the week and more heavily used for recreation on the weekends. Figure 11.1 below illustrates the basic components of the project that compose a smart system based on

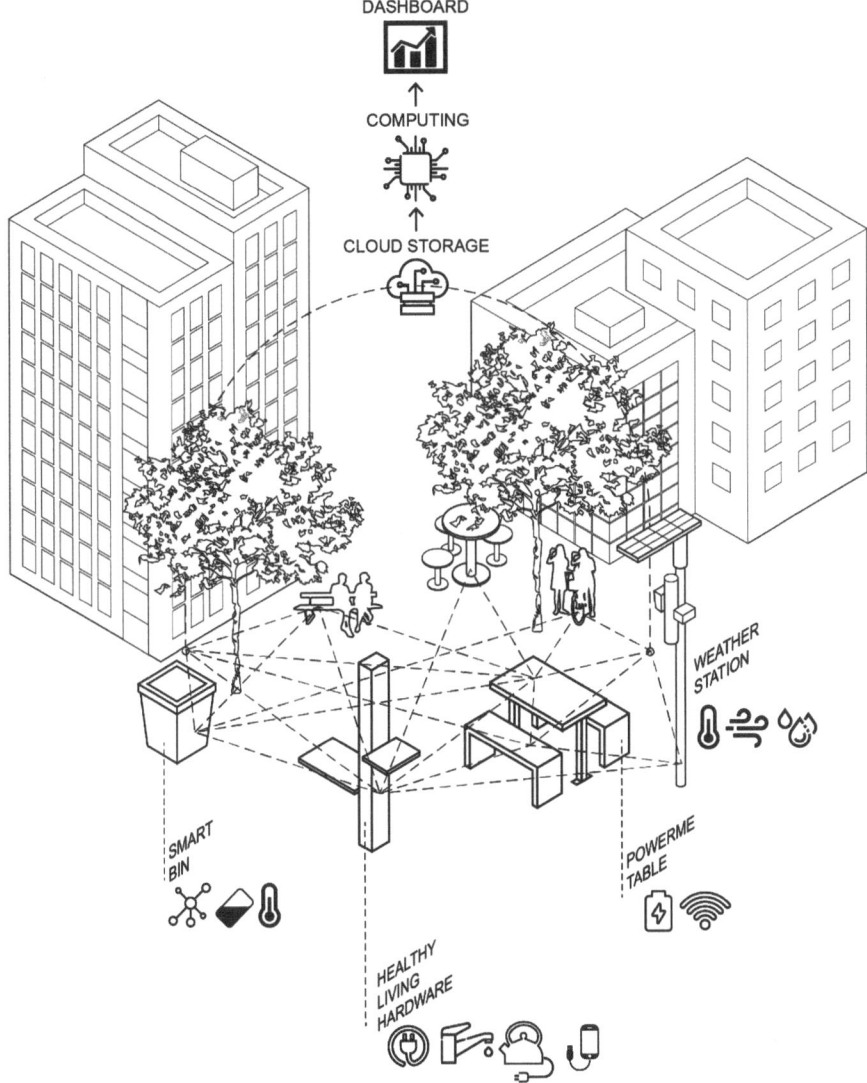

Fig. 11.1 Schematic diagram of Smart Social Spaces. (Authors 2018)

IoT (Internet of Things) technology. The IoT technology as a network of sensors embedded in devices and physical objects provides a means to monitor and manage urban assets and their use. The 'Smart Social Spaces' project has been aimed to collect data from environmental sensors, store it in cloud servers, process and analyse it in real-time, convert it into useful forms and insights, and displays it on a city dashboard. This provides local councils with live user data about public open space and the use and performance of their public furnishings.

A key component of the project was fitting out new and existing street furniture installations with a range of digital wireless sensors (Figs. 11.2 and 11.3) that operate on several different transmission platforms to provide data about the use of these urban furnishings. For example, it is possible to determine how long people spend in

Fig. 11.2 Vibration sensor installed under the seats, Olds Park. (Authors 2018)

Fig. 11.3 Sound level meter installed in the Plaza, Memorial Square. (Authors 2018)

the vicinity of the site; how often a piece of equipment is used; and whether the equipment needs servicing. Sensors also provide data on how long people sit on the benches, and whether they are alone or with someone. This data can be cross referenced with sound and location-specific weather data that is also collected. This gives GRC a rich and detailed picture of the patterns of use for the assets in these settings. The project has introduced a novel data system provided by IoT sensors.

This project has been an opportunity for the industry partner, SFA, to test a new product: the 'PowerMe' table that includes two General Purpose Power Outlets, USB ports, wireless charging and inbuilt power monitoring integrated into a new range of seating. This is a first in Australia. Also a (cigarette) ash receptacle was invented that measures the heat inside the unit. Sound sensors and a weather station were installed in both case study areas to measure those environmental factors. These products, alongside smart bins (measuring fill levels, heat and passing pedestrians), are being combined into an IoT of diverse and extensive data. This evidence base is the start of GRC's new smart asset management system.

In addition, an entirely new product is provided, the Healthy Living Hardware (HLH). This is an innovative street furniture product that aims to improve public health by providing power and water in proximity to heavily used public areas. The HLH includes WiFi, General Power Outlet [GPO], USB and power outlets (Fig. 11.4). The units allow people to recharge their phones, wash their hands, or make a cup of tea. They also include a timed water tap, a grate for drainage at the base and side counter tops for food preparation. With the HLH, people can bring a cooking appliance like an electric BBQ, wok, or hot plate, extending the range of activities that can take place in a public setting, introducing an element of domestic activity into the public space. The HLH has been adapted by SFA to fit their existing product range by integrating elements of their products' design materials and

Fig. 11.4 Healthy Living Hardware and smart bins installed in Olds Park. (Authors 2018)

features, such as timber fittings and other fixtures. The result is that this unit, albeit a prototype of a new typology of street furniture, now looks and feels like it belongs to the existing range of an established repertoire of street furniture options.

Key stages of the project were as follows (also represented in Fig. 11.2):

- **Proposal Development** was mostly conducted by UNSW and SFA. This included identifying sites, preparing the plan for the refurbished sites, identifying opportunities and constraints, refining the initial proposal, developing concepts of smart furniture and turning concepts into products. Feasibility planning and cost assessment was undertaken to ensure the project components are aligned with the budget and schedule;
- **Innovation Legality** included drawing up an agreement between UNSW and GRC to address liability issues associated with the installation of the HLH as a piece of untested in the public realm;
- **Installation**, conducted through the collaboration of SFA and GRC, included site preparation, furniture installation, ensuring water flow and power supply through GPO and USB outlets, and setting up a network connection via 4G modems to enable sending data to cloud servers.

11.6.1 Roles of the Collaborators in the Smart Social Spaces Project

Figure 11.2 illustrates the three contributors to the project and the dynamic nature of their relationships. It also shows the iterative nature of the design thinking behind the project and amongst the players. As well, Fig. 11.2 summarises the major roles of each collaborator.

A summary of tasks by collaborator is as follows:

UNSW

- Wrote and submitted the grant (driver and now leader)
- Led discussions about legalities of the innovative part of the project
- Operationalised the data collection and management systems
- Positioned the location of street furniture in the plaza and park
- Documenting knowledge through journal articles and conference publications (ongoing)
- Promoted the project through social and other media outlets
- Managing the project overall to be completed within time and budget

SFA

- Designed and developed smart street furniture
- Prepared engineering and structural drawings for new furniture
- Managed subcontractors for technical aspects of the project
- Produced, assembled and delivered the smart street furniture

GRC

- Agreed to be formal grant partner and host the project
- Identified the public domain case study sites which have some current social and place-based issues requiring research
- Participated in discussions about legalities of the innovative part of the project
- Prepared the site and installed new street furniture
- Communicated regularly with their local community

Subcontractors

- Developed IoT sensors to install on street furniture
- Advised on the selection of technology to be used
- Provided the data platform for data capture and data visualisation

11.6.2   The Role of the Disciplines in the Smart Social Spaces Project

Collaboration across the various disciplines (planning, landscape architecture, industrial design) and scales (from individual furniture to a public space, street, neighborhood and beyond) in the Smart Social Spaces project is essential to its success. There are always opportunities and challenges collaborating across different disciplines that work at 1:1//1:200//1:2500 scale drawings from objects to site plans to maps, and of course, all involving interactions with people. The UNSW team is made up of industrial designers, city planners, a healthy built environment expert, a landscape architect, an environmental psychologist and an architect who work at these different scales. The SFA team includes industrial designers, engineers, graphic designers, and construction/production staff. The Council team includes project managers, an infrastructure manager, engineering operations staff, electricians, tradespeople, parks crews and the communications coordinator. Collectively, the complexity of the project can be successfully addressed by the expertise in the multidisciplinary teams from across the partners.

11.7   Lessons Learnt

Beyond developing technical and design solutions, there has been a highly cooperative and productive partnership between local government, academia and industry. By working collaboratively, we have been able to provide a better understanding of the use of public space, so that Council can make informed decisions that support and optimise the use of shared public space, encourage social interaction and ultimately, improve public health. The critical insights (concerning processes, outputs and outcomes) presented below can be applied to develop an effective model of research, practice, and local government collaboration that stimulates urban innovations to address complex problems of twenty-first century cities:

1. Processes

    • Start developing the MOUs and legal agreements necessary for the project at the outset or during planning phases
    • Appreciate that knowledge creation can happen through experiential learning or 'learning by doing' which means that new knowledge and innovation cannot be rushed
    • Innovation emerges as collaboration occurs
    • Communication between parties is vital
    • Decide early in the process which collaborator owns the Intellectual Property, including the data generated
    • Enable urban entrepreneurism through supporting start-ups and tech-based companies
    • Identify what you do not know which could interfere with the project and seek solutions

2. Outputs

    • Aim to commercialise new products to ensure the private industry partner benefits
    • Allow the industry partner to experiment/test and improve its products in a real-world setting for both social and economic benefit
    • Ensure these innovative projects provide the industry partner with opportunities for publicity
    • Be aware that new products take time to be designed, tested and launched into the marketplace
    • Improve the public amenity or services by using smart technologies with the end user in mind

3. Outcomes

    • Develop proposals for future projects that can benefit from the established collaboration
    • Assess and reflect on what has been done in order to learn from mistakes or run future projects more effectively
    • Do not underestimate the potential for 'scaling up' or the transferability of findings
    • Set future research agendas
    • Continue to invest in collaboration (with partners and their relationships)
    • Look for non-traditional partnerships

11.8   Innovations and Smart City Projects: Reflections on Collaboration

A smart city is committed to innovation in management and policy as well as innovation in the adoption of technology (Nam and Pardo 2011). The Smart Social Spaces project has been an opportunity to deliver 'technological innovation' by developing new products, as well as 'organisational innovation' by enabling cross-sectoral collaboration. The formation of new partnerships between university, industry, and government represented one of the innovative aspects of the project. Although the Smart Social Spaces project, like most smart city initiatives, started as a pilot project, as the partnerships formed and trust between the actors was built up over the course of conducting the project, additional projects have been identified and the collaboration has been extended.[1]

It was through the sharing and integrating of information and knowledge that this 'smart city innovation' has been fostered in the above case study. Triple helix collaboration enabled the development of a cutting-edge project by bridging efforts that were otherwise isolated: the university analysed the problems and generated creative solutions; industry devised these solutions to improve their services and products; and the local government implemented them in real urban settings to better serve their community. In this way, one partner's activity contributed to another's to ensure the effective use of technology for overall community benefit – ultimately a goal for all three organisations.

Getting involved in collaborative projects is a great opportunity for universities to boost innovation and knowledge. As noted by Boyer (1990), knowledge can be acquired not only through research, but also synthesis and practice. In the Smart Social Spaces project, the university brought different disciplines together and proposed the application of smart technology in a real-life setting. This enabled the scholarship of integration and application that are often seen as a university's functions besides teaching and discovery (Boyer 1990).

Finally, the case study in this chapter provided an example of a dynamic collaboration taking place in a non-linear way in which partners are jointly involved in multiple stages of the project. Flexibility of partners in taking charge of different tasks has been critical in the success of this project; it would not have been possible to define these roles precisely at the beginning of developing an innovative proposal. Also, for smart city projects to include the element of innovation, the involvement of start-up companies and tech-based contractors are essential – these too were central to the success of this project (see Fig. 11.5). This is of great advantage for start-ups as it enables strengthening their link with academia as well as established industry practices.

---

[1](The same set of partners applied for and were successful in Round 2 of Smart Cities and Suburbs grant program in 2018).

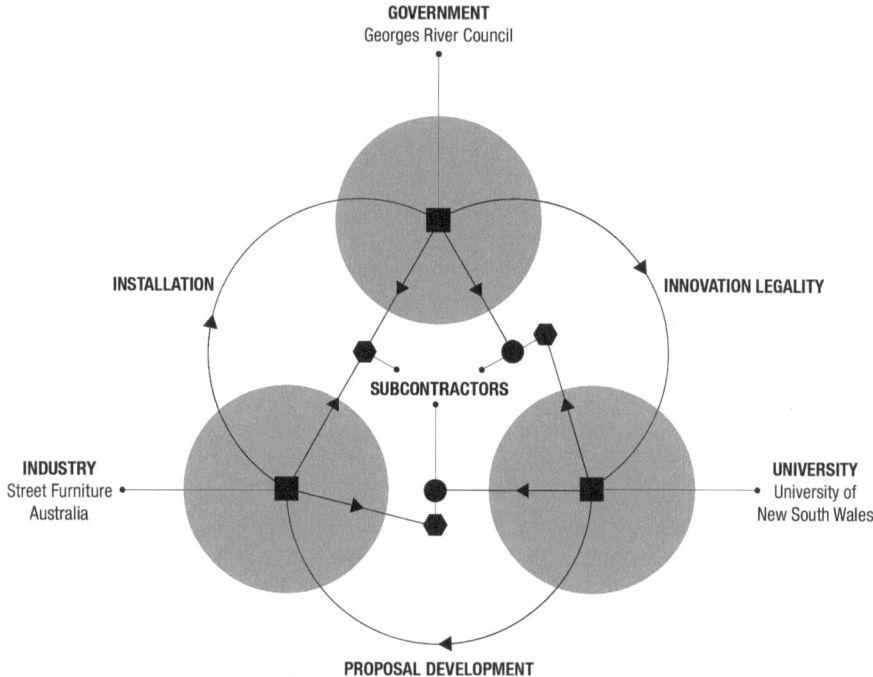

Fig. 11.5 Cross-sectoral collaboration in Smart Social Spaces project. (Authors 2018)

11.9 Conclusion

The Smart Cities Movement is here to stay and can be a catalyst and driver of meaningful partnership formation. Local councils, private industry and academia need to realise the power of collaboration at large or very small scales. Connecting all three groups of collaborators can increase the effectiveness of their individual efforts and initiatives that aim to make cities smarter. Despite *"the project partners having differing objectives and cultures"* (Hatzelhoffer 2012, p. 190) there is much to be gained from working together.

This chapter highlights one case study and discusses the many lessons learnt about the process, outputs and outcomes of smart city initiatives across a range of scales. Mutual trust amongst the players can help ensure the overall success of these projects as innovation often 'tests' and ideally strengthens these relationships.

The effective use of technology for improving infrastructure, precincts, services, planning and design needs an interdisciplinary collaboration. Roles and responsibilities may be blurred or overlap at times, but it makes the triple helix a stronger and more dynamic model of collaboration to deliver a smart city.

References

Adams R, Daly S, Mann L, Dall'Alba G (2011) Being a professional: three lenses into design thinking, acting, and being. Des Stud 32:588–607

Albino V, Berardi U, Dangelico RM (2015) Smart cities: definitions, dimensions, performance, and initiatives. J Urban Technol 22(1):3–21

Allwinkle S, Cruickshank P (2011) Creating smart-er cities: an overview. J Urban Technol 18 (2):1–16

Australian Government Smart Cities and Suburbs Program (2017) Smart cities and suburbs program guidelines. Available at: https://www.business.gov.au/~/media/business/smart-cities-and-suburbs/smart-cities-and-suburbs-program-guidelines-round-one-PDF

Batty M, Axhausen K, Fosca G, Pozdnoukhov A, Bazzani A, Wachowicz M, Ouzounis G, Portugal Y (2012) Smart cities of the future. Eur Phys J Spec Top 214:481–518

Boyer EL (1990) Scholarship reconsidered: priorities of the professoriate. Princeton University Press, Princeton

Brown T (2008) Design thinking. Harv Bus Rev 86:84–92

Brown T, Katz B (2009) Change by design: how design thinking transforms organizations and inspires innovation. Harper Collins, New York

Charles D (2011) The role of universities in building knowledge cities in Australia. Built Environ 37(3):281–298

Cruickshank P (2011) SCRAN: the network. J Urban Technol 18(2):83–97

Dameri RP, Negre E, Rosenthal-Sabroux C (2016) Triple Helix in Smart cities: a literature review about the vision of public bodies, universities, and private companies. In: System Sciences (HICSS), 2016 49th Hawaii International Conference on, pp 2974–2982

Dorst K (2010) The nature of design thinking. In: Proceedings of the 8th Design Thinking Research Symposium (DTRS8), Sydney, October 19–20, 131–139

Etzkowitz H, Zhou C (2017) The triple Helix: university–industry–government innovation and entrepreneurship. Routledge, New York

Hatzelhoffer L (ed) (2012) Smart City in practice: converting innovative ideas into reality: evaluation of the T-city Friedrichshafen. Jovis Verlag GmbH, Berlin

Hollands RG (2008) Will the real smart city please stand up? Intelligent, progressive or entrepreneurial? City 12(3):303–320

Kitchin R (2014) The real-time city? Big data and smart urbanism. GeoJournal 79(1):1–14

Leydesdorff L, Deakin M (2011) The triple-helix model of smart cities: a neo-evolutionary perspective. J Urban Technol 18(2):53–63

Nam T, Pardo TA (2011) Smart city as urban innovation: focusing on management, policy, and context. In: Proceedings of the 5th international conference on theory and practice of electronic governance, pp 185–194. ACM

Oren H (2017) Tel Aviv, 2010–2016: the start-up city of the start-up nation. Collar Venture Rev 5:118–119

Patton MQ (2002) Qualitative research and evaluation methods, 3rd edn. Sage, Thousand Oaks

Tomitsch M (2018) Making cities smarter: designing interactive urban applications. Jovis Verlag GmbH, Berlin

Chapter 12
Enabling Smart Participatory Local Government

Tooran Alizadeh, Somwrita Sarkar, Sandy Burgoyne, Alex Elton-Pym, and Robyn Dowling

Abstract Social media and online communication have changed the way citizens engage in all aspects of their lives, from shopping and education to how their communities are planned and developed. It is no longer one-way or two- way communication. Instead, via networked all-to-all communication channels, our citizens engage in urban issues in a more complex and connected way than ever before. Therefore, the government needs new ways to listen to its citizens.

This paper comprises three components. Firstly, we will build on the growing discussions in literature focused on smart cities, on one hand, and social media research, on the other hand, in order to capture the diversity of citizen voices and better inform decision-making. Secondly, with the support of the Australian Federal Government and in collaboration with case study local governments, we will collect citizen voices from social media platforms, on selected urban projects. Thirdly, we will present preliminary findings in terms of quantity and quality of publicly available online data representing citizen concerns on urban matters. Analyzing the sentiments of the citizen voices captured online and clustering them into topic areas, we will elaborate the scope and value of technologically-enabled opportunities in terms of enabling participatory local government decision making processes.

Keywords Citizen engagement · Crowdsourcing · Social capital · Social media

12.1 Introduction

Amidst the speedy growth of smart city promises and practices, there is an urgent need to take a critical approach and offer an integrated vision for an otherwise fragmented and sectoral concept. In particular, the literature warns about the lack of

T. Alizadeh (✉) · S. Sarkar · S. Burgoyne · A. Elton-Pym · R. Dowling
School of Architecture, Design and Planning, University of Sydney, Camperdown, NSW, Australia
e-mail: tooran.alizadeh@sydney.edu.au

© Springer Nature Switzerland AG 2020
R. Roggema, A. Roggema (eds.), *Smart and Sustainable Cities and Buildings*,
https://doi.org/10.1007/978-3-030-37635-2_12

citizen voices in smart city decision making processes and projects (Alizadeh 2018; Lara et al. 2016; Niaros 2016). Indeed, the current debates around smart cities are full of contradictions. On one hand, there are near-utopian notions around the concept of "smart", with big tech-companies projecting visions of perfectly optimized and smooth-running lives in perfectly organized cities, a return to concepts of the "city as a machine". On the other hand, the same utopia-based vision turns dystopic, when it is pitched against the worst that it could become: the city turning into a sort of an omnipresent, omniscient Truman show or an Orwellian 1984, with, ironically, the big-smart-tech companies running the show (Valverde and Flnn 2018). In the middle of both extremes is a worrying empty space: there is no space in which the everyday messiness and infinite capacity that a "good", "fair", "just", or "sustainable" city must have for adaptability to collective actions, choices, and lives of the citizens, is discussed, conceptualized, acknowledged, or captured. In other words, a framework for civic engagement in design and planning of a city is an essential need in a fully functioning healthy democracy (Dong et al. 2013). In reality, such a city would likely be a middle ground between being completely top-down, system-driven, versus completely bottom-up, citizen-driven. This paper begins by acknowledging this middle space between the two extremes, a space where citizen voices may provide a bridge between the top-down, system-driven city, and the bottom-up, people-driven city.

There are, however, serious questions about how collective citizen voices can be accounted for in the urban development processes already in place, using smart technological advances already at hand. In developing a response to such questions, this paper takes up a challenge to empower citizen voices in smart cities, with special attention to the potential of passive crowdsourcing based on the mostly untapped and unutilized available data in the public domain of social media. This is proposed to enable public engagement in smart city debates and decision making – especially at the local government level where the resources to actively build higher levels of citizen engagement are scarce.

The paper also builds upon the existing links between the two concepts of 'smart cities' on one hand, and 'sustainable cities' on the other. It shifts the focus of attention away from the over- simplified technological focus on smart city concepts, and puts the spotlight on smart cities empowered by smart governance, smart environment, smart living, and smart people (Papa et al. 2013; Yigitcanlar et al. 2015). In doing so, the smart city concept reaches out to the building of social sustainability, as accounting for people's voices in urban decision-making processes is targeted as a means of social empowerment; and ultimately increases collective social capital (Dale and Onyx 2010; Dillard et al. 2009).

The paper starts with a portrayal of the bifurcated smart city landscape, then presents the ways in which citizen voices can be captured via social media using crowdsourcing techniques, and identifies the shortcomings that motivated our study. This leads to the second part of the paper, in which the methodological details of our study – collecting citizen voices in collaboration with local government partners – are included. In the third part of the paper, we present some of the tentative findings in terms of quantity and quality of publicly available online data, representing citizens' voices and their concerns on local urban matters. The paper concludes by elaborating the scope and value of technologically-enabled opportunities in terms of enabling participatory local government decision making processes. Lessons learned contribute to the fast-growing and yet understudied fields of empirical smart cities studies and social media research. Moreover, findings inform governments, on all levels, of the opportunities and challenges involved in capturing meaningful public insights online.

12.2 Broader View: Citizen Voices in Smart Cities

12.2.1 Corporate Smart Cities vs. Alternative Smart Cities

Over the last few years, we have witnessed a spread of smart city projects around the world (Alizadeh 2017) involving cities of all sizes (Kavta and Yadav 2017; Watson 2015) and a diversity of socio-economic statuses. (Sanseverino et al. 2016). Smart city projects cover an incredibly broad range of topics, including e-governance (Bertot et al. 2014; Kumar 2015), smart transport (Bodhani 2012; Debnath et al. 2014), efficient urban services (Lee et al. 2015), and open data (Al-Ani 2017).

Despite the heterogeneous smart city practices and projects worldwide, the critiques seem to be quite focused on what is labelled as 'corporate smart cities' (Hollands 2015a; McNeill 2016; Söderström et al. 2014; Vanolo 2014). From a critical perspective, it is argued that the smart city has crystallized into an image of a technology-led urban utopia permeated with centrally controlled technological infrastructure (Albino et al. 2015; Niaros 2016). In fact, in the corporate vision of smart cities, citizens are often seen as barriers in the race towards smartness; and that they need to be educated as to the benefits smartification can bring (McNeill 2015; Vanolo 2014).

On a positive note, however, the absence of real citizen involvement and participation has encouraged a push for an alternative version of smart cities, to provide a counter-point to the corporate vision (Hollands 2015b; Kostakis et al. 2015). This alternative vision has emanated from small-scale and fledgling examples of participatory community-based type smart initiatives (Chatterton 2013; Niaros 2016; Radywyla and Biggs 2013). Previous studies (Alizadeh 2018), in search of common ground for the growing number of alternative smart initiatives, have put the below list as the core elements that tie together the alternative smart city vision:

- An emphasis on citizen engagement beyond the simple delivery of services
- A democratic bottom up approach: to promote participatory urban technologies, greater social inclusion, and a substantial shift in power from corporations to ordinary people and their communities
- Reliance on dynamic public-private partnership: with an emphasis on participatory governance rather than an entrepreneurial one

- A tendency to identifying the urban problem first, and only then reaching out for the relevant technological solution: with emphasis on the capacities of each city, and its distinct cultures, histories, and political economies
- Is associated with the free software and open access movement and
- Is in the preliminary phase: far from being mature; and mainly exist in seed form.

12.2.2   Power of the Crowd Via Social Media

An essential element of the alternative smart city vision is an emphasis on citizen engagement by empowering their voices (Lee et al. 2014; Papa et al. 2013). However, participatory planning literature acknowledges the difficulties involved in gathering people's voices in urban decision-making processes (Davies et al. 2012; Umemoto 2001). There are, however, two relatively new phenomena – social media and crowdsourcing – which provide opportunities for smart technologies and techniques to capture citizen voices via alternative channels (Kleinhans et al. 2015). Below, we briefly discuss these two phenomena. It should be highlighted that the aim is not to offer an all-inclusive account but rather to provide the foundation for the further empirical parts of the paper.

12.2.3   Social Media

Social media is an umbrella term for many different online platforms which mainly introduced and gained momentum in the last decade, including but not limited to: Twitter, Facebook, and Instagram; each with the key feature of allowing users to connect (Carr and Hayes 2015; Fieseler and Fleck 2013). Social media is broad reaching and allows dispersed groups and individuals to connect and share or promote information relating to common interests, concerns, or causes (LaRiviere et al. 2012; Minton et al. 2012; Walther and Jang 2012). It has played an important role in a number of civic uprisings around the word including but not limited to the Arab Spring, the Occupy Movement, and recent presidential election campaigns in the US (Farro and Demirhisar 2014; Gleason 2013; Morozov 2009). This has prompted a new line of scholarship, focusing on the role of social media in enabling participation, creating collective voice, and facilitating socio-political change (Comunello and Anzera 2012; Howard and Parks 2012; Kavada 2015).

Social media is participatory and interactive, which is also what separates it from the traditional forms of media (Fieseler and Fleck 2013; Minton et al. 2012; Walther and Jang 2012). The user-generated content on social media provides avenues for building bottom-up movements and empowering collective voices (Juris 2012; Linders 2012; Scott and Liew 2012; Willems and Alizadeh 2015). In contrast to these possibilities of empowerment, there is also growing skepticism around the quality of online voices, the sheer size of distracting noises online, claims of inherent bias towards the tech savvy citizen, and the legitimacy of citizen voice captured online (Ferrara et al. 2016; McCafferty 2011; Vitak et al. 2011).

Nevertheless, the complexity of the social media debate is, partially, due to its growing ability to provide an alternative voice (Fieseler and Fleck 2013; Walther and Jang 2012). Its dynamic nature allows for both bottom-up and top-down community involvement. From a bottom-up perspective, there is growing research on the use of community led groups to organize and coordinate via social media in opposition to planning, policy, and manufacturing or development processes (Alizadeh et al. 2018; Maireder and Schwarzenegger 2012; Shav-Ami 2013). In a top-down engagement perspective also, there is a growing line of literature (Afzalan and Evans-Cowley 2015; Evans-Cowley 2012) that argues planners can greatly utilize social media-based opportunities to mobilize and organize citizens.

12.2.4 Crowdsourcing

Howe (2006) first coined the term crowdsourcing in a Wired Magazine article as "the act of a company or institution taking a task once performed by employees and outsourcing it to an undefined (and generally large) network of people in the form of an open call". Since then, there have been many attempts to revise and redefine crowdsourcing based on the diversity of its practices (Estellés-Arolas and González-Ladrón-De-Guevara 2012; Zhao and Zhu 2014). Some of the latest revisions have been proposed in response to the emergent crowdsourcing based on the eminence of social media (Kietzmann 2017; Thapa et al. 2015).

However, the most significant evolution of the crowdsourcing concept stems from a shift, from the original 'task-oriented' approach to what can be described as 'crowdsourcing of opinions' (Alizadeh 2018; Noveck 2015). In this second approach, crowdsourcing is no longer about getting a certain task done by the help of the crowd. Instead, crowdsourcing of opinions is used to gauge opinions, ideas, or perceptions of the public in different forms of polling, sentiment analysis, and opinion mining. Sentiment analysis uses language processing and machine learning to identify which topics different groups talk and care about the most. Social media in general, and twitter, in particular, are rich sources of opinions; and have been used in crowdsourcing of opinions. There are, indeed, numerous examples of companies using crowdsourcing of opinions – via social media – in their marketing efforts (Dowson and Bynghal 2011; Willems and Alizadeh 2015).

Crowdsourcing of opinions, in turn, is then categorized into two broad categories, active and passive. In terms of the difference between active and passive crowdsourcing, Loukis and Charalabidis (2015) argue that active crowdsourcing of opinions is more like mainstream private sector crowdsourcing, which actively stimulates discussions and content generation by citizens on specific topics. The passive crowdsourcing approach, however, is more compatible for the public sector; it passively collects information, knowledge, opinions, and ideas concerning hot topics of the day and important public policies created by citizens without any initiation, stimulation or moderation from government postings (Charalabidis et al. 2012, 2014; Loukis and Charalabidis 2015). Social Media Monitoring (SMM), as a systematic continuous observation, and analysis of the data already available and mostly untapped, is the main source of passive crowdsourcing in the public sector (Loukis et al. 2017).

12.2.5 Shortcomings: Crowdsourcing in Urban Decision-Making Processes

In principal, crowdsourcing has great potential in participatory urban planning; promotes many elements of smart cities including open government; and can be used as an expansion of e-governance to we-governance by facilitating citizen-to-government support, citizen reporting, and citizen-government coproduction of cities (Castelnovo 2016; Linders 2012; Schmidthuber and Hilgers 2017).

Nevertheless, the problem is the scale of uptake of crowdsourcing in urban governments (Alizadeh 2018; Berst et al. 2014; Bertot et al. 2014; Norris and Reddick 2013). Indeed, the small, but growing number of crowdsourcing in urban governments mostly falls into the category of active crowdsourcing; a special question is posed to the public (e.g. in times of emergency responses in disaster management (Liu 2014; Poblet et al. 2017)), or a new application/platform is introduced to reach the crowd (e.g. crowdsourcing of real-time data from the residents about the conditions of local roads (Harford 2014)). This slow uptake may be further impacted by the perceived inherent bias that social media users are the domain of the tech savvy, and not representative of the entire population.

This paper, motivated by the gap in the practice of crowdsourcing, takes a step towards using passive crowdsourcing to inform local urban planning processes. Below, parts describe the ins and outs of the study behind this paper, and unfold some of the preliminary findings.

12.3   Our Study

Our study builds upon the alternative smart city vision, discussed earlier in the paper, and puts citizens' voices at the center of smart city thinking. The paper is part of a project funded in the first round of Smart Cities and Suburbs Program initiated by the Australian Government in 2017. The overall project involves investigating new ways of collecting citizen opinions from a range of online sources, including social media, about four urban infrastructure projects in two Australian cities: Sydney and Brisbane, based in the jurisdiction of two Local Government Areas (LGAs): Canada Bay (Sydney) and Logan (Brisbane). The paper, however, only focuses on two projects located in the Canada Bay revealing some preliminary findings.

Below outlines the scope of the study, and the methods adopted for data collection and analysis.

12.3.1 Scope of the Study

As stated above, the broader study (funded by the federal government) focuses on four urban projects in two Australian metropolitan regions: Sydney and Brisbane, based in the jurisdiction of two Local Government Areas (LGAs): Canada Bay (Sydney) and Logan (Brisbane).

Following consultation with local government partners and preliminary data analysis the following pilot projects were selected. Consideration was given to the:

1. Scale and impact of the project; What was the financial investment, project duration, physical scale and potential disruption to the built environment and therefore citizen use?
2. Diversity in type and location of project; A master-planned development or major infrastructure? Inner city or peri-urban fringe?
3. Role of governments in the urban project; Who is the lead proponent? Local government or state government?
4. Viability of data; Is there sufficient volume of data for analysis? Is the data providing meaningful insights to urban issues?

The selected urban projects ranged from a state-led peri-urban master-planned smart community; local government led master-planned recreation park and urban revitalization precinct, to a state- led major transport infrastructure project. For the purpose of this paper, the projects based in Sydney (City of Canada Bay) are further introduced in Table 12.1.

12.3.2 Methods

12.3.2.1 Data Acquisition

In the first phase, we mined Twitter feeds for targeted data on each of the projects discussed above. Meetings and informal interviews were conducted by the team members with council representatives, who outlined the key pieces of information and background described above for each project. Using this as the basis, and a pilot study of Twitter streams, the team developed a set of hashtags and keywords related to each project.

It should be noted here that using some of these keywords and hashtags returned data that may not be concerned directly to the projects, but could nonetheless reveal interesting and relevant information. For example, "Five Dock" could bring out the café culture activities, biking, and community meeting activities that seem common

Table 12.1 Summary of the Urban Projects at the core of the study – Located in the City of Canada Bay, NSW

Project name	Description
Parramatta Rd Transformation project	The Parramatta Road corridor is described as an important transport and movement route for people who live, work and travel in the area; is characterized by chronic traffic congestion, noise and as the connector of Sydney CBD to Parramatta, is a priority area for the long term growth of Sydney.
	Three renewal areas are identified within the City of Canada Bay LGA, including Homebush, Burwood and Kings Bay Precincts. These precincts are expected to provide an additional 17,000 dwellings to house approximately 36,100 people and provide up to 19,600 new jobs (Landcom 2018).
Five Dock Urban Renewal	The Five Dock Town Centre Urban Urban Renewal project sets out a vision for Five Dock to ensure that the centre continues to provide a strong focus for the community, is a better place to live and work, creates improved opportunities for investment, is easy to get around and provides an enhanced built environment (City of Canada Bay 2018).

Table 12.2 Tweet counts

Project	13/08/18–20/08/18 (Snapshot of 7 days)	Complete tweet repository
Parramatta Rd	68	2015
Five Dock	143	2392

in the Five Dock area in Sydney, which in itself could be a sign of an active community that participates.

Repeated tweets were removed, and only unique tweets were preserved, though a record and count was kept of what was being removed. Retweets that appeared genuine were not removed. Further, any tweets that were filled with random data and appeared to be sourced from bots were also removed. This part of the processing made sure that a maximum amount of meaningful information was being captured.

12.3.2.2 Twitter: Data Processing and Cleaning

The quality of Twitter data depends on multiple factors: primarily, the query used and timing. For example, the query "Five Dock" sometimes returns incorrect results regarding boat docks, or for some content that may not be related to Sydney at all. Timing is another factor as certain events may cause a burst of tweets, followed by spans of silence. For example, "Parramatta road" would become particularly active during events of traffic congestion or accidents. The fluctuating popularity of topics is particularly prevalent with Twitter due to its "trending topics" feature. For this paper, we have collected tweets over a period of 7 years.

Table 12.2 shows counts of the numbers of tweets and the time frames for which tweet data was captured. As is evident, the volume of information and flow of information over time is relatively low. Yet, surprisingly most of the tweet content was very meaningful (further information on this follows). Below table also shows a snapshot of 7 days in order to provide an idea of how many relevant tweets are made in a week, though, this can vary depending on bursts of event-related tweets.

12.3.2.3 Sentiment Analysis

Tweets are then analyzed to assess their overall sentiment, either positive or negative. We first remove meaningless words such as Twitter handles, URLs and stopwords. The remaining, meaningful words are then individually assigned a sentiment score. Words such as "happy", "good" and "sun" are given a positive score while words such as "angry", "traffic" or "lost" are given negative scores. The sentiment analysis is performed using a standard Python based library that uses a Naïve Bayes and bag-of-words approach. These individual scores are combined using the Pattern library's PatternAnalyzer to give a combined sentiment score between -1 and 1. An additional feature of the Python library is that it provides a measure of subjectivity versus objectivity of information: if a tweet contains factual information, it is classified as less subjective, whereas tweets with more opinion-based information are rated more on the subjectivity score. It is to be noted here that given any algorithmic limitations (too technical and out of scope of discussion for this paper), the results would not be perfect, though they are reliable.

12.3.2.4 Clustering Analysis

The final stage in analyzing the data was clustering the tweets, that is, putting the tweets together in clusters that brought out information on a particular recurrent topic that was getting repeated attention. We used a technique called Latent Semantic Analysis (LSA), similar to graph clustering (Sarkar and Dong 2011). First, we created a term-by-document (TDM) matrix. As mentioned above in the case of sentiment analysis, all stop words were already removed from the tweets, and all meaningful non-stop words extracted. Each of these unique m words formed a row in the TDM matrix, $i = 1, 2, \ldots m$. Each column formed one of the n tweets, $j = 1, 2, \ldots n$. Each matrix entry is a record of the number of times the word i occurs in tweet j. We then perform the Singular Value Decomposition based algorithm to cluster the data (Sarkar and Dong 2011). This algorithm extracts a lower-dimensional representation for the high-dimensional tweet data. In this lower-dimensional representation, words and tweets that frequently co-occur with each other lie "close to each other" in this abstract mathematical space. The extracted clusters were then examined by counting the top words in each cluster in order to identify the topics.

12.3.2.5 Preliminary Findings

Figures 12.1, 12.2, 12.3 and 12.4 show the preliminary results of the analysis. Figures 12.1 and 12.2 show the results of the sentiment analysis for Parramatta Road and Five Dock, respectively. Each dot represents a tweet, the x-axis plots the

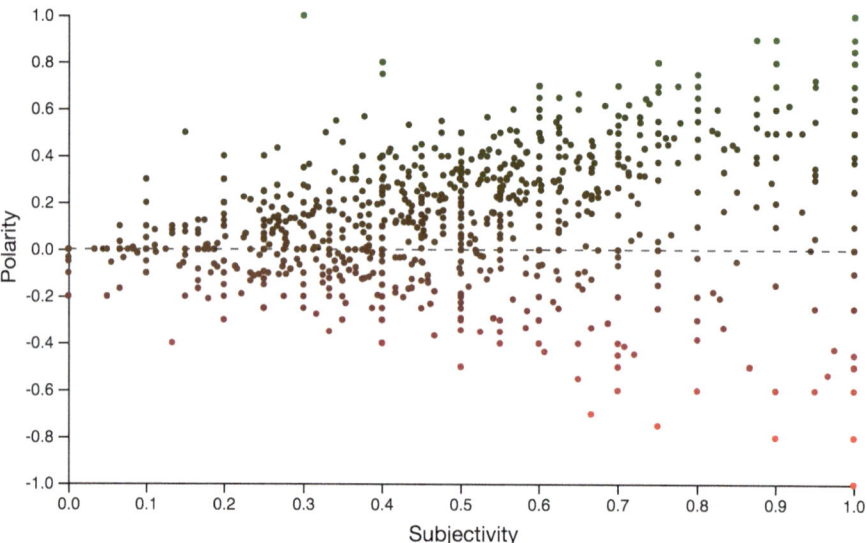

Fig. 12.1 Sentiment analysis for the Parramatta Road tweets

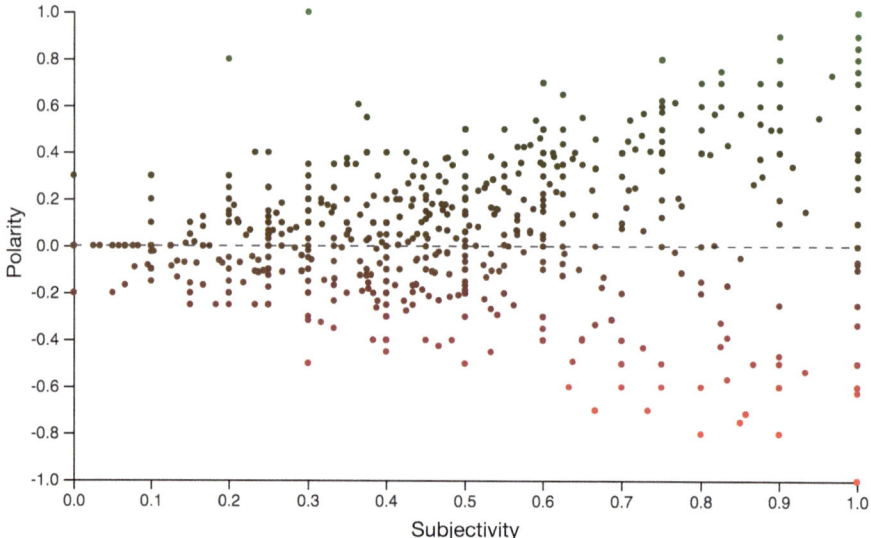

Fig. 12.2 Sentiment analysis for the Five Dock tweets

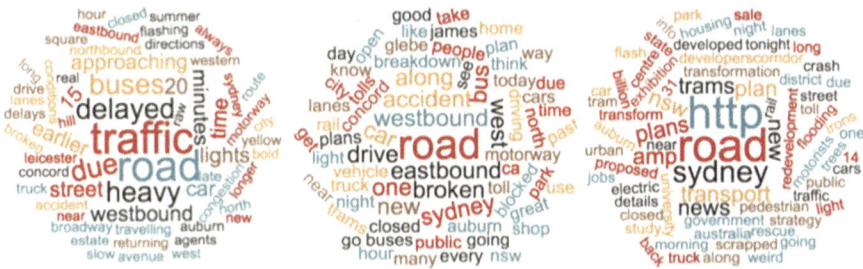

Fig. 12.3 Cluster analysis for the Parramatta Road Tweets (**a**) Delayed traffic, (**b**) Accidents and break downs, and (**c**) Planning and government related tweets

Fig. 12.4 Cluster Analysis for the Five Dock Tweets (**a**) Five Dock popular culture, (**b**) Games and sports events, and (**c**) Traffic delays and congestion

subjectivity of the tweet, and the y-axis measures the polarity (or the positive-negative sentiment). The greener the color of a dot, the more positive the tweet, and the redder the color of a dot, the more negative the tweet.

It is interesting to observe the spread of the data across both urban projects (Parramatta Rd., and Five Dock). It is important to remember that, here, data refers to the diversity of topics discussed by the citizens in their tweets about the two urban projects. First, there is an equal distribution of negative vs. positive opinions expressed in the tweets. This is perhaps surprising; as there is a line of literature which argues that people mostly use Twitter (or social media in general) to vent out, to complain, and basically to be negative about urban issues (Resch et al. 2016). Our sentiment analysis, for both Parramatta Rd. and Five Dock, however, shows a fair distribution of opinions. More interestingly, the less subjective the information shared via each tweet, the lower the spread or dispersion of the polarity. On the contrary, the higher the subjectivity of the information, the higher also the dispersion of the polarity. This shows that more subjective information has higher extremes of emotional expression of positivity or negativity in-built into it. It is also interesting to observe from Figs. 12.1 and 12.2 that the overall distribution "shapes" for both the Parramatta Road and the Five Dock tweets are quite similar.

Table 12.3 Examples of positively and negatively classified tweets

Tweets	Parramatta Road	Five Dock
Positively classified	"Best I've seen that stretch of Parramatta road look" "The joy of Parramatta Road this evening" "NOW is the perfect time to invest in Homebush and become part of a new growing community"	"Looking forward to #Ferragosto2017 today in #fivedock Come on #innerwesties it's a great day for food, sun and fun!" "Good morning #fivedock Happy Tuesday twitterverse! #happydays #lovinglife #canadabay #innerwestisbest"
Negatively classified	"There is a bad pothole round a man-hole/access cover on #ParramattaRoad,westbound,ouside #VictoriaPark just b4 SydneyUniversity. Others closer" "That's so very terrible, @davidtickle_. I've always found Parramatta Road to be a traffic funnel, not good for business or people!"	"Hey @Telstra. Is there something wrong with the broadband service in Five Dock 2046? The service has been slow for 3 days." "#innerwest #sydney be careful of your children! Kidnapping attempt Belfield burwood Five Dock Drummoyne Haberfield"

Regarding the content of the tweets, most of the positive tweets focused on good personal experiences, social events, or at times marketing related information. Most of the negative tweets focused on topics such as traffic and congestion, poor citizen or driver behaviors, or complaints against government or poor infrastructure (e.g. road conditions, telecommunication failure). Table 12.3 provides a few examples of positively and negatively classified tweets for the Parramatta Road and Five Dock.

Figures 12.3 and 12.4 show the results of the cluster analysis for Parramatta Road and Five Dock. The size of the word corresponds to the highest frequency words of each cluster. Figure 12.3 shows word clouds for the top 3 topic clusters identified for Parramatta Road. The clusters can be broadly classified as: (a) Delayed traffic, (b) Accidents and breakdowns, and (c) Planning and government related tweets. Figure 12.4 shows word clouds for the top 3 topic clusters identified for Five Dock. The clusters can be broadly classified as (a) Five Dock popular culture, (b) Games and sports events, and (c) Traffic delays and congestions.

In terms of the content, each cluster identified is rich and meaningful as everyday citizens share their observations, frustration, and happy moments. For instance, a broad range of topic areas are discussed with direct relevance to local government decision making as part of the planning and government related tweets (cluster (c) for Parramatta Rd.): including but not limited to street parking, new constructions' noise pollution, public transport, and housing. Tweet examples follow:

What's wrong with people? This is Parramatta Road, not a quiet Annandale back street!! They should address the parking issues
Somethings needs to be done to deal with noise emanating from the building sites in Parramatta, but sheesh... if you can't build there, where can you build?

Councils across the country are spending more than 1bn putting people in temporary
accommodation because we don't have enough houses and I'm sure there's
nothing better we could spend that money on.

The worst part of Sydney is definitely the overabundance of 413 and 431 buses. I'm
just trying to get down Parramatta road

Very cool that a 40min bus ride can take me between two European capital cities,
while it takes 40 mins just to get down f**** Parramatta road

There is a broader diversity of topic discussed in the Parramatta Road tweets in
comparison to the patterns observed in the Five Dock tweets. This is not unexpected
and relates to the intensity and complexity of issues experienced in Parramatta Road
versus Five Dock, which is a low-profile local project. Having said that, there is still
specific and detailed information captured among Five Dock tweets that could be
valuable – especially if presented and responded to in 'real-time' or 'near-real-time'
basis. Below is an example, which reports a potential break down in the road system:

Some buses travelling through Five Dock and Haberfield are delayed up to
20 minutes due to heavy traffic on Parramatta Rd and Ramsay Rd!!

In sum, the preliminary findings show great potential for further investigations of
topic areas discussed by citizens in their tweets and to better understand the topics of
concern that require responses from government. Especially real-time analysis of
tweets has the potential to inform the government on the burning issues in need of
immediate action. Moreover, we also see potential for longitudinal analysis as it may
reveal more information on the triggers of positive or negative reactions from
citizens or explain how public opinions shift over time about a certain urban project
or issue.

12.4 Conclusion: What We Learned and Where to Go from Here

To date, we have used machine learning methods to extract overall sentiments on the
urban projects at the core of our study. While sentiment analysis is able to provide an
overall idea of positive or negative opinions on the projects, a deeper look at the data
for its content was required. Thus, we used a second set of machine learning methods
(sentiment analysis and cluster analysis) to extract primary latent topics in the data.
These reveal, along with the positive and negative sentiments, the areas of primary
concern for the public.

The analysis reveals that the rate of participation is low, but meaningful. A very
small portion of population have so far participated in the online conversations on
the urban projects at the core of our study. However, those who do participate often
leave meaningful observations that have the capacity to inform the decision-making
process. Our future steps for the study are three folded.

First, we are in the process of advancing our data analytics. The next step is to re-apply sentiment analysis to the tweets classified by topics. The aim is to unravel specific positive and negative sentiments on different dimensions of the projects. This next analysis step will also enhance our understanding of the specific topic areas of concerns for the citizens which need immediate response from the government. Second, we are in the process of a dashboard design to feed back our analysis of citizen voices and concerns, and at time suggestions to the local government to inform decision making processes and outcomes. The proposed interactive dashboard will visually present the 'real time' or 'near real time' data to citizens and local government. Indeed, the next steps of the research project funded by the Australian Government include consultations with the local government partners as part of the dashboard design process. The proposed interactive dashboards will provide further opportunities to test the reliability and quality of social media data and to explore the role that it can play as a representative data source for capturing public opinions on urban projects in 'real time' or 'near-real time'.

Last but not least, we are hoping to expand our study to a larger network of local governments, including a wider diversity of urban projects in different contexts. If anything, our preliminary findings have presented an exciting potential for passive crowdsourcing via social media platform to enhance our understanding of what matters to citizens in terms of each urban project. Building a larger network of local governments will enable responsive urban decision making based on an informed understanding of citizens' concerns and priorities. This will then enable participatory planning which is socially responsible and respectful to the diversity of citizen voices captured online.

References

Afzalan N, Evans-Cowley J (2015) Planning and social media: Facebook for planning at the neighbourhood scale. Plan Pract Res 30:270–285

Al-Ani A (2017) Government as a platform: services, participation and policies. In: Friedrichsen M, Kamalipour Y (eds) Digital transformation in journalism and news media. Springer, Berlin, pp 179–196

Albino V, Berardi U, Dangelico RM (2015) Smart cities: definitions, dimensions, performance, and initiatives. J Urban Technol 22:3–21

Alizadeh T (2017) An investigation of IBM's smarter cites challenge: what do participating cities want? Cities 63:70–80

Alizadeh T (2018) Crowdsourced smart cities versus corporate smart cities. In: Paper presented at the 4th Planocosmo international conference, Bandung, Indonasia

Alizadeh T, Farid R, Willems L (2018) The role of social media in public involvement: pushing for sustainability in international planning. In: Silva CN (ed) New approaches, methods and tools in urban E-planning. IGI Global, New York, pp 310–342

Berst J, Enbysk L, Williams C (2014) Smart cities readiness guide: the planning manual for building tomorrow's cities today. Smart Cities Council, Seattle

Bertot JC, Gorham U, Jaeger PT, Sarin LC, Choi H (2014) Big data, open government and e-government: issues, policies and recommendations. Inf Polity 19:5–16

Bodhani A (2012) Smart transport. Eng Technol 7:70–73

Carr CT, Hayes RA (2015) Social media: defining, developing, and divining. Atl J Commun 23:46–65

Castelnovo W (2016) Co-production Makes Cities Smarter: Citizens' Participation in Smart City Initiatives. In: Fugini M, Bracci E, Sicilia M (eds) Co-production in the Public Sector. Springer, Milan, pp 97–117

Charalabidis Y, Triantafillou A, Karkaletsis V, Loukis E (2012) Public policy formulation through non-moderated crowdsourcing in social media. In: Paper presented at the international conference on electronic participation, Kristiansand, Norway

Charalabidis Y, Loukis EN, Androutsopoulou A, Karkaletsis V, Triantafillou A (2014) Passive crowdsourcing in government using social media. Transforming Gov: People Process Policy 8:283–308

Chatterton P (2013) Towards an agenda for postcarbon cities: lessons from LILAC, the UK's first ecological, affordable, cohousing community. Int J Urban Reg Res 37:1654–1674

City of Canada Bay (2018) City of Canada Bay. Retrieved 2018, Aug 8, from http://www.canadabay.nsw.gov.au/

Comunello F, Anzera G (2012) Will the revolution be tweeted? A conceptual framework for understanding the social media and the Arab Spring. J Islam Christian–Muslim Relat 23:453–470

Dale A, Onyx J (2010) A dynamic balance: social capital and sustainable community development. UBC Press, Vancouver

Davies R, Selin C, Gano G, GuimarãesPereira Â (2012) Citizen engagement and urban change: three case studies of material deliberation. Cities 29:351–357

Debnath AK, Chin HC, Haque MM, Yuen B (2014) A methodological framework for benchmarking smart transport cities. Cities 37:47–56

Dillard J, Dujon V, Kin MC (2009) Understanding the social dimension of sustainability. Routledge, New York

Dong A, Sarkar S, Nichols C, Kvan T (2013) The capability approach as a framework for the assessment of policies toward civic engagement in design. Des Stud 34:326–344

Dowson R, Bynghal S (2011) Getting results from crowds: the definitive guide to using crowdsourcing to grow your business. Advanced Human Technologies Inc., San Francisco

Estellés-Arolas E, González-Ladrón-De-Guevara F (2012) Towards an integrated crowdsourcing definition. J Inf Sci 38:189–200

Evans-Cowley J (2012) There's an app for that: Mobile applications for urban planning. Int J E-Plann Res (IJEPR) 1:79–87

Farro A, Demirhisar E (2014) The Gezi Park movement: a Turkish experience of the twenty- first-century collective movements. Int Rev Sociol 24:176–189

Ferrara E, Varol O, Davis C, Menczer F, Flammini A (2016) The rise of social bots. Commun ACM 59:96–104

Fieseler C, Fleck M (2013) The pursuit of empowerment through social media: structural social capital dynamics in CSR-blogging. J Bus Ethics 118:759–775

Gleason B (2013) Occupywallstreet: exploring informal learning about a social movement on twitter. Am Behav Sci 57:966–982

Harford T (2014) Big data: A big mistake? Significance 11:14–19

Hollands R (2015a) Beyond the corporate smart city: glimps of other possiblities of smartness. In: Marvin S, Luque-Ayala A, McFarlane C (eds) Smart urbanism: utopian vision or false Dawn? Routledge, London, pp 168–184

Hollands RG (2015b) Critical interventions into the corporate smart city. Camb J Reg Econ Soc 8:61–77

Howard PN, Parks M (2012) Social media and political change: capacity, constraint, and consequence. J Commun 62:359–362

Howe J (2006) The Rise of Crowdsourcing. Wired Mag 14:1–5

Juris J (2012) Reflections on #occupy everywhere: social media, public space, and emerging logics of aggregation. J Am Ethnol Soc 39:259–279

Kavada A (2015) Creating the collective: social media, the occupy movement and its constitution as a collective actor. Inf Commun Soc 18:872–886

Kavta K, Yadav PK (2017) Indian smart cities and their financing: a first look. In: Seta F, Sen J, Biswas A, Khare A (eds) From poverty, inequality to smart city. Springer, Singapore, pp 123–141

Kietzmann JH (2017) Crowdsourcing: a revised definition and introduction to new research. Elsevier, London

Kleinhans R, Ham MV, Evans-Cowley J (2015) Using social media and mobile technologies to foster engagement and self-organization in participatory urban planning and neighbourhood governance. Plan Pract Res 30:237–247

Kostakis V, Bauwens M, Niaros V (2015) Urban reconfiguration after the emergence of peer-to-peer infrastructures: four future scenarios with an impact on smart cities. In: Araya D (ed) Smart cities as democratic ecologies. Palgrave Macmillan, New York, pp 116–124

Kumar TMV (2015) E-governance for smart cities. In: TMV K (ed) E-governance for smart cities. Springer Singapore, Singapore, pp 1–43

Landcom (2018) About the parramatta road program. Retrieved Aug 8, 2018, from https://www.landcom.com.au/places/parramatta-road/

Lara AP, Costa EMD, Furlani TZ, Yigitcanlar T (2016) Smartness that matters: towards a comprehensive and human-centred characterisation of smart cities. J Open Innov: Technol Market Complexity 2:8

LaRiviere K, Snider J, Stromberg A, O'Meara K (2012) Protest: critical lessons of using digital media for social change. About Campus 17:10–17

Lee JH, Hancock MG, Hu M-C (2014) Towards an effective framework for building smart cities: lessons from Seoul and San Francisco. Technol Forecast Soc Chang 89:80–99

Lee SW, Sarp S, Jeon DJ, Kim JH (2015) Smart water grid: the future water management platform. Desalin Water Treat 55:339–346

Linders D (2012) From e-government to we-government: defining a typology for citizen coproduction in the age of social media. Gov Inf Q 29:446–454

Liu SB (2014) Crisis crowdsourcing framework: designing strategic configurations of crowdsourcing for the emergency management domain. Comput Supported Coop Work 23:389–443

Loukis E, Charalabidis Y (2015) Active and passive crowdsourcing in government. In: Janssen M, Wimmer MA, Deljoo A (eds) Policy practice and digital science. Springer International Publishing, Cham, pp 261–289

Loukis E, Charalabidis Y, Androutsopoulou A (2017) Promoting open innovation in the public sector through social media monitoring. Gov Inf Q 34:99–109

Maireder A, Schwarzenegger C (2012) A movement of connected individuals. Inf Commun Soc 15:171–195

McCafferty D (2011) Activism vs. slacktivism. Commun ACM 54:17–19

McNeill D (2015) Global firms and smart technologies: IBM and the reduction of cities. Trans Inst Br Geogr 40:562–574

McNeill D (2016) IBM and the visual formation of smart cities. In: Marvin S, Luque-Ayala A, Mc Farlane C (eds) Smart urbanism: utopian vision or false dawn? Routledge, London, pp 34–52

Minton E, Lee C, Orth U, Chung-Hyun K, Lynn K (2012) Sustainable marketing and social media: a cross-country analysis of motives for sustainable behaviors. J Advert 41:69–84

Morozov E (2009) Iran: downside to the "twitter revolution". Dissent 56:10–14

Niaros V (2016) Introducing a taxonomy of the "Smart City": towards a commons-oriented approach? tripleC 14

Norris DF, Reddick CG (2013) Local E-government in the United States: transformation or incremental change? Public Adm Rev 73:165–175

Noveck B (2015) Smart citizens, smarter state: the technologies of expertise and the future of governing. Harvard University Press, Cambridge

Papa R, Gargiulo C, Galderisi A (2013) Towards an urban planners' perspective on Smart City. TEMA J Land Use Mobil Environ 6:5–17

Poblet M, García-Cuesta E, Casanovas P (2017) Crowdsourcing roles, methods and tools for data-intensive disaster management. Inf Syst Front 20(6):1–17

Radywyla N, Biggs C (2013) Reclaiming the commons for urban transformation. J Clean Prod 50:159–170

Resch B, Summa A, Zeile P, Strube M (2016) Citizen-centric urban planning through extracting emotion information from twitter in an interdisciplinary space-time-linguistics algorithm. Urban Plan 1:114–127

Sanseverino ER, Sanseverino RR, Vaccaro V, Macaione I, Anello E (2016) Smart cities: case studies. In: Sanseverino IR, Sanseverino RR, Vaccaro V (eds) Smart cities atlas. Springer, London, pp 47–140

Sarkar S, Dong A (2011) Community detection in graphs using singular value decomposition. Phys Rev E 83:1–16

Schmidthuber L, Hilgers D (2017) Unleashing innovation beyond organizational boundaries: exploring citizensourcing projects. Int J Public Adm 20:1–16

Scott K, Liew T (2012) Social networking as a development tool: a critical reflection. Urban Stud 49:2751–2767

Shav-Ami A (2013) Social protest: the Israeli case. J Enterp Commun People Places Glob Econ 7:373–382

Söderström O, Paasche T, Klauser F (2014) Smart cities as corporate storytelling. City: Anal Urban Trends Cult Theory Policy Action 18:307–320

Thapa BE, Niehaves B, Seidel C, Plattfaut R (2015) Citizen involvement in public sector innovation: government and citizen perspectives. Inf Polity 20:3–17

Umemoto K (2001) Walking in another's shoes: epistemological challenges in participatory planning. J Plan Educ Res 21:17–31

Valverde M, Flnn A (2018) More buzzwords than answers - to sidewalk labs in Toronto. J Landscape Archit Front 6:115–123

Vanolo A (2014) Smartmentality: the Smart City as disciplinary strategy. Urban Stud 51:883–898

Vitak J, Zube P, Smock A, Carr CT, Ellison N, Lampe C (2011) It's complicated: Facebook users' political participation in the 2008 election. Cyberpsychol Behav Soc Netw 14:107–114

Walther JB, Jang J (2012) Communication processes in particapatory websites. Comput-Mediated Commun 18:2–15

Watson V (2015) The allure of 'smart city' rhetoric: India and Africa. Dialogues Hum Geogr 5:36–39

Willems LG, Alizadeh T (2015) Social Media for Public Involvement and Sustainability in international planning and development. Int J E-Plann Res (IJEPR) 4:1–17

Yigitcanlar T, Inkinen T, Makkonen T (2015) Does size matter? Knowledge-based development of second-order cityregions in Finland. disP-Plann Rev 51:62–77

Zhao Y, Zhu Q (2014) Evaluation on crowdsourcing research: current status and future direction. Inf Syst Front 16:417–434

Chapter 13
Data Management Using Computational Building Information Modeling for Building Envelope Retrofitting

Taki Eddine Seghier, Mohd Hamdan Ahmad, Lim Yaik Wah, and Muhamad Farhin Harun

Abstract Computational Building Information Modelling (BIM) is a design paradigm grounded on the use of BIM-based rules and algorithms for data extraction and management, to meet the design objectives and user needs. This design approach has been applied in many aspects and disciplines within the building industry, such as spatial, geometrical, and structural design/optimization, building energetic performance analysis, acoustical optimization, and more.

This article discusses the application of computational BIM for data management within the scope of a research project entitled 'the retrofitting of existing buildings through BIM (RBIM)'. The main goal of this project is to optimize the overall thermal transfer value (OTTV) of the building envelope of existing buildings against the cost of investment. OTTV criteria have been adopted by several green rating tools in different countries, to evaluate building envelope thermal performance and thus assess buildings' cooling loads. Building envelope design under OTTV requirements often requires a cumbersome data collection process and complex design decision-making regarding building envelope component selection.

In the RBIM project, Dynamo was used as the computational BIM design software, to create scripts for the automation of data extraction and data push back from/to the BIM model. Meanwhile, MATLAB was used to perform multi-objective optimization of the OTTV/Cost, using the data extracted from the BIM model as an input.

This paper focuses only on the data management part of the RBIM project. Thus, the development process of the Dynamo scripts is explained and limitations are

T. E. Seghier (✉)
School of Architecture and Built Environment, Faculty of Engineering, Technology & Built Environment, UCSI University, Kuala Lumpur, Malaysia

M. H. Ahmad · L. Yaik Wah
Faculty of Built Environment and Surveying (FABS), Universiti Teknologi Malaysia (UTM), Skudai, Malaysia

M. F. Harun
Faculty of Engineering, School of Computing, Universiti Teknologi Malaysia (UTM), Skudai, Malaysia

© Springer Nature Switzerland AG 2020
R. Roggema, A. Roggema (eds.), *Smart and Sustainable Cities and Buildings*,
https://doi.org/10.1007/978-3-030-37635-2_13

205

discussed. Moreover, delicate tasks during the development of the scripts, such as automating the detection of building envelop elements orientation, are further explained. For validation purpose, the whole RBIM workflow was implemented on a case study building (BIM model), where the applicability of the developed Dynamo scripts for data management was demonstrated and tested.

The findings show a great potential of the implemented workflow in automating data extraction for building envelope performance optimization, as well as reducing working time. Besides this, pushing back the optimized data to the BIM model using Dynamo scripting seems to be applicable.

Keywords Retrofitting · OTTV · Optimization · Revit · Dynamo · Visual programming · MATLAB · Sustainability

13.1 Introduction

A huge investment is devoted globally to environmentally friendly buildings that can provide both high performance and long-term cost saving (Jrade and Jalaei 2013). In order to efficiently guide the design and construction of green buildings, several green building rating systems (GBRS) have been developed around the world, such as LEED (US), BREEAM (UK), Green Mark (Singapore), and GreenRE (Malaysia). When a project team is seeking a green certification under these rating systems, often, design decision-making becomes very time consuming due to the fact that collecting, managing, and documenting the relevant data is a very laborious process (Wu 2010; Wong and Kuan 2014; Jalaei and Jrade 2015; Ilhan and Yaman 2016; Lim et al. 2016; Seghier et al. 2018). Accordingly, architects and designers tend to rely on their previous experience outcomes to make "what they think it is the best design decision". These issues can occur during the design of a new building as well as the retrofitting of existing ones. It has been argued by Biswas et al. (2013) that taking the appropriate steps to automate the process of gathering the necessary information for building environmental analysis is very crucial.

The Building Information Modelling (BIM) design process is based on one data-rich digital model that can be used to perform numerous analyses through building life cycle. It has been argued by many scholars that BIM can support design Decision-Making and sustainability analysis in the very early design stages (Azhar et al. 2011; Jalaei and Jrade 2015; Ilhan and Yaman 2016). Yet, BIM is still not being effectively utilized in improving the sustainability of existing buildings, such as building retrofitting and refurbishment activities. "While the use of BIM for asset management has been acknowledged by researchers and practitioners, BIM is still not being effectively utilized for refurbishment activities" (Alwan 2016). Additionally, Computational BIM tools (i.e. Dynamo, Grasshopper) have broadened opportunities for design optimization and increased the automation of data management throughout the building design process. Several studies have implemented computational BIM tool for building performance analysis such as Energy Efficiency and daylighting optimization (Asl et al. 2011), structural analysis (Makris et al. 2013), acoustical analysis (Andrea Vannini 2015) and building envelope performance assessment (Seghier et al. 2017).

In new green building design and existing building retrofitting, heat gain or loss through the building envelope is usually controlled by specific standards and regulations. For instance, the overall thermal transfer value (OTTV) is one of the mandatory requirements under several green building certifications and standards of several countries including Malaysia, Indonesia, and Singapore. It has been developed in order to guide project teams during the process of building envelope design, thus cut down on the external heat gain, and hence reduce the cooling load of the air-conditioning system (BCA 2008). It is a mandatory requirement for both new and existing buildings. The complexity of OTTV assessment resides in the process of collecting the required data to perform the required calculations (Inhabitgroup 2016; Lim et al. 2016; Seghier et al. 2017). In the Malaysian context, OTTV computation is based on the methodology specified in the Malaysian Standards (MS-1525) (Department of Standards Malaysia 2014). The equation of OTTV applied to commercial buildings is given as follows:

$$OTTVi = 15 \, \alpha \, (1 - WWR)Uw + 6 \, (WWR)Uf \\ + (194 \times OF \times WWR \times SC) \tag{13.1}$$

Equation 13.1 OTTV equation
Where:

WWR is the window-to-gross exterior wall area ratio for the orientation under consideration;
α is the solar absorptivity of the opaque wall;
Uw is the thermal transmittance of the opaque wall (W/m^2 K);
Uf is the thermal transmittance of the fenestration system (W/m^2 K);
OF is the solar orientation factor; as in Table 1; and
SC is the shading coefficient of the fenestration system.
SHGC is solar heat gain coefficient where SHGC = SC $\times$ 0.87.

This paper aims to develop a computational BIM-based data management workflow for building envelope retrofitting. This workflow is implemented within a research project called RBIM, which focuses on the optimization of the thermal performance of the existing building envelope. Different from the existing workflows, which are often based on the usage of Industry Foundation Class (IFC) format, this workflow relies mainly on Dynamo visual scripting and the embedded information within the BIM model components. Thus, three Dynamo scripts are developed to extract automatically the required data from the BIM model in order to be used as an input during the optimization process. Finally, the optimization output is pushed back to the BIM model through another Dynamo script, developed for that purpose. Accordingly, the BIM model is updated automatically with the optimized design option based on two criteria: OTTV performance, and cost of investment. In this paper, first, the overall RBIM framework is explained, then the development process of the Dynamo scripts is discussed, and its usability is demonstrated through case study building.

13.2   RBIM Framework Overview

The RBIM workflow is mainly established to optimize the OTTV of an existing building against the cost of investment. As shown in Fig. 13.1, the RBIM framework is developed to perform a cycle of data, starting by extracting the relevant data from the BIM model using Dynamo. Dynamo scripts are designed according to OTTV requirement and its assessment procedures, which are stipulated in the Malaysian Standards (MS) 1525:2014. Accordingly, all OTTV data requests were interpreted to rules for data management. These rules are compatible with Autodesk Revit and Dynamo current functionalities. Following the extraction of OTTV related data from the BIM model, this data is exported to MATLAB for optimization purpose. Finally, the optimization output data is pushed back again to update the configuration of the building envelope automatically.

Technically speaking, the first stage focuses on the preparation of the BIM model for data extraction. The preparation process requires having a BIM model of the building in question. In the case where no BIM model is available, it is necessary to remodel the existing building (at least the envelope), based on its CAD drawings using a BIM authoring tool (i.e. Autodesk Revit). The choice of selecting Autodesk Revit software in this study is because it is a widely used software in both BIM related research and the practice, not to mention its availability as a student version and its integration with Dynamo (open source visual programming platform for computational BM). The BIM model should be modelled correctly. For instance, the exterior walls should be directed to the outside of the building. Moreover, all building envelope elements (walls and windows) should contain all the required parameters for OTTV assessment, such as U-value of wall, and windows' glazing and cost. As a result, it is recommended to have a BIM model with a level of development (LOD) equal to 300.

On the other hand, data management in the RBIM project is mainly performed using Dynamo scripting. Similar to the other visual programming tools, Dynamo scripting is accomplished by combining the different nodes available in Dynamo

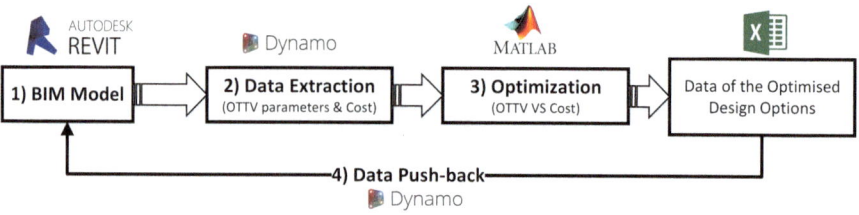

Fig. 13.1 Logic workflow of the RBIM framework for OTTV/Cost optimization

library, in order to create a logic flow for data management according to OTTV requirement. The first stage of data management consists of extracting the relevant OTTV data from the BIM model, as well as the data related to the Revit library, which contains walls and windows types. Following the data extraction process, the extracted data is hosted in an Excel template designed for that purpose. Next, this data is used as an input during the optimization process, which uses a Non-Dominated Sorting Genetic Algorithm II (NSGA II) customized using MATLAB to perform Multi-objective optimization of OTTV against the cost of investment. Finally, another Dynamo script is developed to push back of the optimized data (MATLAB output) to the BIM model. Thus, updating the BIM model automatically with the optimized design option configurations. It is worthy to note that the RBIM framework is open for customization; the user can add additional types of elements to the library and create constraints to the optimization output according to its need.

At the end of the development process, the RBIM workflow was tested and demonstrated on an actual existing building (BIM model), and the generated optimization results were compared to manual calculations, for validation purposes. However, in this paper, the demonstration part focuses only on the applicability of the developed Dynamo scripts to extract the relevant data and push back this data to automatically update the BIM model with optimum design option (MATLAB output).

13.3 Dynamo Scripting Development

Data management scripts in the RBIM framework are classified into four categories, as follows:

- Script A1: Auto data extraction for opaque walls,
- Script A2: Auto data extraction for glazing (windows),
- Script B: Data extraction for Revit library, and
- Script C: Data push back.

Script A1 and A2 are developed to extract building envelope data related to OTTV requirement. This data is mainly related to the exterior opaque walls and exterior windows of the project. It covers the data request of OTTV equation, which includes windows and walls U-value, windows and walls area, windows and walls orientation, shading coefficients (SC1, SC2), wall pitch angle, WWR (%), and the solar correction factor (OF). Script B is designed to extract the library data of Autodesk Revit (Walls and windows elements). This data will be used to create the new design alternative during the optimization process in MATLAB. In order to get accurate optimization results, the Revit library should contain different types of walls, windows of different sizes, and a variety of thermal material properties. More importantly, the cost parameter of each element in the library and the case study model should be assigned with its relevant value. This is because the RBIM framework is basically developed to optimized building envelop OTTV against the cost of investment (cost of the new design option). Script C has been developed to push back the data of the optimum design option (MATLAB output) to the BIM model in Autodesk Revit. This will automatically update the BIM model with the new solution, which consists of the best design alternative in terms of OTTV performance against cost investment. It is important to mention that the scope of this paper includes only the development of script A1, A2, and C.

13.4 Data Extraction

As shown in Fig. 13.2, the process of extracting building envelope's data goes through multiple rules and functions, which have been created and embedded within two Dynamo scripts (A1 and A2) based on OTTV requirements. It is important to note that OTTV is applicable only to air-conditioned buildings with an air-conditioned space ≥ 1000 m^2. Since one building could contain both air-conditioned (AC) spaces and naturally ventilated (NV) spaces (i.e. store, toilet, corridors), in the meantime, it is a crucial step to automate the detection of only the building envelop elements that enclose air-conditioned spaces during the assessment process of OTTV. This has been implemented by adding a new shared parameter called "Is NV space" to the rooms' category within the Revit template. Then, the Dynamo script was developed to filter only the rooms with an unchecked "Is NV space" parameter (see Fig. 13.2a–b–c). The next part (c–d–e) of the script was

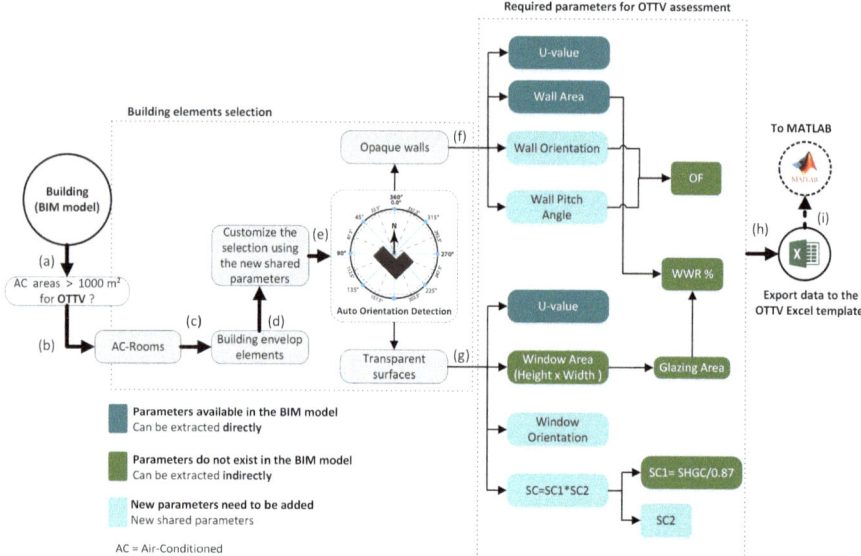

Fig. 13.2 Data extraction logic flow for OTTV requirement (wall and window)

created in order to select all the wall elements of the air-conditioned rooms. Then, only the exterior walls were filtered from this selection, by using the function parameter of the wall elements as a Boolean parameter (Exterior $=1$, interior $= 0$). Moreover, this part included the addition of some new shared parameters to assist the user in manually excluding the walls that do not belong to any room in the project (i.e. walls in the staircases).

The orientation of building elevation has a direct impact of the final measure of OTTV. This occurs because several coefficients in the OTTV equation, such as the solar orientation factor (OF) and shading coefficient of shading devices (SC2) vary according to the orientation of each building elevation. Thus, it is crucial, in the preliminary stages, to automate the detection of building elements orientation of each elevation of the building envelope in order to automate the whole assessment of OTTV. In Autodesk Revit, each wall has a vector perpendicular on its exterior surface. This vector is called "the normal", it is used as a key element in the automation of wall orientation in this study. As shown in the central part of Fig. 14.2 (part; e–g, e–f), script A1 detects the normal vectors of the envelope walls in the project. Then, the normal vectors coordinates (x, y) are assigned with conditions related to the angles defined by the cardinal and intermediate directions (N, E, S, W, NE, NW, SE, and SW) of the design environment of Autodesk Revit. For example, West direction is defined by a normal vector which belongs to an angle that ensures the following condition; $67.5° \geq$ angle $\geq 112.5°$. Therefore, when executing script A1, each wall element is automatically assigned its specific orientation parameter; walls with the same parameter value (same orientation, i.e. North) are grouped together to defined the gross area of each elevation. On the other hand, one of the fundamental modelling rules related to window modelling in Autodesk Revit is that each window should be hosted in a wall element. Based on this rule, data extraction script for windows (A2) has been developed to assign the orientation parameter of each window based on the orientation parameter of its host wall (assigned using the previous process). For instance, if a window is hosted in a wall oriented north, automatically the orientation of the window will be assigned as north as well, and vice versa. Next, in the f–h and g–h parts (see Fig. 14.2), all the relevant walls and windows are categorized according to their orientation parameter, and all the required parameters are extracted. Finally, the list of the extracted data is exported to an Excel template (h–i) which in turn will be used as an input in MATLAB to perform the optimization of OTTV against the cost of investment (i).

13.5 Data Push Back

Data pushback is another key step in the RBIM framework. It opposes the data extraction process discussed previously. In contrast, data pushback consists of feeding the BIM model with the optimized data that will ensure an optimized building envelope design in term of OTTV performance and cost of investment. As shown in Fig. 13.3, the data output of MATLAB optimization is exported to an

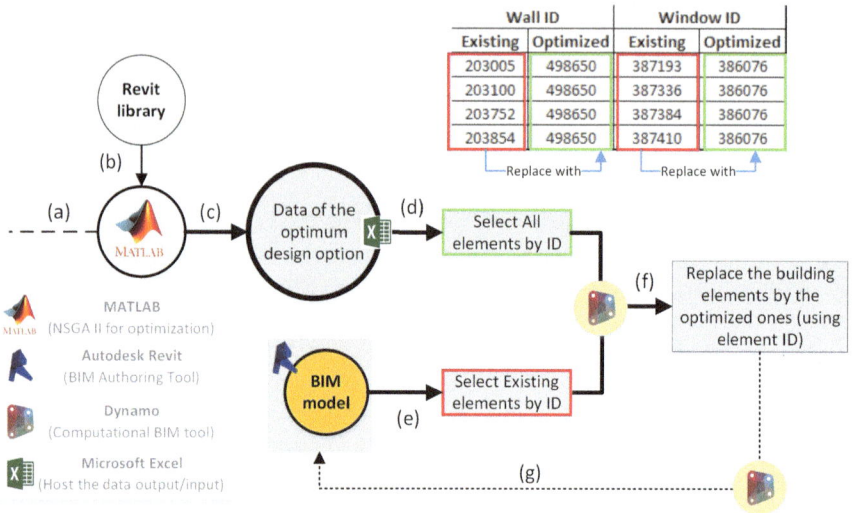

Fig. 13.3 Logic workflow of the data push-back script

Excel template (a–c). This data is pushed back to the BIM model using Dynamo Script C, which was developed for that purpose. Script C will select the optimized design alternatives of walls and windows using their ID parameter (d–f). This selection is performed by reading the available data within the green columns of the Excel file (MATLAB output), see Fig. 13.3. It is worthy to mention that all the design variables generated by MATLAB are linked to the library of Autodesk Revit (b). In the meantime, script C will select the existing walls and windows of the building envelope using their ID parameter (e–f). Finally, after reading both data of the existing and optimised building envelop elements, script C (f–g) will replace each existing wall/window element with the optimized option based on their ID parameter. As result, the model will be updated automatically to the optimum design configuration based on MATLAB data output.

13.6   Case Study

The whole RBIM framework applicability is tested and demonstrated through case study building (BIM model), though in this paper the validation will focus only on the implementation of Dynamo scripts (A1, A2, and A3).

The selected case study consists of an existing office building of four levels with a gross floor area of 7500 m². This building is located in the Faculty of Built Environment and Surveying, Universiti Teknologi Malaysia (UTM). The BIM model of this building was modelled using Autodesk Revit based on CAD drawings. As shown in Fig. 13.4, the building has four facades (NW.NE.SW.SE); each façade

Fig. 13.4 Case study building for the RBIM framework testing; (**a**): photo, (**b**): BIM model

consists of opaque walls built using single brick material and transparent surfaces that consists of windows with single glazing.

The RBIM was tested on the case study building through three optimization scenarios, as follows: optimization which considers only window elements (scenarios one), optimization for both Wall and window elements (scenarios two), and optimization using an additional constraint for window area (scenarios three). All the different scenarios have shown a significant reduction in OTTV, however, the cost of investment varied from case to another. In this paper, only the second scenario and the North-East elevation of the case study are taken as an example.

The optimization process starts by running the data extraction scripts (A1 and A2), which takes only 35 s to extract all the required data. The extracted data consist of two lists of data related walls and windows. These lists cannot be presented in this paper because of their length. At this stage, to prevent including wrong walls, any walls included in OTTV optimization are assigned a specific colour, according to the elevation orientation in the BIM model (i.e. orange colour for North-East orientation). Next, the extracted data is exported to MATLAB to run the multi-objective optimization. The optimization results have shown a reduction in OTTV value from 49.47 to 25.96 W/m^2 (19.91% of reduction) with an additional cost of investment related to building materials that equals to RM 92832 (the cost of the initial building envelop design is RM 101291.95). The data of the optimized design option is then pushed back to the BIM model using script C. At this point, the BIM model of the initial building is updated automatically with the optimized data generated through MATLAB. It is worth mentioning that the time of data push back took around 3 min, however, this period may vary dependently on the BIM model size and its design complexity. Figure 13.5 illustrates the status of the BIM model before and after optimization. It can be seen that the element types of building envelop have been changed to new types, which have more thermal performing materials. For example, windows with single glazing (U-value = 6.7 w/m^2.k) have been changed to new window types with double glazing (U-value = 3.12 w/m^2.k). Meanwhile, single brick walls have been changed to double brick walls with additional insolation layers. As mentioned previously, these optimization results are highly dependent on the possible design options available in the library, and the constraint under of the second scenario.

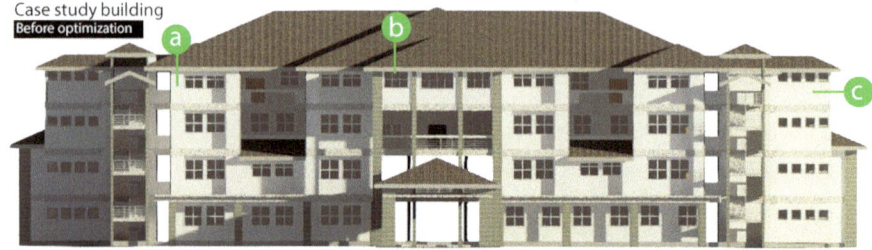

OTTV = 49.47 W/m2; Material cost RM 101291.95

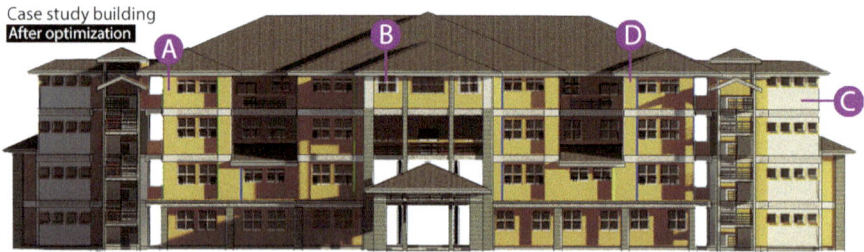

OTTV = 25.96 W/m2; Cost of investment RM 92832

Fig. 13.5 Case study building (North-East elevation) before and after optimization
Before optimization:
   (a): all wall types 12 cm single brick with 2*15 cm plaster
   (b): all windows with single glazing (U-value = 6.7 w/m^2.k)
   (c): walls that enclose naturally ventilated spaces (i.e. WC) have been excluded before running script A1
After optimization:
   (A): all walls: 2*12 cm double brick, 10 cm air cavity with 2*15 cm plaster
   (B): most of the windows with double glazing (U-value = 3.12 w/m^2.k)
   (C): the excluded walls remained with their initial color
   (D): All envelop walls included in the optimization are assigned with the new color (for visualization purpose)

13.7 Discussion and Conclusion

This paper has presented the development of a computational BIM-based workflow for data management using visual §programming (Dynamo). This workflow has been implemented within a research project called RBIM, which seeks the optimization of existing building envelope thermal transfer value (OTTV) against the cost of investment. Two main Dynamo scripts, able to extract/push back data automatically from/to the BIM model, were developed for this purpose. The results of the case study have shown a great potential for using Dynamo to automate data management (extraction and push back) for building envelope optimization. The data extraction process using script A1 and A2 took only 35 s. On the other hand, the data push back of the optimized data to the BIM model took 3 min. More importantly, heat transfer of building envelope (North-East elevation) has been reduced by 19.91% with an additional cost equals that to RM92832.

Despite these achievements, this study is only a preliminary investigation of the usage of Dynamo to automate data management within the RBIM project. Further investigation on how the transparent curtain walls data can be integrated to the developed scripts is required. In addition, an assessment model for shading coefficient of shading devices (SC2) must be integrated. Finally, in order to make the developed scripts more user-friendly, exploring how the developed scripts can be developed further as a Revit plug-in is recommended. It is believed that the RBIM framework and data management scripts in this study can be adapted to optimize other sustainability criteria, such as Co2 emissions and building material selection.

Acknowledgement The authors would like to acknowledge the research funding by Universiti Teknologi Malaysia (UTM), Ministry of Higher Education, Malaysia (MOHE) through the Research University Grant (GUP), project no. 13H40, titled "Retrofitting Building Information Modelling (RBIM) for Sustainable Buildings".

References

Alwan Z (2016) BIM performance framework for the maintenance and refurbishment of housing stock. Struct Surv 34(3):242–255. https://doi.org/10.1108/SS-03-2015-0018

Andrea, Vannini (2015) Andreaarch | Architecture & Computation. Available at: https://andreaarch. wordpress.com/. Accessed 17 Feb 2017

Asl MR et al (2011) Optimo : a BIM-based multi-objective optimization tool utilizing visual programming for high performance. In: Proceedings of the 33rd International Conference on Education and Research in Computer Aided Architectural Design in Europe, pp 1–10

Azhar S et al (2011) Building information modeling for sustainable design and LEED ® rating analysis. Autom Constr 20(2):217–224. https://doi.org/10.1016/j.autcon.2010.09.019

Biswas T, Wang T-H, Krishnamurti R (2013) From design to pre-certification using building information modeling. J Green Build 8(1):151–176. https://doi.org/10.3992/jgb.8.1.151

Building and Construction Authority (BCA) (2008) Code on envelope thermal performance for buildings. BCA. Available at: http://www.bca.gov.sg/PerformanceBased/others/RETV.pdf

Department of Standards Malaysia (2014) Malaysian standard 1525:2014 – energy efficiency and use of renewable energy for non-residential buildings – code of practice (2nd revision)

Ilhan B, Yaman H (2016) Green building assessment tool (GBAT) for integrated BIM-based design decisions. Autom Constr.. Elsevier B.V. 70:26–37. https://doi.org/10.1016/j.autcon.2016.05. 001

Inhabitgroup (2016) ETTV Façade designer released to Singapore market – Inhabit Group. Available at: http://inhabitgroup.com/ettv-facade-designer-released-to-singapore-market-2/. Accessed 10 Feb 2017

Jalaei F, Jrade A (2015) Integrating building information modeling (BIM) and LEED system at the conceptual design stage of sustainable buildings. Sustain Cities Soc.. Elsevier B.V. 18:95–107. https://doi.org/10.1016/j.scs.2015.06.007

Jrade A, Jalaei F (2013) Integrating building information modelling with sustainability to design building projects at the conceptual stage. Build Simul 6(4):429–444. https://doi.org/10.1007/ s12273-013-0120-0

Lim Y et al (2016) Building information modelling for building energy efficiency evaluation. In: Ace, pp 42–48

Makris M et al (2013) Informing design through parametric integrated structural simulation. In eCAADe 2013: computation and performance–Proceedings of the 31st International Conference on Education and research in Computer Aided Architectural Design in Europe, 1, pp 69–77

Seghier TE et al (2017) Building envelope thermal performance assessment using visual programming and BIM, based on ETTV requirement of green mark and GreenRE. Int J Built Enviro Sustain 4(3):227–235. https://doi.org/10.11113/ijbes.v4.n3.216

Seghier TE et al (2018) Integration models of building information modelling and green building rating systems: a review. Adv Sci Lett 24(6):4121–4125

Wong JK-W, Kuan K-L (2014) Implementing "BEAM plus" for BIM-based sustainability analysis. Autom Constr.. Elsevier B.V. 44:163–175. https://doi.org/10.1016/j.autcon.2014.04.003

Wu W (2010) Integrating building information modeling and green building certification: the BIM – LEED application model development, Vasa. Available at: http://medcontent.metapress.com/index/A65RM03P4874243N.pdf

Part V
Urban Ecology

What are spatial ways to integrate ecology and nature in urban environments. How can a city be designed in which the natural quality can be realized.

Chapter 14
Australia's Urban Biodiversity: How Is Adaptive Governance Influencing Land-Use Policy?

Hugh Stanford and Judy Bush

Abstract Green space in cities is valuable in improving the quality of life of urban residents. This occurs through the provision of ecosystem services within areas of open space. Biodiversity plays an important role in enhancing the resilience of urban ecosystems, ensuring the continued supply of ecosystem services.

Urban changes in Australia are expected to result in significant changes in urban biodiversity. Climate change is one factor considered likely to exacerbate many existing urban threats to biodiversity. Where cities are located in areas of high conservation significance, such as Melbourne's temperate grassland communities, these rare and threatened habitats are particularly vulnerable.

Biodiversity planning policy in Australia largely focuses on the retention of existing areas of high native biodiversity. While this is important, a sole reliance on this approach likely results in inflexibility when dealing with uncertain urban changes. This paper questions whether land-use planning policy in Australia has due consideration in addressing the uncertain threat posed by urban changes such as climate change.

The paper investigates the current biodiversity land-use planning policy in Melbourne and Sydney through the lens of contemporary sustainability and resilience theory. The paper uses an analytical framework to assess land-use planning policies' adoption of adaptive governance elements. We conclude that, while some policy measures provide adaptability in the face of uncertainty, more needs to be done to better protect Australia's unique and valuable biodiversity from the unpredictability of urban threats such as climate change.

Keywords Biodiversity · Land-use planning · Policy · Adaptive governance · Climate change · Resilience

H. Stanford (✉) · J. Bush
ICON Science Research Group, Centre for Urban Research, RMIT University, Melbourne, VIC, Australia

© Springer Nature Switzerland AG 2020
R. Roggema, A. Roggema (eds.), *Smart and Sustainable Cities and Buildings*,
https://doi.org/10.1007/978-3-030-37635-2_14

14.1 Introduction and Background

In the face of urban change, the resilience of urban ecosystems underpins the ongoing provision of biodiversity habitats, as well as quality of life for urban residents. Adaptive governance provides a policy framework that is focused on increasing resilience in the face of uncertainties such as climate change. This research investigates to what extent recent urban biodiversity land-use planning within Australia has incorporated an adaptive governance approach. The paper investigates this question by constructing an assessment framework derived from the key elements of adaptive governance, as outlined in the literature, and applying this framework to the assessment of current land-use planning policies in Melbourne and Sydney.

14.1.1 Biodiversity and Ecosystem Services

Throughout history, humanity's success has depended on the health of the environments in which we live. The management of biodiversity, the abundance and variety of genetic traits, species composition, and functional groups within an ecosystem (Müller et al. 2013) are important factors in the ongoing provision of ecosystem services, the benefits humans receive from the environment (Bolund and Hunhammar 1999). These services, whether providing resources, regulating and maintaining ecosystems, or providing cultural opportunities (Harrison et al. 2014; Langemeyer and Gomez-Baggethun 2018) are made more efficient, productive, and stable through enhancing biodiversity, in-turn supporting the humans who rely on them (Cardinale et al. 2012).

As the world increases in population and becomes more urbanised, it is likely that humanity's reliance on ecosystem services will continue to increase (Seto et al. 2012; Bolund and Hunhammar 1999; Wiedmann et al. 2015). Unfortunately, urbanisation has, to date, occurred at the expense of biodiversity, and consequently, the function of ecosystems globally (Seto et al. 2012; Hoornweg et al. 2016). For example, this has been seen in Australia, where Melbourne's urban expansion has resulted in significant loss of temperate grassland communities and the species that rely on them (Ives et al. 2013). However, cities are not 'un- natural' places, with urban environments providing many opportunities for the conservation and enhancement of biodiversity and ecosystem-services (Ives et al. 2016). To ensure an appropriate quality of life for the world's growing urban population, it is increasingly important that planning policy prioritises the provision of ecosystem services through policy that maximises biodiversity within cities (Guo et al. 2010).

14.1.2 The Influence of Urban Changes and Uncertainty

Urban biodiversity faces significant challenges, due to urban changes such as expansion and fragmentation processes (Borgström et al. 2006), and environmental changes. Climate change is one such change which is anticipated to have severe and uncertain impacts on urban biodiversity and the provision of ecosystem services (Coffey and Wescott 2010). This is largely through uncertain developments, such as increasing likelihood of extreme heat events and significant uncertainty regarding future rainfall patterns (Reisinger et al. 2014; Roggema 2016), where the conditions required to retain healthy ecosystems are being altered with a changing climate (Root et al. 2003; Kendal 2011). This has been compounded with a history of inappropriate urban growth and habitat loss, resulting in increasingly fragile urban ecosystems, as seen by the higher proportion of threatened species existing in cities compared to non-urban environments (Ives et al. 2013).

Biodiversity both provides resilience to the uncertainty of climate change through the increased stability of ecosystem services (Cardinale et al. 2012), while also becoming increasingly vulnerable to it (Root et al. 2003). As such, urban biodiversity both protects and needs to be protected in the face of a changing climate. In a climate change future, where social-ecological systems are expected to become less predictable (Roggema 2016), retaining biodiversity for the ongoing provision of ecosystem services is vital to assist in human and non-human adaptation (Green et al. 2016b).

14.1.3 Adaptive Governance

When managing the preservation and enhancement of urban biodiversity in an uncertain future, adaptive governance has been proposed as a policy framework for improving and adapting policy mechanisms to address uncertainty and changing conditions (Olsson et al. 2006). As it is impossible to fully predict future social-ecological changes due to the constant flux of social and ecological variables (Tyre and Michaels 2011), dynamic adaptive governance structures are designed to allow for and utilise disturbances to build knowledge and capacity for responding to future uncertainty (Roggema 2016; Folke et al. 2005). As such, this flexible approach has been shown to result in improved environmental outcomes for the management of biodiversity and the provision of ecosystem services in an unpredictable system (Kenward et al. 2011; Olsson et al. 2006; Gunderson et al. 2016; Chaffin et al. 2016).

A review of the literature has identified the core elements of adaptive governance. Firstly, adaptive governance encourages learning through innovation and experimentation, requiring the ongoing iteration of policy through monitoring and feedback cycles (Karkkainen 2008; Folke et al. 2005; Green et al. 2016a; Tyre and Michaels 2011). Secondly, it encourages a nested governance hierarchy, with leadership coming from central government bodies that provide vision and meaning for action amongst the wider stakeholders. This in turn empowers local and non-government bodies to undertake management and monitoring activities with a level of autonomy (Folke et al. 2005; Green et al. 2016a; Evans 2011; Tyre and Michaels 2011; Roggema 2017). Additionally, adaptive governance requires an increased involvement of stakeholders in creating policy, requiring significant collaboration by non-government bodies (Green et al. 2016a; Heller and Zavaleta 2009; Roggema 2017). Thirdly, adaptive governance relies on institutional networks to be established, connecting all bodies involved in the governance of biodiversity assets, from national government bodies through to local and non-government entities. This is achieved through the use of bridging organisations, where an impartial actor, not directly related to implementing biodiversity policy, acts as a facilitator between different governance bodies (Green et al. 2015; Folke et al. 2005), sharing knowledge evenly throughout the network to build capacity of all governance bodies (Green et al. 2015, 2016a; Folke et al. 2005). The existence of a modest overlap in the responsibilities of governance bodies prevents management duties being unfulfilled and allows for the shifting of governance responsibilities and spreading of risk, should alternative actions be required (Folke et al. 2005).

14.1.4   The Australian and International Context

While methods of biodiversity land-use planning in Australia have, to date, largely not followed an adaptive governance framework, biodiversity management globally is increasingly adopting alternative planning approaches. In Australia, biodiversity policies have largely focused on the protection of native and threatened flora and fauna within a defined protection area, aiming to retain existing environmental conditions and preventing further degradation and loss of native species (Borgström 2018; Erixon et al. 2013; Barr et al. 2016). This demonstrates a static governance approach, which seeks to fix a disturbance event and return a system to its pre-disturbed state (Roggema 2017). This approach fails to build resilience in the face of a dynamic environmental context, where the underlying ecological conditions, required for the persistence of an ecosystem, change (Folke et al. 2004). While a static governance approach has dominated biodiversity policy-making approaches, there is a growing trend of alternative planning approaches, such as the rise of eco-urbanism and nature-based solutions (Duvall et al. 2017). There is growing research interest as to the extent that biodiversity planning within Australia has adopted these changes (Byrne et al. 2014). As such, this paper seeks to determine whether recent biodiversity land-use planning policy within Australia has incorporated a more dynamic approach such as the adaptive governance approach.

14.2   Method

We used a case study approach to assess how current biodiversity land-use planning within Australia incorporates adaptive governance elements. We selected two policy case-studies: The Greater Sydney Commission's Regional and District Plans (GSC), and the Melbourne Strategic Assessment Program (MSA). The GSC and MSA were selected as they are significant planning policies which guide land-use planning and urban development within Sydney and Melbourne, the two fastest growing cities in Australia (ABS 2018). The policies provide the overarching definitions for the structure and design of management activities. The substantial growth and urbanisation within these cities is likely to significantly impact biodiversity outcomes (Ives et al. 2013; McDonald et al. 2013). As such, evaluating the measures taken by these cities to address these issues was considered valuable.

The documents analysed for the GSC case study included the Regional Plan, which set the overarching policy direction, and the four District Plans, which applied the policy direction to locally relevant information (GSC 2018). The MSA policy documents included the Program Report, which outlines the MSA and details how it will be implemented (DPCD 2009), and the Biodiversity Conservation Strategy (BCS), which provides further detail on the management of important biodiversity areas (DEPI 2013). While there are many documents within the MSA, these two most thoroughly outline the objectives and actions of the program.

To investigate how the case study policies utilise adaptive governance in responding to future climate uncertainties, an assessment framework was developed. It brings together the key elements of adaptive governance identified in the previous section (shown in Table 14.1). The analysis did not focus on each policy's biodiversity planning merit, instead focusing on how the policies address adaptive governance.

We used the adaptive governance analysis framework to assess elements that directly related to the biodiversity policy domain of each document. Content was selected due to its use of biodiversity-related terms (biodiversity, ecosystem, habitat, flora, fauna). Policy areas that weren't directly related to biodiversity planning were outside the scope of this analysis. For example, the provision of public open space for recreational purposes or the social impacts of climate change detailed in the GSC plans were excluded from the assessment, as they were considered insufficiently linked to biodiversity planning.

14.3   Results

This section presents the findings of our analysis of how adaptive governance is addressed within the MSA and the GSC policies. The three key elements of the analysis framework are presented separately. The extent to which the two case studies address each of the elements are presented, and findings summarised in the

224   H. Stanford and J. Bush

Table 14.1 Policy analysis framework – adaptive governance

1. A focus on learning through experimentation and feeding lessons back into policy	(a) Experimentation and innovation encouraged in policy (b) Models and initial understandings are treated as provisional and to be replaced by learned knowledge (c) Provision for careful monitoring included (d) Findings feed back into policy

2. A reliance on non-government and local government actors, and networks to undertake implementation	(a) Hierarchy of governance responsibility:
	  (i) State provides vision and meaning for action
	  (ii) On-ground management and monitoring undertaken by local government
	(b) Policy builds capacity for management and monitoring to be undertaken by non-government bodies
	(c) Provision for local and non-government bodies to act autonomously in designing implementation activities
	(d) All stakeholders are significantly involved in creation and review of policy

3. Vertically and horizontally interconnected systems of governance	(a) Use of bridging organisations (b) Mechanisms for sharing knowledge between all levels of governance (c) Modest overlap in policy and responsibility

Adapted from Folke et al. (2005) and others as noted in previous section

following tables. The results will then be further discussed in the following section, to explore how biodiversity land-use planning within each policy reflects or adopts adaptive governance principles.

14.3.1 Element 1: Learning Through Experimentation and Feeding Learning Back into Policy

Both policies demonstrated elements of experimentation, though these were limited to revising existing actions (Table 14.2). The MSA's Adaptive Management component provided an elaborate process of experimentation, learning, and feedback, leading to changes in the way conservation outcomes were achieved. The GSC policy included strategic content relating to monitoring and feedback; however, this was not followed with operational content, such as actions indicating how policy will be refined. The elements of experimentation did not extend to iterations of the policy itself, with objectives and conservation outcomes largely unable to be modified. One exception of this is within the MSA program where conservations outcomes were able to be amended if "...the outcomes are agreed to be technically improbable" (pg. 47 DPCD 2009).

Table 14.2 Element 1 results

Framework section	Melbourne strategic assessment	Greater Sydney Commission
1.a – Experimentation and innovation encouraged in policy?	Adaptive management process outlined in the policy addressed through the inclusion of progressive iteration and experimentation.	Establishment of a Green Grid includes elements of progressive iteration.
1.b – Models and initial understandings are treated as provisional and to be replaced by learned knowledge?	Multiple examples of incomplete information, at times utilising formal models, being replaced by learned knowledge.	Objective 40 of the Regional Plan states: "Plans refined by monitoring and reporting" (GSC 2018). Actions relating to the refinement of the plans were not found.
1.c – Provision for careful monitoring included?	Monitoring processes are widely detailed throughout both MSA policy documents. Monitoring is divided into two parts. Firstly, monitoring is intended to "ensure compliance with the endorsed program" (DPCD 2009). Secondly, adaptive management component of the policy is intended to "... monitor whether the outcomes envisaged for each matter of national environmental significance is being effectively achieved..." in the face of "... changing circumstances and procedures and/or new information relating to matters of national environmental significance" (DPCD 2009).	Monitoring is often referenced within the GSC policy with the principal intention to identify abnormalities in the implementation of the policy and to assist future decision making to ensure the initial objectives and actions of the policy are achieved. Actions relating to this include the development of performance indicators relevant to each Council's context to assess the implementation of the policy.
1.d – Findings feed back into policy?	The MSA's focus on adaptive management creates many avenues to feed learnings back into policy. Firstly, through changes to management practices to achieve the stated conservation outcomes. Secondly, to amend these outcomes if they are deemed "...not achieved or are unlikely to be achieved" (DPCD 2009).	Learnings not fed back into policy but intended to be used to influence future decision making to achieve policy objectives. Actions relating to the delivery of the Green Grid are intended to be progressively refined from learnings gained through the delivery of past Green Grid projects.
2.a – Hierarchy of governance responsibility (i) State provides vision and meaning for action	(i) The Program Report, through outlining outcomes and activities to achieve outcomes, provides a detailed	(i) The GSC policies provide significant vision and meaning for action to be undertaken.

(continued)

Table 14.2 (continued)

Framework section	Melbourne strategic assessment	Greater Sydney Commission
(ii) Operationalization of actions undertaken by local government	vision and meaning for action. (ii) Local government bodies are included in a marginal number of conservation actions compared to state government bodies.	(ii) Local government plays a significant role in undertaking actions outlined in the policy. Local council is the lead organisation for certain actions such as the delivery of the Green Grid.
2.b – Policy builds capacity for operationalization of actions to be undertaken by non-government bodies	Minimal involvement of non-government bodies in undertaking actions outlined in the policy. Land-owners wanting to maintain land ownership after urban development need to enter into an agreement with the Victorian government requiring them to undertaken management activities.	Minimal involvement of non-government bodies in undertaking actions outlined in policy. Regional Plan sets a vision for incorporating non-government action to build ecological resilience. This was largely absent in the policy's actions. A single action strategically incentivises land-owners within rural areas to protect and manage biodiversity assets through the offset market.
2.c – Provision for local and non-government bodies to act autonomously	Local government has limited autonomy. Most actions undertaken by local government involve implementing plans and statutory requirements prepared by state government bodies.	Local Government is provided autonomy as the lead actor in delivering the Green Grid action. Local Governments also have autonomy in reviewing local environmental plans to incorporate the actions outlined in the District Plans.
3.a – Use of bridging organisations	No evidence of bridging organisations.	No evidence of bridging organisations.
3.b – Mechanisms for sharing knowledge between all levels of governance	Monitoring and reporting obligations were the only mechanism for sharing knowledge between different governance bodies.	Monitoring and reporting obligations were the only mechanisms for sharing knowledge between different governance bodies.
3.c – Modest overlap in policy and responsibility	Federal government has delegated authority to the Victorian government. Victorian and local government bodies have some overlap through collaborating on preparation of documents associated with greenfield planning. No overlap between government and non-government bodies.	The delivery of the Green Grid demonstrates overlap between State and local government bodies

14.3.2   Element 2: Non-Government and Local Government Actors and Networks for Implementation

The policies demonstrated marginal involvement of non-government bodies. The GSC policy provided several examples of building capacity and autonomy for local government bodies to undertake actions outlined in the policy, predominantly within the Green Grid project where local governments were identified as the lead agency to operationalise the project (Table 14.2). The MSA program provided little capacity and autonomy to local governments. Both policies demonstrated a submission and feedback process in-line with traditional consultation practices (Table 14.2). There was no evidence found of a more robust collaborative involvement of stakeholders during the creation and review of the policies.

14.3.3   Element 3: Vertically and Horizontally Interconnected Systems of Governance

Both the MSA and GSC demonstrated limited vertical and horizontal interconnections between governance bodies. The GSC demonstrated slightly higher levels of interconnectedness through the Green Grid implementation, a series of interconnected but independent projects, able to be implemented by all levels of government (Table 14.2). While the GSC demonstrates some capacity to act as a bridging organisation between different levels of government, this is heavily constrained, as the GSC exercises authority over local government authorities through its role in overseeing their planning scheme amendments.

14.4   Discussion

The analysis of the two strategies demonstrates limited adaptive governance capacity. While both strategies included some elements, there is significant potential to increase the involvement of local and non-government bodies in decision making, and increase incorporation of feedback loops and policy learning to improve implementation and outcomes. The results of analysis highlight key opportunities for increasing adaptive governance in biodiversity planning approaches. The discussion concludes by reflecting on the analysis framework and its efficacy for policy analysis.

14.4.1   Learning and Feedbacks

Neither strategy demonstrated substantial elements of policy experimentation to influence key policy objectives, limiting the adaptability of both policies to future climate uncertainties. Within the MSA and GSC, capacity for iteration was only found to extend to actions already under the policies' objectives. This is either through rectifying mistargeted policy application, or, as seen in the MSA, undertaking experimentation of management actions to better achieve the stated objectives. The MSA's approach exhibits some adaptive governance elements, which enhance the policy's ability to dynamically respond to uncertain disturbances (Karkkainen 2008; Folke et al. 2005). However, as experimentation only applies to modifying management actions within rigid objectives ('Feedback loop A', Fig. 14.1) rather than revising the underlying policy objectives ('Feedback loop B', Fig. 14.1), this possibly limits the use of adaptive and innovative policy solutions in the face of future and uncertain disturbances. Even when the MSA includes details regarding the ability to amend core objectives, this only applies to objectives that are considered unachieved or unachievable (DPCD 2009). This approach focuses on reducing failure instead of looking at opportunities for innovation and improvement through establishing new objectives or amending objectives in a lateral direction (Green et al. 2016a). As such, the ability for either policy to fully respond to a significant and unpredictable disturbance is reduced (Tyre and Michaels 2011; Green et al. 2016a).

14.4.2   Reliance on Local and Non-Government Actors

Under-reliance on local and non-government bodies demonstrates a missed opportunity for both the GSC and MSA policies, by not incorporating a dynamic biodiversity land-use planning approach to the extent possible. While both policies demonstrate a significant level of strategic content, creating a vision and meaning for action critical in guiding other governance bodies (Folke et al. 2005), the operational components of both policies demonstrate an overreliance on state agencies either directly undertaking actions, or heavily influencing actions to be undertaken by local bodies. Neither policy takes a robust approach towards collaborating

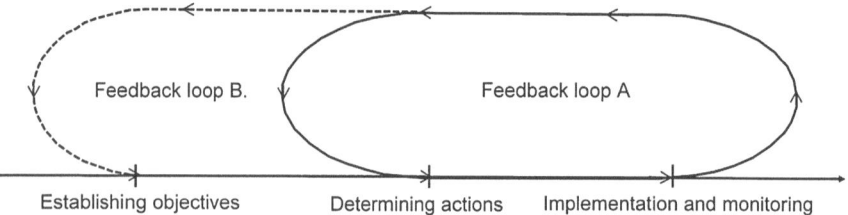

Fig. 14.1 Policy feedback loops

or building capacity with non-government groups to undertake actions or create and review the policy. Consultation processes are limited to standard submission and feedback processes and do not sufficiently incorporate more meaningful forms of engagement (Green et al. 2015). These findings are consistent with research on public participation with climate change adaptation policy (Sarzynski 2015; Green et al. 2015). These factors limit ownership and buy-in from local communities, potentially impacting long-term political and public support, while restricting the flow of information critical for generating improved ideas and decision making (Green et al. 2015; Bodin and Crona 2009; Tyre and Michaels 2011).

14.4.3   Interconnected Governance

Limited vertical and horizontal interconnectedness reduces each policy's adaptability in the face of uncertainty. The omission of an external bridging organisation, independent from the implementation of the policy, limits the capacity for coordination and information flow between different levels of governance. This in turn impacts the generation of new and innovative ideas to enhance decision making (Green et al. 2015). As there are no mechanisms for knowledge sharing other than highly vertical and structured monitoring and reporting processes, elaborate horizontal and vertical interconnections are not able to develop, impacting the ability for the policies to respond to uncertain disturbances (Green et al. 2015, 2016a; ; Folke et al. 2005).

The MSA demonstrates policy and responsibility overlap through delegated authority from federal government to the Department of Planning and Community Development (now DELWP). While there is collaboration with local government and external state agencies, this is limited in scope and characterised by an unbalanced and centrally concentrated power relationship (Fig. 14.2a). The resulting lack of horizontal overlap between responsibilities of governance bodies as defined

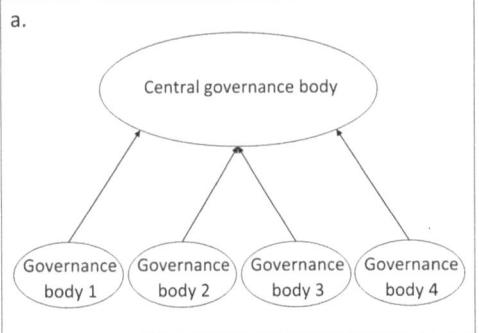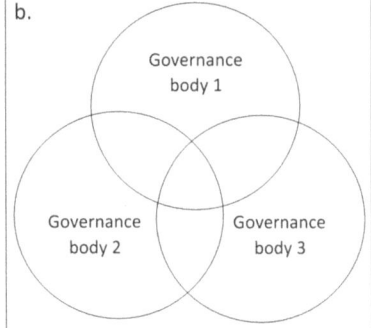

Fig. 14.2 Responsibility overlap and power relationships: the MSA (**a**) and GSC's Green Grid program (**b**)

within the policy, likely results in a reduced ability for governance structures to re-organise and respond meaningfully in the face of short-term and long-term disturbances (Folke et al. 2005). The Green Grid program within the GSC policy demonstrated an alternative model of integration and is discussed further below.

14.4.4 The Green Grid: Innovation in Governance and Green Space?

The GSC's Green Grid program demonstrates the greatest potential for adaptive governance within the two policies, including the possibility of improved vertical and horizontal interconnectedness and greater reliance of local and non-government bodies. The delivery of the Green Grid demonstrates overlap between responsibilities and power relationships of the multiple governance bodies (Fig. 14.2b), as individual groups are tasked with implementing independent projects which together contribute to a broader network of green spaces. This arrangement builds capacity and provides autonomy for both state and local bodies, contributing to the Green Grid, in turn generating buy-in from local government and opening up opportunities to access the institutional knowledge held within local Councils (Green et al. 2016a; Roggema 2017). The overlap enhances the ability for individual bodies to change their field of focus to absorb unexpected disturbances, while ensuring the broader green network is maintained through the implementations of different government bodies (Folke et al. 2005). Together these factors may underpin an enhanced ability to respond to uncertainty.

While the Green Grid program includes multiple beneficial adaptive governance attributes, there are areas where its adaptive capacity can be further enhanced. Greater involvement of non- government bodies would further enhance on-going public support for biodiversity management (Green et al. 2015). This could be achieved through the inclusion of incentive-based policy tools for private land, an important but under-utilised land tenure type within the green network (Goddard et al. 2010; Miller 2008; Feinberg et al. 2015). Additionally, including provision for knowledge sharing through bridging organisations would enhance the collective knowledge pool, improving decision making (Green et al. 2015). Finally, including experimentation of the policy management actions (as demonstrated within the MSA) would allow feedback learning and policy iterations to better adapt to unforeseen disturbances (Folke et al. 2005; Green et al. 2016a).

14.4.5 Future Research Directions in Adaptive Governance

The method used to assess the MSA and GSC was an appropriate way to assess each policies' consistency with adaptive governance theory. We consider the analysis framework suitable for use in a broad range of policy assessments outside the field of biodiversity planning and climate change. The methodology used was best suited for analysing a small number of policies in a high level of detail. To analyse and compare a broader number and range of policies, a scoring mechanism could be incorporated, quantifying and standardising the assessment results. Incorporating additional elements within the assessment framework, such as key policy attributes: strategic, tactical, operational, and reflexive elements (Bush and Hes 2018; Frantzeskaki and Tilie 2014), would provide greater insights into the transformative areas of each policy and where adaptive governance is most significantly reflected. This would allow for a wider scale assessment of biodiversity land-use policies to determine whether this research's findings are reflected in other Australian jurisdictions' policies.

Additionally, given biodiversity's central role in the provision of ecosystem services, it is vital that policy mechanisms are developed to enhance urban biodiversity. As such, further work should be undertaken to investigate various policy mechanisms and their potential impact on urban biodiversity outcomes.

This research has focused on policy content elements and is predicated on the premise that, in practice, adaptive governance provides a better capacity to respond to uncertainty. While there has been some research in this field (for example Kenward et al. 2011), this area requires further investigation into the practicality of adopting a more dynamic policy framework in a policy making context (Chaffin et al. 2016; Karpouzoglou et al. 2016; Orach and Schlüter 2016).

14.5 Conclusion

Urban changes, such as those expected as a result of climate change, are anticipated to have uncertain and severe impacts on biodiversity in cities, likely impacting provision of ecosystem services and the quality of life for those living in urban environments. Adaptive governance provides a useful policy framework to enhance a system's ability to respond to uncertain disturbances and would benefit the resilience of biodiversity planning in the face of climate change. This paper has investigated two Australian land-use policies to determine to what extent Australian urban biodiversity planning adopts an adaptive governance approach. The research found that, while the two policies demonstrated some elements of adaptive governance, the overall inclusion of adaptive governance elements is limited. This likely leads to an increased vulnerability of Melbourne and Sydney's biodiversity values in the face of unpredictable disturbances associated with climate change. This paper contributes to the understanding of Australian biodiversity policies' adaptive capacity and presents an analysis framework suitable for assessing the adoption of adaptive governance within and beyond biodiversity land-use policy.

References

ABS (2018) Regional population growth, Australia: population estimates by significant urban area, 2007–2017, table, 3218. Australian Bureau of Statistics, Canberra

Barr L, Watson J, Possingham H, Iwamura T, Fuller R (2016) Progress in improving the protection of species and habitats in Australia. Biol Conserv 200:184–191

Bodin Ö, Crona B (2009) The role of social networks in natural resource governance: what relational patterns make a difference? Glob Environ Chang 19(3):366–374

Bolund P, Hunhammar S (1999) Ecosystem services in urban areas. Ecol Econ 29(2):293–301

Borgström S (2018) Governance perspectives on urban biodiversity. In: Ossola A, Niemelä J (eds) Urban biodiversity: from research to practice. Routledge, New York, pp 150–166

Borgström S, Elmqvist T, Angelstam P, Alfsen-Norodom C (2006) Scale mismatches in management of urban landscapes. Ecol Soc 11(2):16

Bush J, Hes D (2018) Urban green space in transition to the eco-city: policies, multifunctionality and narrative. In: Hes D, Bush J (eds) Enabling eco-cities. Springer, Singapore, pp 43–64

Byrne J, Sipe N, Dodson J (2014) What is environmental planning. In: Byrne J, Sipe N, Dodson J (eds) Australian environmental planning: challenges and future prospects. Routledge, New York, pp 3–8

Cardinale B, Duffy J, Gonzalez A, Hooper D, Perrings C, Venail P, Narwani A, Mace G, Tilman D, Wardle DA, Kinzig A, Daily G, Loreau M, Grace J, Larigauderie A, Srivastava D, Naeem S (2012) Biodiversity loss and its impact on humanity. Nature 489(7415):326–326

Chaffin B, Garmestani A, Gunderson L, Benson M, Angeler D, Tony C, Cosens B, Craig R, Ruhl J, Allen C (2016) Transformative environmental governance. Annu Rev Environ Resour 41:399–423

Coffey B, Wescott G (2010) New directions in biodiversity policy and governance? A critique of Victoria's land and biodiversity white paper. Aust J Environ Manag 17(4):204–214

DEPI (2013) Biodiversity conservation strategy for Melbourne's growth corridors, Victorian Department of Environment and Primary Industries, Melbourne

DPCD (2009) Delivering Melbourne's newest sustainable communities: program report, Victorian Department of Planning and Community Development, Melbourne

Duvall P, Lennon M, Scott M (2017) The 'natures' of planning: evolving conceptualizations of nature as expressed in urban planning theory and practice. Eur Plan Stud 26(3):480–501

Erixon H, Borgström S, Andersson E (2013) Challenging dichotomies – exploring resilience as an integrative and operative conceptual framework for large-scale urban green structures. Plan Theory Pract 14(3):349–372

Evans J (2011) Ch 10: conclusions. In: Environmental governance. Taylor & Francis, London, pp 209–219

Feinberg D, Hostetler M, Reed S, Pienaar E, Pejchar L (2015) Evaluating management strategies to enhance biodiversity in conservation developments: perspectives from developers in Colorado, USA. Landsc Urban Plan 136:87–96

Folke C, Carpenter S, Walker B, Scheffer M, Elmqvist T, Gunderson L, Holling C (2004) Regime shifts, resilience, and biodiversity in ecosystem management. Annu Rev Ecol Evol Syst 35 (1):557–581

Folke C, Hahn T, Olsson P, Norberg J (2005) Adaptive governance of social-ecological systems. Annu Rev Environ Resour 30:441–473

Frantzeskaki N, Tilie N (2014) The dynamics of urban ecosystem governance in Rotterdam, The Netherlands. Ambio 43(4):542–555

Goddard M, Dougill A, Benton T (2010) Scaling up from gardens: biodiversity conservation in urban environments. Trends Ecol Evol 25(2):90–98

Green O, Garmestani A, Allen C, Gunderson L, Ruhl J, Arnold C, Graham N, Cosens B, Angeler D, Chaffin B, Holling C (2015) Barriers and bridges to the integration of social– ecological resilience and law. Front Ecol Environ 13(6):332–337

Green O, Garmestani A, Albro S, Ban N, Berland A, Burkman C, Gardiner M, Gunderson L, Hopton M, Schoon M, Shuster W (2016a) Adaptive governance to promote ecosystem services in urban green spaces. Urban Ecosyst 19(1):77–93

Green T, Kronenberg J, Andersson E, Elmqvist T, Gómez-Baggethun E (2016b) Insurance value of green infrastructure in and around cities. Ecosystems 19(6):1051–1063

GSC (2018) Greater Sydney region plan: a metropolis of three cities – connecting people. Greater Sydney Commission, Sydney

Gunderson L, Cosens B, Garmestani A (2016) Adaptive governance of riverine and wetland ecosystem goods and services. J Environ Manag 183:353–360

Guo Z, Zhang L, Li Y (2010) Increased dependence of humans on ecosystem services and biodiversity. PLoS One 5(10):e13113

Harrison P, Berry P, Simpson G, Haslett J, Blicharska M, Bucur M, Dunford R, Egoh B, Garcia-Llorente M, Geamănă N, Geertsema W, Lommelen E, Meiresonne L, Turkelboom F (2014) Linkages between biodiversity attributes & ecosystem services: a systematic review. Ecosyst Serv 9:191–203

Heller N, Zavaleta E (2009) Biodiversity management in the face of climate change: a review of 22 years of recommendations. Biol Conserv 142(1):14–32

Hoornweg D, Hosseini M, Kennedy C, Behdadi A (2016) An urban approach to planetary boundaries. Ambio 45(5):567–580

Ives C, Beilin R, Gordon A, Kendal D, Hahs A, McDonnell M (2013) Local assessment of Melbourne: the biodiversity & social-ecological dynamics of Melbourne, Australia. In: Elmqvist T (ed) Urbanization, biodiversity & ecosystem services: challenges & opportunities. Springer, Dordrecht, pp 385–407

Ives C, Lentini P, Threlfall C, Ikin K, Sanahan D, Garrard G, Bekessy S, Fuller R, Mumaw L, Rayner L, Rowe R, Valentine L, Kendal D (2016) Cities are hotspots for threatened species. Glob Ecol Biogeogr 25:117–126

Karkkainen B (2008) Bottlenecks and baselines: tackling information deficits in environmental regulation. Tex Law Rev 86:1409–1444

Karpouzoglou T, Dewulf A, Clark J (2016) Advancing adaptive governance of social-ecological systems through theoretical multiplicity. Environ Sci Pol 57:1–9

Kendal D (2011) Potential effects of climate change on Melbourne's street trees and some implications for human and non-human animals. Proceedings of the State of Australian Cities Conference

Kenward R, Whittingham M, Arampatzis S, Manos B, Hahn T, Terry A, Simoncini R, Alcorn J, Bastian O, Donlan M, Elowe K, Franzen F, Karacsonyi Z, Larsson M, Manou D, Navodaru I, Papadopoulou O, Papathanasiou J, von Raggamby A, Sharp R, Soderqvist T, Soutukorva A, Vavrova L, Aebischer N, Leader-Williams N, Rutz C (2011) Identifying governance strategies that effectively support ecosystem services, resource sustainability, and biodiversity. Proc Natl Acad Sci 108(13):5308–5312

Langemeyer J, Gomez-Baggethun E (2018) Urban biodiversity and ecosystem services. In: Urban biodiversity: from research to practice. Routledge, New York, pp 36–53

McDonald R, Marcotullio P, Güneralp B (2013) Chapter 3: urbanization and global trends in biodiversity and ecosystem services. In: Elmqvist T, Parnell S, Fragkias M, Schewenius M, Güneralp B, Seto K, Marcotullio P, Wilkinson C, McDonald R (eds) Urbanization biodiversity and ecosystem services: challenges and opportunities. Springer, Dordrecht, pp 31–52

Miller J (2008) Conserving biodiversity in metropolitan landscapes: a matter of scale (but which scale?). Landsc J 27(1):114–126

Müller N, Ignatieva M, Nilon C, Werner P, Zipperer W (2013) Chapter 10: patterns and trends in urban biodiversity and landscape design. In: Elmqvist T (ed) Urbanization, biodiversity and ecosystem services: challenges and opportunities. Springer, Dordrecht, pp 123–174

Olsson P, Gunderson L, Carpenter S, Ryan P, Lebel L, Folke C, Holling C (2006) Shooting the rapids: navigating transitions to adaptive governance of social-ecological systems. Ecol Soc 11(1):18

Orach K, Schlüter M (2016) Uncovering the political dimension of social-ecological systems: contributions from policy process frameworks. Glob Environ Chang 40:13–25

Reisinger A, Kitching R, Chiew F, Hughes L, Newton P, Schuster S, Tait A, Whetton P (2014) Australasia. In: Barros V, Field C, Dokken D, Mastrandrea M, Mach K, Bilir T, Chatterjee M, Ebi K, Estrada Y, Genova R, Girma B, Kissel E, Levy A, MacCracken S, Mastrandrea P, White L (eds) Climate change 2014: impacts, adaptation, and vulnerability. Part B: regional aspects. Contribution of Working Group II to the Fifth Assessment Report of the Intergovernmental Panel on Climate Change. Cambridge University Press, Cambridge/New York, pp 1371–1438

Roggema R (2016) The future of sustainable urbanism: a redefinition. City Territory Archit 3:22

Roggema R (2017) The future of sustainable urbanism: society-based, complexity-led, and landscape-driven. Sustainability 9(8):1442

Root T, Price J, Hall K, Schneider S, Rosenzweig C, Pounds J (2003) Fingerprints of global warming on wild animals and plants. Nature 421(6918):57–60

Sarzynski A (2015) Public participation, civic capacity, and climate change adaptation in cities. Urban Clim 14:52–67

Seto K, Guneralp B, Hutyra L (2012) Global forecasts of urban expansion to 2030 and direct impacts on biodiversity and carbon pools. PNAS 109(40):16083–16088

Tyre AJ, Michaels S (2011) Confronting socially generated uncertainty in adaptive management. J Environ Manag 92:1365–1370

Wiedmann T, Schandl H, Lenzen M, Moran D, Suh S, West J, Kanemoto K (2015) The material footprint of nations. Proc Natl Acad Sci 112(20):6271–6276

Chapter 15
Mapping the Permeability of Urban Landscapes as Stepping Stones for Forest Migration

Qiyao Han and Greg Keeffe

Abstract Large-scale urbanisation has become a significant barrier to the migration of trees, which is being exacerbated by accelerated climate change. Maintaining and increasing landscape permeability is expected to be an effective strategy to facilitate the process of forest migration through. This study develops a new methodology to map the permeability of urban landscapes as stepping stones for the movement of seed dispersal agents. Since seed dispersal agents experience their landscapes as hierarchical mosaics of patches, two spatial scales—habitat and home-range scales—are simultaneously considered in the study. The proposed method combines a least-cost path model and a graph theory-based approach. The least-cost path model is applied to map the potential movements of seed dispersal agents, based on which two graph theory-based indices—the probability of connectivity index and the integral index of connectivity—are used to quantify the accessibility of the landscape at habitat and home-range scales, respectively. This method is demonstrated by a case study in the Greater Manchester area, UK. The Eurasian jay, Eurasian siskin, coal tit and the grey squirrel are selected as main dispersal agents in the study area. The results compare the permeability of urban landscapes for different seed dispersal agents, and identify key areas likely to facilitate the process of forest migration through Greater Manchester. Recommendations regarding landscape design and management to improve permeability are also discussed.

Keywords Forest migration · Permeability · Stepping stones · Seed dispersal · Climate change

Q. Han (✉) · G. Keeffe
School of Natural and Built Environment, Queen's University Belfast, Belfast, UK

© Springer Nature Switzerland AG 2020
R. Roggema, A. Roggema (eds.), *Smart and Sustainable Cities and Buildings*,
https://doi.org/10.1007/978-3-030-37635-2_15

15.1 Introduction

As a response to global climate change, many tree species are moving to higher latitudes or elevations with more suitable climate conditions (Hampe 2011). However, modern urbanisation means that they will have to overcome substantial anthropogenic barriers (e.g., agricultural land, buildings, and highways), which may impede their ability to keep pace with the rapidly changing climate or even modify their migration patterns (Tomiolo and Ward 2018). Within this context, assessing and increasing the permeability of urban landscapes is expected to be an effective strategy to facilitate this ecological process. Here, "permeability" refers to the capacity of a landscape to support species' movement.

Several methods have been proposed for assessing landscape permeability. Most of them focused on specific landscape features related to habitat quality or human modification, such as land cover type, road density, and housing density (e.g., Gray et al. 2016; Littlefield et al. 2017). Other studies estimated permeability by modelling or experiments (Shimazaki et al. 2016; Gastón et al. 2016; Cline et al. 2014; Caryl et al. 2013). Additionally, Anderson et al. (2015) utilised genetic data to infer the permeability of landscape features to the movement of chipmunks.

Although these methods provide spatially explicit estimates of landscape permeability, they may be less useful for the study of forest migration. On one hand, they focus on the movement of active dispersers (animals) and may be unsuitable for tree species that depend on passive seed dispersal. Successful forest migration depends on effective seed dispersal between forest fragments, which is affected by the ways in which seed dispersal agents move and interact with the landscape (Clobert et al. 2012). Therefore, the movement of dispersal agents should be considered in the efforts to assess landscape permeability. Moreover, since different dispersal agents may respond very differently to the landscape (Saunders et al. 1991), a sound understanding of their dispersal abilities is also required. On the other hand, in human-dominated environments where landscapes are highly modified and fragmentated, the dominant function of landscape patches is serving as a series of stepping stones that form dispersal paths and transmit ecological flows, rather than providing habitats (Boscolo and Metzger 2011). In this respect, the spatial pattern of urban landscapes might be of great importance to tree migration, because it directly influences the accessibility of stepping stones for dispersal agents, whereas landscape features related to habitat quality or human modification might be of limited value.

Accordingly, this study proposes a new method for mapping landscape permeability based on a measure of landscape accessibility, assuming that landscapes with higher accessibility for seed dispersal agents might have a higher probability of seed dispersal and therefore are more permeable to the migration of trees. Since the focus of this study is on seed dispersal, the behaviour of dispersal agents is mainly considered; other biotic or abiotic factors such as soil type, habitat quality, plant diversity, or interspecific competition are excluded. In addition, to account for the movements of animals at multiple scales, the habitat and home-range scales of dispersal agents are simultaneously considered in this study. The proposed method combines a least-cost path (LCP) model and a graph theory-based approach. The LCP model is applied to map potential movement pathways of dispersal agents, based on which graph theory-based indices are used to quantify landscape accessibility. The Greater Manchester area, UK, is used as a case study to demonstrate this mapping method.

15.2 Method

15.2.1 Data

We use the 2010 topography layer in the Ordnance Survey Master Map as the landcover data, which gives a comprehensive view of 13 land-cover types in the study area (http://digimap.edina.ac.uk/). At the same time, to compare the degree of permeability with the intensity of human modification of the landscape, the greenspace layer (with detailed land use categories which captures the major aspects of human modification) in the Ordnance Survey Master Map is used to classify urban landscapes as (1) natural, with a low intensity of human modification (e.g. natural woodland); (2) semi-natural, with an intermediate intensity of human modification (e.g. camping park, cemetery, golf course, public park or garden); or (3) manmade, with a high intensity of human modification (e.g. transport, bowling green, sports facility).

According to a research by the Forestry Commission (https://www.forestry.gov.uk/fr/infd-837f9j), there are a number of tree species that need to migrate through Greater Manchester in this century, including the European larch (*Larix decidua*), the Sitka spruce (*Picea sitchensis*), the sweet chestnut (*Castanea sativa*), the lodgepole pine (*Pinus contorta*), the Scots pine (*Pinus sylvestris*), the sessile oak (*Quercus petraea*), and the beech (*Fagus*). Most of them are dispersed by frugivorous birds. The acorns and nuts of lodgepole pine, sweet chestnut, sessile oak, beech, and Scots pine are moved by the Eurasian jay (*Garrulus glandarius*), while the Eurasian siskin (*Spinus spinus*) and the coal tit (*Periparus ater*) are the principal dispersal agents for European larch and Sitka spruce. Besides this, the grey squirrel (*Sciurus carolinensis*) is also considered as a main dispersal agent in the study area, given that this small mammal is highly mobile and can disperse chestnuts and acorns readily through fragmented urban landscapes (Rushton et al. 1997).

The spatial records of the four dispersal agents are obtained from the UK's NBN Atlas. For Eurasian jays and grey squirrels, their dispersal distances are obtained from literature, as well as other key parameters (see Table 15.1). However, for the remaining two species, direct observation records of their daily dispersal are not available. In this study, we use the model developed by Sutherland et al. (2000) to estimate their daily dispersal distances based on their body masses. The minimum

Table 15.1 The key parameters of seed dispersal agents (Conway and Fuller 2011; Gómez 2003; Dyer 1995; De Montis et al. 2016; Rolando 1998; Nupp and Swihart 2000; Pascual et al. 2014; Senar et al. 1992; Sellers 2011)

Dispersal agent	Habitat size	Home-range size	Daily dispersal distance	Long-distance dispersal	Body mass
Eurasian Jay	≥ 4 ha	≥ 10.7 ha	≤ 1 km	1–5 km	161.7 g
Eurasian Siskin	≥ 4 ha	≥ 8 ha	≤ 0.5 km	0.5–3 km	13.8 g
Coal Tit	≥ 1 ha	≥ 3 ha	≤ 0.4 km	0.4–5 km	9.25 g
Grey Squirrel	≥ 0.0625 ha	≥ 0.5 ha	≤ 0.15 km	0.15–2 km	510 g

home-range sizes of these species are then derived from the estimate of dispersal distances (Jenkins et al. 2007).

15.2.2 Landscape Accessibility at Habitat Scale

The assessment of landscape accessibility starts with an identification of landscape networks for dispersal agents. Land-cover types are reclassified, as either a habitat or non-habitat area for dispersal agents. For the aim of this study, broadleaved, coniferous and mixed forests are selected as suitable for habitat. After that, we use the minimum habitat size (see Table 15.1) as grain size to change the resolution of the habitat map for each dispersal agent, aggregating small, scattered habitat fragments into large, contiguous habitat patches. Since the dispersal probability between habitats is inversely related to the least-cost distance between them (de la Pena-Domene et al. 2016), a least-cost path (LCP) model is applied to map the paths between habitats (as shown in Fig. 15.1a). The LCP model uses a raster-based optimisation algorithm to identify the optimum path between patches, in terms of cumulative land-cover resistance (Watts et al. 2010). In the study of Greater Manchester, the resistance value of each land-cover type is obtained by habitat suitability modelling using the MaxEnt software (Phillips et al. 2017). The spatial records of each species and the land-cover map are used as input data. For the following analysis of accessibility, each set of interconnected patches is defined as a component (an isolated patch makes up a component itself).

A graph theory-based index, probability of connectivity (PC) (Saura and Pascual-Hortal 2007) is used to transform the habitat network into a node-link graph and calculate the accessibility of each habitat (Fig. 15.1b). The PC index is a probabilistic index that integrates both patch area and inter-patch distance in one measure. It has been shown to relate well to actual species movement and occurrence patterns (Awade et al. 2011). We evaluate the accessibility of each habitat patch based on a quantification of its contribution to the overall PC value of the component that the patch belongs to, using the Graphab software.

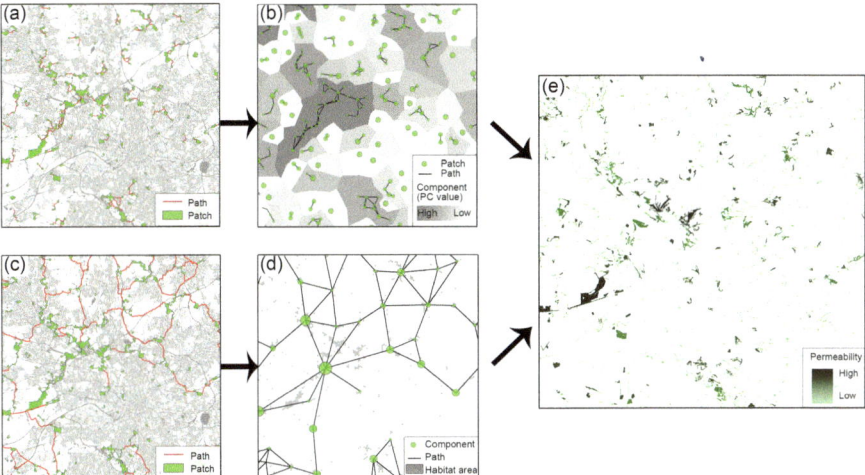

Fig. 15.1 Illustration of The Mapping Method. (**a**) Identify Habitat Networks, (**b**) Evaluate Landscape Accessibility at Habitat Scale, (**c**) Identify Home-range Network, (**d**) Evaluate Landscape Accessibility at Home-range Scale, and (**e**) Assess Landscape Permeability

15.2.3   Landscape Accessibility at Home-Range Scale

The home range of an animal is composed of a cluster of connected habitat patches, which could support its minimum resource requirement. Different home ranges are connected by the paths of long-distance dispersal, using the LCP model (Fig. 15.1c). The distance threshold of the paths is determined by the maximum distance that the animal could move in its search for new home ranges.

At home-range scale, the integral index of connectivity (IIC) (Pascual-Hortal and Saura 2006) is applied for the assessment of landscape accessibility rather than the PC index, because IIC has been shown to better relate to the functional connectivity among home ranges (Decout et al. 2012). The accessibility of each home range is evaluated by a measurement of its contribution to the overall connectivity (IIC value) of the landscape, using the Graphab software (Fig. 15.1d).

15.2.4   Landscape Permeability to Forest Migration

We calculate the permeability of each habitat area by multiplying the result of PC and IIC, given the interactions between landscapes at different scales (Fig. 15.1e). Habitat areas are then classified into three categories, high-, medium-, and low-permeability, using the method of natural breaks in ArcGIS. Finally, we combine the resulting permeability map with the map of human modification to identify areas for improvement.

15.3   Results

As shown in Table 15.2 and Fig. 15.2, the aggregation of habitat areas yields 498, 498, 1677, and 7240 habitat patches for Eurasian jays, Eurasian siskins, coal tits, and grey squirrels, respectively. After that, the potential paths between habitats are identified using the LCP model, based on the land-cover resistances values derived from the habitat suitability modelling (Table 15.3). The connected habitats are then divided into 91, 171, 248, and 1255 home ranges for the four dispersal agents, respectively.

Figure 15.3 illustrates the relative accessibility of individual habitat and home-range nodes. Nodes with high values are critical for maintaining landscape connectivity, and therefore can be regarded as key stepping stones for seed dispersal. As shown in the figure, for all the four dispersal agents, only a handful of home-range patches are responsible for a disproportionate share of seed dispersal events in the landscape network.

After the calculation of permeability in Sect. 15.2.4, each habitat area is assigned a value representing its permeability to forest migration, with higher values indicating greater ease of movement. The permeability values are categorised into four classes: high-permeability, medium- permeability, low-permeability, and impermeable. The impermeable class is the areas that are suitable for habitat but cannot be identified as patches for dispersal agents. Table 15.4 shows the range of permeability values for each class. The percentage of each class regarding both patch number and habitat area is presented in Fig. 15.4. The percentage of habitat patches with high or

Table 15.2 Landscape elements for dispersal agents

Scale	Number	Eurasian Jay	Eurasian Siskin	Coal Tit	Grey Squirrel
Habitat	Patch	498	498	1677	7240
	Path	423	192	1726	6546
	Component	192	347	533	2609
Home-range	Patch	91	171	248	1255
	Path	182	273	1900	7938

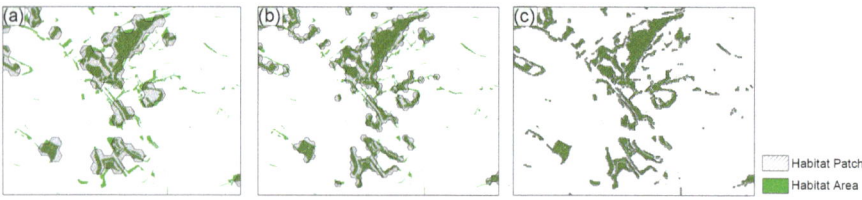

Fig. 15.2 Example of Habitat Patches Identified for Different Dispersal Agents: (**a**) Eurasian Jays and Siskins, (**b**) Coal Tits, and (**c**) Grey Squirrels

Table 15.3 Habitat Suitability Index (HSI) scores obtained from the MaxEnt software and the land-cover resistance (R) values for the four dispersal agent

Land Cover	Eurasian Jay		Eurasian Siskin		Coal Tit		Grey Squirrel	
	HSI	R	HSI	R	HSI	R	HSI	R
Building	0.46	54	0.45	55	0.51	49	N/A	1000
Health	0.43	57	0.45	55	0.43	57	0.2	80
Marsh	0.43	57	0.45	55	0.43	57	0.2	80
Residential	0.47	53	0.58	42	0.61	39	0.71	29
Agricultural	0.4	60	0.45	55	0.39	61	0.3	70
Orchard	0.43	57	0.45	55	0.43	57	0.2	80
Roads	0.69	31	0.45	55	0.43	57	0.63	37
Rock	0.5	50	0.45	55	0.43	57	N/A	1000
Grassland	0.41	59	0.58	42	0.43	57	0.2	80
Scrub	0.67	33	0.57	43	0.57	43	0.37	63
Urban	0.43	57	0.45	55	0.43	57	0.28	72
Water	0.85	15	0.78	22	0.86	14	0.35	65
Woodland	N/A	1	N/A	1	N/A	1	N/A	1

medium permeability is higher for Eurasian jays and siskins than for the other two dispersal agents. Nevertheless, the total percentage of the high and medium class between the four dispersal agents is not very different, in terms of habitat area. This is because most of the low-permeability and impermeable areas are small patches. Figure 15.5 shows the spatial distribution of the four permeability classes for different dispersal agents. The differences in spatial distribution indicate that land-scape permeability to forest migration is influenced by the dispersal capabilities of local species.

We integrate the results of the four dispersal agents to obtain a permeability map of Greater Manchester (Fig. 15.6a). In summary, around 13% of the total habitat area is very permeable to forest migration when all the four dispersal agents are considered, while the areas corresponding to the medium-permeability class account for 24%. Those low-permeability and impermeable areas cover more than 60% of the total habitat area, although most of them (95%) are smaller than 1 ha.

Figure 15.6b describes the percentages of permeable areas in natural, semi-natural and manmade greenspaces in Greater Manchester. Landscapes showing high- or medium-permeability occupy 30% of natural, 34% of semi-natural, and 19% of manmade greenspaces. These relatively permeable natural and semi-natural greenspaces are very important for forest migration, because they can support both seed dispersal and plant establishment, whereas the high- and medium-permeable manmade greenspaces are potential locations where habitat quality should be improved to increase their contributions to forest migration. At the same time, 69% of natural and 65% of semi-natural greenspaces appear low permeable to forest migration, indicating that they are isolated habitats where permeability could be improved by adding new stepping stones to increase their accessibility.

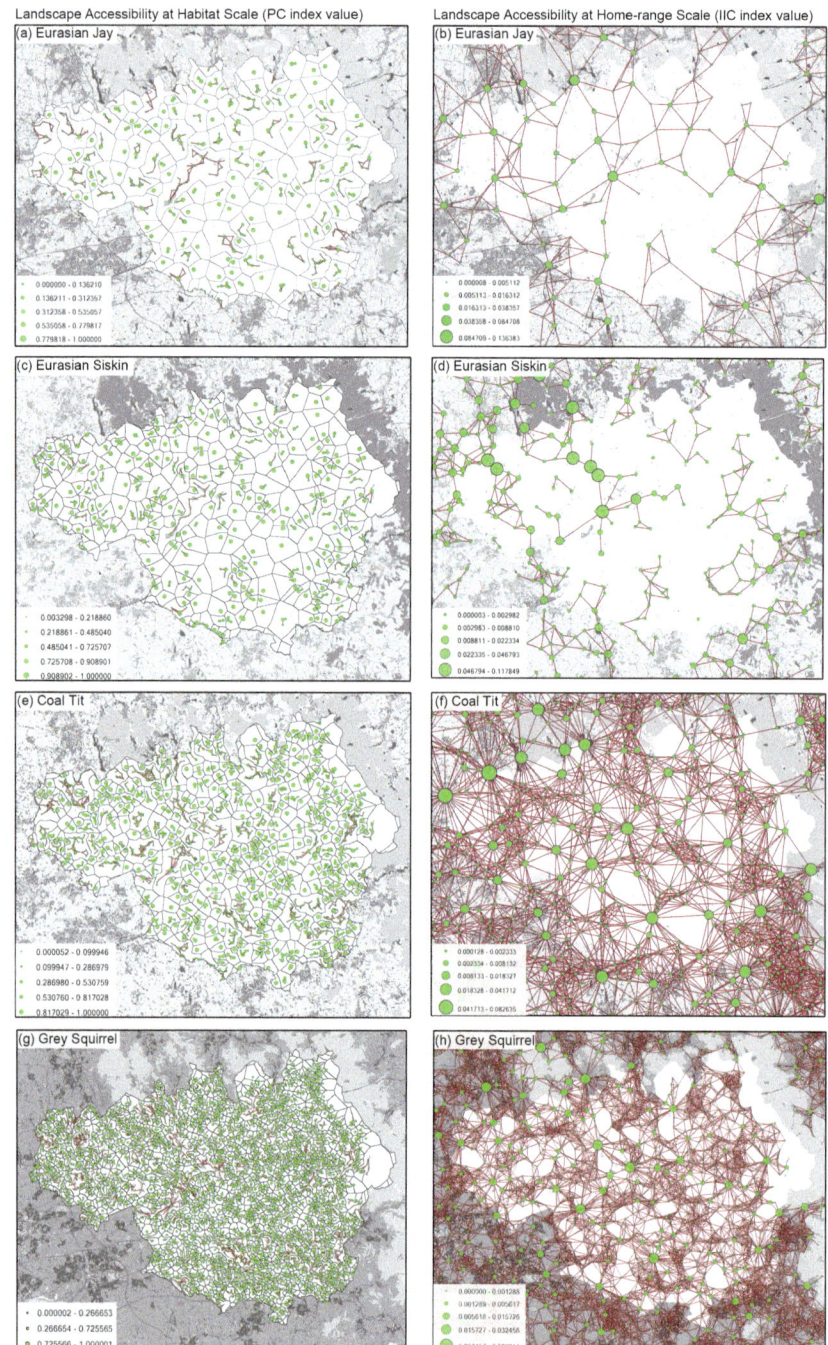

Fig. 15.3 Landscape Accessibility for Dispersal Agents

Table 15.4 Classification of landscape permeability

Permeability	Eurasian Jay	Eurasian Siskin	Coal Tit	Grey Squirrel
High	0–0.0039	0–0.0047	0–0.0017	0–0.0024
Medium	0.0040–0.0135	0.0048–0.0212	0.0018–0.0081	0.0025–0.0108
Low	0.0136–0.0364	0.0213–0.0617	0.0082–0.0265	0.0109–0.0231

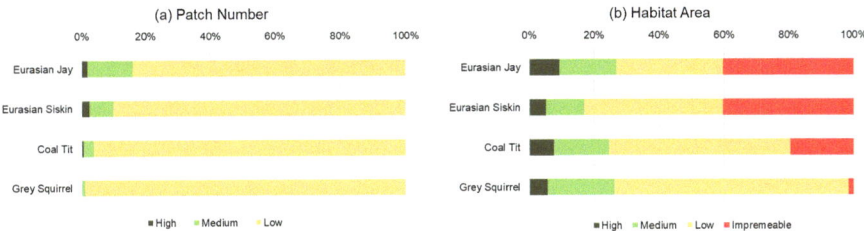

Fig. 15.4 Percentages of the Four Permeability Classes

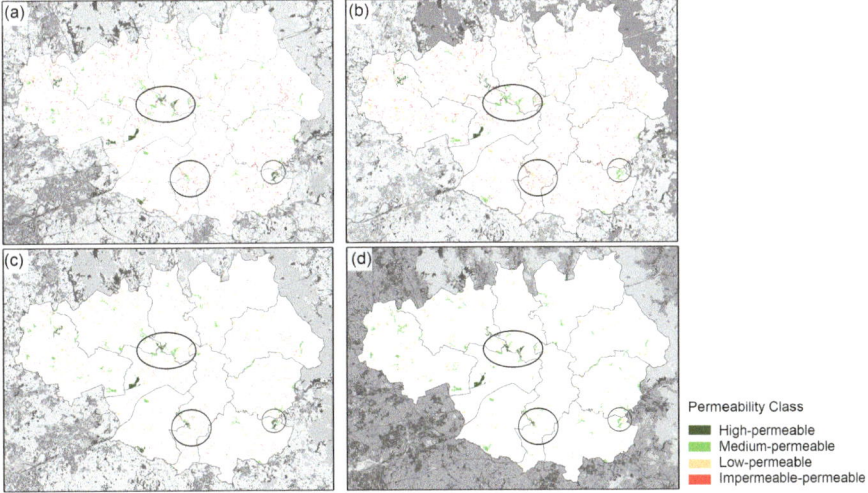

Fig. 15.5 Spatial distribution of the four permeability classes for (**a**) Eurasian Jays, (**b**) Eurasian Siskins, (**c**) Coal Tits, and (**d**) Grey Squirrels

15.4 Conclusion

This study develops a novel method to map the permeability of urban landscapes to forest migration. It combines an LCP model that identifies landscape networks for dispersal agents, and a graph theory-based approach which evaluates landscape accessibility at multiple scales. This allows designers to re-visualise highly modified and fragmented urban landscapes as stepping stones for seed dispersal, which in turn

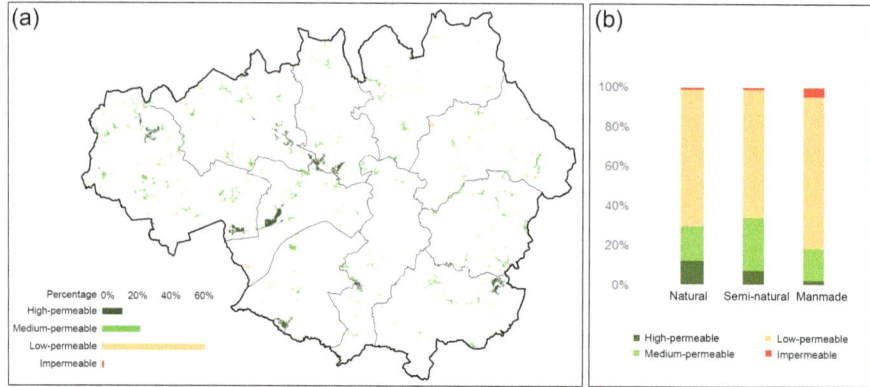

Fig. 15.6 (**a**) Permeability Map of Greater Manchester Considering Four Dispersal Agents, (**b**) Percentages of Areas of Each Permeability class in Natural, Semi-natural and Manmade Greenspaces

allows for a more piecemeal form of landscape design to have strategic benefit by engaging as part of a larger system.

The proposed method is applied to the case study of the Greater Manchester area, UK. The results identify urban green spaces with the potential to facilitate the climate-driven migration of trees through the city and provide a comparison of the permeability maps generated from different dispersal agents. Moreover, this study combines the map of permeability with the map of human modification to illustrate the application of the proposed method to incorporate other considerations and analytical possibilities. It is believed that this method would be especially important for landscapes where human activity is intense and implementing large continuous reserves is not possible. Future research should explore how to add new patches (through forestation or restoration programs) in the urban matrix to favour the movement of dispersal agents and thereby to increase the permeability of urban landscapes to forest migration.

Acknowledgments The first author is funded by the China Scholarship Council of Chinese government. We would like to thank Paul Caplat, Helen Roe, Gary Archibald Boyd and Gul Kacmaz Erk for their comments and suggestions.

References

Anderson SJ, Kierepka EM, Swihart RK, Latch EK, Rhodes OE Jr (2015) Assessing the permeability of landscape features to animal movement: using genetic structure to infer functional connectivity. PLoS One 10:e0117500

Awade M, Boscolo D, Metzger JP (2011) Using binary and probabilistic habitat availability indices derived from graph theory to model bird occurrence in fragmented forests. Landsc Ecol 27:185–198

Boscolo D, Metzger JP (2011) Isolation determines patterns of species presence in highly fragmented landscapes. Ecography 34:1018–1029

Caryl FM, Thomson K, van der Ree R (2013) Permeability of the urban matrix to arboreal gliding mammals: sugar gliders in Melbourne, Australia. Austral Ecol 38:609–616

Cline BB, Hunter ML, Banks-Leite C (2014) Different open-canopy vegetation types affect matrix permeability for a dispersing forest amphibian. J Appl Ecol 51:319–329

Clobert J, Baguette M, Benton TG, Bullock JM (2012) Dispersal ecology and evolution. Oxford University Press, Oxford

Conway GJ, Fuller RJ (2011) Multi-scale relationships between vegetation pattern and breeding birds in the upland margins (ffridd) of North Wales. British Trust for Ornithology, The Nunnery/ Thetford/Norfolk

de la Pena-Domene M, Minor ES, Howe HF (2016) Restored connectivity facilitates recruitment by an endemic large-seeded tree in a fragmented tropical landscape. Ecology 97:2511–2517

De Montis A, Caschili S, Mulas M, Modica G, Ganciu A, Bardi A, Ledda A, Dessena L, Laudari L, Fichera CR (2016) Urban–rural ecological networks for landscape planning. Land Use Policy 50:312–327

Decout S, Manel S, Miaud C, Luque S (2012) Integrative approach for landscape-based graph connectivity analysis: a case study with the common frog (Rana temporaria) in human-dominated landscapes. Landsc Ecol 27:267–279

Dyer JM (1995) Assessment of climatic warming using a model of forest species migration. Ecol Model 79:199–219

Gastón A, Blázquez-Cabrera S, Garrote G, Mateo-Sánchez MC, Beier P, Simón MA, Saura S, Wintle B (2016) Response to agriculture by a woodland species depends on cover type and behavioural state: insights from resident and dispersing Iberian lynx. J Appl Ecol 53:814–824

Gómez JMa (2003) Spatial patterns in long-distance dispersal of Quercus ilex acorns by jays in a heterogeneous landscape. Ecography 26:573–584

Gray M, Wilmers CC, Reed SE, Merenlender AM (2016) Landscape feature-based permeability models relate to puma occurrence. Landsc Urban Plan 147:50–58

Hampe A (2011) Plants on the move: the role of seed dispersal and initial population establishment for climate-driven range expansions. Acta Oecol 37:666–673

Jenkins DG, Brescacin CR, Duxbury CV, Elliott JA, Evans JA, Grablow KR, Hillegass M, Lyon BN, Metzger GA, Olandese ML, Pepe D, Silvers GA, Suresch HN, Thompson TN, Trexler CM, Williams GE, Williams NC, Williams SE (2007) Does size matter for dispersal distance? Glob Ecol Biogeogr 16:415–425

Littlefield CE, McRae BH, Michalak JL, Lawler JJ, Carroll C (2017) Connecting today's climates to future climate analogs to facilitate movement of species under climate change. Conserv Biol 31:1397–1408

Nupp TE, Swihart RK (2000) Landscape-level correlates of small-mammal assemblages in forest fragments of farmland. J Mammal 81:512–526

Pascual J, Senar JC, Domènech J, Herberstein M (2014) Are the costs of site unfamiliarity compensated with vigilance? A field test in Eurasian siskins. Ethology 120:702–714

Pascual-Hortal L, Saura S (2006) Comparison and development of new graph-based landscape connectivity indices: towards the priorization of habitat patches and corridors for conservation. Landsc Ecol 21:959–967

Phillips SJ, Dudík M, Schapire RE (2017) Maxent software for modeling species niches and distributions (Version 3.4.1) [Online]. Available: http://biodiversityinformatics.amnh.org/ open_source/maxent/. Accessed 2018

Rolando A (1998) Factors affecting movements and home ranges in the jay (Garrulus glandarius). J Zool 246:249–257

Rushton SP, Lurz PWW, Fuller R, Garson PJ (1997) Modelling the distribution of the red and grey squirrel at the landscape scale: a combined GIS and population dynamics approach. J Appl Ecol 34:1137–1154

Saunders DA, HOBBS RJ, MARGULES CR (1991) Biological consequences of ecosystem fragmentation: a review. Conserv Biol 5:18–32

Saura S, Pascual-Hortal L (2007) A new habitat availability index to integrate connectivity in landscape conservation planning: comparison with existing indices and application to a case study. Landsc Urban Plan 83:91–103

Sellers RM (2011) Movements of coal, marsh and willow tits in Britain. Ringing Migr 5:79–89

Senar JC, Burton PJK, Metcalfe NB (1992) Variation in the nomadic tendency of a wintering finch Carduelis spinus and its relationship with body condition. Ornis Scand 23:63–72

Shimazaki A, Yamaura Y, Senzaki M, Yabuhara Y, Akasaka T, Nakamura F (2016) Urban permeability for birds: an approach combining mobbing-call experiments and circuit theory. Urban For Urban Green 19:167–175

Sutherland GD, Harestad AS, Price K, Lertzman KP (2000) Scaling of natal dispersal distances in terrestrial birds and mammals. Conserv Ecol 4:16

Tomiolo S, Ward D (2018) Species migrations and range shifts: a synthesis of causes and consequences. Perspect Plant Ecol Evol Syst 33:62–77

Watts K, Eycott AE, Handley P, Ray D, Humphrey JW, Quine CP (2010) Targeting and evaluating biodiversity conservation action within fragmented landscapes: an approach based on generic focal species and least-cost networks. Landsc Ecol 25:1305–1318

Chapter 16
Contemporary Urban Biotopes: Lessons Learned from Four Recent European Urban Design Plans

Martin Knuijt

Abstract Increasing urbanisation requires rethinking what liveable cities are about. Within the next decade, both climate change and the shortage of resources on water, energy and nutrients will have strong effects on the urban environment. The challenge is to create 'healthy cities'. Bringing landscape to the cities can strongly contribute to making cities healthier, more resilient, and more vibrant, to accommodate all its citizens. The key to providing this new perspective is discovering how to create healthy cities in densely built areas, and strengthen the urban metabolism, while also addressing externalities, such as urban heat island effects, increased storm events, and sea level rise.

The objective of this paper is to gain insight in the new complexity that arises from the increasing relevance of landscape and planting in dense urban environments, in order to set a contemporary agenda for urban green space design. The increasing need to develop healthy, circular, and climate adaptive cities leads to new demands on urban green spaces (Knuijt 2013). A journey along recent urban plans for Rotterdam, Athens, London, and Utrecht demonstrates that, by rebalancing traffic in cities, a vibrant and green public realm can be realised. Re-balancing transportation and a shift to multimodal mobility have a large impact on spatial qualities of cities and provides creating access for all. The public realm can be transformed into a green and blue network, and will set the scene to activate public realm for vibrant city centres.

The complexity that arises from the new demands on green space in dense urban environments is explained through the analyses of four case studies in Rotterdam, Athens, London, and Utrecht and literature reviews. These four 'research by design' projects are discussed and evaluated to reveal the increasing (societal) relevance of urban green space.

These projects offer new perspectives on the integration of climate adaptive design and circularity. Toolboxes for heat mitigation and for water sensitive design are developed and applied in the designs. Today's complexity to increase the degree

---

M. Knuijt (✉)
OKRA Landscape Architects, Utrecht, The Netherlands
e-mail: martinknuijt@okra.nl

© Springer Nature Switzerland AG 2020
R. Roggema, A. Roggema (eds.), *Smart and Sustainable Cities and Buildings*,
https://doi.org/10.1007/978-3-030-37635-2_16

of self-sufficiency within city limits and regions as well as climate adaptation requires continuous monitoring of the level of incorporation of the different aspects of 'healthy living' into the realized development and assessment of the standards each year. Adding today's aspirations on including biodiversity calls for the idea of 'urban biotopes', turning the green into an urban ecosystem that can evolve over time. Creating a circular economy within the dense urban development, plus climate adaptive design and aspirations on creating an urban biotope, make urban development complex. It requires careful consideration about how to balance energy production and green and how to integrate underground infrastructure, in order to make sure that the proper conditions for urban green are set.

Keywords Urbanisation · Healthy cities · Resiliency · Climate adaptation · Circular economy · Biodiversity

16.1   Introduction

An increasing number of people are living within city limits; the world is becoming rapidly urbanised, and adjacent natural resources are being exhausted at an unsustainable rate. The quality of our lakes, rivers and streams is decreasing due to run-off. Ecological and agricultural areas are being developed to accommodate the population shift, resulting in a loss of 'green.' At the same time, we are facing climate change due to which contemporary cities are faced with the task of providing fresh water and access to restorative naturalised areas, while protecting citizens from natural disasters like flash floods, coastal storm surges, heatwaves, and droughts.

In the European densely built mega polis and metropolitan areas, there are urgent questions on how to develop healthy (peri-)urban environments with integrated water and drought management, and heat island effect mitigation solutions. Within the cities, the fundament for change lies in creating efficient transportation systems, connected to the urban network of spaces. A decade ago, the agenda for improving the quality of the public was set. Successful transformations of infrastructure into green structures, such as the Cheonggyecheon River in Seoul (Kodukula 2011), and urban transformations of the industrial to post industry cities in Bilbao (Areso 2009) and Melbourne (Adams 2009), have shown that large scale transformations are possible.

The challenge is to create 'healthy cities'. Following up to the change towards a direction of creating green landscape cities, our contemporary challenge is to incorporate the 'metabolism of the city' within the context of the densely built urban environment. The challenge is to manage the exchanges of energy, material, and population in a responsible and sustainable way—considering and strengthening the urban metabolism. Being aware of the growth of the world's population and the ongoing increase of percentage of people living within city limits, it will be imperative to find answers to unhealthy living conditions and to increase the degree of self-sufficiency within city limits and regions.

Creating a healthy living environment requires the reduction of distances between working and living, plus rebalancing traffic into a shift towards public transport and slow movement. Furthermore, it includes climate-proof cities and landscapes, addressing the issues of storm water management (Lenzholzer 2015). Other topics can be related to strategies for energy transition, the shift from fossil fuels to sustainable energy management, and for re-thinking waste, nutrients, and food production within the city limits (Troy 2012). The increasing need to develop healthy, circular, and climate adaptive cities leads to new demands for urban plans. In that context, the role of green spaces in the development of cities will change and become more relevant (City of Copenhagen 2016). To gain insight in the new complexity regarding urban green space design, four recent urban plans for Rotterdam, Athens, London, and Utrecht were analysed on how different demands on green space were integrated in the designs. This paper describes lessons learned from these four urban design projects, in order to set a contemporary agenda for dense urban environments as 'urban biotopes'.

16.2   Methodology

The new perspective on creating healthy cities can strongly contribute to make cities healthier, resilient, and more vibrant, in order to accommodate all of its citizens. Via 'research by design projects', new perspectives arise on integrating climate-sensitive design and circularity, and insights are developed on how to achieve substantial improvements in the public realm. Three recent urban design projects in Rotterdam, Athens, and London are discussed and evaluated as to the way in which they offer integrated solutions for urban challenges, regarding climate change, circular economy, and mobility. The contribution of these three case studies to these challenges is evaluated through assessment of the design projects' output and through literature reviews. Moreover, a currently ongoing urban design project in the Dutch city of Utrecht is reviewed as a pilot project for the application of the concept 'urban biotope'.

16.3   Results

16.3.1   Rotterdam, the Connected City Centre

The first step to the transformation of cities is to change mobility. The strategy for urban change in Rotterdam, and later for London and Athens was based on three pillars: creating access for all and rebalance transportation, transforming the public realm into a green and blue network, and activate the public realm to create vibrant city centres.

Re-balancing transportation has a large impact on spatial qualities of cities. With increased mobility, the quality of life of our cities is under pressure. Mobility is an issue in contemporary metropolitan regions and in rapidly growing cities. In many cities, auto-centric city planning has led to vehicular movement exceeding maximum capacity of road space, resulting in congestion and low-quality anonymous space, lacking identity. Prioritizing automobile transportation results in a lack of safe pedestrian and cyclist networks. It has a negative impact on quality of life; air pollution takes a heavy toll on our health. Being aware that about a third of the total amount of air pollution comes from motorised traffic, development of clean transport to improve air quality will help to increase livability of cities.

The shift from vehicular mobility to multimodal is part of the strategy for Rotterdam city centre, focusing on expansion of pedestrian networks, public transit, and cycling. The idea of space needed to change, not by simply removing lanes of traffic on the main boulevards and roads, but by shifting from vehicular orientated space to pedestrian orientated space. For the city centre, an area of four by four kilometers, the switch from hard traffic (car-oriented) space to 'soft traffic' (pedestrian, cyclist and public transport oriented) space allows the centre to become accessible in a different way—the public realm itself to become a catalyst for urban revitalisation. Focal points in the city were connected, and the old pre-WWII streets were activated, so that pleasant and attractive networks for slow traffic could be regenerated. Resulting into a connected city, the ground is set for greening the space, and to create inviting places to stay (Knuijt 2008).

16.3.2 Re-think Athens and a Toolbox for Heat Mitigation

Creating space for pedestrians and cyclists sets the ground for a resilient and climate proof public realm (Bulleri 2018). The proposed transformation of Athens' city centre interlinked infrastructural change with the built environment, and created a basic framework for a blue-green network (Salles 2013). Changing the heart of Athens into a true contemporary metropolitan city centre required transformation of the city triangle into a lively part of the city. Newly gained space, as a result of the major step towards a walkable city by reducing car traffic in this area, will transform it into a vibrant, green, and accessible heart of the city (Knuijt 2013). The combination of water solutions was key to make the city more resilient, adaptable, and dynamic. The blue-green network served a multifunction purpose—storm-water and drought management and heat island effect mitigation (Figs. 16.1 and 16.2).

To mitigate urban heat island effects urban measures to improve the urban microclimate, energy consumption and thermal comfort of citizens are required and integrated into the design principles of the public realm. A heat mitigation design toolkit was part of the proposal of OKRA landscape architects: the addition of greenery, use of light materials, and integration of open water helps reduce the urban heat island effect. A contextual approach defines where the tools for different categories can be applied. Trees and other vegetation provide shade and allow of

Fig. 16.1 Heat mitigation toolbox Athens—measures and spatial distribution. (Knuijt 2013)

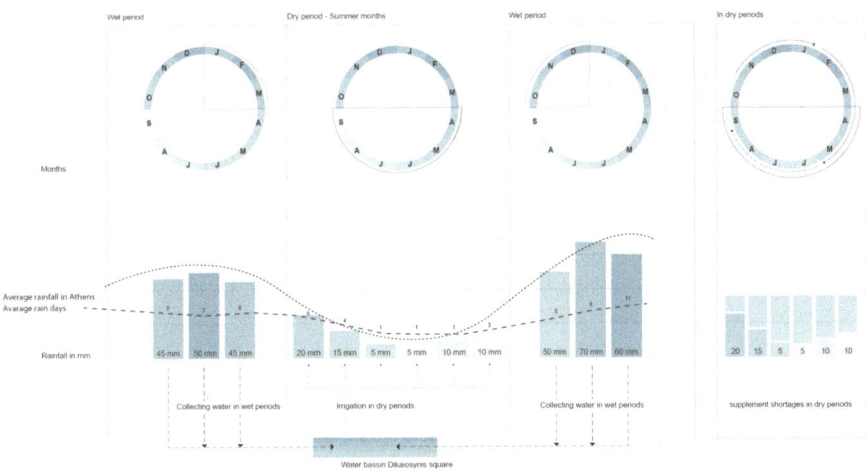

Fig. 16.2 Rainfall in Athens, seasonal change. (Knuijt 2013)

evapotranspiration, both having a cooling effect. Providing water for the trees stimulates transpiration, and contributes to cooling as well. Europe's largest rainwater retention system allows the area to be self-sufficient on water for irrigating green areas.

Parallel to the design process, monitoring of the results has taken place. Similar to the technical aspects, such as traffic modelling, aspects of climate adaptive design can also be calculated (Santamouris et al. 2012). It is not quantities indicating the amount of green and water of the design; it is the performance that can be indicated in figures. The broad notion 'sustainability' gets precision. For the Re-Think Athens project, the heat mitigation toolbox was evaluated and translated into the design for public realm. Measurements on site were executed (University of Athens 2013) and during the design stages the outcome on heat reduction of the proposals were calculated via using ENVI-met simulations A cooling of 1.5 °C plus 20% of the thermal comfort index on a typical summer day was the aim. Results of 1.5 up till 3.0° was; the outcome. (Werner Sobek GT 2013).

16.3.3   London Meridian Water and a Toolbox for Water Sensitive Urban Design

To adapt to climate change and prevent urban areas from the negative effects of flooding and drought, consideration of these aspects needs to be integrated at an early stage of planning for urban developments. Regenerating the water system is key to a contemporary and healthy relationship between nature and culture (Hoyer et al. 2011).

For London's largest building development, Meridian Water, situated on a brownfield area, OKRA landscape architects developed a water sensitive urban design toolbox. This was a result of considering that water solutions and green in public realm in the River Lee valley could work as an 'urban water machine'. A coherent set of tools was developed, contributing to healthy cities by integrating (Figs. 16.3 and 16.4).

The first category of tools is about preventing areas from flooding. The second category is about ensuring that integrated water management capturing and reuse is of essence. The green areas are designed to act as flood storage system within the built environment. Additional water storage can be designed into parks and squares serving dual purposes—leisure, recreation, and sport most of the year, and storm-water storage during intense storm events. Storm-water can be stored in above or below ground containers, used as an alternative for irrigation. Storm-water ponds can be seen as attractive natural elements that also serve as habitat to urban flora and fauna. Canals and pools within the public realm can hold water while not only providing something beautiful to look at, but also allowing for evapotranspiration to occur, thus lowering the urban temperature. Capturing as much rainwater on site as possible is one way in which we can ensure climate proof cities. The third category related to sound water management includes tools to filter, to infiltrate and to

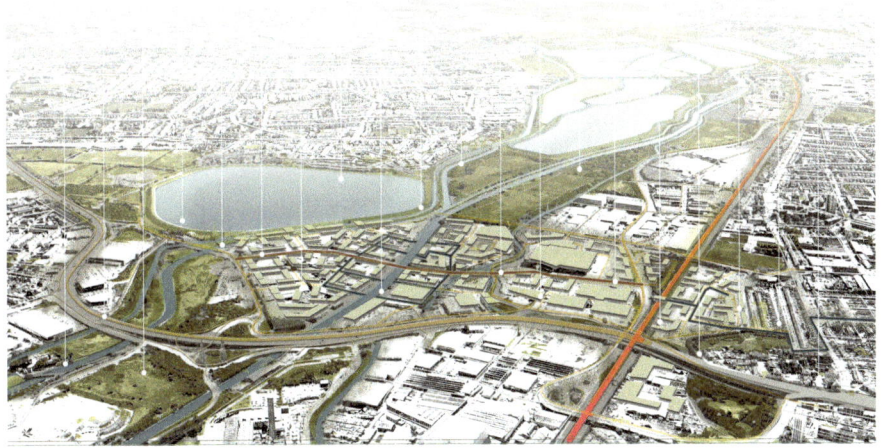

Fig. 16.3 Toolbox for water sensitive urban design, London Meridian Water (Knuijt 2016a, b) water into the public realm by storing, filtering, and infiltrating (Knuijt 2016a, b)

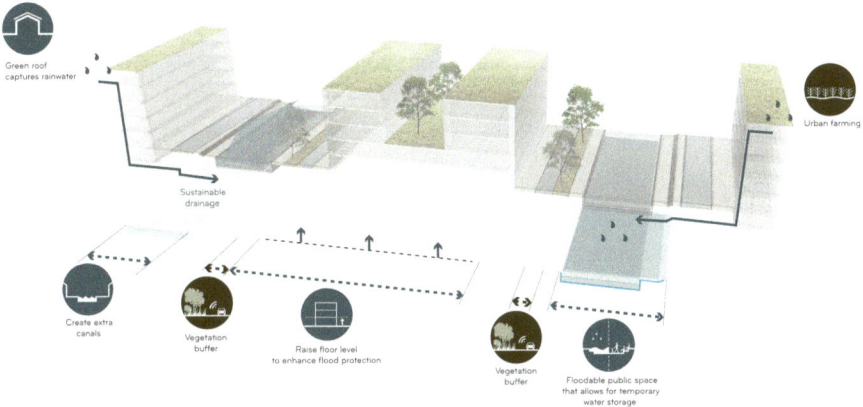

Fig. 16.4 Toolbox for water sensitive urban design translated into typical section for one of the waterways, London Meridian Water. (Knuijt 2016a, b)

recharge. The last category of tools is about the educational aspect of water. In addition to rain gardens, swales, ponds, streams, and channels, other water elements within public areas such as water-squares, fountains and interactive play elements can serve as protection, by providing additional temporary storage and/or reuse. Moreover, creating hydro-centric recreational opportunities is a way to bring people into contact with water; providing these opportunities is essential for education and awareness. Within the design of public realm, water and green can be integrated in multiple scales, resulting in attractive public realm.

16.3.4   *Merwedekanaalzone, Utrecht: A New Horizon*

Improving biodiversity in the urban context and preventing diseases brings nature close to people living in the city. In the most recent urban design project of OKRA landscape architects, the Merwede area in Utrecht, a large brownfield development along the Merwedekanaal, 'urban biotopes' are designed in the public realm and on inner courts plus rooftops. The concept of urban biotope, in this case, means to design a resilient urban planting plan with a balanced nutrient and water supply system for planting. Based on the existing trees in the area, a selection of additional species is made to create variety in size of trees, understory planting, groundcover and perennials. Part of the planting is edible green, providing nuts and fruits for birds and insects.

Within the context of high density urban development, improved water sensitive solutions, waste management, and energy neutral developments are included. Creating a circular economy within the dense urban development, plus a climate adaptive design and aspirations of creating an urban biotope make the situation complex. It is so complex, that choices have to be made where energy production can take place and at what place green on roofs has priority. It even requires re-thinking and integrating underground infrastructure, to avoid that at a late design stage it becomes clear that part of the green can't be realised.

The project itself is regarded as a 1:1 design lab and requires monitoring of the results, executed by research institutes. Most likely, today's highest standards will be normal standards in a few years, and might be outdated standards within 10 years. This requires continuous monitoring of the level of incorporation of the different aspects of 'healthy living' into the realised development and assessment of the standards each year. Results during construction will be set against today's baseline, and include achievements during the next 10 years, thus being able to evaluate the different aspects on water sensitive solutions, heat mitigation, carbon reduction, energy circles, and waste management for a longer period (Fig. 16.5).

16.4   Conclusion: Towards Healthy Cities

The base of the above mentioned urban strategies is to enhance green and connected cities. Firstly, fundamental change is possible via rebalancing a city's mobility system, creating access for all, transforming public realm into a green-blue network and activating public realm to create vibrant city centres.

Secondly, the role of planting in this green-blue network is beyond aesthetics and has increased societal relevance. Planting is becoming increasingly important to tackle big societal, urban challenges, such as climate adaptation, biodiversity, and heat mitigation. While traditionally, cities can be regarded as petrified landscapes, the integration of landscape and city can result in the creation of holistic cities.

Moreover, increasing urban densification and complexity caused by the need to design and organise cities in a circular way, demands holistic green strategies. In

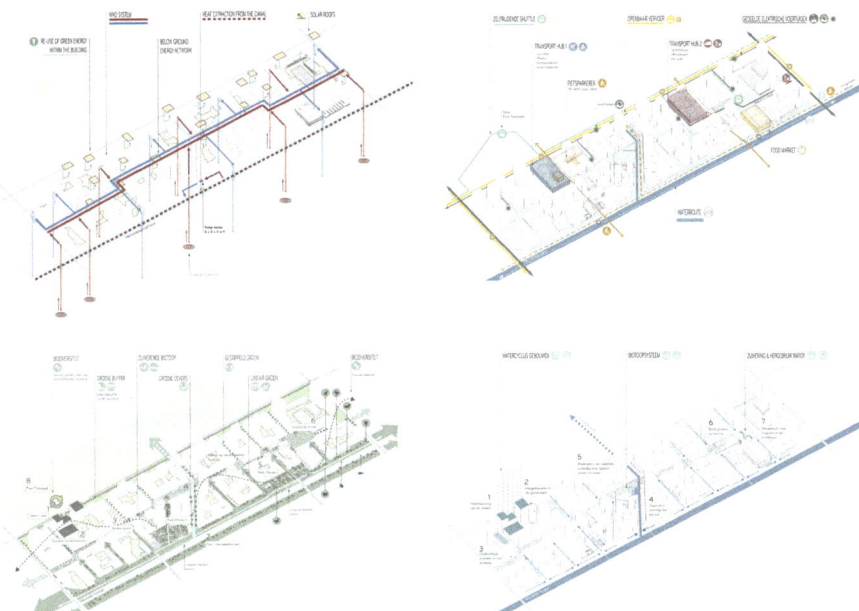

Fig. 16.5 Climate adaptive solutions and circular integrated in the plans for Merwede. (Owners collective Merwede, 2017)

fact, the city should be regarded as a system, connecting places with multi-modal and multi-functional corridors. Cities, like organisms, require inputs, such as water, energy, people, and produce outputs, such as waste. The continuous exchange of energy, material, and population is essential to the way a city functions, and can be regarded as a metabolism.

Finally, today's aspirations on including biodiversity in highly urban environments calls for the concept of 'urban biotopes'. The concept of urban biotopes regards a holistic approach in which planting, nutrients, and water together form new urban ecosystems. Urban biotopes seem promising for designing cities, in such a way that a balanced and holistic green system can evolve over time. Although urban biotopes require careful consideration about how to integrate energy production, underground infrastructure, and mobility in such a way that proper conditions for urban green are set, the concept is very promising for the necessary development of long lasting green networks and resilient cities.

The design task today is to manage those exchanges in a responsible and sustainable way—considering and strengthening the urban metabolism and creating healthy cities. Entering a new era will require that once more the interaction to other fields of expertise will be key to find strategic solutions to new challenges. Mixing up and integrating all disciplines seems a fairly logic approach when working in this context. To find answers to today's complex challenges, a clear vision and a strong collaboration between dedicated people on different fields of expertise is required.

Universal are the qualities that emerge from the landscape and that do connect us with mother earth. The qualities of change are the new programs that will be drivers for spatial adaptation. The creation of landscape cities today, being healthy and sustainable leads to interesting cities of tomorrow. It goes beyond blue-green networks: creating green that enhances urban biodiversity and brings nature to cities even in the densely built urban environment will be the next step in ensuring that cities are resilient enough to be a good place for working and living tomorrow.

References

Adams R (2009) From industrial cities to eco-urbanity; the Melbourne case study. In: Radovic D. (ed) Eco-urbanity: towards well-mannered built environments. Routledge, New York

Areso I (2009) Bilbao's strategic evolution. From the industrial to the post-industrial city. In: Instituto Cervantes. 'Restoring, regenerating, rethinking: the urban transformation of Madrid. In: Barcelona and Bilbao

Bulleri A (2018) Esercizi di riscatto urbano; Considerazioni sulla riqualificazione degli spazi aperti a Atene da Piazza Monastiraki a "Re-Think Athens" – Exercises in urban redemption; Considerations on the redevelopment of open spaces in Athens, from Monastiraki Square to 'Re-Think Athens'. In: Paesaggio Urbano (urban design) 2018-2, pp 124–133

City of Copenhagen (2016) Climate adaptation and urban nature. Copenhagen

Group of Building Environmental Studies, Physics Department, University of Athens (2013) Thermal and wind measurements performed in the Panepistimioy street in the city of Athens (final report). Athens

Hoyer J, Wolfgang D, Kronawitter L, Weber B (2011) Water sensitive urban design; principles and inspiration for sustainable stormwater management in the city of the future. (Research in the context of the European research project SWITCH.) Jovis, Berlin

Knuijt M (2008) The connected city, the metamorphosis of Central Rotterdam's public space. Topos (64) pp 50–55

Knuijt M (2013) One step beyond, a new city center for Athens. In: Topos, vol 85, pp 62–69

Knuijt M (2016a) Stadtentwicklung in Wassersensibelen Bereichen / Urbanisation en zone aquatique sensible. In Anthos 4-2016, pp 20–23

Knuijt M (2016b) Stadt als Stoffwechsel; Gemischte Netzwerke aus Stadt und Landschaft, TU Berlin, Heft 35, 2016

Kodukula S (2011) Reviving the soul in Seoul: Seoul's experience in demolishing road infrastructure and improving public transport. A joint case study by GIZ and KOTI. In: Case studies in sustainable transport, volume 6. GIZ, Deutsche Gesellschaft für Internationaler Zusammenarbeit, Eschborn

Lenzholzer S (2015) Weather in the city; how design shapes the urban climate. nai010, Rotterdam

Salles A (2013) Athènes rêve d'une coulée verte pour lutter contre la pollution et les embouteillages. Le Monde, Paris

Santamouris M, Gaitani N, Spanou A, Saliari M, Gianopoulou K, Vasilakopoulou K (2012) Using cool paving materials to improve microclimate of urban areas – design realisation and results of the Flisvos project. Build Environ 53, pp 128–136

Troy A (2012) The very hungry city; urban energy efficiency and the economic fate of cities. Yale University Press, London

Werner Sobek Green Technologies (2013) Final report 'Microclimatic simulation II design conclusions'. Athens

Chapter 17
The Influence of Landscape Architecture on Landscape Construction Health and Safety

John Smallwood

Abstract The influence of design on construction Health and Safety (H&S) is well documented in literature, as the concept and practice of 'designing for construction H&S'. However, there is a paucity of literature relative to landscape construction H&S, and none relative to the influence of landscape architecture on landscape construction H&S. Furthermore, no research has been conducted in this subject area, despite landscape construction entailing exposure to numerous hazards and risks. Given the status quo, a quantitative study was conducted among members of the Institute of Landscape Architects South Africa (ILASA), the objectives of the study being to determine, among others, perceptions, and practices of landscape architects in terms of landscape construction H&S. The salient findings include: site handover, site meetings, and site inspections/discussions predominate in terms of the frequency landscape construction H&S is considered/referred to on various occasions; method of fixing predominates in terms of the frequency construction H&S is considered/ referred to relative to design related aspects; position of components predominates in terms of the extent design related aspects impact on landscape construction H&S; tertiary landscape architecture education addresses landscape construction H&S to a minor extent; respondents rate their knowledge of landscape construction H&S and 'design for landscape construction H&S' skills as poor, and experience predominates in terms of respondents' acquisition of knowledge of landscape construction H&S. Conclusions include: respondents are committed to landscape construction H&S, however they are lacking in knowledge; the extent to which respondents perceive landscape architecture impacts on landscape construction H&S indicates inadequate knowledge; the ratio of action in terms of consideration of/reference to landscape construction H&S as a percentage of perceived impact of landscape design thereon indicates that there is potential to enhance such consideration/refer- ence; respondents appreciate the potential of interventions to improve landscape construction H&S, and the extent to which landscape architectural programmes

J. Smallwood (✉)
Department of Construction Management, Nelson Mandela University, Port Elizabeth, South Africa
e-mail: john.smallwood@mandela.ac.za

© Springer Nature Switzerland AG 2020
R. Roggema, A. Roggema (eds.), *Smart and Sustainable Cities and Buildings*,
https://doi.org/10.1007/978-3-030-37635-2_17

address landscape construction H&S reflects inadequate commitment thereto on the part of the related stakeholders. Recommendations include: tertiary education landscape architectural programmes should include appropriate 'designing for landscape construction H&S' modules as a component of a subject – probably design. The ILASA should develop practice notes relative to landscape construction H&S, and the South African Council for the Landscape Architectural Profession (SACLAP) should include construction H&S in their six work stages as per the identity of work (IoW) in a more comprehensive manner.

Keywords Construction · Health and safety · Influence · Landscape architects

17.1 Introduction

The definition of 'designer' in the South African Construction Regulations (Republic of South Africa 2014) includes, inter alia, a landscape architect. The Construction Regulations require designers to, among other things, modify the design or make use of substitute materials where the design necessitates the use of dangerous procedures or materials hazardous to H&S, and consider hazards relating to subsequent maintenance of the structure and make provision in the design for that work to be performed to minimise the risk. This alludes to the term 'designing for safety', which Behm (2006) defines as "The consideration of construction site safety in the preparation of plans and specifications for construction projects." Thorpe (2006) in turn contends that design is an important stage of projects, as it is at this stage that conceptual ideas are ideally converted into constructable realities. Thorpe (2006) further states that designing for safety is one of a range of considerations that need to be balanced simultaneously during design.

However, there is a paucity of literature pertaining to landscape construction H&S, and the role of landscape architecture in landscape construction H&S. Given this paucity, a study was undertaken, the objectives being to determine the:

- frequency at which landscape architectural practices consider landscape construction H&S on various occasions, and relative to various design related aspects;
- extent to which various design related aspects impact on landscape construction H&S;
- potential of various interventions to contribute to an improvement in landscape construction H&S;
- means by which landscape construction H&S knowledge is gained, and
- landscape architects' rating of their knowledge of landscape construction H&S and 'design for landscape construction H&S' skills.

17.2 Review of the Literature

17.2.1 Health and Safety Legislation and Recommendations Pertaining to Designers

In terms of the South African Construction Regulations (Republic of South Africa 2014), clients and designers have responsibilities with respect to construction H&S.

Clients are required to, prepare an H&S specification based on their baseline risk assessment (BRA), which is then provided to designers. They must then ensure that the designer takes the H&S specification into account during design, and that the designers carry out their duties in terms of Regulation 6 'Duties of designers'. Thereafter, clients must include the H&S specification in the tender documentation, which in theory should have been revised to include any relevant H&S information included in the designer report as discussed below.

Designers in turn are required to, inter alia: consider the H&S specification; submit a report to the client before tender stage that includes all the relevant H&S information about the design that may affect the pricing of the work, the geotechnical-science aspects, and the loading that the structure is designed to withstand; inform the client of any known or anticipated dangers or hazards relating to the construction work, and make available all relevant information required for the safe execution of the work upon being designed or when the design is changed; modify the design or make use of substitute materials where the design necessitates the use of dangerous procedures or materials hazardous to H&S, and consider hazards relating to subsequent maintenance of the structure and make provision in the design for that work to be performed, to minimize the risk. To mitigate design originated hazards, requires hazard identification and risk assessment (HIRA) and appropriate responses, which process should be structured and documented.

Despite the requirements of H&S legislation relative to clients and designers, in general, the related statutory councils' respective identities of work (IoW) record limited H&S deliverables (Deacon 2016). The point with respect to the respective IoW is that in general, the deliverables inform with respect to the competencies of practitioners and graduates. In the case of the South African Council for the Landscape Architectural Professions (SACLAP) (2011), the reference to H&S is as follows: Stage 1 'Project Initiation and Briefing' - nil; Stage 2 'Concept and Feasibility' - advise the client regarding the appointment of an H&S consultant where necessary; Stage 3 'Design and Development'; Stage 4 'Tender Documentation and Procurement' – nil; Stage 5 'Construction Documentation and Management' - where the compliance of landscape contractors could be monitored in accordance with the requirements of the H&S consultant, and Stage 6 'Project Close Out' – nil.

Furthermore, the International Labour Office (ILO) (1992) recommends that designers should: receive training in H&S; integrate the H&S of construction workers into the design and planning process, and not include anything in a design which would necessitate the use of dangerous structural or other procedures or hazardous materials, which could be avoided by design modifications or by substitute materials.

17.2.2 Landscape Construction H&S

The Occupational Safety and Health Administration (OSHA) in the United States of America (USA) identified landscape and horticultural services, which include landscape construction, as one of the most hazardous industries in the USA. Potential hazards include: motor vehicle and other equipment accidents; ergonomic injuries such as back strains; exposure to noise, heat, cold, chemicals, and insects; amputations; slips, trips and falls; eye injuries, and electrocutions (Integrity Insurance 2013).

17.2.3 Statistics

Table 17.1 provides an overview of landscape gardening injury statistics for the years 2008 to 2014 (Federated Employers Mutual 2015). The mean accident frequency rate of 1.71 indicates that 1.71 per 100 workers experience a disabling injury, which results in a loss of a shift or more after the day of the injury. The highest is 2.04 relative to 2010. The mean for all classes is 3.13, the highest being 3.86 relative to 2008. Then, the fatality rate for the year 2010 equates to 25.8 per 100,000 workers [(100,000 / 7755) x 2], which is high.

Table 17.1 Landscape gardening injury statistics for the years 2008–2014

Year	Accident frequency rate	Employees (No.)	Accidents (No.)	Fatal accidents (No.)
2008	1.83	8898	163	0
2009	1.80	9025	162	2
2010	2.04	7755	158	2
2011	1.57	8010	126	0
2012	1.66	8749	145	0
2013	1.61	8466	136	2
2014	1.44	8430	121	0
Mean	1.71			

17.3 Research Method

A previous study conducted among engineers in South Africa to determine their perceptions and practices with respect to construction H&S investigated the: frequency at which construction H&S is considered on various occasions, and relative to various design related aspects; extent to which various design related aspects impact on construction H&S; sources of H&S knowledge, and the potential of various aspects to contribute to an improvement in construction H&S (Smallwood 2004). The study reported on constitutes a replication of this prior study, which study in turn constitutes the origin of the occasions, aspects, and sources. Given that it is landscape oriented, some amendments were necessary.

The questionnaire consisted of primarily closed ended five-point Likert scale type questions – 10/13 questions were closed ended. The questionnaire, accompanied by a cover letter explaining the rationale for the study, was forwarded per e-mail to 139 members of the ILASA. 21 Responses were included in the analysis of the data, which equates to a net response rate of 15.1%. A follow up e-mail was sent after a few weeks in an endeavour to enhance the response rate, but with limited success. Possible reasons for the response rate include the subject relative to the practice of landscape architectural design, namely landscape construction H&S.

Descriptive statistics in the form of frequencies and a measure of central tendency in the form of a mean score (MS) were computed to present the findings of the empirical study. The MS is based upon a weighting of the responses to the five-point Likert scale type questions, and ranges from a minimum of 1.00 to a maximum of 5.00. The MS thus enables the range of percentage responses to be interpreted, and parameters, occasions, aspects, and interventions to be ranked. Due to the number of responses, inferential statistical analysis was not possible.

17.4 Research Findings

Table 17.2 presents the importance of eleven project parameters to respondents in terms of percentage responses to a range of 1 (not) to 5 (very), and a MS ranging between 1.00 and 5.00. It is notable that all eleven MSs are above the midpoint of 3.00, which indicates the parameters are more than important, as opposed to less than important.

It is notable that 6/11 (54.6%) parameters' MSs are > 4.20 ≤ 5.00 – between near major to major/major importance. The environment (natural), ranked third, and public H&S, ranked fifth, are within this range. The remaining 5/11 (45.4%) MSs are > 3.40 ≤ 4.20 – between important to more than important/more than important. Project H&S ranked eighth is in this range. It is notable that the MS of public H&S is 4.45, and for project H&S 3.90 – an absolute difference of 0.55, and the former is 19% more important than the latter.

Table 17.2 Importance of project parameters to respondents

Response (%)

Parameter	Unsure	Not Very					MS	Rank
		1	2	3	4	5		
Client satisfaction	0.0	0.0	0.0	4.8	9.5	85.7	4.81	1
Project quality	0.0	0.0	0.0	0.0	23.8	76.2	4.76	2
Environment (natural)	0.0	0.0	0.0	5.0	20.0	75.0	4.70	3
Designer satisfaction	0.0	0.0	0.0	4.8	38.1	57.1	4.52	4
Public H&S	0.0	0.0	0.0	5.0	45.0	50.0	4.45	5
Project cost	4.8	0.0	0.0	4.8	38.1	52.4	4.29	6
Project schedule	0.0	0.0	4.8	28.6	28.6	38.1	4.00	7
Project H&S	0.0	0.0	4.8	28.6	38.1	28.6	3.90	8
Contractor satisfaction	0.0	0.0	5.0	25.0	60.0	10.0	3.75	9
Worker satisfaction	4.8	4.8	4.8	28.6	47.6	9.5	3.55	10
Labour productivity	4.8	4.8	0.0	33.3	42.9	14.3	3.48	11

Table 17.3 Frequency of consideration / reference to landscape construction H&S on various occasions

Response (%)

Occasion	Unsure	Never	Rarely	Sometimes	Often	Always	MS	Rank
Site handover	9.5	0.0	9.5	9.5	28.6	42.9	4.16	1
Site meetings	4.8	0.0	4.8	14.3	38.1	38.1	4.15	2
Site inspections / discussions	0.0	0.0	4.8	14.3	42.9	38.1	4.14	3
Preparing project documentation	0.0	9.5	0.0	23.8	38.1	28.6	3.76	4
Pre-tender meeting	4.8	0.0	23.8	19.0	23.8	28.6	3.60	5
Evaluating tenders	9.5	4.8	14.3	33.3	23.8	14.3	3.32	6
Constructability reviews	9.5	4.8	23.8	23.8	28.6	9.5	3.16	7
Pre-qualifying contractors	4.8	9.5	23.8	23.8	23.8	14.3	3.10	8
Detailed design	0.0	14.3	14.3	33.3	28.6	9.5	3.05	9
Working drawings	0.0	19.0	14.3	38.1	14.3	14.3	2.90	10
Client meetings	0.0	0.0	52.4	19.0	19.0	9.5	2.86	11
Design coordination meetings	0.0	9.5	23.8	52.4	9.5	4.8	2.76	12
Deliberating project duration	4.8	23.8	23.8	28.6	14.3	4.8	2.50	13
Concept (design)	0.0	9.5	57.1	14.3	19.0	0.0	2.43	14

Table 17.3 presents the frequency at which landscape architects consider/refer to landscape construction H&S relative to fourteen occasions, in terms of a frequency range, never to always, and a MS ranging between 1.00 and 5.00. It is notable that

Table 17.4 Comparison of the frequency at which landscape architects consider / refer to landscape construction H&S relative to design related aspects and the impact of the aspects on landscape construction H&S

Design related aspect	Consider		Impact		
	MS	Rank	MS	Rank	Con./ Imp.
Position of components	3.33	3	3.62	1	0.89
Method of fixing	3.55	1	3.24	2	1.14
Content of material	2.88	9	3.24	3	0.84
Specifications e.g. hard surfaces	3.29	4	3.19	4	1.05
Details	2.95	7	3.10	5	0.93
Edge of materials	3.35	2	2.95	6	1.21
Position of vegetation / features	2.95	6	2.86	7	1.05
Design (general)	2.55	13	2.85	8	0.84
Finishes	2.95	8	2.70	9	1.15
Mass of materials	2.82	10	2.67	10	1.09
Plan layout	2.48	16	2.58	11	0.94
Elevations	2.38	18	2.53	12	0.90
Method of planting	2.60	12	2.50	13	1.07
Schedule	2.50	15	2.47	14	1.02
Texture of materials	2.50	14	2.45	15	1.03
Surface area of materials	3.19	5	2.41	16	1.55
Mass of vegetation / features	2.80	11	2.35	17	1.33
Texture of vegetation / features	2.26	19	2.28	18	0.98
Content of vegetation	2.47	17	2.21	19	1.21

9/14 (64.3%) MSs are above the midpoint of 3.00, which indicates consideration of/reference to landscape construction H&S relative to these occasions can be deemed to occur.

It is notable that no occasions are > 4.20 ≤ 5.00 – between often-always, however, 5/14 (35.7%) are > 3.40 ≤ 4.20 – between sometimes-often. It is notable that the top three ranked occasions are downstream during Stage 5, namely site handover, site meetings, and site inspections/discussions. Then fourth ranked preparing project documentation, and pre-tender meeting are Stage 4 occasions. Those occasions ranked sixth to twelfth (50%) have MSs > 2.60 ≤ 3.40 – between rarely to sometimes. Evaluating tenders, and pre-qualifying contractors are Stage 4 occasions, whereas constructability reviews, detailed design, working drawings, client meetings, and design coordination meetings are Stage 3 occasions. Client meetings also occur during Stage 1 and 2. Deliberating project duration and concept (design) have MSs > 1.80 ≤ 2.60, and thus the frequency is between never to rarely. The former occurs during Stages 1, 2, 3 and 4. The latter is a Stage 2 occasion.

Table 17.4 provides a comparison of the frequency at which landscape architects consider/refer to landscape construction H&S, relative to design related aspects and the impact of the aspects on landscape construction H&S in terms of MSs, ranks, and the ratio of consider (Con.) to impact (Imp.). The table reflects action as a percentage of perceived impact. The 'impact' MSs are greater than the 'consider' MSs in 12/19

Table 17.5 Potential of interventions to contribute to an improvement in landscape construction H&S

Response (%)

Intervention	Unsure	Minor.........................Major					MS	Rank
		1	2	3	4	5		
Awareness (D & C)	0.0	0.0	0.0	9.5	47.6	42.9	4.33	1
Safe working procedures (C)	0.0	0.0	4.8	14.3	28.6	52.4	4.29	2
Contractor planning (C)	0.0	4.8	4.8	19.0	38.1	33.3	3.90	3
Design of equipment (construction) (C)	23.8	4.8	4.8	14.3	28.6	23.8	3.81	4
Constructability (general) (D)	0.0	4.8	9.5	28.6	33.3	23.8	3.62	5
Specification (D)	0.0	0.0	14.3	33.3	38.1	14.3	3.52	6
Workshop facilities on site (C)	9.5	9.5	9.5	28.6	28.6	14.3	3.32	7
Design of tools (C)	19.0	14.3	0.0	23.8	33.3	9.5	3.29	8
Mechanisation (D & C)	14.3	9.5	9.5	28.6	23.8	14.3	3.28	9
Reengineering (D & C)	23.8	4.8	9.5	38.1	9.5	14.3	3.25	10
Details (D)	0.0	9.5	9.5	38.1	38.1	4.8	3.19	11
General design (D)	0.0	14.3	23.8	33.3	19.0	9.5	2.86	12
Prefabrication (D)	9.5	14.3	9.5	47.6	14.3	4.8	2.84	13

(63.2%) cases, and lower in 7/19 (36.8%) of cases. In terms of the lowest ratio, the greatest difference is relative to content of material (0.84) and design (general) (0.84), followed by position of components (0.89). In terms of the highest ratio, the greatest extent is relative to surface area of materials (1.55), followed by mass of vegetation/features (1.33), and edge of materials (1.21).

Table 17.5 presents the potential of interventions to contribute to an improvement in landscape construction H&S in terms of percentage responses to a range of 1 (minor) to 5 (major), and a MS ranging between 1.00 and 5.00. It is notable that 11/13 (84.6%) MSs are above the midpoint of 3.00, which indicates the interventions have major as opposed to minor potential to contribute to an improvement. 'D' denotes design, and 'C' construction.

It is notable that only 2/13 (15.4%) interventions' MSs are > 4.20 ≤ 5.00 – between near major to major / major potential – awareness (D & C), and safe work procedures (SWPs) (C). 4/13 (30.8%) MSs are > 3.40 ≤ 4.20 – between potential to near major/near major potential - contractor planning (C), design of equipment (construction) (C), constructability (general) (D), and specification (D).

The remaining 7/13 (53.9%) interventions have MSs > 2.60 ≤ 3.40 – between near minor potential to potential. 5/7 MSs are in the upper half of this range - workshop facilities on site (C), design of tools (C), mechanisation (D & C), reengineering (D & C), and details (D).

In summary: one 'D&C' and one 'C' have between near major to major/major potential; two 'C' and two 'D' have between potential to near major/near major potential, and two 'C', two 'D & C', and three 'D' have between near minor potential to potential/potential.

Table 17.6 Form in which landscape construction H&S should be addressed in landscape architecture programmes

Form	Yes (%)
A subject 'construction H&S'	42.9
Included in a range of subjects	23.8
Included in the subject 'design'	23.8
Unsure	9.5
Not at all	0.0

Table 17.7 Respondents' source of landscape construction H&S knowledge

Source	Yes (%)
Experience	95.2
Magazine articles	38.1
Tertiary education	23.8
Workshops	19.0
Practice notes	14.3
Postgraduate qualifications	14.3
CPD seminars	9.5
Other: H&S act	4.8
Journal papers	4.8
Conference papers	4.8

In theory and practice, all the interventions have the potential to contribute to an improvement in landscape construction H&S.

Respondents were required to rate their knowledge of landscape construction H&S and 'design for landscape construction H&S' skills in terms of percentage responses to a scale of 1 (limited) to 5 (extensive). The MS of 3.00 indicates the rating is between near limited to average as the MS is $> 2.60 \leq 3.40$.

Respondents were required to indicate the extent tertiary landscape architecture education addresses landscape construction H&S in terms of percentage responses to a scale of 1 (minor) to 5 (major). The resultant MS of 2.15 indicates the extent is between a minor extent to near minor/near minor extent, as the MS is $> 1.80 \leq 2.60$.

It is notable that 42.9% of respondents identified 'A subject construction H&S', and 23.8% each of 'Included in a range of subjects', and 'Included in the subject design' (Table 17.6).

In terms of respondents' source of landscape construction H&S knowledge, 95.2% identified experience, followed by magazine articles (38.1%), and tertiary education (23.8%) (Table 17.7). The seven other sources were identified by less than 20% of the respondents.

17.5   Conclusions

The traditional project parameters of cost, quality, and time are more important than project H&S, which indicates that landscape architects' perceptions reflect those of built environment designers, and other stakeholders. However, public H&S is substantially more important than project H&S, which reflects an awareness of the impact of the landscaped environment on the users in the form of the public.

Landscape construction H&S is considered/referred to on various occasions mostly to a major as opposed to a minor extent, and relative to design related aspects mostly to a minor as opposed to a major degree, which indicates a degree of commitment to landscape construction H&S, yet inadequate knowledge. There is minor as opposed to major appreciation in terms of the extent design related aspects impact on landscape construction H&S, which indicates inadequate knowledge. Then, the ratio of action in terms of consideration of/reference to landscape construction H&S as a percentage of perceived impact of landscape design thereon indicates that there is potential to enhance such consideration/reference.

There is major as opposed to a minor degree of appreciation of the potential of interventions to contribute to an improvement in landscape construction H&S.

Landscape architectural programmes address landscape construction H&S to a limited extent, which indicates that: the presenters of such programmes are likely not committed thereto; ILASA and SACLAP are not engendering the inclusion thereof in such programmes; ILASA and SACLAP are not interrogating the degree to which it addressed in such programmes during accreditation panel visits, and ILASA and SACLAP are likely not commitment thereto.

Respondents' self-rating of their knowledge of landscape construction H&S and 'design for landscape construction H&S' skills, and the level of acknowledgment relative to experience as the source of landscape construction H&S knowledge further confirms that landscape architectural programmes address landscape construction H&S to a limited extent.

17.6 Recommendations

Recommendations include that tertiary education landscape architectural programmes should include appropriate 'designing for landscape construction H&S' modules as a component of a subject – ideally design. The ILASA should develop practice notes relative to landscape construction H&S, and the SACLAP should make more comprehensive reference to construction H&S in their six work stages of their IOW. Furthermore, SACLAP accreditation reviews of tertiary education landscape architectural programmes should interrogate the extent to which landscape construction H&S is addressed, or rather embedded in such programmes. Both ILASA and SACLAP should actively promote landscape construction H&S-related continuing professional development (CPD).

References

Behm M (2006) An analysis of construction accidents from a design perspective, The Center to Protect Workers' Rights, Silver Spring

Deacon CH (2016) The effect of the integration of design, procurement, and construction relative to health and safety. Unpublished PhD (Construction Management) Thesis. Nelson Mandela Metropolitan University, Port Elizabeth

Federated Employers Mutual (2015) FEM's Accident Stats as at March 2015. http://www.fem.co.za/accidentStatistics.aspx

Integrity Insurance (2013) Landscape industry manual, integrity insurance, appleton

International Labour Office (ILO) (1992) Safety and health in construction, ILO, Geneva

Republic of South Africa (2014) No. R. 84 Occupational Health and Safety Act, 1993 Construction Regulations 2014. Government Gazette No. 37305, Pretoria

Smallwood JJ (2004) The influence of engineering designers on health and safety during construction. J South Afr Inst Civil Eng 46(1):2–8

South African Council for the Landscape Architectural Professions (SACLAP) (2011) Identification of work for the South african council for the landscape architectural professions. SACLAP, Ferndale

Thorpe B (2006) Health and safety in construction design. Gower Publishing Limited, Aldershot

Part VI
Space and Place

How can the quality of places be increased using new functions and uses such as urban agriculture. Can these places play a role in developing regenerative cities? And how do we measure the quality of these places.

Chapter 18
A Multi-Criteria Decision Analysis Based Framework to Evaluate Public Space Quality

Peijun He, Pieter Herthogs, Marco Cinelli, Ludovica Tomarchio, and Bige Tunçer

Abstract Good public space is an inherent part of liveable cities. However, due to the complexity and ambiguity of the concept of public space quality, capturing essential characteristics to assess the quality of public space in a quantitative framework is not straightforward. In this paper, we introduce the Public Space Quality Index (PSQI), a Multi-Criteria Decision Analysis (MCDA) based framework to measure the quality of public space in a systematic manner. This paper discusses the development of our research methodology, which includes four key phases: criteria selection, criteria ranking and weighting, criteria quantification, and criteria aggregation. We also present the results of the first phase, and discuss the main findings from an expert workshop we conducted with Singapore-based urban design professionals, as a validation and learning moment for our past and future work, respectively.

Keywords Public space quality · MCDA · Urban design · Evidence-based design

P. He (✉) · P. Herthogs (✉)
Future Cities Laboratory, Singapore ETH Centre, ETH Zürich, Singapore, Singapore

Singapore University of Technology and Design, Singapore, Singapore
e-mail: peijun@arch.ethz.ch; herthogs@arch.ethz.ch

M. Cinelli
Future Resilient Systems, Singapore ETH Centre, ETH Zürich, Singapore, Singapore
e-mail: marco.cinelli@frs.ethz.ch

L. Tomarchio · B. Tunçer
Singapore University of Technology and Design, Singapore, Singapore

Future Cities Laboratory, Singapore ETH Centre, ETH Zürich, Singapore, Singapore
e-mail: ludovica_tomarchio@mysutd.edu.sg; bige_tuncer@sutd.edu.sg

© Springer Nature Switzerland AG 2020
R. Roggema, A. Roggema (eds.), *Smart and Sustainable Cities and Buildings*,
https://doi.org/10.1007/978-3-030-37635-2_18

18.1 Introduction

Public spaces are essential in the creation of vibrant and sustainable cities: they are our platforms for social interaction, social mixing and social inclusion, facilitate the exchange of ideas, culture, and skills, support trade, leisure and tourist activities, and have many other functions. As they define the fabric of urban life, there has been continuous interest in studying public space use and the design of qualitative public space, from various perspectives, in many different disciplines.

However, evaluating public space quality is not straightforward. A wide range of characteristics contribute to how humans perceive open spaces, from diverse dimensional properties such as geometry, topology and other physical attributes of the place, to the number and profiles of users of the space, the flexibility and multi-functionality of the space and its boundaries, and so on. Moreover, as is often the case in design, some of these characteristics are inherently and intricately interrelated, making it difficult to assess the contributions of each individual part to the quality of the whole.

Despite these difficulties, scholars in the field have always tried to inform design by means of guidelines and heuristics, requiring them to qualitatively and quantitatively describe and measure properties of public space. Canonical works of urban design, such as Jacobs (1961), Whyte (1980) or Gehl (1987), describe systematic examinations of (the reasons behind) different levels of liveliness in urban public spaces from a social use perspective, and how these relate to particular characteristics and design features of these spaces. In order to evaluate public spaces and inform design practice, the findings – or, if lacking, narratives – of urban design studies are often distilled into guidelines and principles, ranging from straightforward qualitative checklists (e.g. Gehl Institute 2017; Carr et al. 1992), over multi-spectral frameworks (e.g. Montgomery 1998; Gehl Institute 2018), to comprehensive manuals (e.g. Llewelyn-Davies 2000). In recent years, more comprehensive models to quantify, assess and score public space qualities have been proposed. Ewing and Handy (2009) developed operational definitions to measure five urban design qualities related to walkability (imageability, enclosure, human scale, transparency, and complexity), applied to (commercial) streets in the United States. Mehta (2014) developed a public space index based on 45 different variables to evaluate the quality of pubic space from five dimensions including inclusiveness, meaningful activities, comfort, safety and pleasurability. Varna and Tiesdell (2010) utilized various descriptors and indicators to evaluate public space based on the concept of publicness such as ownership status, walking opportunities, means of control and others. Other studies focus on the impact of a limited set of easily measurable criteria on human activity in public space; Beirão and Koltsova (2015), for example, put forward that a street's liveliness potential can be expressed in terms of the number of entrances in a street, the street length, and the territorial depth of building entrances. For a comprehensive overview of public space quality guidelines, measures and evaluations, consult Koltsova (2017).

While literature shows an increased interest in evidence-based, measurable guidelines and criteria for "good" public space design, there are still important research gaps to be bridged on our road towards informed urban design approaches and tools. We put forward four important limitations. Firstly, the level of validity is often unclear (in part due to the necessary reliance on case study research) and the range of applicability is often limited (because validation studies, due to their time-consuming nature, are restricted along public space types, locales, times, etcetera). As both limitations could be considered inherent to the topic, this implies that a robust methodology to evaluate public space quality ought to include extensive validation and cater towards extendibility, respectively. Secondly, while most public space design guidelines and evaluation methods aim towards the same goal (i.e. "better public space"), and there is a high degree of overlap between the criteria, characteristics or design ambitions they prescribe, the wide range and variety of used terms, descriptions and structuring categories makes it difficult to compare different methods and combine lessons or evidence provided by each individually. Together, point one and two have led to a fragmented landscape of results. Thirdly, the majority of evaluations and guidelines only qualitatively mention the important aspects for assessing public space quality, but seldom illustrate how to quantitatively measure each aspect in detail. Fourthly, although some researchers may have attempted to quantify each characteristic of public space, they lack a systematic way to combine all important features into a score that can measure public space quality in a more objective manner. Evaluation methods that do propose a single score (e.g. Mehta 2014) apply a simple weighted sum approach; for a complex topic like public space quality, relative weights can be easily criticised and should not be considered robust (Narula and Reddy 2015). Moreover, a single score will not manage to capture the diverse differences in opinions and appreciations different users will have about the same public space. Carmona (2015) argued that one of the defining characteristics of the public spaces in a global city is the sheer diversity of spaces and overlapping typologies, and that more and more urban design professionals acknowledge that there is no such thing as an "ideal pubic space" that suits all users, implying a paradigm shift from a "public realm that is equally appealing to all" towards "an acceptance that users are diverse and will seek different things from their spaces".

With these limitations in mind, we are developing the Public Space Quality Index (PSQI), a Multi- Criteria Decision Analysis (MCDA) based framework to quantitatively measure the quality of public space. It is a systematic attempt to integrate a comprehensive list of criteria of public space into scores that can be used to compare and rank different public spaces. Our ambition is to create a methodological framework that is:

1. Incremental, so a sub-set of criteria and related design guidelines can be developed at a time;
2. Extendible, so more criteria can be added at a later stage;
3. Adaptable, so others can use it in their contexts, using their particular design guidelines;

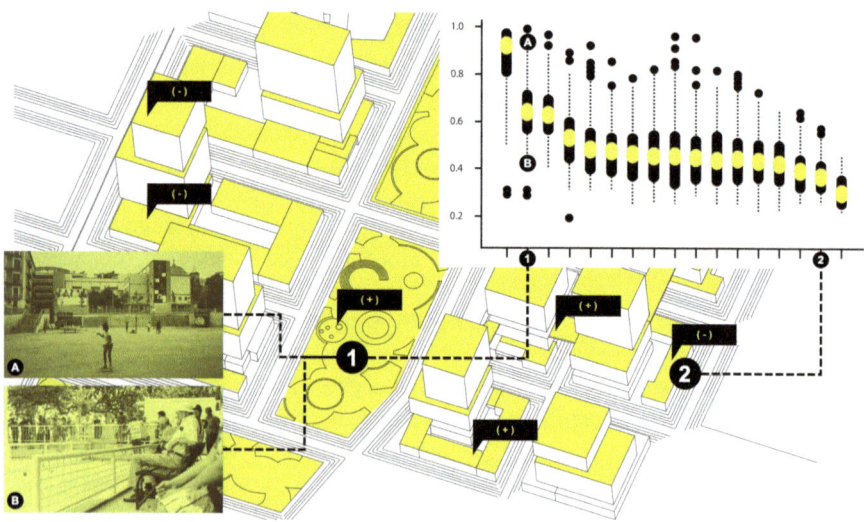

Fig. 18.1 Illustration of the goal we are working towards: a framework that can demonstrate of how well a space supports diverse user needs, as well as a robust, general score that indicates if a space does well for many

4. Coherent under change, so it will be possible to compare scores for different user groups or expert types, but at the same time provide a single, robust score for per public space, based on criteria that are crucial for everyone, including a degree of uncertainty.

Figure 18.1 illustrates the goal we are working towards. Each entry in the top-right chart shows the range of scores of public spaces in the neighborhood below, under various assessment and weighting scenarios that represent the preferences of particular user groups. In this example, we highlighted two user groups, school children (A) and seniors (B), but any group could be included in the scoring model, be it climate comfort experts, government agencies, or parents. Because each group values criteria differently, they score each space differently. Assessing many user groups (or many individuals) will allow us to plot all public space quality scores; if a score is consistently high among different users, it is likely that this public space has a high general quality – this is indicated by the yellow dots of the box plot graph in the illustration. What we end up with is both an indication of how well a space supports diverse user needs, as well as a robust, general score that indicates if a space does well for many. This can in turn inform the design of networks of diverse public spaces, some with general appeal, others catering to particular users. Note that, although the general framework we are developing can be expanded to different countries, we are focusing on the PSQI for use in Singapore in this work.

In this paper, we present the results of the first phase of the development of the PSQI. We start by describing the four phases of our methodology to build this systematic MCDA framework (i.e. criteria selection, ranking and weighting elicitation, criteria quantification, and criteria aggregation). We continue by discussing the selection of our initial list of criteria and our first test: an expert workshop to both get feedback on our set of selected criteria, and try out a number of rank and weight elicitation methods in preparation for phase two. Concluding remarks and possible future directions are given and discussed in the last section.

18.2   A Methodology to Derive a Public Space Quality Index

Multi-Criteria Decision Analysis (MCDA) represents a collection of instruments and methods to understand the structure of a decision-making problem and the multitude of dimensions that characterise it. MCDA helps the decision makers investigate complex problems with qualitative- quantitative attributes, using open and explicit judgement criteria to obtain better informed and justifiable choices. It has been widely employed to solve problems in different areas, such as environment (Soltani et al. 2015), sustainability (Zavadskas et al. 2015), and urban planning (Curwell et al. 2005). Our proposed MCDA-based framework to evaluate public space quality has four consecutive phases, each of which will be explained in more detail in the following subsections.

18.2.1   Phase 1: Selecting Public Space Quality Criteria

In the first phase, researchers decide which characteristics of a public space (such as the shape, the number of trees and benches, . . .) are relevant in describing its quality. Defining and selecting which public space characteristics will be quantified as criteria is done based on a literature review and discussions with urban design experts. The aim is to get a clear understanding of the problem that needs to be assessed using MCDA, the interrelations between its multiple criteria, and how these criteria are perceived (by experts, users, or other stakeholders).

Effective problem structuring and criteria selection are fundamental steps of the MCDA framework. We followed a set of guidelines for MCDA criteria selection developed by Akadiri and Olomolaiy (2012) and Akadiri (2013):

- Comprehensiveness: all the possible aspects of public space should be covered in order to assure a comprehensive evaluation list, i.e. criteria should be ideally selected based on what is desirable to measure not on which criteria is available;
- Applicability: the criteria should be generally applicable to assess public space quality as precisely as possible;
- Transparency: the criteria have to be easily understandable and selected in a traceable manner, so as to avoid misunderstandings and misinterpretations;
- Practicability: the criteria must be implementable and operational.

After developing an initial set of chosen criteria that together could define public space quality, this initial set of criteria is discussed by a panel of urban design and public space experts. It also enables testing a number of ranking and weighting elicitation methods in preparation for phase two. Section 18.3 discusses the development of our criteria list and the expert panel (under the form of a workshop).

18.2.2 Phase 2: Ranking and Weighting Criteria

When the criteria that determine public space quality have been selected, the next thing to determine is how much each one contributes to the overall quality, i.e. determining their rank and the relative distance between each rank (which reflects their weighting). This is commonly done based on a large sample survey. In our study, we plan to survey several different expert and user groups, in order to map out their different of the relative importance of particular public space characteristics.

As long as the rankings are obtained, there are various methods to elicit the weights, such as the rank sum weight method, rank exponent weight method, reciprocal method, or rank-order centroid weight method. Moreover, the original ordinal ranks can also be used as weight by simple normalization. In this work, we use multiple weighting methods to build different weighting scenarios based on the results from the workshop, as we aim to explore the impact of different weightings on the overall public space quality score.

18.2.3 Phase 3: Quantifying the Public Space Characteristics

This phase determines how to translate each criterion into a 0–1 score (i.e. how to quantitatively measure them). In addition, the value function of each criterion is determined, which is a mathematical presentation of people's judgement of the criterion. For example, a measured score of 0.5 for a criterion might already represent 90% of the value that criterion adds to the total score. This value function elicitation is again based on a survey of the same groups surveyed in phase two (and could be part of the same survey).

There are several common approaches in building the value function, such as direct rating technique, curve fitting, bisection techniques and many others (Stewart 2005). However, the construction of a value function for every evaluation criterion in the framework is not an easy task. On one hand, finding a value function (monotonic or not, continuous or discrete, linear or convex, etc.) to correctly describe the essential property of one criterion is demanding, especially for those criteria that are hardly measurable (e.g. human scale). On the other hand, different user groups or experts can have different perceptions of the same criterion and thus it is possible to have multiple value functions per criterion – though this is not commonly done in MCDA, it is necessary to explore variations in user group appreciation of public space. To tackle the first difficulty, we will not only tailor the value function for each criterion based on a thorough understanding of its property using traditional methods such as direct rating technique, but also plan to embrace new technologies to create and represent different virtual public spaces as a parametric 3D model. In this way, survey participants can change criteria and directly see how it is represented in a virtual representation. For example, suppose we want to obtain the value function of criterion Secure, we can check the participants' perception of "feeling safe" by showing them visualisations with different numbers of cameras, asking the question: "to what extent do you feel protected against harm: not all – could be better – good enough – good?"

To tackle the second challenge, we will obtain different value functions from different user profiles to understand the various needs and requirements among different groups of people. Moreover, it is possible that even the participants in the same user group will provide different values functions, thus we will adopt a similar strategy as the Stochastic Multi-Objective Acceptability Analysis (SMAA) developed by Tervonen and Figueira (2008) to explore the whole space of parameters (and related criteria and valuation).

18.2.4   Phase 4: Criteria Aggregation and Public Space Quality Index

The final step aggregates all the criteria into a public space quality score using their respective weights and value functions obtained from surveys. We will use a weighted-sum aggregation method in this work, commonly used in MCDA literature. Unlike most methods, because we will elicit criteria weights and value functions from many different user groups, we will also be able to map how public spaces score for different user groups, and determine a single, robust score representing the "general appeal" of these spaces.

There are a range of alternative aggregation methods in MCDA, but the simplest and most used one is the additive model, as shown in Eq. (18.1):

$$V_{k,j}(p) = \sum\nolimits_{i=1}^{m} w_{ik} v_{ij}(p) \tag{18.1}$$

where $V_{k,j}(p)$ is the quality score of public space p, w_{ik} is the k-th weighting scenario of criterion i, and is the i-th value function of criterion which reflects experts' preference of public space p's performance on criterion i.

However, when using multiple criteria to model the quality of a complex subject, such as public spaces, the criteria will likely have some interdependency, i.e. some might positively or negatively reinforce others. Interactions between criteria (Stewart 2005) is an important issue in MCDA and will be factored in to obtain a more accurate quality score. In order to do this, we revise Eq. (18.1) as follows:

$$V_{k,j}(p) = \sum_{i=1}^{m} w_{ik}v_{ij}(p) - \frac{1}{2} \sum_{i=1}^{m} \sum_{t=1,t\neq i}^{m} \beta_{ti}\sigma_{ti}(p) \qquad (18.2)$$

where $\sigma_{ti}(p)$ represents the interaction between the i-th and t-th criterion with a value of $\{0, 1, -1\}$, meaning no interaction, positive interaction, and negative interaction, respectively; β_{ti} demonstrates the extend of such interaction which can be measured qualitatively, e.g. a "weak-mild-medium-strong-very strong" scale has corresponding value of 0.1, 0.3, 0.5, 0.7, 1.

18.3   Results and Discussion

18.3.1   Initial Set of Public Space Quality Criteria

Following the guidelines discussed in Sect. 18.2.2, we have done an extensive literature review of topics on public space, public space quality, publics space assessment and related sources (in part reflected in the introduction). We developed an initial list of 19 public space criteria in a first attempt to develop a comprehensive set (with comprehensive implying it should capture the factors commonly influencing quality). Part of this 19-criteria list and its corresponding descriptions are shown in Table 18.1, per illustration.

Table 18.1 Public space quality criteria

ID	Criteria	Description
1	Flexible use	The space use is not strictly predefined or regulated regarding what activities can take place, or if users can improvise uses or rearrange urban furniture.
2	Interactive boundaries	Interactive boundaries support active, passive, and social engagement (e.g. facade transparency, shopfronts, visually interesting boundaries, …).
3	Maintenance and cleanliness level	The maintenance and cleanliness level of the space. Also refers to surface materials, finishes, correct choice of materials, ease of maintenance, accessibility for maintenance.
16	Supports active engagement	The space can invite people for physical activities, which may include: areas for active activities (sports, exercise, play), interactive water features, or interactive green features (grass, sand, climbing trees, …).
17	Supports passive engagement	The space can invite people to observe the space, which may include: presence of water feature; trees; green feature; interesting views; opportunities for "people watching".
18	Supports social engagement	The space can invite people to interact socially, which may include: grouping of seating that allows conversations; opportunities for communal/group activities, opportunities to eat/drink in group; opportunities for the exchange of goods (e.g. market).

18.3.2 The Expert Workshop

The expert panel to assess our initial list of criteria and prepare for the second phase of the PSQI development took the form of a 90-min expert workshop. We selected and invited several Singapore-based experts in urban design and public space use, resulting in a group of 14 experts (practitioners and academics) with various backgrounds (small and large offices, specialists in mobility, thermal comfort, human cognition, . . .) and a balanced gender and (non-)Singaporean ratio.

The goals of this workshop were: firstly, to discuss the soundness and completeness of the list of criteria we proposed; secondly, to get an initial set of results that rank and weight this list; thirdly, to understand the extent to which different experts differ in opinion on public space quality, and how comfortable they are quantifying something of a complex and qualitative nature. Note that while we obtained ranks and weights for our criteria list, these results are only used to inform the development of Phase Two of this study.

Part 1: discussing and extending the criteria list. After a short introductory presentation of our research goals, we introduced our criteria list to the experts, describing and explaining each criterion individually. We then opened the floor for discussion. The experts generally agreed on the proposed list, but suggested adding two more criteria to the list – "Inclusion and Social Justice" and "Visual Attractiveness". With these added, there was general consensus on the resulting list of 21 criteria (with the caveat that the phrasing would need more clarity).

Part 2: individual direct scoring of the list of 21. In the first exercise, we asked experts to directly score each criterion in our extended list, using a pre-made electronic survey. Experts answered the question "How important do you think the criteria are for good public space design?" on a 0–6 scale, where 0 meant "not important at all" and 6 meant "extremely important". This short task gave them time to familiarise themselves further with our list of criteria, and put an initial – individual – value on the importance of each criterion in terms of its contribution to public space quality. The survey results were then aggregated, and the criteria list ranked by descending average score (see Fig. 18.2, right panel). The three most highly rated criteria were "Human Scale and Reference", "Pedestrian Accessibility"

 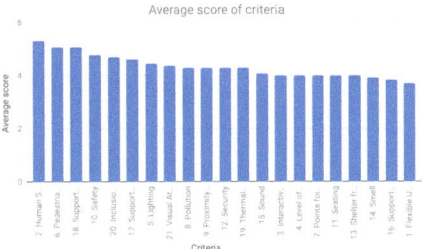

Fig. 18.2 Left pane: workshop participants voting on criteria in the co-creative Swing method. Right pane: ranked average scores for 21 public space quality criteria (direct scoring)

and "Supports Social Engagement"; perhaps surprisingly, rated least important to public space quality were "Smell", "Supports Active Engagement", and "Flexible Use". Aside from testing a direct scoring method to elicit ranks and weights, done individually as if it were an online survey, this part of the workshop also gave us an average rank of importance for our list of 21 criteria.

Part 3: co-creative Swing method. The second, most important and final exercise of the 90-min workshop required the experts, now split into two representative groups with a moderator each, to rank and weight criteria together, reaching consensus for each decision. This co-creative exercise can also enable the researchers to observe and analyze the differences in opinion between experts. The format is an adapted version of the traditional Swing method (e.g. Montibeller and Franco 2007), which is an individual weight elicitation method. For the exercise, the experts were required to follow the following steps:

1. Consider the first four criteria on the ranked list resulting from the previous exercise. Now imagine a very bad public space, one that performs poorly on all these four criteria. (The moderator would put the four considered criteria in the center of the table and ask the questions.)
2. Decide which of the four criteria you would like to change from their worst to their best condition in order to make this a better public space. This first, most important criterion gets 100 points as a score.
3. Add the next highest criterion from the ranked list to the table, to again have four criteria to compare. Choose the next most important criterion to be improved, assuming all the criteria are at their worst. This criterion needs to be scored (0–99), always lower than the previously selected criterion. The participants are asked to first write their score on a post-it note and paste it under the criterion they think is most important; this brings forward their individual opinions before having to reach consensus (see Fig. 18.2, left pane). They are then asked to debate their choices, and reach consensus on which of the four criteria to select, and how to score it. The recorded difference between individual and group opinions can shed light on group dynamics in decision making.
4. Repeat Step 3 until all the criteria have been scored.

The ranked list of 21 criteria resulting from the previous exercise allowed us to develop a much shorter Swing method exercise (required due to time constraints): firstly, we limited this exercise to the top 12 criteria from the list; secondly, we introduced a moving bracket of 4 criteria to choose from (as explained above), rather than all criteria as per the standard Swing method – this is only possible because we start from a pre-ranked list. In addition, at any time an expert could nominate an additional criterion into the selection bracket if they thought it was equally important than the current four.

The experts were divided into two groups in order to compare differences between groups. Table 18.2 shows final ranking results of group A and B, the scores per criterion, and the rank differences (A to B). As can be seen, the experts in both groups emphasized the importance of Inclusion and Social justice, Human Scale and Reference, Pedestrian Accessibility, and Supports Passive Engagement, which almost

Table 18.2 Results of the Co-creative Swing Method

Group A result		Group B result		Rank difference
Rank. criteria	Score	Rank. criteria	Score	|A-B|
1. Inclusion and social justice	100	1. Human scale and reference	100	2
2. Human scale and reference	99	2. Pedestrian accessibility	95	1
3. Supports social engagement	89	3. Inclusion and social justice	80	3
4. Pedestrian accessibility	84	4. Thermal comfort	79	2
5. Supports passive engagement	83	5. Supports passive engagement	75	0
6. Visual attractiveness	82	6. Supports social engagement	72	4
7. Proximity to diverse and vibrant city life	79	7. Proximity to diverse and vibrant city life	71	0
8. Pollution	78	8. Pollution	60	0
9. Thermal comfort	77	9. Security	55	5
10. Lighting	74	10. Visual attractiveness	52	2
11. Security	71	11. Safety	50	2
12. Safety	69	12. Lighting	35	1

aligns with the Top 5 criteria from the first exercise. This may reflect the preference of the experts for more human-oriented criteria. Low priorities for Lighting, Security, and Safety may be related to the context of Singapore, which is highly secure by default. The rank difference (last column) shows a relatively similar ranking between both groups. Surprisingly, Thermal Comfort has the largest rank difference. This is likely because it was only the 12th and last criterion in the initial list (despite its importance in a tropical city), but Group B (which included a thermal comfort expert) decided to nominate "thermal comfort" into the selection bracket early on. The large difference in the range of scores given by both groups is due to a (spontaneous) decision by Group B to scale their allocated scores over a wider range after finishing their exercise.

The differences in ranks and scores demonstrate that different experts have quite varying preferences when it comes to public space quality. Nevertheless, we also observe that the rank difference between the groups is not that big, with strong outliers more likely caused by the exercise format than anything else. Moreover, the relatively small difference in score between many criteria, and the discussions they emerged from, seems to suggest that the criteria on the list of 21 are indeed deemed crucial to public space quality.

18.4   Conclusion and Future Work

In this paper, we presented a systematic Multi-Criteria Decision Analysis framework to evaluate the quality of public spaces. We discussed the essential steps of our methodology in detail and illustrated the first steps of the methodology: a set of public space quality criteria, and the results from an expert workshop. The workshop serves as an important linkage between the first phase of the research and second phase. Some relevant lessons learnt were:

- Our list of 21 criteria was deemed comprehensive by these 14 experts; within the list, there were clearly more and less important criteria.
- While expert opinion varied strongly in individual scoring exercises, consensus (and changes in opinion) where not difficult to reach; this is something to take into account when eliciting weights and values from individual surveys.
- It is difficult to rate the importance of very qualitative, context-based criteria, let alone quantify them or elicit a value function. Hence, for Phase 2, we decided to develop a better, context-based survey using a parametric 3D model for easier value function elicitation.
- Clear language is crucial when describing criteria. Nevertheless, even clear language is open to fuzzy interpretation. It is also strongly tied to particular (urban design) ambitions that criterias describe. In order to reduce the risk of interpretation, we are exploring a criteria list that only uses design guidelines to inform the incremental selection of criteria, but directly generates total scores from individual indicators.
- The participants lauded the workshop format, and the majority spontaneously asked to remain updated on the work, which illustrates interest in the topic of quantifying public space quality.

Integrating the lessons learnt, we are now in the process of finalising the criteria list and the ranking method to build our survey for the final case study, in which will be the scoring of a set of public spaces in Singapore. We have already gathered different types of data, such as temperature, humidity, and people counts, from sensors deployed in Jurong East in Singapore. We are also designing methods to collect data for the other public space design characteristics in our framework. All of this data actually captures the unique Singapore-based context and culture, and hence it would be important to collect and use different data when applying our framework to other countries with different cultures.

Our proposed methodology could eventually serve as a tool for various stake-holder groups, such as the urban designers and policy makers, to evaluate and compare public space designs in a more quantitative way. It will enable them to explore expected qualities and user appreciations of public space, both in terms of a diverse range of needs, and in terms of general quality. Nevertheless, it is important to stress that the PSQI is not an attempt at developing a "perfect recipe" for public space design; our ambition is to inform design by exploring how a comprehensive set of relevant criteria influences the appreciation (and possibly behavior) of different user groups. We believe this tool cannot be exhaustive in its quantification, or in its influence on design or materialization.

Finally, note that while the PSQI is developed as an independent tool to assess public space quality in this work, it has also been designed to be incorporated into an integrated model to simulate the potential presence of people in public spaces. We refer interested readers to Herthogs et al. (2018) for details on this model.

References

Akadiri PO, Olomolaiye PO (2012) Development of sustainable assessment criteria for building materials selection. Eng Constr Archit Manag 19(6):666–687

Akadiri PO, Olomolaiye PO, Chinyio EA (2013) Multi-criteria evaluation model for the selection of sustainable materials for building projects. Autom Constr 30:113–125

Beirão JN, Koltsova A (2015) The effects of territorial depth on the liveliness of streets. Nexus Network Journal 17(1):73–102

Carmona M (2015) Re-theorising contemporary public space: a new narrative and a new normative. J Urban Int Res Placemaking Urban Sustain 8(4):373–405

Carr S, Francis M, Rivlin LG, Stone AM (1992) Public space, Cambridge series in environment and behavior, vol XV. Cambridge University Press, Cambridge, 400 S. ISBN: 0-521-35148-0 0-521-35960-0 PBK

Curwell S, Deakin M, Symes M (eds) (2005) Sustainable urban development Volume 1. Routledge, London. https://doi.org/10.4324/9780203299913

Llewelyn-Davies (2000) Urban design compendium. English Partnership and the Housing Corporation, London

Ewing R, Handy S (2009) Measuring the unmeasurable: urban design qualities related to walkability. J Urban Des 14(1):65–84

Gehl J (1987) Life between buildings: using public space. Van Nostrand Reinhold, New York

Gehl Institute (2017) Twelve quality criteria. Technical report. https://gehlinstitute.org/wp-content/uploads/2017/08/TWELVE-QUALITY-CRITERIA.pdf

Gehl Institute (2018) Inclusive healthy places. Technical report. https://gehlinstitute.org/wp-content/uploads/2018/07/Inclusive-Healthy-Places_Gehl-Institute.pdf

Herthogs P, Tunçer B, Schläpfer M, He P (2018) A weighted graph model to estimate people's presence in public space – the visit potential model. Proc eCAADe 2:611–620

Jacobs J (1961) The death and life of great American cities. Jonathon Cope, London

Koltsova J (2017) Inverse urban design support: attribute extraction from the local context. PhD thesis, ETH Zurich

Mehta V (2014) Evaluating public space. J Urban Des 19(1):53–88

Montgomery J (1998) Making a city: urbanity, vitality and urban design. J Urban Des 3(1):93–116

Montibeller G, Franco LA (2007) Decision and risk analysis for the evaluation of strategic options. In: Supporting strategy: frameworks, methods and models. Wiley, Chichester, pp 251–284

Narula K, Reddy BS (2015) Three blind men and an elephant: the case of energy indices to measure energy security and energy sustainability. Energy 80:148–158

Soltani A, Hewage K, Reza B, Sadiq R (2015) Multiple stakeholders in multi-criteria decision-making in the context of municipal solid waste management: a review. Waste Manag 35:318–328

Stewart TJ (2005) Dealing with uncertainties in MCDA. In: Multiple criteria decision analysis: state of the art surveys. Springer, New York, pp 445–466

Tervonen T, Figueira JR (2008) A survey on stochastic multicriteria acceptability analysis methods. J Multi-Criteria Decis Anal 15(1–2):1–14

Varna G, Tiesdell S (2010) Assessing the publicness of public space: the star model of publicness. J Urban Des 15(4):575–598

Whyte WH (1980) The social life of small urban spaces. Washington, DC: The Conservation Foundation, 125 p. ISBN: 0-89164-057-6

Zavadskas EK, Turskis Z, Bagočius V (2015) Multi-criteria selection of a deep-water port in the eastern Baltic Sea. Appl Soft Comput 26:180–192

Chapter 19
Factors Influencing Urban Open Space Encroachment: The Case of Bloemfontein, South Africa

Lindelwa Sinxadi and Maléne Campbell

Abstract Rapid changes in land use and occupancy patterns of urban open spaces have led to value conflicts in terms of the quest for sustainable neighbourhoods. Urban open spaces are becoming extinct due to rapid urbanization, hence affecting the spatial patterns of urban land use. Such gradual disappearance has resulted from intensity of land use for residential, business, and community facilities, among others. This has created challenges in terms of the value and sustainability of open spaces, land use management, and preservation. This study seeks to explore, in its entirety, the incidence of open space encroachment in the Mangaung Township, Free State Province of South Africa. Adopting a case study approach, this study deploys a variety of techniques such as focus group discussions, face-to-face semi-structured interviews, and personal observation for data elicitation at different intervals. Individual semi-structured interviews were conducted with purposively recruited town planning, land invasion, and environmental management professionals from local and provincial levels of government. Discussants for the focus group discussions consisted of community members who have encroached upon open spaces and those occupying properties around open spaces. Events within the study area were observed at intervals and memos drawn therefrom. The accruing data was analyzed thematically, relying on qualitative content analysis (QCA). The study's findings, besides the provision of an insight into the drivers of this malaise, chronicles the plethora of strategies which have been adopted and implemented to curtail its continued occurrence, highlights the strengths and weaknesses of these strategies, and proffers recommendations on how to optimally surmount this imbroglio. Findings indicate that the high cost of the available land for housing, poor sustenance and management of the available housing stock by municipality officials,

L. Sinxadi (✉)
Department of Built Environment, Central University of Technology, Free State, South Africa
e-mail: TobaM@cut.ac.za

M. Campbell
Department of Urban and Regional Planning, University of the Free State, Bloemfontein, South Africa
e-mail: campbemm@ufs.ac.za

© Springer Nature Switzerland AG 2020
R. Roggema, A. Roggema (eds.), *Smart and Sustainable Cities and Buildings*,
https://doi.org/10.1007/978-3-030-37635-2_19

non-participation of community members in planning processes, and poor enforcement of land use regimes remain salient contributors to the preponderance of open space encroachment. This study's findings hold immense implications for planning practitioners as well as other professionals and policy makers working within the urban planning and socio-economic development praxes both within the province and beyond.

Keywords Participation · South Africa · Strategy · Sustainability · Urban open spaces

19.1 Introduction

Urban open spaces form integral aspects of land use planning. From a sustainability perspective, they are viewed as critical to the environment and quality of life. Campbell (2001) states that open spaces make cities attractive and viable places for people to live, work, and play. Accordingly, town planning guidelines dictate that developments must comprise of open spaces, even though no criteria on the number, location, and usage of these spaces is provided. During land use planning, different land uses are allocated to built-up and non-built up environments. According to Maruani and Amit-Cohen (2007), built-up environments include residential, industrial, and commercial areas, whereas the non-built environments comprise of open spaces. Turner (1992) reiterates the importance of the usage of the term "open space" which was employed for the first time by the London 1833 Select Committee on Public Walks. The Open Space Act 1906 (1906:11) defines an open space as "any land, whether enclosed or not, on which there are no buildings or of which not more than one-twentieth part is covered with buildings, and the whole or the remainder of which is laid out as a garden or is used for purposes of recreation or lies waste and unoccupied". Mashalaba (2013:40) defines an urban open space as "a piece of land, either developed or pristine, that is either existing or planned to maximize the ecological integrity of an urban area by sustaining both urban and natural ecosystems; while improving the quality of human life in both social and economic terms". Accordingly, parks, gardens, wetlands, allotments, trees and forests or grasslands in urban areas can be described as urban open spaces.

Rapid changes in land use and occupancy patterns on urban open spaces have led to value conflicts in terms of the quest for sustainable neighbourhoods. Urban open spaces are becoming extinct due to rapid urbanization, hence affecting the spatial patterns of urban land use. The gradual disappearance of urban open spaces is evident in the increasing emergence of informal settlements and urban sprawl. Also, intensity of land use for residential, business, community facilities, etc. have contributed to this situation. Mensha (2014a, b) states that African countries such as Kenya, Nigeria, Ghana, Sierra Leone, and Senegal have lost urban open spaces, due to rapid urbanization and urban sprawl. This has created challenges in terms of the value and sustainability of open spaces and land use management. Urban open spaces, as part of the urban planning process, need to be controlled by the urban planning regulations, yet, planning jurisdictions in different countries still utilize outdated documents for this purpose. For instance, the Mangaung Metropolitan Municipality in Free State, South Africa still relies on the Town Planning Scheme of 1969 in regulating land uses. Decision-making processes also take long, and this affects the development negatively and contributes to the encroachment of open spaces.

To achieve its objective, the rest of the paper is structured accordingly: theoretical perspectives; a review of literature on the following aspects: urban open spaces and sustainable neighbourhoods; encroachment of open spaces as a societal malaise; the incidence of open space encroachment in Mangaung; justification of research methodology deployed; presentation and discussion of findings section, and the conclusion.

19.2 Theoretical Perspective

The contents of this section emphasize the significance of open spaces in urban areas and its contribution thereof to the concept of sustainable neighbourhoods.

19.2.1 Sustainable Neighbourhoods and Open Space Planning

Urban open spaces form an integral component of sustainable neighbourhoods. Socially, a neighborhood is seen as a "community", whereas the ecological stance focuses on the unique qualities of a target property. According to Al-Hagla (2008:2), the social perspective states that sustainable neighborhoods are communities that "meet the diverse needs of existing and future residents, their children and other users, contribute to a high quality of life and provide opportunity and choice". From this definition, Campbell (2001) highlights different components, namely, governance, transport and connectivity, services, environment, economy, housing and built environment, sociology, and culture. In terms of ecological perspective, a neighborhood refers to an ecosystem created for humans and entails the microclimatic conditions. This area must provide comfort and sustenance to the target populace.

UN-Habitat (2015) identified principles for planning sustainable neighborhoods. These include (i) adequate space for streets and an efficient street network; (ii) high density; (iii) mixed land use; (iv) social mix; and (v) limited land use specialization. All these main principles of planning play a key role in fostering sustainable neighborhoods as they draw three key features, namely, vibrant street life which provides safe and vibrant neighborhood life, encourage walkability and affordability

(Dehghanmongabadi 2014). Urban open spaces play a significant role in achieving sustainable neighborhoods. In achieving sustainable neighborhoods, the key issues to be considered include space management, space function, and landscape. Space management refers to the sustainable lifestyle, community participation, sense of space, and resource management. The space function focuses on car reliance and the need to travel, while sustainable landscape promotes self-sustaining and regulatory systems (Al-Hagal 2008).

19.2.2 Encroachment of Open Spaces as a Societal Malaise

19.2.2.1 Causal Factors of Urban Open Space Encroachment

Urban open spaces are faced with challenges that include rapid urbanization, poor enforcement of land use regimes, poor sustenance and management, and low prioritization. These challenges have led to gradual disappearance of urban open spaces and have threatened sustainable planning of urban areas. Rapid urbanization is a major contributor towards urban open space extinction, and a human rights-based approach to this can have a major impact to environmental sustainability policies (UN-Habitat 2016). In planning, the implementation of land use regimes has been lacking, especially in land earmarked for urban open spaces. Outdated land use schemes and delays in approval of land use change application hinder effective implementation for planning and development. Personnel (shortage and unskilled), financial constraints, and political interference also contribute to poor enforcement. Poor sustenance and management of urban open space is another causal factor for encroachment. Arguably, this has resulted from the lack of involvement and education of the community about the value of open spaces. Mismanagement of these spaces has also led to underwhelming functionality performance, thus resulting in environmental degradation. In addition, proper planning for open spaces is not highly prioritized, hence encroachment. Evidence indicates a neglect of these spaces by the relevant authorities whilst focusing mostly on other social amenities (Mensha 2014a, b).

19.2.2.2 Measures to Curb the Incidence of Encroachment

Effective planning for urban open spaces contributes to curbing the incidence of encroachment. Strategic and holistic plans, as well as legal frameworks, need to be formulated and implemented by authorities. Planners need to integrate urban open space infrastructure needs to their urban frameworks. Involvement of different stakeholders and community participation have also played a crucial role in planning for urban open spaces, thus managing encroachment in that it increases the sense of place and ownership among the residents (Haaland and van den Bosch 2015; World Health Organisation 2017). When planning for urban open spaces, the social dimension of sustainability must be considered. This can be achieved by linking neighbourhoods, urban open spaces, and community assets to address accessibility issues/challenges. The community members must be educated about protection and conservation of urban open spaces, in order to address the ecological dimension of sustainability. Urban open spaces must fulfil the goal for environmental sustainability, and create opportunities for recreation and health, thus indirectly achieving the economic dimension of sustainability (Lindsey 2003). In this instance, planners play a critical role in evaluation of the environmental and social impact on the environment. The plans done by planners during the planning process must support the principles of sustainability and must be shared with the community during the community participation process with the aim of promoting community sustainability (Berke and Conroy 2000).

Other countries, like China, have a variety of policies for the management of urban spaces. This includes policies like "Afforestation Project for the Plains of Beijing" and the "Urban Green System Planning of Beijing" (2004–2020). These policies failed to achieve the goal of controlling urbanization. Other policies used for the use and management of open spaces include the Urban Green Space Work Plan. This was also used in Beijing to increase urban open space area through implementation of the 2016 Urban Greening Work Plan. Beijing also uses the "Green Line Management System" to control the changing of urban open space into other land uses (Li et al. 2016).

19.2.2.3 Understanding the Incidence of Open Space Encroachment in Mangaung

Mangaung is one of the metropolitan municipalities in South Africa, with a population of approximately 747,431. It covers 9887 km^2, and comprises three prominent urban centres, which are surrounded by an extensive rural area. It consists of Bloemfontein, Botshabelo, Thaba Nchu, Soutpan, Dewetsdorp, Wepener and Vanstadersrus (Fig. 19.1). It is centrally located within the Free State and is accessible via National infrastructure including the N1 (which links Gauteng with the Southern and Western Cape), the N6 (which links Bloemfontein to the Eastern Cape), and the N8 (which links Lesotho in the east and with the Northern Cape in the west via Bloemfontein) (MMM Reviewed Integrated Development Plan 2015–16:65, Mangaung Metropolitan Municipality Integrated Development Plan 2017/18).

Mangaung is experiencing a high rate of urban open space encroachment, which is mostly caused, among other factors, by rapid urbanization. UN-Habitat (2016) states that urbanization plays a key role in eradicating poverty when it is planned properly and managed. Highly urbanized countries are always associated with low levels of poverty, as a lot of people have escaped from poverty due of urbanization. This is due to higher levels of productivity, employment opportunities, improved quality of life, and access to improved infrastructure and services (UN-Habitat 2016). The municipality took a resolution in August 1998 in one of its Council

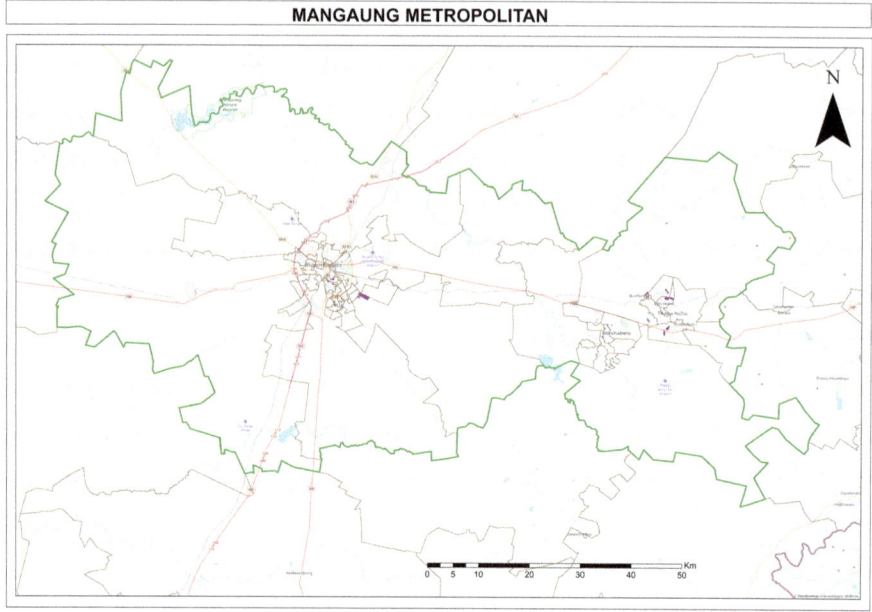

Fig. 19.1 Map of Mangaung Metropolitan Municipality, Free State, South Africa

Meetings that families illegally residing on erven (properties) with land uses other than "Residential" in areas for which the town planning has already been finalized as well as other existing or future townships areas, will not be accommodated in terms of town planning, surveying and the provision of services. However, this is now not the case because an exception is made on each land use application lodged for informal settlement upgrading on urban open spaces. Urban open space encroachment on municipal land has become very common in Mangaung. In such instances, both the Planning and Human Settlements Directorates will follow all the planning processes to formalize those areas.

19.3 Methodology

The main objective of the study is to explore, in its entirety, the incidence of open spaces encroachment in Mangaung Townships, Free State Province of South Africa. The study took a qualitative stance and adopted a case study approach with a variety of techniques, such as focus group discussions, face-to-face semi-structured interviews, and personal observation for data elicitation at different intervals. Creswell and Poth (2018) state that a case study approach explores a contemporary bounded system through detailed, in-depth data collection through the deployment of a coterie of techniques like observations, interviews, documents, etc. The selected cases are subsequently described, understudied and analyzed. Individual semi-structured interviews were conducted with purposively recruited town planning, human settlements, land invasion and environmental management professionals from local and provincial levels of government. Discussants for the focus group discussions included community members who have encroached upon open spaces and, those occupying properties around open spaces. The selected members form part of the ward committees governing the study area. Personal observations were conducted around the study area to get first-hand information on the state of urban open spaces in Mangaung. The observations were also undertaken from memos drawn thereof. Due to personnel constraints at local government, 10 individual semi-structured interviewees were involved in the study. The accruing data was analyzed thematically relying on a set of pre-set themes that evolved from literature.

19.4 Presentation and Discussion of Findings

The findings from the study will be discussed concurrently according to the thematic areas. These themes form part of the causal factors influencing urban open space encroachment. Two themes were selected for discussion: rapid urbanization and poor enforcement of land use regimes.

19.4.1 Rapid Urbanization

Rapid urbanization has resulted in extinction of open spaces. Urbanization is caused by a variety of factors that include rural-urban migration, natural population increase, expansion of the metropolitan periphery, and illegal occupation of land (Cohen 2006). Rapid urbanization has refocused attention on planning. This has led to different perceptions or value conflicts by different urban stakeholders regarding urban open spaces. These perceptions include categories such as the economic, recreational, environmental, and housing values. These value conflicts around urban open space make it extremely difficult for planners to achieve the objectives of planning. Literature highlights that some West African countries like Nigeria, Ghana, Senegal, and Sierra Leone have lost urban open spaces due to rapid urbanization. Beijing, for instance, has experienced a similar challenge in 2014 where it lost some urban open spaces due to rapid urbanization (Mensah 2014a, b; Li et al. 2016). This is also evident in Mangaung as the interviewees and the discussants (people residing around the urban open spaces) support the notion that their townships are losing urban open spaces due to rapid urbanization. The residents and municipal officials indicated that most people encroaching urban open spaces are non-South African citizens and people from all over the country in search for better quality of life. Some of the discussants indicated that they have lost the urban open spaces surrounding their properties because they have not been used for the initial land use. This is affecting the spatial patterns of urban land use.

The actual perpetrators of the encroachment indicated that they understand that there is value for urban open spaces, but their major concern is shelter, that is, proper housing with basic services. Solution to their challenge is encroachment on urban open spaces, with the perception that local government will provide adequate or formal housing for them with basic services. Speak and Tipple (2006) argue that people living in informal settlements where there are no proper basic services may be included on the category of homelessness. Homelessness can also be classified as rooflessness, houselessness, insecure accommodation, and inferior or substandard housing. Speak (2004) states that homelessness manifests itself as destitution, which is mostly due to extreme poverty, isolation, and shortage of societal resources. Homelessness can be defined in terms of categorisation, that is, supplementation, survival and crisis management. Some of the people who encroached these urban open spaces are family men who left their rural homes in search for better job opportunities. They stay in informal settlements and send their income to their families. This is referred to as supplementing rural livelihood. Some fall under the survival strategy of homelessness which originate from the supplementation strategy in that the survival homeless people also migrate from rural to urban areas in search for jobs. Most of these people left their homes with the intent to work for their families but they are often unable to send enough money to their families for better survival. Due to the lifestyle of staying in informal settlements, it becomes impossible for them to go back to rural areas and connect with their families (Speak 2004). During the discussion, some of them indicated that they have more than 20 years staying in shacks without water and electricity. They are promised proper housing with basic services on yearly basis but are unsuccessful.

Another argument of the encroachers on urban open spaces is inability to afford proper housing hence choosing informal settlements. Goal 11 of Sustainable Development Goals (SDGs) is to "Make cities inclusive, safe, resilient and sustainable" (United Nations 2016). This goal includes ensuring access for all to adequate, safe and affordable housing (SDG11.1). This is fundamental to inhibiting and addressing the problem of homelessness or residing in informal settlement. Homelessness is part of inadequate housing that remains a global sustainability challenge. Provision of adequate and affordable housing to citizens is a major challenge of urbanization and can result into homelessness. People who have encroached urban open spaces also lack access to basic services, which is a constraint and a cause for homelessness. Homelessness represents one of the extreme forms of deprivation and exclusion. It is a critical factor for persistent poverty and exclusion globally. This is a challenge for sustainable and inclusive urbanization (UN- Habitat 2016a, b, c, 2005). Lastly, some of the discussants are unable to secure an initial foothold in the economy. They have migrated from rural to an urban area in search for better job opportunities but be unsuccessful, due to acute poverty, unsustainable dependency and lack of alternatives. They opt not to return back home and settle to encroach urban open spaces because they believe they will be allocated adequate housing (Cross and Seager 2010; Somerville 2013; Nooe and Patterson 2010; UN-Habitat 2000).

19.4.2 Poor Enforcement of Land Use Regimes

The Spatial Planning and Land Use Management Act (2013) states that the municipality must have a land use scheme to enforce law to all stakeholders. Issues that lead to poor enforcement include the dysfunctional nature of land use regimes and delays in decision-making processes in planning applications. Mangaung is challenged with poor enforcement of land use regimes, that led to encroachment in urban open spaces. It was confirmed by the interviewees and discussants of focus groups. Areas that were earmarked for "parks" were encroached for residential purposes. Other urban open spaces are already rezoned for residential purposes, whereas others are still in the process of being rezoned for residential purposes or are left open because they are inhabitable. Rezoning by replacement of informal shacks with formal housing in planned and serviced townships reveals that there is a gap between town planners' assumptions and reality (Watson 2002). The discussants affected by the loss of urban open spaces argue that the municipality is failing them in that they allow encroachment by infill development. Infill development refers to the residential development occurring within the existing residential area (Rowley and Phibbs 2012). Some of these spaces for infill development are referred to as opportunistic spaces. Maruani and Amit-Cohen (2007) classify opportunistic spaces under the opportunistic planning model which refers to the pattern where urban open spaces emerge due spaces left over after systematic planning process. These spaces are usually left over because they are either small, irregular, or inaccessible for other land uses and can be poorly suited to be used as urban open spaces. Also, due to shortage of personnel in the municipality, interviewees confirmed they are challenged with poor enforcement of land use regimes. Should there be any encroachment, they cannot act promptly to evict land invaders of municipal land.

Olufemi (2004) confirms that land invasion and eviction is a challenge for land, housing, and planning policy makers. Land invasion occurs due to desperation of space for shelter, and is persistent because people lack resources to build their shelter formally and legally. Land invasion is led by poor enforcement of land regimes because urban open spaces are not used for their initial land use. People encroaching urban open spaces in Mangaung indicated that they invade these spaces because there is no proper enforcement of land regimes. They have seen properties that were zoned for recreational purposes or schools being rezoned for residential purposes. Some indicated that they have been on the "waiting list" for more than 10 years without provision of proper housing. The only way for housing allocation is continuous land invasion. Currently, some of the actual perpetrators of encroachment have been allocated properties with basic services on the land they have encroached.

19.5 Conclusion

Rapid changes in land use and occupancy patterns of urban open spaces remain a major concern in planning and development. This study seeks to proffer patterns for avoiding the incidence of open spaces encroachment in South Africa. Focus of the study is on one of the townships in Mangaung, Free State, because there has been recurring gradual disappearance of urban open space due to encroachment for residential purposes. Semi-structured interviews, focus groups, and personal observations were used for data elicitation. Findings indicated that the incidence of urban open space encroachment is prevalent. Mangaung townships has lost some urban open spaces due to rapid urbanization and poor enforcement on land use regimes. The different stakeholders involved in planning have different perceptions on the uses and values of urban open spaces, and these include economic, recreational, environmental, and housing. This indicates that there are different value conflicts and it makes it difficult for the planners and other professionals to achieve the planning objectives for urban open space and for other issues, such as housing and land use management. In addition, collaboration between municipal officials during planning processes will assist in facilitating sustainable environments. As such, planners have a critical role to play in terms of promoting sustainable neighbourhoods and this includes formulation of policies that would curb the encroachment of urban open spaces. Planners, other professionals involved in the planning practice and the community members, must consider sustainable development concept in shaping sustainable neighborhoods.

References

Al-Hagal KS (2008) Towards a sustainable neighborhood: the role of open spaces. Int J Archit Res 2(2008):162–177
Berke PR, Conroy MM (2000) Are we planning for sustainable development? An evaluation of the 30 comprehensive plans. APA J 66(1, Winter):21–33
Campbell K (2001) Rethinking open space, open space provision and management: a way forward, report presented by Scottish executive central research unit. Edinburgh, Scotland
Cohen B (2006) Urbanization in developing countries: current trends, future projections, and key challenges for sustainability. Technol Soc 28:63–80
Creswell JW, Poth CN (2018) Qualitative inquiry and research design: choosing among the five approaches, 4th edn. Sage, Thousand Oaks
Cross C, Seager JJ (2010) Towards identifying the causes of South Africa's street homelessness: some policy recommendations. Dev South Afr 27(1):143–158
Dehghanmongabadi A (2014) Introduction to achieve sustainable neighbourhoods. Int J Arts Commer 3(9)
Haaland C, van den Bosch CK (2015) Challenges and strategies for urban green-space planning in cities undergoing densification: a review. Urban For Urban Green 14(2015):760–771
Li F, Sun X, Li X, Hao X, Li W, Qian Y, Liu H, Sun H (2016) Research on the sustainable development of green-space in Beijing using the dynamic systems model. Sustainability 8965:1–17

Lindsey G (2003) Sustainability and urban creenways: indicators in Indianapolis. J Am Plann Assoc 69(2, Spring). © American Planning Association, Chicago

Mangaung Metropolitan Municipality. Reviewed integrated development plan 2015–16

Maruani T, Amit-Cohen I (2007) Open space planning models: a review of approaches and methods. ScienceDirect Dep Geogr Environ Israel Landsc Urban Plann 81(2007):1–13

Mashalaba YB (2013) Public open space planning and development in previously neglected townships. Unpublished dissertation. Department of Urban and Regional Planning. University of the Free State

Mensah CA (2014a) Urban green spaces in Africa: nature and challenges. Int J Ecosyst 4(1):1–11

Mensah CA (2014b) Destruction of urban green spaces: a problem beyond urbanization in Kumasi City (Ghana). Am J Environ Protect 3(1):1–9

Nooe RM, Patterson DA (2010) The ecology of homelessness. J Hum Behav Soc Environ 20 (2):105–152

Olufemi O (2004) Socio-political imperatives of land invasion and eviction: revisiting the Bredell case, Johannesburg, South Africa. Centre for Urban and Community Studies. International conference. Toronto, June 2004

Rowley S, Phibbs P (2012) Delivering diverse and affordable housing on infill development sites. AHURI final report no.193. Australian Housing and Urban Research Institute, Melbourne

Somerville P (2013) Understanding homelessness. Hous Theory Soc 30(4):384–415

South Africa (2013) Spatial planning and land use management act, Act 16 of 2013. Department of Rural Development & Land Reform. Government Press, Pretoria

Speak S (2004) Degrees of destitution: a typology of homelessness in developing countries. Hous Stud 19(3):465–482

Speak S, Tipple G (2006) Perceptions, persecution and pity: the limitations on interventions for homeless in developing countries. Int J Urban Reg Res 30(1):172–188

Turner T (1992) Open space planning in London (from standards per 1000 to green strategy). Town Plan Rev 63(4):365–386

UN-Habitat. Global Report on Human Settlements (2009) Planning sustainable cities. Nairobi, Kenya

UN-Habitat. World Cities Report (2016) Urbanization and development: emerging features. Nairobi, Kenya

UN-Habitat (2000) Strategies to combat homelessness. United Nations Centre for Human Settlements, Nairobi

UN-Habitat (2005) Financing urban shelter: global report on human settlements 2005. Nairobi

UN-Habitat (2015) Discussion note 3: urban planning. A new strategy of sustainable neighbourhood planning: five principles. Nairobi, Kenya

UN-Habitat (2016a) Fundaments of urbanization. Evidence base for policy making

UN-Habitat (2016b) World Cities Report. Nairobi

UN-Habitat (2016c) Habitat III policy paper 10 – housing policies (unedited version). Nairobi

United Kingdom (1906) Open Space Act, Act of 1906. Chapter 25. UK

Watson V (2002) Do we learn from planning practice? The contribution of the practice movement to planning theory. J Plan Educ Res 22(2):178–187

World Health Organization (WHO) (2017) Urban green spaces: a brief for action. Europe

Chapter 20
Urban Agricultural Practices in the Megacities of Dhaka and Mumbai

Tazy Sharmin Momtaz

Abstract Dhaka city in Bangladesh, with a population density of approximately 27,700 people per square kilometre, is one of the densest cities in the world. The urban morphological characteristics and urban design and planning provisions in the high-density city restrict land availability for growing food locally. As a result, the food, especially vegetables and fruits available in the local market of Dhaka, are transported from the rural regions of the country, with high food miles, mostly adulterated with preservatives for longer shelve life. Evidence shows that this megacity is adopting useful pathways to integrate safer local food production or urban agriculture within the built environments. Unsafe available food is not the only problem of this city. In more than 30 years, Dhaka city has lost enormous amounts of open spaces, decreasing from 44.8% to 24.1%. As a result, the city has lost its agricultural land, other food producing spaces such as home and community gardens, small vegetable farms, and recreational spaces such as neighbourhood parks. Local food production on rooftops can be a gateway for safer food production, as well as adding extra green spaces to the city.

The aim of this research is to review existing and emerging urban agricultural practices and current planning policies in the city of Dhaka and selected high-density cities of the world, using an exploratory literature review and analysis of two case studies from these cities. One innovative case study from Dhaka would be analysed and compared to one relevant case study from Mumbai, India, considering productive spaces, current performance, potential to grow local food, and ability to make social awareness which would cause other people of the city to adopt urban agriculture as a part of their urban life.

The outcome of this research will provide a realistic view of the status of Dhaka's urban agricultural system and practice, in comparison to another high-density city's food production practices. This will give a clear idea about current research and planning, and urban design policy of urban agriculture. Mainly, it will identify the

T. S. Momtaz (✉)
Faculty of Design, Architecture and Building, University of Technology Sydney, Ultimo, Australia
e-mail: tazysharmin.momtaz@student.uts.edu.au

© Springer Nature Switzerland AG 2020
R. Roggema, A. Roggema (eds.), *Smart and Sustainable Cities and Buildings*,
https://doi.org/10.1007/978-3-030-37635-2_20

research gap this practice creates, and would provide some recommendations for integrating local food production within the city to improve and to build a resilient future.

Keywords Urban agriculture · Rooftop garden · Dhaka · Mumbai · Food security

20.1 Introduction and Context

Dhaka is one of the major cities and the capital of Bangladesh. It has a population of 12 million and a population density of approximately 27,700 people per square kilometres. It is predicted that the population of Dhaka will be over 20 million people by the year 2020. Although this population is contributing to the economy of the country, but it is simultaneously causing unbearable pressure on the city's infrastructure. Due to land scarcity and topography, the land price of Dhaka is very high even compared to other developed countries (Baker 2007). Figure 20.1 presents photos from Dhaka and Mumbai.

On the other hand, Mumbai is situated almost 2300 km away from Dhaka and is regarded as the economic capital of India. Similar to Dhaka, this city is highly populated, with a population of 23.5 million people, almost the double as Dhaka's current population. The population density of Mumbai is about 20,482 persons per square kilometer, making it the ninth most crowded city of the world.

This paper focuses on the urban agriculture practices in these two dense cities, which are located in the same region of the world. Although agriculture is not one of the principal activities of most of the cities in the world, in a number of specific countries, there are many urban dwellers who rely solely on urban agriculture for food and livestock production for their supply of nutrition and food security (Zezza and Tasciotti 2010). Urban agriculture can play a major role in making these cities self-sufficient through local food production. Many cities in the world are currently

Dhaka, Bangladesh   Mumbai, India

Fig. 20.1 Two high density cities. (Photos by M. Islam and K. Faiz)

practicing urban agriculture within city limits. Cuba showcases a good example of how urban agriculture practices can flourish in a country to tackle food supply crisis. In 1989, when Cuba was cut off from the Soviet Block and was also under US embargos, a food crisis began to take place on the small island nation. Hunger became a part of Cubans' daily life. Not only the food import stopped, but also import of other agricultural materials such as animal feed, fertiliser, and industrial equipment were also stopped. This crisis led the government to take initiate urban organic farming in Cuban cities such as Havana (Clouse 2014). Among South East Asian countries, Dhaka and Mumbai are the two high-density cities that have been impacted by rapid urbanisation. But when it comes to formal urban agriculture, both cities have a culture of growing food within the city as a hobby, but not as a strong urban component. Compared to Bangladesh, India has more major cities, and urban agriculture is slowly getting its position in the urban fabric. For example, urban agriculture is a new concept in the city of Hyderabad in India. More than 400 house-holds of this city practice urban agriculture, and the horticulture department of the government has a subsidy system for citizens who want to initiate farming of their own. Despite having this promising initiative from the government, there is hardly any noticeable urban agriculture in the core areas of the city. This is because the city is very dense and tenants do not have access to spaces where they could grow food. Some of the residents have received this subsidy from the government and applied it to practices of growing food in the outskirts of the city (Awasthi 2013).

20.2 A Review of Urban Agricultural Practices in Dhaka and Mumbai

Islam's (2002) research established that almost 78% of rooftops in Dhaka had already incorporated some forms of gardens. This city has not introduced any formal 'Urban Agriculture' (UA) system and there is very limited literature available on this specific topic (Das 2017; Sajjaduzzaman et al. 2005; Shariful Islam 2002). Although approaches to UA practices in Mumbai are different from Dhaka, limited work has been conducted on UA research for this city. The only notable urban agriculture typology that exists in the Dhaka city is a rooftop garden. The community garden concept did not progress, because communities could not afford to invest in a land area for community garden due to high land price in Dhaka, and this is not a common cultural practice (Zinia and McShane 2018). However, in Mumbai, three typologies of UA system are widely practised such as (1) farming along the railway tracks, (2) community garden, (3) rooftop garden (Satterlee 2015).

Historically, UA in Mumbai was informal and had only existed in the form of a terrace or balcony garden (Satterlee 2015). These are popular as part of a food production system in Mumbai, especially to meet one's personal recreational need (Vazhacharickal 2014). Simultaneously, Mumbai's community gardens are run by a group of people or non-government organisation with various goals, such as creating awareness on local food production with local people and to teach them gardening techniques so that they are inspired to start gardening themselves. Produces from these community gardens are usually distributed between the garden members, and sometimes get donated to nearby churches and hospitals (Satterlee 2015).

The Indian Railway plays a major role in Mumbai in implementing a government-initiated food production system in the city. This organisation has allocated their unoccupied land to employees with lower socio-economic profiles to farm and grow vegetables. The project is called "Grow more food", and has protected Indian Railway's unoccupied lands from illegal encroachment. The huge farming culture in the city is also linked to cultural practices of many agricultural migrants from the adjacent rural areas (Vazhacharickal 2014). Produces from these farms are usually consumed by the farmers, and they sell the extra vegetables to the local markets (Satterlee 2015). Sadly, this innovative approach towards urban agriculture farming is threatened by the overuse of fertilisers and pesticides by the poor farmers who lack formal training and knowledge about organic farming and are motivated only to increase the production. Lack of knowledge on organic farming and its benefits have led to the high chemical contamination in vegetables that farmers grew in this project (Vazhacharickal 2014). Nevertheless, this is the only formal approach from the government's side to encourage urban farming for food production in this city (Satterlee 2015).

Similar to Mumbai, there is little to no involvement from the government's side to promote urban agriculture in Dhaka. As discussed earlier, community gardens or gardening on land is difficult in the city, due to the unaffordable land price and unavailability of land. However, rooftop gardening has significant potential in the city. The dense built environment of the city provides a larger roof space area available, and a huge opportunity for city dwellers to grow their own food on the rooftop gardens. Also among the many green adaptation methods, rooftop gardening is one of the most affordable methods for the Dhaka dwellers to increase the number of green spaces in the city (Zinia and McShane 2018). When Shariful (2000) interviewed the current food growers of Dhaka, he found out that, usually, homeowners prefer to have gardens on the rooftop but they are not very willing to let their tenants practice gardening on the rooftops of their houses. Each rooftop gardener in Dhaka spends around $50–$80 Canadian dollars annually as an expense on the garden and half of the gardeners have prior experiences of gardening and in agriculture (Shariful Islam 2002).

Several types of planters are used in Dhaka's rooftop gardens and are made out of permanent to temporary materials. Using of Permanent structures such as cemented platforms are less (approximately 5%). Up to 83% of these planters are clay pots, and the rest of the planters are plastic drums (Sajjaduzzamanet al. 2005). Unlike Dhaka, Mumbai's UA farmers mostly upcycle un-biodegradable waste by using them as planters and other gardening materials (Satterlee 2015). Where Dhaka's UA is hobby-driven, Mumbai's urban agriculture aims to help improve waste management practices within the city. Every community garden or rooftop garden has a provision for using waste as compost for the plants (Satterlee 2015).

75% of gardeners who grow food on their rooftop gardens in Dhaka's context are from middle-class backgrounds. Rich and poor classes are less involved in rooftop gardening in Dhaka (Sajjaduzzaman et al. 2005). The economic conditions of Mumbai's urban farmers present variations of social and economic statuses who participate in urban food production. While rooftop and community gardeners are usually the upper-middle class, railway farmers fall directly below the poverty line in Mumbai (Satterlee 2015).

20.3 Analysis of Roof Garden Case Studies from Dhaka and Mumbai

20.3.1 Rooftop Gardening Project in Dhaka

The Food and Agriculture Organization of the United Nations (FAO), in collaboration with Department of Agricultural Extension (DAE), Ministry of Agriculture, Bangladesh, had funded a rooftop gardening project in two major cities of Bangladesh, Dhaka and Chittagong. The project was implemented for two years, from 2015 to 2017. On record, this is the only formal urban agricultural project in Bangladesh where a global organisation worked solely on food production in a rooftop garden. The main objective of this project was to increase awareness about rooftop gardens as a medium to achieve food security and nutrition in growing agricultural produces; reducing air pollution and if possible to generate income, strengthening the local economy. In this project, in two cities, around 250 demonstration rooftop gardens were established, and the same number of house owners received training and knowledge on latest rooftop garden technologies from the experts from three non-government organisations, one public university and a research institute. Compared to the total population of the city, the share of householders included was very minimal. Still, under this project, almost 800 people received basic training on rooftop gardens, and 45 young people with limited educational backgrounds received training to be an aspiring gardener (Das 2017). Apart from this personal training provision, this project also took an initiative to educate school children. The result of this project commenced an incentive from the government and two city corporations declared 10% rebate on holding tax for any building which has a green roof on it (Das 2017).

20.3.2 A Rooftop Garden Case Study in Dhaka

A rooftop garden was selected as a case study for Dhaka is situated in Uttara, Sector 06, at the fringe of Dhaka. The area is comparatively new, so it is more planned and organised than other parts of the city. The selected rooftop garden attracted

Fig. 20.2 A rooftop garden case study in Dhaka (Photo by the author) and the plan of a planter box. (Source: Drawn by the author)

significant media coverage, a TV Show entitled 'Rooftop farming' was hosted by a celebrity agriculture enthusiast, Shykh Seraj has showcased this garden in their TV show. Figure 20.2 presents a photograph of the rooftop garden case study in Dhaka, and a plan of the planter box placed on the rooftop garden.

The selected rooftop garden is set upon the roof of a five-story privately-owned house. The total area of the house is about 229.9 m^2; 50% of the total rooftop is dedicated to growing food and includes mainly fruit trees planted in concrete boxes. It has permanent structures to accommodate trees on the rooftop. Structures of the garden are planned and constructed accordingly to sustain the extra loads of the trees in the garden. The garden is a notable example; not common in the city of Dhaka and also compared to other urban agricultural practices of Dhaka. Preplanning of this garden at the design stage before construction began and the owner's enthusiasm for gardening and recreating a green space in a dense urban area, make this garden an outstanding example in Dhaka. The built-up areas, roads and green spaces on the three adjacent blocks of the selected case study show that all the plots are almost of the same size; have a good road connection with each other, but the setback between buildings are very narrow. Satellite imagery of this planned part of this dense city indicates that this city lacks in green or open spaces. From this image, it is also noticeable that the rooftop garden on the building of the selected case study contributes a reasonable amount of green space in the neighbourhood. An aesthetically designed artificial fountain and seating areas to relax on the rooftop garden indicate that the garden functions as a social space for family and friends.

This garden has two sizes of planters to grow trees. The larger planter is 101 cm × 101 cm with a depth of 50 cm. The smaller planter is a permanent structure and is 66 cm × 53 cm with a depth of 50 cm. There are altogether 25 planters on the rooftop, out of which 19 are of a large size, and the remaining six are of a smaller size. However, on this rooftop garden, the owner has comparatively larger size trees, such as mango or star fruit trees, which generally grow on the ground. On the site visit to the rooftop garden case study, 12 types of fruit trees such as mango, star fruit, lemon, guava, lychee, tamarind, etc., and exotic fruit trees, such as avocado and Thai longan were recorded. There were some vegetables such as spinach, mint, pumpkin, and others growing in the garden.

They have created two types of visible drainage system for the garden. The first type is linked to the planter to drain out excess water after watering the plants from the planter to the rooftop garden. The second type of drainage system is to drain the water draining out of the planters and rainwater from the rooftop garden to the ground level.

One of the main problems was that the movement paths were narrow and were only 38 cm wide, although pre-designed. The spreads of the large trees at lower heights at some points make it difficult to walk through these paths, and this also poses a problem for regular maintenance of the garden. During the site visit, it was observed that drainage was blocked at several points making these paths water-logged and slippery.

20.3.3 A Rooftop Garden Case Study in Mumbai

Mumbai Port Trust's central kitchen rooftop garden is selected as the case study. It is a celebrated institutional garden in Mumbai and is different from a private residential rooftop garden case study in Dhaka. It was founded by the catering officer, Preeti Patil as a way of recycling enormous food waste created by the thousands of meals prepared in this kitchen (Satterlee 2015).

The rooftop of the central kitchen of Mumbai Port Trust Authority building is about 229.9 m^2 in size (Marielle Dubbeling 2012). In 2002, this garden started with only five plants, but it now contains about 150 different types of plants in the garden (Marielle Dubbeling 2012; Pendharkar 2008). The garden not only has fruits and vegetables, it also has a separate section for growing medicinal plants and herbs (Pendharkar 2008). The photos and research studies indicate that the garden was not pre-designed, rather it was designed organically at the post-construction phase when the building was already operating. Similarly, the materials used as planters are mostly recycled waste. 90% of the waste generated by 30,000 employee's meal preparation are recycled in this garden (Marielle Dubbeling 2012). Figure 20.3 shows the sketch of the organic organisation of the garden, the use of different types of material as planters, and plants of different types and sizes.

This garden had started as a small initiative, just to solve an immediate problem. However, over time, it has become a model garden for many urban gardeners in Mumbai and in India. After the success of this garden, Preeti Patil founded one of the most successful organisations named 'Urban Leaves' to inspire and help people with farming knowledge and techniques ('MbPT Terrace Garden and the philosophical and practical base of Urban Leaves').

Both the rooftop gardens in Dhaka and Mumbai in two different settings have similarities, as they are creating food forests in two high-density megacities. Table 20.1 compares the two selected rooftop garden case studies in Dhaka and Mumbai.

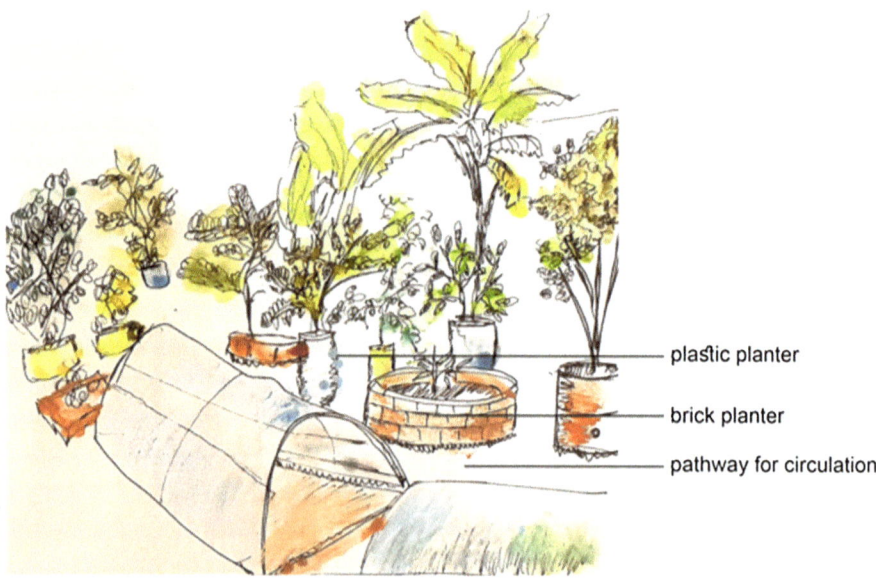

Fig. 20.3 Sketch of a rooftop garden case study in Mumbai (drawn by the author). (Source: http://cityfarmer.info/mumbai-port-trust%E2%80%99s-%E2%80%98wild%E2%80%99-kitchen-garden-india/)

Table 20.1 A comparison of rooftop garden case studies in in Dhaka and Mumbai

Category	Rooftop garden, Dhaka	Rooftop garden, Mumbai
Land use	Residential	Institutional
Area	278.7 m^2	229.9 m^2
Concept	Pre-planned	Organic
Purpose	Hobby garden	Waste management
Produce type	Mainly fruits, partially vegetables	Mainly vegetables, partially fruits and herbs
Planter type	Permanent, cemented	Recycled waste, temporary

20.4   Discussion

A literature review of urban agricultural practices in two megacities, Dhaka and Mumbai, from the same geographical region, has been conducted in this paper. An critical comparison of two selected roof garden case studies has also been completed. Dhaka and Mumbai have very similar urban conditions and are considered as comparatively very dense cities of the world. Although the scarcity of the available land areas exists, these cities are still putting efforts to grow their own food locally. Balcony and rooftop gardens are popular forms of urban agriculture in both cities. Community gardens play a huge role as a small-scale urban agricultural typology for Mumbai, whereas in Dhaka, community garden typology is absent in the local food growing scenario. In Dhaka, the majority of urban farmers in rooftop gardens are from middle-class backgrounds, but in Mumbai, members of upper-middle-class society are more involved in growing food in community and rooftop gardens within the city.

The literature on quantitative data and core aspects of the garden, such as the amount and types of produce, growing seasons, expenditure, quality of food produced, and how much food is distributed at the outlets, is not available for Mumbai nor Dhaka. Limited and almost no planning or food policy that aims to improve the uptake of urban agriculture in that region is available. Urban agriculture is appreciated for its potential to grow food for the city dwellers worldwide. However, the potential of urban agriculture in this specific region is yet to be explored. Future research should explore and compare how cultural practices, affordability, and people's behaviour impact climate and other issues and provide varied urban performance linked to food for moving towards a resilient food future.

From both case studies, apart from their differences, one aspect is common: that a fully functioning food-producing rooftop garden is a possibility in both Dhaka and Mumbai. In both case studies in Dhaka and Mumbai, one individual and a private group, respectively, dedicated themselves to grow food voluntarily. However, their small initiatives have contributed towards increasing two small green spaces and providing access to nutritious food in the city. If the governments of these two cities are able to implement intervention techniques and planning and food policies for urban agriculture, it could have broader and more effective benefits to offer. Building capacities in people to use efficient farming techniques; developing guidelines on the types of suitable plants that could be grown on the rooftop gardens and measuring how rooftop local food production could improve micro-climate of cities would be very beneficial. All these benefits collectively could make a positive difference in the overall performance of cities.

20.5 Conclusion

This paper analysed literature references from two selected cities from neighbouring countries, Dhaka and Mumbai, to study the urban agriculture practice in both places. Analysis of two case studies of rooftop gardens were also conducted, to comprehend their similarities and differences in practice methods and to portray these urban agricultural practices as examples of functioning food production systems on the rooftops of these dense cities. Literature and case studies both indicate possibilities of implementing successful urban agriculture practices in these cities, but more research is needed on this field along with efficient and effective policies from government's side to promote urban agriculture in Dhaka and Mumbai.

Acknowledgements The author is very thankful to the referees for their comments.

References

Awasthi P (2013) Urban agriculture in India and its challenges. Int J Environ Sci Develop Monit (IJESDM) 4:48–51

Baker JL (2007) Dhaka: improving living conditions for the urban poor. World Bank Office, Dhaka

Clouse C (2014) Cuba's urban farming revolution: how to create self-sufficient cities. https://www.architectural-review.com/essays/cubas-urbanfarming-revolution-how-to-create-self-sufficient-cities/8660204.article. Accessed 8 Sept

Das AK (2017) Enhancing urban horticulture production to improve food and nutrition strategy, FAO Bangladesh, Bangladesh

Dubbeling M, Massonneau E (2012) Rooftop agriculture – a climate change perspective. RUAF Foundation, Katajelaan

'MbPT Terrace Garden and the philosophical and practical base of Urban Leaves'. http://purvita10.wixsite.com/urbanleaves/beginning. Accessed 13 Sept

Pendharkar A (2008) Mumbai Port Trust's 'Wild' Kitchen Garden India. http://cityfarmer.info/mumbai-port-trust%E2%80%99s-%E2%80%98wild%E2%80%99-kitchen-garden-india/. Accessed 8 Sept

Sajjaduzzaman M, MASAO K, Muhammed N (2005) An analytical study on cultural and financial aspects of roof gardening in Dhaka metropolitan city of Bangladesh', Int. J Agric Biol 7:184–187

Satterlee K (2015) Cultivating sustainable cities: a comparative study of urban agriculture in Mumbai, India and New York City, USA. Digital Commons @ Connecticut College. Available at: http://digitalcommons.conncoll.edu/cgi/viewcontent.cgi?article=1012&context=envirohp. Accessed 5 Dec 2016

Shariful Islam KM (2002, September) Rooftop gardening as a strategy of urban agriculture for food security: the case of Dhaka City, Bangladesh. In: International Conference on Urban Horticulture, vol 643, pp 241–247

Vazhacharickal PJ (2014) Urban and peri-urban agricultural migration: an overview from Mumbai Metropolitan Region (MMR), India. Int J Soc Sci 3:347

Zezza A, Luca T (2010) Urban agriculture, poverty, and food security: empirical evidence from a sample of developing countries. Food Policy 35:265–273

Zinia NJ, Paul M (2018) Significance of urban green and blue spaces: identifying and valuing provisioning ecosystem Services in Dhaka City. Eur J Sustain Develop 7:435–448

Chapter 21
Re-imagining Urban Leftover Spaces

Jasim Azhar, Morten Gjerde, and Brenda Vale

Abstract In most developed cities, leftover spaces in the urban fabric can be seen both as having potential and as threatening. Researchers have pointed out the issues, conditions, and importance of the positive utilisation of leftover spaces. However, there is insufficient information available on how to go about using such spaces. The revitalisation and aesthetic quality of leftover spaces could expand the dynamism of a city through strategic design interventions. This study seeks to understand the potential of different types of urban leftover space to be used in more effective ways than they are present. This paper, therefore, examines how such leftover spaces are defined and can be redesigned to become part of a built environment. The paper reports on affective and aesthetic responses that could lead to a better understanding of human perceptions of such spaces. A visual preference study, utilising semantic differentials for reimagined leftover spaces, was carried out to understand the differences between participants' preferences. A further comparison between participants who had occupations in built environment areas, and those who did not, showed that for both groups, the most preferred spaces were those that included vegetation. T-test analyses of the correlation results confirmed that participant professional expertise is not a preference factor when it comes to the design of leftover spaces, and in this respect, the study contradicts theories that hold that there are differences in the ways experts and non-experts perceive the environment.

Keywords Urban leftover spaces · Environmental perception and aesthetics · Visual preference approach

J. Azhar (✉) · M. Gjerde · B. Vale
School of Architecture, Victoria University of Wellington, Wellington, New Zealand

© Springer Nature Switzerland AG 2020   307
R. Roggema, A. Roggema (eds.), *Smart and Sustainable Cities and Buildings*,
https://doi.org/10.1007/978-3-030-37635-2_21

21.1   Introduction

The capital of New Zealand, Wellington, expects a significant population growth of 200,000–250,000 inhabitants from 2015 to 2040 (DIA 2015). This rapid population growth is a problem numerous cities face worldwide. New Zealand's population increased by 1.9% in 2015 (Statistics New Zealand 2015), outpacing Australia, which had a 1.4% increase in the same year (ABS 2016). Of the New Zealand population, 87% resides in 138 recognised urban centres, ranging in size from around 1000 people to more than 1000,000 (DIA 2015). Globally, the shift from rural to city living has increased the demand for resources, including water, food, and energy for urban populations (Satterthwaite et al. 2010). The growth and quality of future urbanisation will, therefore, have a massive impact on international resource availability and sustainability, affecting the quality of life for many people. There is an urgent need to design, test, and implement effective policies to address these issues. Like many other places, Wellington's development growth plan has been set up with a focus on achieving sustainable solutions, conserving natural environments, maintaining livability, keeping the city compact, and achieving maximum affordability.

Urban growth varies from area to area, making it almost impossible to follow only one development model (Turok and McGranahan 2013). To that end, any development will depend on the current infrastructure, traditional and cultural desires, topography, financial resources, and the institutional scope for planning and political stability for growth management. As cities expand across productive arable land, it is essential to investigate the potential value and usage of unused land or leftover spaces in currently developed areas of cities. An urban setting exists not only as a physical environment, but also as a shared space for personal perceptions and experiences, such that a city can be studied as an episode resulting from a continuing relationship between the built environment, civic processes, and human experience. When considering urban development, it is vital to understand the potential purposefulness of leftover spaces. The redevelopment process, including space assessment, has layers and structures from reading space to interpreting it and generating meanings through diverse activities. Capturing and engaging with the qualities of the intermediate, often invisible phenomena of the city, suggests the need for an alternative approach to utilizing leftover spaces efficiently and productively. Urban leftover spaces invite many possibilities for the integration of new techniques from integrating natural attributes to tactical solutions for the built environment.

This paper investigates the potential value of redesigning semi-public urban leftover spaces in more creative ways within the Wellington urban fabric. By 2040, the Wellington city council aims to reduce the presence of cars and car parks in the city centre by making it more compact, and leftover spaces have a great potential to be utilised strategically as part of this process. What is required are practical design solutions for the usage of semi-public (privately owned but the public can use that) urban leftover spaces. The visual preference study reported in this paper investigates design initiatives that could be more compelling for the public in terms of preferred design solutions. The key concern of this study is to explore and evaluate different design options with different attributes that influence people's perceptions of the usage of such outdoor spaces.

21.2 Urban Leftover Spaces

Trancik (1986) started theoretical research into urban leftover spaces approximately 30 years ago. He investigated their aspects and referred to them as 'lost space', as such spaces were ill-defined, had no significant outlook, and had a negative impact on the built environment. Furthermore, he argued such spaces had no definite or measurable boundaries and created division in use through policies or zoning. Urban leftover spaces are a fundamental part of an urban system and can occur next to planned development, along with and under highways or railway lines, are often stumbled upon unnoticed, and are known as no man's space, or land set aside for development. These are spaces of uncertainty (Muller and Busmann 2002), which are considered to be meaningless by a large segment of the community (Akkerman and Cornfeld 2009). Lacking officially assigned uses, leftover spaces are abandoned spaces that lie outside the rush and flow, as well as the control regulations of a city (Qamaruz-Zaman et al. 2012). In the name of 'progress', they are commonly considered as places devoid of function (Doron 2008). These spaces are vacant, unkempt, and underutilised with no defined function, often being between stages of formal development but indefinitely waiting for future use. Leftover spaces have been given different names throughout history, but the scale, spatial quality, and usability remain the real parameters with which to describe them. The literature employs different names for leftover spaces, often with varying scales, but no authors have dealt with or tested possible solutions for future regeneration from within. As some of the names suggest, these spaces seem vague and unloved. The critical issue of time and temporality is entirely excluded from the official definitions of leftover spaces. Azhar and Gjerde (2016) have categorised urban leftover spaces into six types that can be seen as having the potential for usage. These are between enclosed by buildings on 2 or 3 sides, in front of a building, at the back of a building, underneath a building, and on a rooftop. These spaces intervene between adjacent objects and often become problematic for the physical and social fabric of the city. There is thus a need to search for transformational opportunities. Urban leftover spaces exist because of several factors and are present in every major city, and often adversely affect the urban centre by disrupting the flow of neighbourhoods and districts, creating visually unappealing places, and reducing pedestrian interest in the surrounding businesses. Moreover, such spaces do not contribute to successful processes within cities, by neglecting to provide significant programmatic and social functions.

21.3   Perception of an Environment

Psychology plays a vital role in investigating the science of interaction between humans and their natural environment (Keniger et al. 2013). The field of psychology evolved in the early nineteenth century. However, i t was not until the mid-twentieth century that the importance of understanding the human-built environment became vital. Wilson and Baldassare (1996) describe the built environment as the relationship of people's needs to their surroundings, but that it also has to provide symbolic and functional needs. Environmental psychology encompasses the natural and constructed setting for human existence.

Furthermore, the main objective of environmental psychology is to enhance and upgrade the physical conditions for humans in the constructed setting. It also encourages the improvement of the human-nature relationship. 'Experience' and 'Perception' are the most commonly used keywords in an understanding of environmental psychology (Gieseking 2014). Experience relates to the transaction between intuition and already assessed knowledge. It differs from person to person and produces social contrasts, whereas perception is about identifying and interpreting the knowledge by using different senses. Rapoport (1982) claimed that the interdependence of a person to his/her environment is most essentially linked to sensual experiences and perception, while Gibson (1997) elaborated the idea of perception more deeply. He claimed that human perception was not just attached to the environment, but also accounted for the potential outcomes of that environment for human benefit and usage. However, Brebner (1998) argued that human thoughts, emotions, and feelings are influenced, both physically and emotionally, by what surrounds them. Wohlwill (1976) stressed the importance of the visual aspect and its effect on human psychology. Taylor et al. (2008) differentiated the two facets of perception: the dimension of sensory passiveness (or the idea of having any sensual experience), and the physical response that involves the action of a body. However, Seamon (2010) contradicted this idea by emphasising that both aspects were intertwined with each other. He said that in a day-to-day routine, both bodily actions and sensory responses are working continuously. The actions are coupled outcomes instead of a separate response and should be viewed as an integrated response.

21.3.1   Aesthetic Assessment

The idea of beauty or beautifying by the processing of human cognition and perception is known as aesthetics (McWhinnie 1968). This also includes emotional behaviour. People react varyingly to different environments around them, depending on their past occurrences and experiences, their closeness to all the views, and their expectations and the duration of exposure. Ulrich (1983) stated that aesthetic response is about the individual preference that provides a feeling of happiness or sadness and works through cognitive activity by visual confrontation. The aesthetic quality of any built or natural environment is a measure of a viewer's visual perception and responsiveness to that area (Company and York 2009). McWhinnie (1968) used aesthetics as a benchmark to explain the responsiveness of people towards a visual stimulus. Whether the stimulus is beautiful or not, it creates an analogy of aesthetics through human cognition. Also, if a specific visual appearance is more beautiful or pleasing, the preference is automatically diverted to it. Beauty rating is a result of this hypothesis. It is argued that visual impacts can be explained through various elements and not just a single factor. These include visual character and quality (e.g., form, line, colour, and texture), visual exposure, the viewer's idealised mental image, and the number of viewers who are expected to see the project (Hagerhall et al. 2008).

21.3.2   Visual Preference Study

Preference understanding is a vital element used to analyse how people judge an environment, including how they characterise and project it. This judgment can be different from person to person based on individual preferences. Habe (1989) confirmed that visual elements in a building are essential for a space preference. His study found evidence that photographs, responsiveness, and multi-dimensional scaling were essential in deriving the dimensions of perception. In visual preferences, photos of an environment are used as substituent agents for the original (Arriaza et al. 2004). Researchers like Kaplan and Kaplan 1989 and Sanoff (1991) have studied the reliability of this procedure. Nasar and Stamps (2009) suggested that showcasing photographs of a scenario or environment induce the same response in people as if the pictures were real. Tversky et al. (2006) found that the visuals of an environment exist in the human cognition, and can be as significant as real expressions. Furthermore, in 1957, Osgood, Suci & Tannenbaum used a linguistic analytical approach. They created a bipolar grouping to testify the efficacy of affective domains. As a result, photomontage became a way of creating altered images, by coupling or omitting elements to form a well-composed picture of future reality (Waldheim 2006).

21.4   Method

This visual preference study used photomontages to represent three alternative design modifications for six different types of leftover spaces in Wellington. All one-point perspective photos were treated in Photoshop to reconstruct the specific visions while emphasising one attribute in each context, and noting that these attributes changed with the context. All leftover spaces were designed without changing the current usage of the site. Concepts of providing more vegetation,

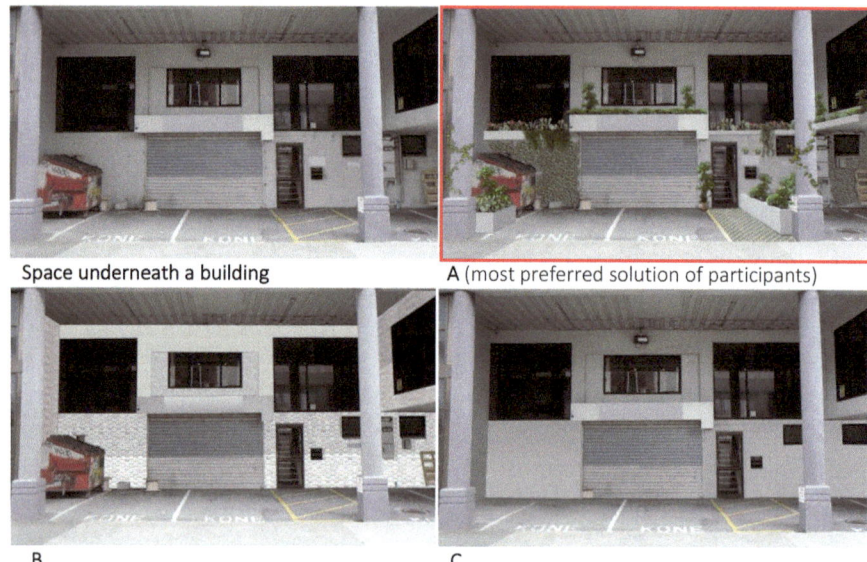

Space underneath a building | A (most preferred solution of participants)

B | C

Fig. 21.1 Example of a redesigned leftover space

creating seating space, improving cleanliness, changing surface materials, removing the boundary walls, creating clear pathways, and installing wind turbines or solar panels were photomontaged for different types of space (Fig. 21.1). These concepts were extracted from a previous study, where participants gave their suggestions for designing such sites. Each photo was rated using a 3-point Likert scale to reveal the differences and to understand the data more efficiently (1 Dislike, 2 Neutral, 3 Like). Benson (1971) recommended using a 3-point Likert scale for its practical convenience, and some claim as few as two response categories might be adequate in practice (Jacoby and Mattell 1971). The second part of the study was related to the semantic differential measures, which sought each participant's reaction to the redesigned space through a series of stimulus words/concepts. The concepts (adjectives) were evaluated through a 5-point bipolar rating scale. This section investigated reactions using the concepts of attractiveness (ugly-beautiful), satisfaction (annoying-pleasing), buildable (impossible-realisable), usability (boring-interesting) and mood (constrained-energetic). The adjectives were chosen according to how they best fitted the research aims and were consistent throughout the study. The Likert-scale reveals how much people agree or disagree with a particular statement, whereas the semantic differential scale decides how much of a trait or quality the item has, as rated through the bipolar scale defined by adjectives (Osgood and Snider 1969).

21.4.1 Participants

The participants responded to an interactive web-based survey made using Qualtrics. The study was initiated after approval by the human ethics committee. Invitations to participate were sent through email and by putting up posters in local cafes. The invitation emails were sent to administrators in the different Schools of Victoria University, Wellington City Council, New-Zealand Institute of Architects, University of 3rd Age[1]1, and Wellington City Library with a view of inviting both adults and students to participate in the study. By the end, data were collected from 121 individuals and imported into the Statistical Package for the Social Sciences (SPSS) software for analysis.

21.4.2 Sample Demographics

Overall, 96 participants completed the survey and 25 partially completed it. In terms of gender, 42.7% of respondents were male, 55.5% female and 1.8% did not answer. Participants with built environment knowledge formed 23.6% of the sample, with the remaining 76.4% being from different fields. Just over half (52.7%) of participants had an NZ European background compared to 47.3% with contrasting cultural ethnicity. In terms of formal education, 68.0% of respondents had a postgraduate qualification.

21.4.3 Procedure

The first step investigated the preferences for the whole sample of 121 respondents and identified the most appealing attributes (semantic differential) for each redesigned leftover space. The second step investigated the subgroups of 26 (23.6%) built environment participants compared to 85 (76.6%) respondents from other occupations, to see if there was any difference in preferences. Arnheim (1977) pointed out that built environment experts not only see what a building or a place looks like but also deconstruct it to understand how it was built and works. It is also claimed that built environment professionals perceive differently from other people and have different preferences (Posner 1973).

---

[1]An international movement founded in 1973. It focuses on improving living standards and helping the personal development of older or retired people (Marcinkiewicz 2011).

21.4.4 Data Analysis

Different methods of analysis were deployed to understand the relationships between the sample groups. The mean preferences for the most and least preferred redesigned photos were measured on the Likert scale (1–3) by using a descriptive frequency test in SPSS. The simple technique of calculating the mean, standard deviation ($\pm$SD) and percentage of the most preferred design was used. Kendall's tau-b (τb) correlation test was conducted between the most liked images with the respondent's attitude (semantic differential scale). Kendall's tau-b (τb) correlation coefficient calculated the strength and direction of association in a nonparametric measure, such as exists between two variables measured on an ordinal scale (Laerd Statistics 2016). A comparison was made to evaluate the alignment of the built environment with the non-built environment participants. The percentages for the most preferred design on a 3-point Likert scale were calculated for participants from different fields of study. The Cronbach alpha (α) reliability test was used to check the internal consistency of several variables for the semantic scale 1–5 before independent sample T-tests were carried out. Cronbach's alpha (α) was 0.81 and indicated a high level of internal consistency. The averages of semantic differential responses were calculated to find the difference in preferences among the respondents through an independent sample T-test. The differences of opinion between the built environment and the non-built environment participants were analysed using independent sample T-tests.

21.5 Results

21.5.1 Whole Sample (n = 120)

The preference value on the 3-point Likert scale fluctuated, but the most preferred design solution among all participants related to adding more vegetation to all leftover spaces (Fig. 21.2). A Kendall's tau-b (τb) correlation test for different images revealed that the most likeable image had a strong, positive association with all affective appraisals (semantic differentials) except for the bipolar category of "impossible to realisable". This suggested this category was independent of the association and was not influenced by the image's likability.

The space in front of a building had a different preference ranking among all participants. The first preference was for removing the boundary wall, whereas the second most preferred design was providing more vegetation. The Kendall's tau-b (τb) correlation test revealed the most likeable image in front of a building had a weak, negative association with the one affective appraisal category of "boring to interesting".

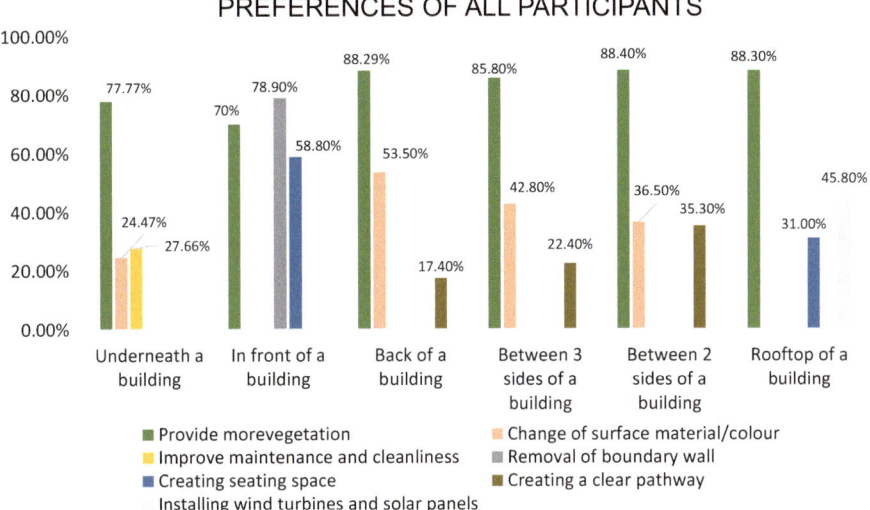

Fig. 21.2 Preferences of all participants

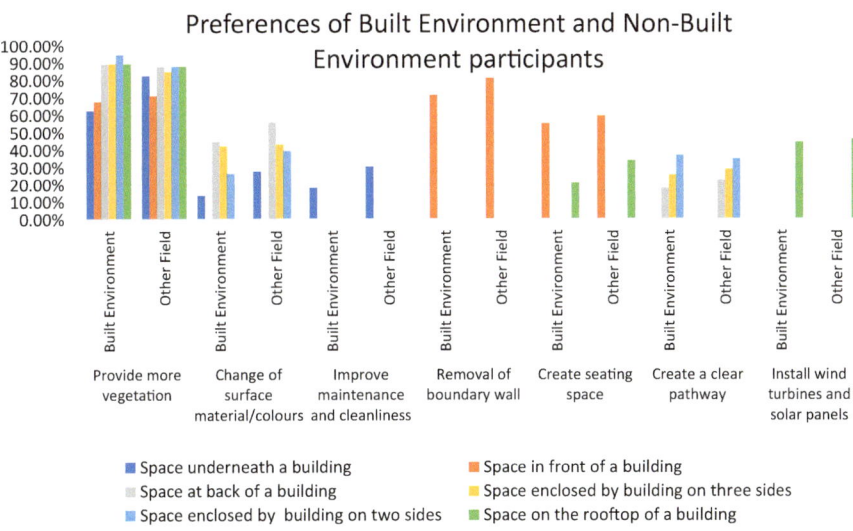

Fig. 21.3 Preferences of Built Environment participants and others

21.5.2 *Built Environment and Non-built Environment Participants*

The most robust agreement for both groups was for spaces that had an element of vegetation in them (Fig. 21.3). The space at the back of a building, enclosed by the buildings on three sides, enclosed by the buildings on two sides, and a rooftop space all with the introduction of vegetation were valued higher by BE than NBE participants.

An independent sample T-test for both the groups confirmed that there was no statistically significant difference in opinions between the built environment and non-built environment group (P > 0.05) for all redesigned leftover spaces. The space in front of a building with the removal of boundary walls was liked best by both groups. The option with vegetation was the second most liked design for both BE and NBE participants. An independent sample T-test for both groups confirmed that there was no statistically significant difference in opinions between the BE and NBE participants (p > 0.05) for all redesigned leftover spaces.

21.5.3 Discussion

The quantitative study demonstrated (1) an overall desire for incorporating nature into built environments and (2) that there was no difference in preferences for the built environment and non-built environment participants. The preferences for all six examples of leftover spaces were similar for each scenario. A study by Ulrich (1981) found that natural environments are usually evaluated as having a high rate of aesthetic quality over built environments, and have relevance to the aesthetic response. At the same time, Ulrich's study also suggests that incorporating vegetation is not always practical and can be expensive since it comes with maintenance, care or stewardship issues. Respondents also exhibited preferences for visual openness, as evidenced by the preference scores for the scene depicting the removal of a boundary wall in front of a building. This result suggests that the entrance to a building should be designed to be open and inviting and creating areas that are perceived as claustrophobic should be avoided. Visual quality can influence a person's experience significantly because people react to what appears before them. Different needs and demands can be catered for to create positive settings for each type of leftover space, and this could enhance people's attitudes and behaviours. The most realisable solutions among all designs were related to vegetation, change of surface material or its colours, improving cleanliness, and creating pedestrian pathways and seating spaces. The comparisons between the preferences expressed by built environment participants and lay people were similar for all design solutions. This suggests both groups perceived the designed spaces similarly, and the sample T-tests for opinions of leftover spaces were the same. Also, the aesthetic judgments of preferences were aligned between all participants. Overall, it seemed that if a leftover space is designed with natural elements, this could induce a spatial preference. However, solutions regarding preferences for installing different types of plants (trees, shrubs, climbers, and ground cover) is nonexistent in the literature for leftover spaces, and this is a possible area for further research. Another issue that needs further exploration is the light exposure and scale when it comes to measuring preferences. In this study, only one view of each space was given, with an attempt to have a similar level of light in each. Ephemeral qualities, such as light level, affect emotional responsiveness. However, it does appear from the analysis that human intervention with natural design solutions is a crucial aspect of how leftover spaces could be improved.

21.6 Conclusion

This study has examined and assessed three different design schemes for urban leftover space, by asking survey participants about their evaluations, to gain a fuller understanding of which solution is preferred. This study also suggests a direction for design schemes in that they should consider enhancing the aesthetic quality of space with the use of natural elements. For example, the use of urban leftover spaces could aid in mitigating climate change through urban food production, if both the owner and the public agree with such changes in use. The importance of leftover spaces between, over, and under buildings as part of the public realm should be realized by city stakeholders, managers and end users. It is not clear yet what future cities will look like with the implementation of sustainable measures that include emerging technologies, but it is vital to develop new strategies to cope with humanity's demands for resources and designing such spaces through stakeholder participation is a need for future cities.

References

ABS (2016) Women outnumbering men in Australia. Australian Demographic Statistics, Aug 2018

Akkerman, Cornfeld A (2009) Greening as an urban design metaphor: looking for the city's soul in leftover spaces. Structurist 2009:30–35

Arnheim R (1977) The dynamics of architectural form. University of California Press, Berkley/Los Angeles

Arriaza M, Canas-Ortega JF, Canas-Madueno JA, Ruiz-Aviles P (2004) Assessing the visual quality of rural landscapes. Landsc Urban Plan 69:115–125

Azhar J, Gjerde M (2016) Rethinking the role of urban in-between spaces. J Archit Sci Rev. ASA conference proceedings, Taylor & Francis, Adelaide

Benson PH (1971) How many scales and how many categories shall we use in consumer research? A comment. J Mark 35:59–61

Brebner J (1998) Personality and individual differences, happiness and personality. Personal Individ Differ 25(2):279–296

Company E, York N (2009) Measuring the perceived aesthetic quality of photographic images. Cathleen Daniels Cerosaletti and Alexander C. Loui. Image, Rochester, pp 47–52

DIA (2015) Setting the scene. The Department of Internal Affairs, Aug 2018

Doron G (2008) Those marvellous empty zones at the edge of cities in Dehaene and De Cauter, Heterotopia and the city: public space in a post-civil society. Routledge, London, pp 203–213

Gibson E (1997) An ecological psychologist's prolegomena for perceptual development: a functional approach. In: Dent-Read C, Zukow-Goldring P (eds) Evolving explanations of development: ecological approaches to organism-environment systems. American Psychological Association, Washington, DC, pp 23–45

Gieseking J (2014) Environmental psychology. In: Teo T, Barnes M, Gao Z, Kaiser M, Sheivari R, Zabinski B (eds) International encyclopedia of critical psychology. Springer, New York, pp 587–593

Habe R (1989) Public design control in American communities: design guidelines/design review. Town Plan Rev 60(2):195–219

Hagerhall CM, Laike T, Taylor RP, Küller M, Küller R, Martin TP (2008) Investigations of human EEG response to viewing fractal patterns. Perception 37(10):1488–1494

Jacoby J, Matell M (1971) Three-point Likert scales are good enough. J Mark Res 7:495–500

Kaplan R, Kaplan S (1989) The experience of nature. A psychological perspective. Cambridge University, Cambridge

Keniger L, Gaston K, Irvine K, Fuller R (2013) What are the benefits of interacting with nature? Int J Environ Res Public Health 10(3):913–935

Laerd Statistics (2016) Kendall's tau-b using SPSS statistics. Statistical tutorials and software guides. Retrieved from https://statistics.laerd.com/

Marcinkiewicz A (2011) The University of the Third Age as an institution counteracting marginalisation of older people. J Educ Cult Soc 2:38–44

McWhinnie H (1968) A review of research on the aesthetic measure. Acta Psychol 28:363–375

Muller and Busmann (2002) Spaces of uncertainty. Kenny, M, Markus, C, Berlin

Nasar J, Stamps A (2009) Infill McMansions: style and the psychophysics of size. J Environ Psychol 29(1):110–123

Osgood C, Snider J (1969) Semantic differential technique: a sourcebook. Aldine, Chicago

Osgood C, Suci G, Tannenbaum PH (1957) The measurement of meaning. University of Illinois Press, Oxford

Posner M (1973) Cognition: an introduction. Scott, Foresman, Oxford, p 208

Qamaruz-Zaman N, Samadi Z, Azhari N (2012) Opportunity in leftover spaces: activities under the flyovers of Kuala Lumpur. Procedia Soc Behav Sci 68:451–463

Rapoport A (1982) The meaning of the built environment: a nonverbal communication approach. Sage, Beverly Hills/London

Sanoff H (1991) Visual research methods in design. Van Nostrand Reinhold, New York

Satterthwaite D, McGranahan G, Tacoli C (2010) Urbanisation and its implications for food and farming. Philos Trans R Soc B Biol Sci 365(1554):2809–2820

Seamon D (2010) Merleau-Ponty, perception, and environmental embodiment: implications for architectural and environmental studies. In: McCann R, Locke P (eds) Carnal echoes: Merleau-Ponty and the flesh of architecture. Routledge, New York

Taylor KI, Salamoura A, Randall B, Moss H, Tyler LK (2008) Clarifying the nature of the distinctiveness by domain interaction in conceptual structure: comment on Cree, McNorgan, and McRae. J Exp Psychol Learn Mem Cogn 34(3):719–725

Trancik R (1986) Finding lost space, theories of urban design, 1st edn. Wiley, New York

Turok I, McGranahan G (2013) Urbanisation and economic growth: the arguments and evidence for Africa and Asia. Environ Urban 25(2):465–482

Tversky B, Agrawala M, Heiser J, Lee P, Hanrahan P, Phan D, Stolte C, Daniel M (2006) Cognitive design principles: from cognitive models to computer models. In: Magnani L (ed) Model-based reasoning in science and engineering. King's College, London, pp 227–247

Ulrich R (1981) Natural versus urban scenes: some psychophysiological effects. Environ Behav 13:523–556

Ulrich R (1983) Aesthetic and affective response to natural environments. In: Altman I, Wohlwill J (eds) Human behavior and the natural environment. Plenum, New York, pp 85–125

Waldheim C (2006) The landscape urbanism reader. Princeton Architectural Press, New York

Wilson G, Baldassare M (1996) Overall 'sense of community' in a suburban region: the effects of localism, privacy and urbanisation. Environ Behav 28:27–43

Wohlwill J (1976) Environmental aesthetics: the environment as a source of effect. In: Altman I, Wohlwill JF (eds) Human behaviour and environment, vol 1. Plenum Press, New York, pp 37–86

Chapter 22
A New Model for Place Development – Bringing Together Regenerative and Placemaking Processes

Dominique Hes, Cristina Hernandez-Santin, Tanja Beer, and Lewis Lo

Abstract This paper describes the bringing together of two practices, placemaking and regenerative development. Placemaking is a relatively recent term, describing a city making movement focusing on the process of developing places through the active participation of the citizens that conceive, perceive, and live in that place. It aims to create place attachment, a foundational concept of environmental psychology linked to positive outcomes in health, community participation, civic behaviour, and perceptions of safety. Regenerative development is an approach to supporting design for place to focus on the delivery of vital, viable, and resilient places, able to evolve over time to support all human and non-human life. In this paper, these two practices are integrated under the 'Place Agency' model. This model harnesses the key strengths from both practices, while providing ways to address their limitations. The research approach used to integrate the models was discursive grounded theory; where each practice, its rhetoric, its tools, and case studies was looked at. The content was analyzed using inductive coding to identify potential synergies. The resulting model indicates that merging these two practices can deliver a place designed for both human and non-human participants, potentially shifting city making from a largely anthropocentric based practice. The combined approach supports the ability to look across history and its attributes to understand a place's potential, while providing a method through which the community can actively participate in the city making process. Placemaking can thus become a strategy to bring forward this potential, test, play, and evaluate regenerative initiatives, in context of spatial, temporal, social, and ecological influences.

Keywords Placemaking · Regenerative development · Cities · Human and non-human agency

D. Hes (✉) · C. Hernandez-Santin · T. Beer · L. Lo
Thrive Research Hub, Melbourne School of Design, Faculty of Architecture, Building and Planning, University of Melbourne, Melbourne, Australia
e-mail: dhes@unimelb.edu.au

© Springer Nature Switzerland AG 2020
R. Roggema, A. Roggema (eds.), *Smart and Sustainable Cities and Buildings*,
https://doi.org/10.1007/978-3-030-37635-2_22

22.1 Introduction

22.1.1 Aims of This Paper

More and more people are moving into cities. As such, it is important to create places
that support the future of cities as places of living, working, creating and contributing. Designing for people is only one aspect of city making, though. Incorporating
strategies for nature integration and the non-human aspects of life is also critical, not
just for their own sake, but for the wellbeing of the whole system, including humans,
who have an innate need to be connected to nature (Wilson 1984). This paper brings
together regenerative development and placemaking to create Living Environments.
In this paper, we define living environment as "a setting that is thriving, healthy, and
resilient because its ecological, social, and economic systems are continually
nourished." (Plaut et al. 2016, p. 2). This paper outlines the two practices in question,
then presents the methods used to bring them together, exploring their key strengths,
concluding with a suggested integrated framework.

22.1.2 What Is Placemaking Practice

Placemaking is a worldwide practice focusing on the process of developing places
through the active participation of the citizens that conceive, perceive and live in that
place (Arefi 2014). It aims to create place attachment, a foundational concept of
environmental psychology linked to positive outcomes in health, community participation (Anton and Lawrence 2014), civic behaviour, and perceptions of safety
(Billig 2006). It is possible to conduct placemaking through formal (i.e. strategic
placemaking) to informal (i.e. tactical urbanism) approaches. The key characteristics
of a placemaking project are: (1) a process which puts emphasis on deep engagement
with the community of an area; (2) the use of relatively small projects to trigger long-
term benefits; and (3) the aim of improving life quality by developing social
cohesion and place attachment that contributes to the planning and investment in
public places (Kyle et al. 2004).

There is strong evidence (over five studies) that placemaking can foster place
attachment in increasingly dense, diverse and mobile communities (Hidalgo and
Hernandez 2001; Lewicka 2010; Scannell and Gifford 2010). The strengths of
placemaking lie in its adaptiveness to context, its ease and often affordable ways
to reimagine spaces (PPS n.d.). Successful placemaking efforts are often
community-led or have undergone extensive community engagement, where the
'placemakers' take the time to build a relationship with the people of that area. In
many ways, the placemaker role is to provide a safe space for the community to
voice their opinions and needs, and subsequently work with them to come up with
key initiatives.

Placemaking is simultaneously a process (of community engagement) and a product (which may or may not be a design). It can be a time-consuming practice in which trained facilitation, communication, and listening skills are critical. Because time is often limited, placemaking projects can easily be superficial in their engagement, and thus fail to achieve the intended long- term benefits and can contribute to inequality and gentrification across communities (Fincher et al. 2016). What is needed is a way to think long term to integrate the ecological, or non-human aspects, and to develop capacity over time for the place; to strengthen not only itself and its stakeholders, but the broader systems on which it relies.

22.1.3 *What Is Regenerative Development*

Regenerative development is an approach that applies the ecological worldview. Plaut et al. (2016, p. 2) defined it as "the process of cultivating the capacity and capability in people, communities, and other natural systems to renew, sustain, and thrive". Simplified, our approach to regenerative development is to:

1. Understand the flows through a system that bring it to life, that create a living system. Flows are the various resources, including 'intangibles' like culture and social cohesion, that interact with the place.
2. Design place-based solutions that create multiple, mutual benefits between these flows by focusing on the opportunities for creating relationships.
3. Operate within the context of the place to ensure its relevance, resilience, and ability to adapt.

Though in its infancy in application, Regenerative Development is informed by systems thinking (see i.e. Meadows 2008), ecological thinking (see i.e. Du Plessis and Brandon 2015), and indigenous thinking (see i.e. Mang and Haggard 2016). Critically, regenerative development is about working within a system to enable the potential of the system to emerge, to co-evolve the aspects of the system so that it can constructively adapt to change and evolve towards increasing states of health and abundance. There are examples of the application of regenerative development ideas internationally, mostly related to reflections on specific projects and their outcomes (Mang and Reed 2012), and case studies found on practitioner pages such as Regenesis and the Institute for the Built Environment (IBE) at Colorado State University. While these provide insights into the outputs of regenerative development projects, there is a need to better understand the process that supports regenerative thinking and contrast it to 'business as usual'. That is: how do we operationalize these abstract concepts of creating ecological, social, and economic benefit within a place? It is in the operationalization that the potential of bringing these two approaches together is born.

22.2 Methods

The research started with a literature review of both scholarly and practice-based publications of placemaking and regenerative development initiatives. The following section presents a summary of the key aspects that were revealed from the literature review. This literature was coded inductively, identifying where their approaches complemented or mirrored each other.

Inductive coding allowed us to convert papers, case studies, manuals, online content, and books into keywords, approaches, and concepts that suggested synergy between the two approaches. Inductive coding supported the research process, with continual revisiting of the codes allowing an unfolding or revealing of how the two practices can work together, and a sense of the synergistic potential. This is unlike deductive coding, where one is trying to prove a hypothesis and has pre- conceived ideas of the outcomes.

We used a 'discursive grounded theory' approach to bring this data together. Grounded theory is a systematic methodology in the social sciences, involving the construction of theory through the gathering and analysis of data (Martin and Turner 1986; Strauss and Corbin 1994). Grounded theory is a research methodology which operates inductively. For this research, we started with the question of the ability for regenerative development to contribute to the ecological potential of placemaking. We continually reviewed the data collected, repeated ideas, concepts, or elements through coding. These were grouped into concepts, categories and themes, resulting in the approach outlined herein. 'Discursive grounded theory' is the term we used, because this was a collaborative and iterative process of discussion, argument, deliberation, and negotiation between researchers and practitioners. This was not a single researcher and a computer using software; the 'codes' and their analysis were developed through consultation, conversations, and testing.

22.3 Outcomes

Much like placemaking, regenerative development is often regarded as a practice – one that requires co-creation between professional regenerative development practitioners and the users of the development project. Unsurprisingly, this is often centered around underlying connotations, experiences, stories, feelings, and values that the stakeholders hold for a place (Mang and Haggard 2016). Co-creating an understanding of place identifies underlying patterns of meaning and interactions, which allows for better integration of human social and economic processes with ecological processes and is something both practices aim for. In placemaking, it is also about co- creating an understanding of place, its values to the stakeholders, and what will contribute to the betterment of the stakeholders and the place. Thus, although ends differ somewhat, practitioners of regenerative development and placemaking often perform many of the same tasks.

Table 22.1 Differences and gaps between placemaking and regenerative development

	Placemaking	Regenerative development
Goals and place	Social	Socio-ecological
	Humans as beneficiaries	Humans as catalysts and co-evolving with environment
	Respecting meaning of place in and of itself place as final product	Meaning of place as means and agent of co-evolution
		Focused on potential-building
Scale of actions	Local, with consideration of wider geographical, social or policy context on occasion	Nested within local, proximal and global spheres of influence
	Cross-scale influence may be important to ensure sustainability of local places	Mutual cross-scale benefits are important
Orientation towards the future	Often neglected or assumed	Co-evolutionary responses shaped by pattern discernment
	Place-keeping to ensure longevity and periodic revitalization of placemaking uncertainty is something to be managed (if not neglected)	Lacks clear guidance on ensuring socio- ecological durability
		Respects uncertainty without guidance on what to do with it

We contend then that both practices can lead to long-term care and evolution of places and promote wellbeing of all stakeholders. Yet they approach a place differently spatially and temporally, in their understanding of the potential of a place and its ability to evolve. It is our contention that combining the two approaches creates the potential of a more holistic framework. This requires understanding and reconciling their different approaches with different practices, goals and visions (Table 22.1).

Importantly, it is the differences between approaches that offer the opportunity to investigate if their integration will provide a greater potential to create living environments. The following sections consider their ability to support the evolution of place, ability to integrate spatial and temporal aspects of place, and the ability to elicit the potential of place.

22.3.1   Evolution of Place

Living environments are resilient and respond to events and opportunities in ways that serve the whole system and make it more vital and viable. Therefore, looking at the way both approaches address evolution is critical. For a place and its stakeholders to be able to evolve constructively through change, it is important to support its ability to identify, respond and adapt to change.

In placemaking, the focus is usually on humans as the beneficiaries of a place, with the important requirement that such benefits should be tailored to how people make sense of the place. Thus, placemaking is most often conceived as activities that develop or increase place attachment and place understanding (i.e. meaning) in addition to enhancing environmental quality and amenities. It is essentially anthropocentric, with the care for the ecosystem aspects of place being strongly related to the perception of the people involved in the value of the ecosystem to their thriving. Further, the support for the non-human aspects of place will reflect the values of the human stakeholders, meaning that sometimes what is perceived to be good for human stakeholders will have priority over the non-human. Examples include the recent push to continue mining and using coal because this results in jobs, or the strong views of wind power. Without a vision for the complexity of a place and all its human and non-human stakeholders, the opportunity to develop and address the place is reduced. Often it is the non-human that are the heralds of change (think the canary in the mineshaft). Therefore, being able to conceive place as creating mutually beneficial relationships between the human and non-human is critical in being able to identify and respond constructively to change.

Placemaking that leads to place attachment without the context of all the stakeholders of the place and its complexity also means that it can result in unhelpful outcomes; it can lead to conscious and unconscious bias that reduces the ability to evolve through change. For example, Devine-Wright (Devine-Wright 2014), suggests strong place attachment attitudes may actually make people more resistant to making changes. In worse cases, place attachment can result in self-segregation and the manifestation of xenophobia, as evidenced by residents in certain white neighbourhoods in South Africa bemoaning the arrival of black neighbors (Dixon and Durrheim 2000). Such negative consequences can become a profound limiting factor for the evolution of places, or otherwise drive places towards degeneration.

Therefore, to create a living environment, the ability to create a strong place attachment is an outcome that needs to be carefully planned into the process of working in that place. Creating a strong story of place that connects people to the benefits of the ecosystems and to their continued thriving in that place is critical, as is the ability to be invested in that place so as to understand it and be able to work with is as circumstance change.

This is the contribution that regenerative development can provide. Regenerative development seeks a co-evolutionary relationship between humans and the ecosystems of that place. As Mang and Reed (2012, p. 5) explain, regenerative development is "not preservation of an ecosystem, nor is it restoration. Instead, it is the continual evolution of culture in relationship to the evolution of life." The philosophy of regenerative development is therefore neither anthropocentric, nor explicitly strictly ecocentric either. Rather, proponents of regenerative development envision a relationship where development and ecosystems are changing in response to one another, such that each ultimately benefits from their relationship with the other. Moreover, they argue that value-generating capacity is only possible by (re)growing such a relationship over time, which is not only distinctive to each entity but itself an agent of evolution within open systems (Mang and Haggard 2016).

To this end, regenerative development focuses on potential-building, feedback loops, and understanding the human and non-human aspects of place. Their relationship beyond the place therefore enables those involved to be more able to observe, plan, and respond to change. Mang and Reed (2012) describe regenerative development as consisting of two central and interrelated endeavors: (1) choosing the right phenomena to work on so as to maximize the system's potential for evolution and (2) building capacity and a 'field of caring and commitment' among stakeholders. Doing so is not only accepting change but committing to a never-ending process of change and openness. Humans' role, then, in this co-evolutionary relationship is to be catalyzers and active participants, creating or contributing to processes with the potential to generate a healthy place without trying to tightly control the direction the system evolves in (Mang and Reed 2012; Hes and Du Plessis 2014).

From this point of view, place – more specifically, the storytelling of place – can be treated variously as a framework, a mechanism and a process, rather than as a final product (Mang and Reed 2012). As a framework, story of place helps humans learn how to understand and co-evolve with their environments better, provided such storytelling explicitly addresses the relationship between humans and the ecology/non-human they are enmeshed in. As a mechanism, story of place helps create the field of caring and commitment necessary for continuous potential building. As a process, place itself acts a change agent after design and construction, calling on human actors to respond and remake meaning as its potential is realized or changes with regards to wider systemic changes through placemaking.

In summary, regenerative development allows stakeholders to understand their place in context, and to think about how to support that place to become more vital, viable, and able to observe and respond to change. Placemaking allows the testing of this with the stakeholders in a point in time, in a specific place. It is these temporal and spatial aspects that are discussed next.

22.3.2 Temporal and Spatial Scale of Place to Enable Adaptation and Resilience

22.3.2.1 Temporal

As discussed above, placemaking is essentially an activity undertaken to connect people to a place with an endpoint in mind, for example, increased activation of a park, activating a dying main street, safety of a laneway, etc. As such, it is mostly short-term, with some placemaking practice such as tactical urbanism being temporary. However, only thinking in endpoints and solving single space related problems can be a limitation when working in an ever-changing and evolving place.

In contrast, regenerative development's focus is on fostering co-evolutionary relationships between humans and places, culture and ecosystems/non-human. It does this while looking across time, in some projects taking the native American Indian concept of design for 7 generations (learning from the past three, designing for the future potential of three and the middle one now). Its aim is to create greater vitality, viability and ability to adapt (Mang and Haggard 2016), by looking at what worked throughout history, and improving the relationships between aspects of the place, its stakeholders, and their ability to reach their potential.

Bringing placemaking in line with regenerative development practice provides a way to understand the place over time, to ensure that the making and management of the place aligns with the essence of the place, based on how it worked throughout history and using placemaking to test if that is still relevant now.

22.3.2.2 Spatial

Placemaking focuses primarily on the site itself, and rarely addresses the reciprocal impact that occurs between the place and its stakeholders across different scales; it merely acknowledges that this reciprocity can occur. Arguably, this may be due to its essential focus on place attachment and meaning. As discussed previously, place attachment can manifest as insularity if the local communities' understanding of their place and place identity do not mesh with external elements around them. Mere apathy, lack of understanding, time, or ability to support external elements may also prevents cross-scale or cross-place considerations (Dempsey and Burton 2012; Mathers et al. 2015).

Regenerative development practice recognizes that development projects are always limited in scale, regardless of how large they are. Thus, regenerative development takes a simple nested approach to the cross-scale interactions (think a village square, within a village, within a watershed). This nested framework considers three layers of influence: the project site itself, the proximate whole and the greater whole. The proximate whole refers to a relatively localized system that is immediately relevant to the project, as defined through an understanding of the natural flows in the system or through cultural and social agreement. The greater whole is the greater system that may affect and be affected by the project in more indirect ways or over longer periods. This may include entities at the city, regional, national, or global scales, such as the international market or global climate patterns. The result of this approach is that the project acknowledges its role to provide positive benefits for aspects beyond the site, and this becomes part of its essence and its story of place. It also means that its capacity to influence the other levels is explicitly part of the development process, it is explicitly integrated into the project. In regenerative terms, it 'does what it can' (Hes and Du Plessis 2014) to contribute beyond its boundaries through its design, so as to create a stronger whole. For example, the design of tracks of greenery in a housing development can provide a potential wildlife corridor.

In the process of bringing placemaking and regenerative development together, it is placemaking that provides place-based projects that can be 'acupuncture points' that catalyze a community engagement with the story of place and its potential. As Mang (2009) write, "Places, as attractor points, therefore, are evolutionary agents in that they become points within a larger system in which new life and new distinct patterns of existence can emerge" (p. 40). Additionally, whereas placemaking efforts can sometimes lead to communities responding to external threats by rejecting their influence, regenerative development's attention to the reciprocal nature of nested cross-scale interactions suggests a different response. Regenerative development gives a way to think of the proximate and greater wholes and their mutual relationships to the project, while placemaking provides a way to test and refine ideas of what happens in the place to manifest these relationships.

22.3.3   The Ability to Elicit Potential

Thus far, this paper has outlined that both placemaking and regenerative development aim to create better places by working with its potential. A challenge lies in the amount of information required to elicit the potential of place, particularly if integrating all the aspects of the site. Placemaking elicits potential through a process of strong community engagement throughout the design and development of the project. This can then be tested through tactical urbanism and other temporary techniques, so that the lived outcome of the ideas can be experienced creating a stronger connection to the potential of the project. Yet, as outlined above, this is an anthropocentric approach, and ultimately may fail to lead to ongoing thriving and the capacity to evolve, because it overlooks the non-human elements of the ecosystem of which humans are an integral part. Therefore, all aspects of the ecosystem need to be incorporated. Yet, limitations on time, resources, and capacity might make the data needed to do this seems to onerous.

Regenerative development practitioners have a counterpoint to this: they argue that though data of a site is important, it is more critical to identify the patterns that this data reveal. The often-used example is that of knowing your life partner, or children: you don't know them by the state of their liver, or blood pressure, or ingrown toenail, you know them by the pattern of who they are, and how that reveals their essence. For a project, Mang and Reed (2012) advocate for the use of story-telling to create a 'story field' to focus practitioners' and stakeholders' attention towards evolving patterns in "the whole system and what [the system] is attempting to become" (p. 12). As the short story above illustrates, it provides for a practical course of action for coming to a decision about what phenomena to work on. It is a way to bring together the complexity of the assessments of physical assets, ecosystems, geology, history, hydrology, and so on. Again, placemaking gives the tools to engage people to work together at a specific point in time, at a specific place, while regenerative development provides the story and context of how this could be done to achieve greater potential.

22.4   Discussion: Integrating Placemaking and Regenerative Development for Continual Co-evolution

Placemaking and regenerative development have strengths and weaknesses that are complementary. Working together, these two frameworks have the potential to harness each other's strengths and use them to minimise their limitations.

Placemaking benefits from regenerative development through the systems approach, ecological considerations, nested thinking beyond the project boundary, long-term visioning and planning, and pattern analysis to identify potential. Regenerative development benefits from placemaking as it provides a way to test, refine, implement, and learn, based on creating relationships to a place in a specific point of time. This is the Place Agency model. When integrating both practices, placemaking then becomes an acupuncture point where the regenerative potential of a place manifests itself.

The active nature of placemaking implementation brings life to a site allowing the community to build a stronger relationship to the place (attachment) and providing opportunities to become active participants of its ongoing growth (stewardship). Meanwhile, the reflective nature of regenerative development allows the community and place experts to continuously analyse the flows that bring the system to life, find the new patterns, design new solutions that lead to socio- ecological benefits, and adapt the regenerative strategy in response to the acupuncture points.

The Place Agency model is a call for regenerative practitioners and placemakers to work together to integrate these complementary place-centric views. The collaboration process will harness the key strengths posed by each framework to deliver Living Environments that are constantly evolving. The process of integration would start as follows: (A) during the conceptual design phase, the Place Agency model proposes the regenerative development approach for site analysis based on system's thinking. This analysis will holistically integrate human and non-human participants in their analysis and identify key patterns of the area. Placemaking is applied as part of this analysis to develop the community engagement strategy suited to the project and to lead a process where the community identifies their values, their needs, and current perspectives of the site. This results not only in a design, but a whole strategy for the ongoing evolution and improvement of the site from a social, ecological, and economic standpoint. (B) During the consultation and detail design process, placemaking is used to deliver some short-term interventions, trialing different aspects of the design (e.g. using tactical urbanism techniques). These interventions constitute a quick and responsive way to observe the community response to the regenerative ideas, while keeping the interest going throughout the design and planning process of any project. (C) Finally, during the construction phase, the community engagement process is used to continue working with the community, design team, and regenerative practitioners to develop a place management strategy which responds to the ever-changing interests of the community, in celebration of the past and present of the place.

22.5 Conclusion

Placemaking and Regenerative Development are two approaches to design and project implementation aiming to deliver healthy built environments that are relevant to the unique attributes of each community and geographical areas. Both approaches are place-centric, with placemaking also being people-centric while regenerative development represents a socio- ecological approach. These practices differ on three key elements:

1. Evolution of place: Placemaking has mostly grown as a social movement focused on delivering temporary or permanent people-friendly shared spaces. Regenerative development is a socio-ecological framework that brings in the importance of both social and natural systems to create vibrant and resilient places.
2. Temporal and spatial scale of place: Both frameworks comprise an ongoing and adaptable process constantly revisiting what is working and what is not. However, regenerative development works on much longer timeframes and detailed understandings of systems, while placemaking poses a much more flexible approach suitable for trial and error interventions, embedded in a specific space at a specific time.
3. Eliciting potential: While placemaking identifies opportunities through community consultation, regenerative development finds potential within a living system through observation of patterns.

These three differences are complementary and can support the alternative framework in moving beyond its limitations. This paper presents a new approach, the 'Place Agency' model, to harness these complementary aspects of placemaking and regenerative development. This model understands placemaking as a point of time, nested, within regenerative development path and allowing the potential of the place to manifest. It proposes placemaking as a fun, quick, and responsive approach to trialing key ideas considered for the long-term outcomes sought by regenerative development. Placemaking providing an ongoing catalyst, or acupuncture point to revisit, adapt and re-align the regenerative development journey. By implementing regenerative development and placemaking together, places can grow from an anthropocentric approach to one that considers the whole system. Together, both frameworks can successfully support a co-evolving process suitable to deliver Living Environments.

References

Anton CE, Lawrence C (2014) Home is where the heart is: the effect of place of residence on place attachment and community participation. J Environ Psychol 40:451–461

Arefi M (2014) Deconstructing placemaking: needs, opportunities, and assets. Routledge, London/New York

Billig M (2006) Is my home my castle? Place attachment, risk perception, and religious faith. Environ Behav 38(2):248–265

Dempsey N, Burton M (2012) Defining place-keeping: the long-term management of public spaces. Urban For Urban Green 11(1):11–20

Devine-Wright P (2014) Dynamics of place attachment in a climate changed world. In: Manzo L, Devine-Wright P (eds) Place attachment: advances in theory, methods and applications. Routledge, London, pp 179–191

Dixon J, Durrheim K (2000) Displacing place-identity: a discursive approach to locating self and other. Br J Soc Psychol 39(1):27–44

Du Plessis C, Brandon P (2015) An ecological worldview as basis for a regenerative sustainability paradigm for the built environment. J Clean Prod 109:53–61

Fincher R, Pardy M, Shaw K (2016) Place-making or place-masking? The everyday political economy of "making place". Plan Theory Pract 17(4):516–536

Hes D, du Plessis C (2014) Designing for hope: pathways to regenerative sustainability. Routledge, New York, NY

Hidalgo MC, Hernandez B (2001) Place attachment: conceptual and empirical questions. J Environ Psychol 21(3):273–281

Kyle G, Graefe A, Manning R, Bacon J (2004) Effect of activity involvement and place attachment on recreationists' perceptions of setting density. J Leis Res 36(2):209–231

Lewicka M (2010) What makes neighborhood different from home and city? Effects of place scale on place attachment. J Environ Psychol 30(1):35–51

Mang NS (2009) Toward a regenerative psychology of urban planning. Saybrook University, San Francisco, California

Mang P, Haggard B (2016) Regenerative development: a framework for evolving sustainability. Wiley, Hoboken

Mang P, Reed B (2012) Designing from place: a regenerative framework and methodology. Build Res Inf 40(1):23–38

Martin PY, Turner BA (1986) Grounded theory and organizational research. J Appl Behav Sci 22 (2):141–157

Mathers A, Dempsey N, Molin JF (2015) Place-keeping in action: evaluating the capacity of green space partnerships in England. Landsc Urban Plan 139:126–136

Meadows DH (2008) Thinking in systems. Chelsea Green Publishing, White River Junction, pp 77–78

Plaut J, Dunbar B, Gotthelf H, Hes D (2016) Regenerative development through LENSES with a case study of Seacombe West. Environ Des Guid 88:1

PPS (n.d.) What is placemaking: what if we built our communities around places? Available from https://www.pps.org. Accessed 20 Mar

Scannell L, Gifford R (2010) Defining place attachment: a tripartite organizing framework. J Environ Psychol 30(1):1–10

Strauss A, Corbin J (1994) Grounded theory methodology. Handb Qual Res 17:273–285

Wilson EO (1984) Biophilia. Harvard Press, Cambridge, MA

Part VII
Inclusivity

The way people can be included in the planning and design process in order to create the best sustainable and smart future for their cities. This looks at a diversity of participatory planning and design approaches.

Chapter 23
Public Participation: A Sustainable Legacy for Olympic Parks

Eveline Mussi, Christine Steinmetz, Catherine Evans, and Linda Corkery

Abstract Since the 1980s, plans for the Sydney Olympic Park (SOP) have been underway to transform the heavily contaminated brownfield site of Homebush Bay into a thriving urban precinct of residential neighbourhoods, business districts and regional parklands. This transformation has been underpinned by Environmentally Sustainable Development (ESD) principles, as established by Sydney Olympic organisers in the bid for the 2000 Olympic Games. The Sydney Olympic Park Master Plan 2030 envisions 23,500 residents and a doubling of office spaces, responding to state goals for the growth of the Western Sydney region. Such population growth challenges SOP's management to remain loyal to sustainable development, maintaining a 'local' level of engagement and participation with residents and balancing competing interests from businesses who view themselves as integral stakeholders in SOP. The importance of public participation to sustainable development and contemporary challenges to its application, especially in public- private mega-projects, is the focus of this chapter. The chapter will discuss findings from a 4-year doctoral investigation of social sustainability in the SOP and how public participation has been considered and implemented in the process of developing it, post-Games. Findings emerged from an analysis of SOP-specific planning documents, in-depth interviews with experts who have been or are working on the development of SOP, and focus groups with residents living in the neighbouring communities of Newington and Wentworth Point. The chapter will first set out the problem of Olympic parks and establish the definition of sustainable development as a social problem, thus bringing attention to the importance of public participation. It briefly introduces the methodology used, the context of SOP and presents the findings, unpacking the reality of public participation in the post-Games development of SOP and how it has been perceived by experts and residents. Findings reveal a 'watered down' approach to social participation strategies, as perceived by residents, despite the increasing importance of social participation in state and local planning instruments. Some of the barriers in applying strategies for

E. Mussi (✉) · C. Steinmetz · C. Evans · L. Corkery
Faculty of Built Environment, University of New South Wales, Sydney, NSW, Australia
e-mail: e.mussi@unsw.edu.au

© Springer Nature Switzerland AG 2020
R. Roggema, A. Roggema (eds.), *Smart and Sustainable Cities and Buildings*,
https://doi.org/10.1007/978-3-030-37635-2_23

social participation are discussed by experts in relation to difficulties in defining local communities, especially in a project of regional importance and the limited legal responsibility of the Sydney Olympic Park Authority for communities around its boundaries. This is taking a toll on how residents engage with SOP and how they connect to an ethos of shared values for sustainability. Novel approaches to social engagement with residents living within and around SOP's boundaries will be critical to the ongoing management of SOP as an icon of sustainability, especially with the expected population growth.

Keywords Sustainable development · Social sustainability · Public participation · Olympic parks · Olympic legacy · Mega-projects

23.1 Literature Review

23.1.1 Sustainable Olympic Parks

Often referred to as "white elephants", Olympic parks can be problematic mega-projects for host cities. In The Olympic Cities (Pack and Hustwit 2013), Jon Pack and Gary Hustwit document a pattern of underutilised or abandoned large urban spaces as defining characteristics of the Olympic legacy, with key examples including Beijing's green Olympic park and the deteriorated aquatic centre at the OAKA site from the Athens 2004 Olympics. The combination of mega-sized architectural projects, onerous costs and the underutilisation of these large infrastructures post-Games suggests that Olympic parks are unsustainable mega-projects.

Given this recurring problem, and with increasing concerns for sustainability and legacy, host cities and the International Olympic Committee (IOC), have rebranded Olympic parks in marketing materials and repackaged them for consumption as sustainable urban-regeneration mega-projects. Sydney Olympic Park, developed for the 2000 Olympic Games, set the benchmark, planned as an environmentally sustainable site, to become a thriving urban centre, long-term. The IOC later directed that Olympic parks should become "accommodation, entertainment and tourism districts", with "new housing, retail and leisure outlets, and extensive sport, recreation, transport and other infrastructure developments" (IOC 2012, p. 88). However, despite promising discourses, problems with sustainability and legacy of Olympic infrastructure persist (Gold and Gold 2013).

The key to a successful Olympic park legacy, according to Shirai (2009, p. 4), "lies in the long-term relationship between the Olympic Park, as a distinct economic and political entity, and it's very different pre-existing surrounding neighbourhood." As with other contemporary mega- projects that have proliferated around the world, Olympic parks are developed and managed to attract investment, tourism, and new populations that can afford their attractions (Lehrer and Laidley 2008). This pattern reflects urban entrepreneurial strategies that emerged in the 1980s in response to the increasing influence of private corporate power in urban development (Harvey 1989) and corresponding changes to cultural planning strategies in redefining cities' landscapes towards international competitiveness (Friedmann 1986). Such mega-projects have a tendency of becoming "Machiavellian formulas" of unattained promises (Flyvbjerg 2005), while spaces are created to the exclusion of local communities, with the undesirable consequence of lacking a vibrant urban life (Woodcraft et al. 2011). Shirai (2009) proposes physical and programmatic attributes, such as better access and social activities, to integrate Olympic parks to their urban context. However, the extent to which Olympic parks, post-Games, are sustainable places and how management considers local voices and needs through public participation strategies, remains an issue.

23.1.2 Sustainable Development as a Social Problem

Public participation has been a recurring theme in the sustainable development debate, especially under the umbrella of social sustainability. The widely accepted definition of sustainable development, announced in 1987 by The Brundtland Commission, "development that meets the needs of the present without compromising the ability of future generations to meet their own needs" (WCED 1987, p. 43), has been criticised as problematic (Lélé 1991; Jacobs 1999; Fergus and Rowney 2005; Shiva 2012). While its breadth might allow irreconcilable positions to find common ground, it also leaves the concept vulnerable to opportunistic interpretations (Lélé 1991; Jacobs 1999), particularly in the built environment (Krueger and Gibbs 2007). Fergus and Rowney (2005) suggest that the unchallenged economic-scientific rationalistic framework through which the term has been defined leaves little space for diverse perspectives, perpetuating the unsustainable condition of contemporary global society (Basiago 1999; Adams 2006). As Keough (2008) concludes:

> [Sustainable development] has been handicapped from birth by its association with "sustained yield resource management" and the international development paradigm, whose practice and philosophy have been eloquently and passionately critiqued (Keough 2008, p. 65).

Thus, because of the narrow interpretation, and despite internationally adopted environmental initiatives, sustainable development has failed to mitigate the deepening global environmental, economic and social crisis (Shiva 2012). An alternative proposition suggests a social interpretation to the sustainable development challenge (Stefanovic 2000; Littig and Grießler 2005; Keough 2008; Vallance et al. 2011) viewing sustainable development as "socio-scientific subject matter, not just a question of natural sciences" (Littig and Grießler 2005, p. 69). From this perspective, achieving sustainable development requires public involvement, to promote open debate and inclusion of different views. It is in this sense, that public participation gains importance as a strategy for sustainability (Fergus and Rowney 2005; Keough 2008).

23.1.3   The Importance of Public Participation
         for Sustainable Urban Development

Public participation, as a collaborative process, brings together experts' and communities' knowledge in order to develop places that are inclusive, diverse, and vibrant. While the application of strategies for public participation in sustainable urban development has been questionable, especially in the context of mega-projects for the Olympic Games, in theory, the importance of public participation is emphasised, to promote the design of vibrant places for people, local wellbeing (Woodcraft et al. 2011), and even a city's prosperity (UN-Habitat 2016).

The campaign for the development of cities for people has a long history. For instance, Jane Jacobs (1961) discussed and actively campaigned against the negative impacts of massive urban redevelopment and top-down urban planning on community life and wellbeing. Jan Gehl (1971) talks about "life between buildings", arguing that it is of greater importance to discuss the social rather than the spatial, given that it is life which must define the physical quality of public spaces. Carr et al. (1992) suggested that the emphasis given by built environment professionals on the physical attributes of spaces often results in a "simplistic, deterministic conception of the functioning of public places, one that has turned out to be limiting in many respects" (Carr et al. 1992, p. 85). Hester (2006) proposed processes of ecological democracy, which would result in very different urban forms from the current emphasis "in extravagant architecture"; there would be a renewed "search for roots, foundations, and fundamentals – the basics of satisfying life" (Hester 2006, p. 7). What is evident from these perspectives is the common theme that vibrant and authentic public spaces depend on understanding the local environmental-social context, the activities taking place, and local values, alongside strategies such as immersion, observation, and public participation.

However, public participation remains largely problematic, especially in the context of public- private partnerships for urban regeneration (i.e. Olympic parks). Research reports that public participation runs the risk of being co-opted by private stakeholders in ways that conveniently legitimise public involvement, especially if challenging the established urban agenda (Kaethler et al. 2017). The conflict between recognition and resistance to implementing appropriate approaches to public participation asks for an examination of how approaches are perceived to result in vibrant, meaningful, and sustainable places.

23.2   Research Method

This chapter draws on a 4-year doctoral study which investigated the social sustainability of Olympic parks, using SOP as an in-depth case-study. Through the in-depth case- study method, it was possible to investigate the views and experiences of different stakeholders involved with SOP, particularly the views of residents and experts responsible for its development. SOP is an ideal case study for examining issues relating to social sustainability of mega projects following key criteria. First, SOP was developed in response to ESD principles, introduced in the documents for the 2000 Olympic bid and crystallised post-Games in the Sydney Olympic Park Authority Act (NSW 2001). Second, it has been well over 15 years since the 2000 Olympic Games, and, as Furrer (2002) argues, it can take at least a decade for an Olympic site to achieve its legacy intentions. Third, SOP is an active urban space. Annual reports from the Sydney Olympic Park Authority (SOPA) show that SOP is attracting an increasing number of visitors annually and generating greater interest in the area as a site of commercial and business development. A PricewaterhouseCoopers report states the economic attractiveness of SOP, ranked twentieth in the local economies in Australia in 2014 (Wade 2014). Thomas Bach, president of the IOC, referred to it as a benchmark for other Olympic cities (SOPA 2015). Despite official reports, there have been persisting criticism of SOP as an underutilised space (Irvine 2012). Given such legacy reports, external conflicting perceptions, and sustainability ideals, SOP is a best case-scenario to investigate if and how local residents are engaging and participate with its planning and development.

The study of SOP was based on a social sustainability framework revolving around three principles: local empowerment, identity, and community. Public participation is a key strategy for local empowerment. The analysis of public participation in the development of SOP involved a triangulation of qualitative methods, to validate the information interpreted from the different sources. First, the research relied on the analysis of planning instruments, official publications and newspaper articles, extracting terms and narratives that pointed to processes of public involvement. The analysis of documents also informed the guide for in-depth interviews with 12 experts who were or have been working in the development of SOP, including planners, managers, and members of SOPA's Board of Directors. Finally, five focus groups were conducted with the residents of the local residential communities adjacent to SOP, Newington and Wentworth Point, to explore the reality of public participation from the perspective and experiences of these local communities. The triangulation of data, and more significantly the data collected from local residents, validated or questioned the approaches proposed in documents and applied by SOPA in the development of SOP as a sustainable place. Newington and Wentworth Point were selected because of their proximity to SOP, longer-term development, and managerial considerations by SOPA. At the time of the research, residents living within the boundaries of SOPA only just moved in, and therefore were not present long enough to experience different processes, changes, and current approaches to public involvement. The data collected were transcribed and analysed using NVivo qualitative data analysis software.

23.3   Background to Sydney Olympic Park (SOP)

SOP is envisioned, in state planning instruments, as a specialised precinct, set to become "a major location for employment and high-density housing while remaining a major sports and entertainment centre" (NSW Government 2014, p. 23). Sydney Olympic Park 2030 Master Plan (MP2030) responds to the metropolitan vision, projecting that the 760 ha of the park, including the 460 ha of the Millennium Parklands conservation area, will become a "vibrant and sustainable township" with 14,000 residents, 31,500 workers and 5000 students, remaining a major events space (SOPA 2010, p. 22). This vision for SOP, former Homebush Bay, has been evolving long before the 2000 Olympics, as part of state planning strategies for Sydney's competitiveness as a global city (Waitt 2003).

Homebush Bay was a large, mostly stated-owned, industrial site, which included factories, brickfields, and abattoirs. With the closing down of industries, Homebush became a dumping ground for residential waste and, more critically, harmful chemical waste from neighbouring industries. As a vacant, large, state-owned site, it became the ideal location for an Olympic park, to direct Sydney's expansion towards its Western suburbs through the development of urban infrastructure and rebranding of a long-stigmatised area. Since the late 1970s, a number of NSW state level planning schemes and policies for Homebush Bay, reflected global entrepreneurial aspirations and relied to various degrees on privatisation and urban aesthetic improvements, including environmental remediation (Freestone and Gibson 2006; Hu 2012). These planning proposals envisioned the removal of most of the previous and obsolete industrial activities, adding new sports, recreation and entertainment facilities, business and technology, housing, and extensive areas of urban parklands. The Sydney Regional Environmental Plan No. 24: Homebush Bay (SREP 24), gazetted in 1989, solidified these visions and new public-private state government institutions were established to deliver the state vision for the site. In 1991, the Sydney Olympics 2000 Bid Limited (SOBL) was formed, to prepare the bid for the 2000 Olympic Games (Cashman 2011). SOBL was constituted of influential figures from the private and public sectors (McGeoch 1994), who worked closely with the Homebush Bay Development Corporation (HBDC), established in 1992 as part of a new centralised public-private planning system.

In a pre-bid design competition for the Olympic Village, one of the winning entries by Greenpeace- allied architects proposed a sustainable Olympic village. This shifted the theme of the Sydney bid to the "Green Games", including an innovative Environmental Guidelines for the Summer Olympic Games (1993), proposed by Greenpeace in collaboration with the NSW Government and SOBL. Despite the promises for sustainability, the alliance of Greenpeace with the government and private organisations raised criticisms for potentially conditioning the use of sustainable development to intrinsic political-economic interests (Beder 1999). After being awarded the Games, the Olympic Coordination Authority (OCA) was established to develop the infrastructure for the Olympics. Accountable to the Minister for the Olympics, this centralised governance system persisted after the games with the OCA replaced by the Sydney Olympic Park Authority (SOPA), a state development corporation committed to the management of SOP.

SOP is now a growing, active urban area, catering for almost 5500 large events each year and a daily working population of more than 19,000 people. Such population growth and new urban functions led to the recognition of SOP as a new suburb in 2009. In 2012, SOP's first residents moved into the newly opened Australia Towers, adding a permanent population of over 600 to the site. The site is also surrounded by other residential areas such as Carter Street, Wentworth Point, and Newington, the former Olympic Village. Wentworth Point and Carter Street are designated by the New South Wales Planning Department as Urban Activation Precincts (UAP), high-density residential and commercial priority growth areas. The development of SOP as a specialised centre, to accommodate a total of 23,500 residents and a doubling of office space, bordering other high-density residential areas, challenges the management of SOP as an events precinct and its ideological commitment to sustainable development, adding pressure to the importance of public participation and engagement with SOP.

23.4 Findings: The Who and How of Public Participation

23.4.1 Expert's View

Expert's views of sustainability, social sustainability, and the principle of public participation are usually directed by official policies and documents. Overall, planning instruments for SOP refer to the Brundtland Commission's interpretation of sustainability, as specified in the literature. However, there is very limited explicit reference to the concept of social sustainability.

Consequently, a series of in-depth interviews with SOPA's executive of major projects, member of the board of directors, and management staff, revealed an uneven understanding of social sustainability, despite the common reference to the Brundtland interpretation of sustainability, as well as the three-pillar model: "it needs to be environmental, economic and social, otherwise it doesn't work", as referred Peter Duncan, Director of Estate Management for the OCA). For some participants, social sustainability is linked to the presence of diverse communities at SOP. For others, it is about promoting the environmental values of SOP. Others assert that limited public participation implied that "the social component was almost not considered at all" during the phase of developing SOP for the Olympic Games, as states Sue Holliday, director general of the NSW Planning Department between 1997 and 2003, firmly linking social sustainability to public participation.

Despite the conceptual issues around social sustainability, public participation has been referred in planning policies as a guiding principle to support the sustainability agenda, specifically in the 1993 Environmental Guidelines. In 2008, public participation was to some extent reinforced and complemented in the revision of the Guidelines. However, differently from the original version, public participation was not directly stated as a planning strategy. The revised 2008 Guidelines use terms such as "community involvement" with the local identity or "participation" in different activities to encourage an appreciation for the place (SOPA 2008, p.10). In terms of planning strategy, there is indirect reference to public participation by highlighting the importance of responsive governance to local community's needs:

> [SOPA] will promote a sustainable place to support the changing business, event, visitation, worker and resident needs of the precinct in the future (SOPA 2008, p. 11).

Paramount to responding to changing social needs is to define strategies for communication and participation. However, the 2008 Guidelines offered limited specifications in that regard, other than "involvement", "responsiveness", and broadly referring to the different users of SOP, as in the above quote. Therefore, the 'who' and 'how' of responsive governance and public participation is left to experts' scrutiny. One of the apparent issues that emerged from interviews with experts is the lack of clarity regarding who are the communities to whom SOPA must be responsive, and how should public participation happen.

In Craig Bagley's planning capacity as executive manager of Major Projects at SOPA, the 'who' of public participation relates to those within SOP's boundary. However, as h e further discussed, there is also "a huge growing population around us and we know that they all see this more like their backyard than a regional open space", making specific reference to the green areas of the Millennium Parklands. There is, therefore, a conflict with managing SOP in response to the growing number of residents around its boundaries, particularly regarding the appropriate environmental management of the parklands. A communications strategist for SOPA adds that it is not only about how communities outside SOP's boundaries use the site, but also how SOP's activities influence the communities around it. As explained, because the influence of SOP's activities reaches beyond its boundaries, "I view the residents inside SOP in the same light as the residents on our boundaries" (Interviewee 2, 2014, pers. comm.). There is an uneven but general view that residential communities adjacent to SOP's boundaries affect and are affected by the places and activities at SOP, conflicting with SOPA's official responsibility for the area within SOP's boundaries.

The lack of clarity regarding who SOPA must include in managing SOP as a sustainable place has affected approaches to participation. Bagley (2013, pers. comm.) explained that in the first few years post-Games, public participation strategies were not considered because "there was no community"; community would arrive during the lifetime of the plan. However, there was already a growing community living in the surrounding areas. Up until 2012, the large majority of SOP's population, meaning those within its boundaries, consisted of growing businesses and its employees. And, as Bagley (2013, pers. comm.) pointed out: "We [at SOPA] have done quite well with the business community and the workers". SOPA has put the effort to participate in the meetings of the Sydney Olympic Park Business Association (SOPBA), and to have representatives present in SOPA's planning meetings, as experts explained. David Baffsky, who has been on the board of SOPA since 2009, explained that the re-assessment of SOP's master plan was largely based on consulting local businesses and understanding their needs and challenges in operating at SOP. Baffsky further commented that there are also differences in how SOPA responds to the needs of different businesses: "Each business is different … different size, different priority … different level of involvement and capacity, but that's not unnatural" (2014, pers. comm.). As businesses can often be an anchor for urban precincts, priorities of the area could largely be influenced by businesses themselves and not the residents. Richard Cashman, professor at the University of Technology Sydney and founding director of the Centre for Olympic Studies in 1996, also noted that business stakeholders have been the "louder voice" in the planning process of SOP; "they are an important group, because they bring money to the park, and they also bring people" (2013, pers. comm.). Not uncommon, the reality of 'voice' within a community can become the issue of contention often overtaking the real priority of the SOP—sustainable values for an entire community.

There is, nonetheless, an effort from SOPA to inform communities around its borders about events and activities at SOP, particularly as it helps to create a harmonious relationship with SOP as an events precinct. The intent to keep local communities informed is for "setting expectation around what this park needs to do, because, at the end of the day, the park was created for major events" (Interviewee 2, 2014, pers. comm.). In addition, those informative tools are also about "building relationships, so that open communication can happen, and we can tell people what we're doing, what we're thinking, what we're planning" (Interviewee 2, 2014, pers. comm.). While such an informative strategy is potentially positive and ensures that residents are accustomed to the nature of SOP as an events' space, it does not facilitate responsive management to local needs. Plans for SOP's development and how SOP is managed as a sustainable place does not involve the participation of local stakeholders beyond its borders, particularly local residents. This approach to participation reflects a corporate view to place management, based on responsiveness to businesses rather than the broader community. Conclusively, James Weirick, professor of landscape architecture at the University of New South Wales and a critical voice of the Olympic Games, remarked that SOP is evolving as a "successful business park" (2013, pers. comm.), rather than an urban precinct with daily local activities, vibrant and promoting sustainable values locally.

23.4.2 Local Residents' View

In five focus group discussions, thirty-seven residents from Newington and Wentworth Point shared their perceptions about the type of communication and engagement they have with SOPA, acknowledging its effect on how local residents' use the park and, importantly, on how they connect, identify and respect SOP's sustainability values.

First, while social sustainability is also an unclear concept among residents, it became evident in the group discussions that the presence of community and shared values for place are elements that support quality of life. Community and value are aspects that emerge from daily experiences and perceptions of the local natural and built environment, services, activities and informal socialization. Governance strategies to promote local engagement and responsibility for place have a significant role to play in promoting community, shared values and, therefore, social sustainability, as evident through the group discussion.

Participants talked about two types of engagement approaches implemented by SOPA. One approach focuses on providing information to residents, through flyers and notes on local newsletter, about management matters such as the events program, road closures and other temporary changes in the area. In general, participants shared a positive view about being kept up-to-date about potential traffic issues in their area. This allowed residents to "plan ahead" (Resident 1, N2) or "work around" (Resident 2, WP2) these changes in their daily routines. As intended, informing adjacent and local communities of events and management issues helps create a harmonious relationship between the reality of SOP as an events place and the everyday routine of people living around it. SOPA's efforts to inform residents about events and programs at SOP also encourages local use of the facilities and offerings of the park, while promoting the presence of residents from adjacent communities.

While participants talked positively about these communication strategies, they also remarked that there is a diminishing frequency of communication from SOPA over time. A participant from Wentworth Point explained: "you would get a letter drop once a month. They used to open the tennis; they used to advertise the music, the movie. You used to get a flyer on your letter drop quite regularly" (Resident 6, WP1). Another added: "I haven't seen that for years. Before, they were very, very keen to advertise what they were doing" (Resident 3, WP1). Others agreed that efforts from SOPA in engaging with residents were better in the early post-Games years. The discontinuation of SOPA's engagement strategies is perceived as a disinterest towards local communities, viewed as a loss to both SOP and local residents. As a participant from Wentworth Point commented: "I think they are doing their businesses a disservice by not actively promoting to us out here" (Resident 3, WP2). Another participant from Newington also said: "I don't think they [SOPA] are sufficiently proactive with us, and perhaps they should be more engaging" (Resident 2, N3). From the residents' point of view, maintaining communication with local residents promotes local engagement, their presence at the site, greater use of local businesses and, consequently, better activation of SOP's urban precinct. Potentially, the limited presence of these local communities in the park has consequences to the everyday diversity and vibrancy of SOP's public urban space, and the external view that SOP lacks vibrancy and social life (Irvine 2012).

Participants also spoke to SOPA's limited efforts to inform residents about planning schemes and the future development of the area. As a resident from Newington explained, "[SOPA] never tells us what they are doing ... what they are going to do in development" (Resident 2, N3). Similarly, Wentworth participants stated: "I don't know how we become aware of future plans and changes (Resident 2, WP1). The limited communication regarding planning matters, and therefore the lack of awareness about future development, was identified as unfavorable to promoting a sense of security regarding residents' expectations and aspirations for the area they live. Another resident from Wentworth Point continued:

> When we went to a meeting, when we first came here, we only knew about this section here being developed. And now they've told us they are going to be developing another section across this side as well (R5, WP1).

The development site the resident is referring to, located in front of Wentworth Point and within SOPA's boundary, was not part of the initial plans for the development of the area. The initial plan was to maintain the site as part of the Millennium Parklands. As other participants agreed, this presents a conflict with what they expect for the area, as it imposes greater number of people and more density in a site reserved for environmental conservation. This also indicates that the new plans are not only not communicated, but, more importantly, they are not based on initial consultation and understanding of contextual information on how residents identify with the place they invested and live in—for them, it is about quality of life.

23.4.3 Loss of the Sustainability Identity

The limited engagement of local residents with SOP is having an impact particularly on their association with the sustainability identity of the area. This is evident by the striking differences on how participants from Newington and Wentworth Point talked about sustainability. Identity with sustainability is associated with the community areas, and not the SOP.

At Wentworth Point, most participants commented that, while they feel a strong sense of community within the residential area, especially as it has been promoted by developers through different strategies, "sustainability was not influential in our decision" to move into the area (Resident 1, WP1). A review of real estate marketing campaigns and design plans makes evident that the site is not being developed based on sustainability principles. Wentworth is characterised by high-rise apartment buildings on a waterfront setting, with small and few community areas, rather showcasing a usual upper-market waterfront development character. The local physical qualities, together with lack of marketing strategies by developers to promote the area's sustainability, and the weak engagement of residents with SOPA, potentially explains the limited shared identity of residents of Wentworth Point with the sustainability values of the broader area.

In contrast, Newington was the former Olympic Village for the Sydney Games; essentially, the original project through which principles of sustainability became integral to the Green Games. Sustainability was "the ethos of creating this suburb", a participant noted (Resident 4, N2). An ethos of a sustainable community both in natural and built form was at the heart of the residential design brief. Other participants shared an appreciation for how the physical attributes of Newington reflected principles for sustainable development; for example, the "blend of the natural environment", referring to the integration of the site with the Parklands vegetation and local topography, "as well as the sustainable environment" (Resident 8, N1), referring to the extensive use of solar panels throughout the area and an integrated water-management system. For many, sustainability was, by large, a determinant in their decision to live in Newington:

> I was drawn here because it was a sustainable house, and I felt this was one unique—at that stage—chance to move into a village that stood for sustainability. That was the thing that attracted me most (Resident 1, N2).

Many of the long-term residents in Newington who moved into the suburb, aware of its sustainability ethos, discussed this as a strong reason for moving there. However, at the same time, the group of longer-term residents also discussed that the local shared identity for sustainability is weakening through time. As remarked, "I think if you called this a green suburb now people might laugh at you" (Resident 1, N3). The noted diminishing interest for sustainability among Newington residents was discussed in relation to changes in residents' behaviour, with less care for the green infrastructure of their units and the physical attributes of Newington as an environmentally sustainable place. For some, this is also having an impact on a shared sense of care for the place, and therefore a sense of community.

The deteriorating shared values for sustainability in Newington was also discussed in relation to the management arrangements in Newington, organised around five precinct committees constituted of volunteering residents. Committees lack authority and expertise to impose established regulations among residents and other stakeholders (i.e. real estate agents). Residents are very concerned that the shared value for sustainability in Newington has become watered down through time. Therefore, governance strategies ensuring residents' engagement and participation with the place's values long-term is having an impact on the area's sustainability, environmentally and socially.

It is possible to highlight that social sustainability for residents, means that in their daily practice of living and experiencing the natural and built surrounds of the area they live, individually and as a community, they would have a more intimate relationship between their lived experience physically and eventually, develop some level, or inkling of a sense of stewardship towards the overall place, both environmentally and socially. Something that is only possible if one is living and doing and being in place. Therefore, strategies for participation and engagement need go beyond informing or just merely engagement, but also tapping into residents' experiences, values and expectations.

Importantly, in the case of Newington and Wentworth Point, the shared identity for sustainability, or lack thereof, is not associated with the SOP. Besides management issues within the residential area, this problem is potentially due to residents' limited engagement and participation with SOP's spaces, activities and limited voice towards the future development of the area.

23.5   Discussion and Conclusion

The analysis of SOP shows that the long-term vision of SOP becoming a feature of the global city—enhanced competitiveness to attract investment and capital—has framed approaches to sustainability. This directed the more usual focus on principles for resource management, including the extensive use of solar power, water catchment system, among other technical measures to mitigate environmental impacts. Sustainability did not imply changes to that global vision and related entrepreneurial strategies. In this context, social sustainability remains a less clear dimension of sustainable development, and public participation strategies are, to some extent, aligned to the main corporate agenda.

Public participation is supporting the development of SOP in response, primarily, to the needs of its local businesses. While this is anchoring the long-term goal of developing SOP as a successful, environmentally sustainable, business park, it is not encouraging the local surrounding communities to engage with the area. In the views of participants from Newington and Wentworth Point, SOPA's diminishing efforts to communicate with local residents has been detrimental to their presence at SOP, to the loss of its local businesses and of generating more diverse, inclusive and vibrant public spaces. Importantly, the limited engagement of adjacent communities with SOP is having an effect on how residents associate with the identity of the area, thus failing to promote a sustainability culture, locally. This will become increasingly problematic for the management of SOP as a sustainable place with the number of residents in the area growing exponentially, as envisioned in planning policies for the densification of the area.

Problems with conceptualising the social dimension of sustainable development persists, especially where its environmental principles continue to support urban entrepreneurial visions and processes (Krueger and Gibbs 2007) and public participation is not applied as a platform for open debate, but instead is used to inform and re-enforce established agendas (Kaethler et al. 2017). If public participation is used as a platform to encourage debate and share views among local communities and experts, perhaps SOP would look different. It would still host large events, but would also have smaller active areas to support daily life of local communities. In this way, SOPA could promote its sustainability ideals by reflecting a renewed search for fundamentals to satisfying local community life, which Hester (2006) and others have long argued is critical the development of places for people.

References

Adams W (2006) The future of sustainability: re-thinking environment and development in the twenty-first century, Report of the IUCN Renowned Thinkers Meeting, 29–31 January 2006, The World Conservation Union (IUCN). Accessed 25 Aug 2014. http://cmsdata.iucn.org/downloads/iucn_future_of_sustanability.pdf

Basiago A (1999) Economic, social, and environmental sustainability in development theory and urban planning practice. Environmentalist 19(2):145–161

Beder S (1999) Through the revolving door: from Greenpeace to big business. PRWatch 6(3):1–4

Carr S, Francis M, Rivlin L, Stone A (1992) Public space. Cambridge University Press, Cambridge

Cashman R (2011) Sydney Olympic Park 2000 to 2010: history and legacy. Walla Walla Press, Sydney

Fergus A, Rowney J (2005) Sustainable development: lost meaning and opportunity? J Bus Ethics 60(1):17–27

Flyvbjerg B (2005) Machiavellian megaprojects. Antipode 37(1):18–22

Freestone R, Gibson C (2006) The cultural dimension of urban planning strategies. In: Monclus J, Guardia M (eds) Culture, urbanism and planning. Ashgate, Hampshire, pp 21–42

Friedmann J (1986) The world city hypothesis. Dev Chang 17(1):69–83

Furrer P (2002) Sustainable Olympic games: a dream or a reality? Bollettino della Società Geografica Italiana, series XII VII(4):795–830

Gehl J (1971) Life between buildings: using public space. Island Press, Washington, DC

Gold J, Gold M (2013) "Bring it under the legacy umbrella": Olympic host cities and the changing fortunes of the sustainability agenda. Sustainability 5(8):3526–3542

Harvey D (1989) From managerialism to entrepreneurialism: the transformation of urban governance in late capitalism. Geografiska Annaler B 71(1):3–17

Hester R (2006) Design for ecological democracy. The MIT Press, London

Hu R (2012) Shaping the global Sydney: the City of Sydney's planning transformation in the 1980s and 1990s. Plan Perspect 27(3):347–368

International Olympic Committee (IOC) (2012) Sustainability through sport: implementing the Olympic Movement's agenda 21. IOC, Lausanne

Irvine J (2012) The winner is . . . white elephants, Sydney Morning Herald, July 27, pp 17

Jacobs J (1961) The death and life of great American cities. Random House, New York

Jacobs M (1999) Sustainable development: a contested concept. In: Dobson A (ed) Fairness and futurity: essays on environmental sustainability and social justice. Oxford University Press, Oxford, pp 21–45

Kaethler M, De Blust S, Devos T (2017) Ambiguity as agency: critical opportunists in the neoliberal city. CoDesign 13(3):175–186

Keough N (2008) Sustaining authentic human experience in community. New Formations Spring (64):65–77

Krueger R, Gibbs D (eds) (2007) The sustainable development paradox: urban political economy in the United States and Europe. The Guilford Press, New York

Lehrer U, Laidley J (2008) Old mega-projects newly packaged? Waterfront redevelopment in Toronto. Int J Urban Reg Res 32(4):786–803

Lélé S (1991) Sustainable development: a critical review. World Dev 19(6):607–621

Littig B, Grießler E (2005) Social sustainability: a catchword between political pragmatism and social theory. Int J Sustain Dev 8(1):65–79

McGeoch R (1994) The bid: how Australia won the 2000 Olympics. William Heinemann Australia, Melbourne

NSW Government (2014) A plan for growing Sydney: a strong global city, a great place to live. NSW Government, Sydney

Pack J, Hustwit G (2013) The Olympic city. Plexi Productions

Shirai H (2009) From global field to local neighbourhood: sustainable transformation of the Olympic Park for the city. The final report, The Cities Programme, The London School of Economics and Political Science, London

Shiva V (2012) The great Rio U-turn. Parkistan Perspect 17(2):161–163

SOPA (2010) Sydney Olympic Park master plan 2030. SOPA, Sydney

SOPA (2015) President of International Olympic Committee Thomas Bach visits Sydney Olympic Park the greatest example of Olympic legacy, SOPA. Accessed 01 May 2015. http://www.sopa.nsw.gov.au/resource_centre/park_news/

Stefanovic I (2000) Safeguarding our common future: rethinking sustainable development. State University of New York Press, Albany

Sydney Olympic Park Authority (SOPA) (2008) Environmental guidelines: Sydney Olympic Park. SOPA, Sydney Olympic Park

UN-Habitat (2016) Global public space toolkit: from global principles to local principles and practices, UN-habitat. Accessed 26 June 2018. In: At. https://unhabitat.org/books/global-public-space-toolkit-from-global-principles-to-local-policies-and-practice/

Vallance S, Perkins H, Dixon J (2011) What is social sustainability? A clarification of concepts. Geoforum 42(3):342–348

Wade M (2014) Sydney Olympic Park: how the west was won, The Sydney Morning Herald, 31 May. Accessed 11 June 2014. http://www.smh.com.au/nsw/sydney-olympic-park-how-the-west-was-won-20140530-399kc.html

Waitt G (2003) Social impacts of Sydney Olympics. Ann Tour Res 30(1):194–215

Woodcraft S, Hackett T, Caistor-Arendar L (2011) Designing for social sustainability: a framework for creating thriving new communities. The Young Foundation, London

World Commission for Environment and Development (WCED) (1987) Our common future. Oxford University Press, Oxford

Chapter 24
Adaptation of "Participatory Method" in Design "for/with/by" the Poor Community in Tam Thanh, Quang Nam, Vietnam

Nguyen Hanh Nguyen and Hung Thanh Dang

Abstract It is likely that many projects related to residential planning and resettlement for low- income people have faced many challenges, not only in Vietnam, but in many countries around the world. Those projects are granted by the governments in order to provide more comfortable and healthier living conditions and better facilities for impoverished communities. Nevertheless, these efforts do not show sufficiency nor effectiveness, and a lack of sustainability. For example, the appearance of "Ghost" cities in some countries, or the desolate villages that are relocated for the farmers in flooding regions in the south of Vietnam. In reality, after the process of resettlement and relocation, people cannot naturally adapt to new living environments. As a result, their new life rapidly becomes boring; without stable jobs and income, the household economic condition is in poverty and they find ways to go back to their comfort zone. Most projects designed for the poor usually indicate an impasse, because they themselves cannot self-revitalise and develop after the projects and supports finish. The insufficiency or failure of several community programs results from lack of understanding or ignorance of the needs of a community. In the field of architectural design and urban planning, a fairly common method known as "participatory design" is a useful tool in designing the well-being of the community. However, the shallow understanding of this method's application can lead to very little achievements in practice. The paper uses a successful project for an impoverished residential neighbourhood in Tam Thanh, Tam Ky, Quang Nam, Vietnam to show how to implement the principles of participatory design in the real environment, through listening and understanding the local people's concep-

N. H. Nguyen (✉)
Department of Architecture, Ho Chi Minh City University of Architecture, Ho Chi Minh City, Vietnam

H. T. Dang
Department of Architecture and 3D Design, School of Art, Design & Architecture, University of Huddersfield, Huddersfield, UK
e-mail: thanh.dang@hud.ac.uk

© Springer Nature Switzerland AG 2020
R. Roggema, A. Roggema (eds.), *Smart and Sustainable Cities and Buildings*,
https://doi.org/10.1007/978-3-030-37635-2_24

349

tions and demands. Furthermore, the paper shows the adaptation of participatory concept: "from passive to active community" which is the key point to achieve successes in this project. Three principal steps were employed, including identifying core issues in the community, creating changes in human perceptions and living conditions, and demanding further operation and maintenance of the community. This inspiring project attracted a positive involvement of local people, volunteers, and experts in all stages: analysis, design, and implementation. Now, the impoverished commune in Tam Thanh is revitalised by a model of a lively ecotourism village, which results in the household income's improvement, a delightful life for everyone, and sustainable social development. The successes of the regenerative project in Tam Thanh are significant and inspirable for many similar poor communities across Vietnam. The participatory designing model is initiated and implemented by the ideas and huge contributions of the local people, in efforts to transform their habitat better and more sustainable in long-term.

Keywords Low-income people · Passive and active community · Participatory design · Tam Thanh · Vietnam

24.1   Introduction

The direct and active participation of communities in local development plans is a recent trend, due to getting positive and far-reaching effects in the long-term. It is argued that the community is the principal object that uses and operates results of the plans after completion. Local people entirely comprehend the issues and needs of the region where they live, so that they have effective contributions to the plans' achievements (improvements in the environment and living quality for themselves). The essence of "Design for the community" means "involvement"; however, this theory is not easy for architects/researchers to approach and apply in the best way. The most common problems are that most practitioners often impose the expert thoughts and the subjective views on the core values of projects. Even most of the projects that won in the competitions related to "design for the community" also show the subjective thinking of the authors without proper "involvement" of the public. In other words, the way in which we are operating may imply with a charity or a gift of a designable product to convey the characteristics and visions of an individual or a group of professionals. Origination of these projects is to serve everyone, but they do not propose the desires of the community; therefore, they lose connection to the public.

   According to current observations for community programs, most architects are focusing on designing buildings based on the logical thinking of facilities and the sparkle appearance of subjectivism. Meanwhile, the real needs of the poor, stabilisation of life and improvement of income, receive fewer concerns. Thus, products titled "design for the community" should be questioned. It is certain that to achieve the aim (improving the living conditions of the poor neighbourhoods) access by "orientation" will be more appropriate than the "installation".

From that point of view, we developed an idea for a case study: "Community-based ecotourism plan in Tam Thanh, Tam Ky". The project goal is to engage the active role of people in freeing themselves from poverty and in building a stable life. We came up with a right approach for right objects: participation of everyone from the beginning steps, and a key role of them in the proposal of the ideas, development, organisation, management, and maintenance of the project's consequences. Professionals and managers only undertook to guide and provide the necessary knowledge/supports in order to help them realise and solve their problems. The project went through three principal stages: Problem detection, Concept building, and Construction development. The "Co-doing" method including Co-analysis, Co-design, and Co-implementation was applied for all phases.

24.2   Literature Review

Since the early twenty-first century, the conception of "liveability" has been used as a measure for quality of the urban life in a city/country around the world (To 2018). Apart from the factor of economy, a great city for the living is a lively enjoyable place which promotes human values, benefits, and inspirations for the citizens. An achievement of a liveable place/city requires a combination of many factors. In this, the organisation of public or community spaces is vital in urban spatial structural planning. However, the approach and implementation of the public/community spaces significantly depend on considerations of location, activities, and everyone's needs and excitement. Their success is ensured by the means for everyone that uses them and they are not violated and abandoned.

Kevin Lynch (1960) pointed out three approaches for making a place, including "top-down", "bottom-up", and "participatory". "Top-down" manner is a traditional perspective; the local people do not contribute to the ideas, efforts, or properties to build up the project. Therefore, that approach's effect is not predictive, while the employment of "bottom-up" way shows creation and innovation. The importance is that the community is involved in generating and developing the concept of public or community places, and then uses them. The methodology of "bottom-up" is adequate; however, there are still some restrictions: application on a larger urban scale, ability of dissemination, and high agreement of the community. In the large cities of Vietnam, we can find the interesting small public spaces for people's daily activities such as a book street, a corner of beer, a little space to enjoy a glass/cup of Vietnamese coffee, playground for kids or dancing.

The last approach is "participation", which combines both above methods. The participatory principle raises as a new approach for the entirely active involvement of the community for the long- term (To 2018) (Lynch 1960). The community participation in public projects implies the involvement of people in a community in the different phases; and they together contribute their thinking, effort, and even money for the project's achievements. Furthermore, they not only propose the detailed plans, but also solve unexpected problems of each stage (Dayaratme 2016) (European Commission 2012). The transformative changes in the urban design show that the design for the community (passive community) is replaced by design with the community (participation of community) and design by the community (community involvement is active in all project phases). However, getting successful projects for the community complies with huge participation and union of local people, patience, and connectivity of the whole community and individuals. Across Vietnam, some of the plans for the community development have been fulfilled since the 2000s, for example, "Regeneration and Preservation of the Ancient Hanoi Quarter" in 2004–2007; "Community Project" with participation of local people in An Giang and Soc Trang (2005–2009); and "Artistic Village" in Tam Thanh, Tam Ky (2016-present) (To 2018).

In 1999, Nguyen T. H. Nguyen studied the self-build houses by people in the ancient town of Hanoi. In the beginning step – site analysis, the participation of households in the examined region was considered as a significant factor for applicability of the project. Although the model of self- built houses was enhanced, due to the limit of the budget along with under-evaluation of people's desires, the housing prototype was likely abandoned and downgraded quickly after the short-term use. Between 2004 and 2007, a pilot project of the participatory design – "Regeneration and Preservation of the Ancient Hanoi Quarter" was proposed and developed, but gained few results (Nguyen 2005). The project was to find a good solution for resettlement of the Hanoi people in the ancient quarter. The failure of that project was due to the neglect of the factor of household income when the families moved to a new residential place. Moreover, the resettlement links to the changes in habits in the living and working environment of people; however, these were not mentioned in the process.

Another scheme for an impoverished community was operated along to the canals in Binh Dong 1, Tan An, Long An, Vietnam in 2014. It was a success, due to understanding and using the appropriate solutions for the real problems and needs of the low-income families there. The wonky shelters and the deprived living environment were replaced by the houses and public spaces for the families. The planning and designing schemes were based on the opinion of all individuals. Moreover, a budget for the construction of the buildings and public facilities was launched and contributed.

From the previous experiences, Le (2016) questioned "Why do we need differences?" for the community development in Tam Ky, Quang Nam. She explained the significance of "participation" and how to approach this method in the community projects. The operation of these projects asks a harmonious and sophisticated combination of two methods of "bottom-up" and "participatory" under a developing economic condition like the one in Vietnam.

24.3   Research Methods

24.3.1   Case Study

Tam Thanh (Tam Ky, Quang Nam) is a poor coastal commune in the central part of Vietnam, with more than 3000 households of 12,000 populations across seven villages. The geographic features here are unique: a 3 km eastern beach from north to south, and the western border facing the Truong Giang River. All villages are connected by only 6 km central road. The primary mean of local people's living is fishing, with an uncertain income.

In early 2016, "Tam Thanh Mural Village" was born in collaboration of Tam Ky Council and The Korea Foundation. In seven villages in Tam Thanh, Trung Thanh village in the middle of the whole commune was selected for development of this project. Many successes of "Tam Thanh Mural Village" are so phenomenal, that a poor village is vividly revitalised, becomes a new attraction for tourists, and delivers inspirations to the other poor communities and the young Vietnamese generation (Duong 2017). However, one restriction seems to be the lack of facilities for visitors when staying here for longer. The first tourism services were very fundamental, including vehicle keepers and some drink kiosks. Some questions for the development of sustainable tourism are addressed: how to create long-lasting benefits for local low-income people not only in Trung Thanh, but also in all seven villages, and how to welcome visitors for longer. Tam Ky Council devises a long-term plan for sustainable tourism based on the successes of the Tam Thanh mural village. An idea for a project: "Community-based ecotourism development in Tam Thanh, Tam Ky" was born (Fig. 24.1).

The basic principle of community design is to channel the weak areas, what the main problems of a poor community are, and what they need for change. The local people are the principal object of the project, so all activities devised relate to them. Profound analysis of the influences of nature and society was developed. Some potential of Tam Thanh was found: 7 km of coastline with quiet and less sloping white sand beaches that are suitable for swimming and resting activities, and unique geographical and ecological characteristics. Villages are wrapped by sea at one side

Fig. 24.1 Mural village in Tam Than.h (Thuy Duong 2017)

and river at another side. Furthermore, human resources are a beneficial factor. Local people are warm and friendly and are conserving many vernacular cultural values. Local cuisine is an identity because of available specialties from sea and surrounding lands.

From the beginning, a volunteer team consisted of more than 30 people: experts in urban and community development/architecture/planning/landscape (UN-Habitat, UNESCO, Cities Alliance) and involvement of students of universities in HCMC, Da Nang, and Singapore. Then, in the later stage of the project, the number of volunteers rose up to 60 members and more than 130 volunteers of various backgrounds. Considering that local people are the main object, their active participation is the key to this project. Therefore, at the very first stages, the leading targets were planned: any tourism products must be simple for people to operate and achieve, the critical goal is benefits for people's livelihoods and improvable income, that the traditional values and local lifestyle do not deform, and the natural environment is not affected.

24.3.2 Phases of the Project

In order to achieve harmony and sustainable development goals, the appropriate approach was that local people were involved in the project from the beginning, and acted as a key role in building the idea, and handling, organizing, and managing the results. The experts should guide and orient, and provide the knowledge and tools to assist them in implementing their "problems". The project went the stages: survey, problem discovery, and workshops (workshop 1: Co-analysis and training; workshop 2: Co-design; and workshop 3: Co-implementing), as shown in Table 24.1.

24.4 Results

24.4.1 Movement of the "Design for the Community" Concept

In the field of community design, the change of design method is gradually shifting from "Design for Community" - a completely passive community, to "Design with Community" – community's participation in phases of the project, towards the final aim of "Design by Community" – ideas' generation and implementation by community actively. The last two operative methods seem to be progressive and corresponding to the sustainable development of philosophy: "Give a man a fish, and he will eat for a day".

Table 24.1 Involvement of local people, authorities, volunteers in all phases of the project (source: adopted from Nguyen 2016)

No	Phases		Local people	Commune authorities	City authorities	Project steering committee	Experts	Students
1	Survey, potential detection		X	X	X	X	X	
2	Project's preparation & students' invitation						X	X
3	Workshop 1: **Co- analysis**	Training				X	X	X
		Field survey	X	X	X	X	X	X
		Needs' survey	X	X		X	X	X
		Presentation	X	X	X	X	X	X
4	Workshop 2: **Co-design**	Planning ideas	X	X	X	X	X	X
		Architecture	X	X	X	X	X	X
		Experience activities	X	X	X	X	X	X
		Media & fundraising	X	X	X	X	X	X
5	Workshop 3: **Co-implementing**	Planning & landscape	X	X	X	X	X	X
		Architecture	X	X	X	X	X	X
		Experience activities	X	X		X	X	X
		Media activities		X	X	X	X	

Table 24.2 Movement in "Design by Community" model

Phases	Step 1	Step 2	Step 3
	Popular in Asia and Vietnam	Less popular, several projects are using and getting high efficiency	Desirable model in the future
Approaching	Design for Community	Design with community	Design by community
Characteristics of involvement	Community is entirely passive	Community involves in design	Community generates ideas And fulfills them
Attitude of attendance	Architects and other parties consider difficulties of community	Architects and other parties share with community and build up the connection and belief in community	Every member of the project understands inner problems of community and be responsible, enthusiastic and cooperative
Effects	Getting immediate impacts but less efficiency for long- term, local people do not see long-lasting values to remain the project after leaving of sponsors	Community maintains products of project because they are their achievements of "co-design"	Community is active, responsible for remaining and developing more to achieve all items of projects for long-term even after leaving of sponsors

This philosophy was introduced to the project in Tam Ky to make significant changes in the lives of regional people. The real evidence showed an inevitable movement in the design method. The changes in that movement were as follows (Table 24.2):

In step 2 and step 3, the authors employed a model of "Co-Design", which is a collaborative and participatory design concept for local people. It is likely that this is an effective and sustainable concept for community development projects, according to the principle that the community is simultaneously the "owner", the "client", the operator, and the manager of a project's categories. Therefore, the transformation of approach from "for the community" to "with the community" achieves effects on the community and real changes in the regeneration of residential neighbourhood.

The process of community participation included five categories: Identification of needs – Proposal of solutions (technique, legality, institution) – Support of means (finance, organization) – Development - Management. The project's decisions were made by all members of the community instead of the acceptance of the individual opinions. As shown in Fig. 24.2, the architects could participate in three out of all activities including I. Determination of needs, II. The suggestion of solutions (the major support to provide design solutions which are based on the needs of the community), IV. Implementation. Listening to the expectations of local people and provoking passion in them was very important, to engage their efforts and involvement in the project. In that way, they correctly perceive their benefits and responsibilities during the whole project process and even after that, so they are ready for changes not only for them but also for the community.

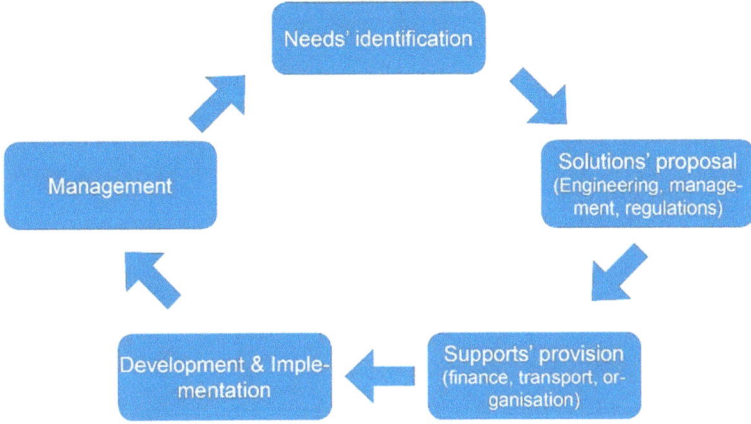

Fig. 24.2 Steps in "Design by the community"

24.4.2 Unique Tourism Concepts

The travelling concepts that formed "in" Tam Thanh and "for" Tam Thanh become the unique products that are positive results of the project. They are as follows:

24.4.2.1 The Art Road of Vernacular Transports

Boats and coracles that are familiar transports of regional people were used to create the artistic works of painting and installation created. Volunteers who are famous artists instructed people to decorate and place 111 boats around Tam Thanh (pathways and riverside) to create the special landscape points for the region. The paintings on the boats introduce lovely images of the motherland and vernacular lives/culture in Tam Thanh. In May 2017, the art road of coracles was certificated in Vietnam Guinness Book of Records.

24.4.2.2 Garbage Free Village

Understanding the impacts of garbage on the environmental quality and long-term development of tourism is that an international volunteer generated an idea of a free rubbish village in Workshop 1. By the observations, much rubbish was disposed around the village (roads and coasts) after starting to welcome tourists. The commitments to using plastic products were formed for both local people and tourists. In particular, the plastic bags are not allowed to use here. The drinking plastic bottles are collected and recycled after use. The small changes in daily awareness and habits of people will inspire and help them understand their responsibilities for their homeland (Figs. 24.3 and 24.4).

Fig. 24.3 Co-analysis stage of Tam Thanh project

Fig. 24.4 Co-Design stage of Tam Thanh project

24.4.3   The Village of Flowers

Cacti are a plant type found in the coastal village. The regional people mix them with some wildflowers to create a naturally fascinating landscape for all visitors when coming here. The model of that landscape can be employed widely within the village by more flowers and cacti. The villagers register to plant trees and flowers, which are consulted by landscape experts and commit to looking after them.

24.4.3.1   Tam Thanh "Village-Stay

The same as the conception of "Homestay", Tam Thanh "Village-stay" is devised for the model of eating, sleeping, and resting services under village scale. This discovery is due to poor condition of local people. They are not able to organise a homestay; therefore, the combination of the services (bathing, sleeping, eating) based on the family circumstances can make a perfect and exciting package accommodation.

24.4.3.2   Amazing Experience Activities in Tam Thanh

The enjoyable rowing and biking experience is attractive the visitors' attention. Two cooperative groups propose these two activities. The coracles and bikes are available in households. Furthermore, they are so familiar with people's life. Thus, they voluntarily register according to a specific team (biking or rowing). The activities and instructions for the visitors are taken over by people in the village. Besides the experience of space and natural landscapes, the short experiential tours associated with local culture and specialties are to promote the traditional values and chic hand-made products, which are gradually lost.

Initially, the project was funded by donors after operating. One part of that fund was deducted for the poor households in order to help them change their career.

24.4.4   Facilities of the Tourism Concept

To support the needs of tourism development, some facilities were planned and designed according to methods of Co-Design and design "with" community. They are as follows:

- A reception space includes greeting gate, parking area, and an information desk.
- The playground is at cultural houses. Seven cultural houses over all seven villages were proposed for the renovation of the existing playground.
- To build a new cultural house in Trung Thanh village
- To arrange the observatories along to river and beaches
- A sustainable concept is for Tam Thanh Community Arts Village. Tam Thanh will be an interesting place of artists and artworks. The artists can come to enjoy and explore new experiments and experiences for their creations to the community (Fig. 24.5).

24.5   Discussion

Under the encourageable results and the positive effects on the principal object of the project, the awareness of local people in Tam Thanh is significantly transformative. It is certain that they want to participate more actively in all actions and become a part of the project. They perceive positive changes in their environment and lives. Previously, when given money or home from the benefactors, the households could

Fig. 24.5 Co-Implementation of Tam Thanh project

not use or preserve it well for long-term and poverty still covered their life later. For the project for the community in Tam Thanh, the people contribute to building up the facilities by their efforts, passions, even their money (even if that money is borrowed). Therefore, they understand how to maintain what they build and to develop new categories. The more important thing is that they find delight and inspiration in the community activities, to serve not only their own lives but also their homeland and visiting tourists. All participants, including the experts and lecturers/students from many universities, also improved the practical experiences in designing the projects for the community. They can hold the core values to get successes of a community project, through communicating, and listening to needs and issues of local people. In collaboration with the community and the successes of the "bottom-up" approach, the participants can open new paths in the field of "design for the community", to undertake more realistic and more effective follow-up projects.

24.6   Conclusion

The practical experience of a design project by the community:

Along with the success of "Tam Thanh Mural Village", the community projects attached later in Tam Thanh initially transform a poor commune by a liveable and attractive place for not only the local families but also tourists. Tam Thanh is recognised on the tourism map and worthy of receiving the Asian Townscape Award 2017.

Collaboration for success: The collaborative support between volunteer experts, architects, and students from different backgrounds provoked the self-confidence of the poor local people to seek the realistic solutions for problems by their resources and efforts.

The local people involved in the community projects are often the most deprived group. Thus, the experts and practitioners need to understand and employ the correct approach, for example, the "bottom-up" method (where people come from). As a result, people transform from the passive condition (people usually wait for external support) to active participation, to hold chances for changes in their life. Holding the active empowerment contributes to reducing the construction costs and promoting the community connection.

The local government is a significant factor in supporting and facilitating the actualisation and success of the project. The supports of authority include an open-minded vision, acceptance for new and better changes, belief in responsibility, and implementation of people in the local area. These factors effectively contribute to the feasibility of the project.

In a collaborative relationship with the community, the patience, comprehension, and sympathy are necessary personalities to build a good relationship and belief in the community, as well as requiring the dedicated and highly responsible participation of each member.

The architects involving the projects for the community should be aware of their role that gives supports and engages the participation of people to come up with the idea instead of proposing strict ideas/solutions. When considering and evaluating the design options with the community, the architects need to consider humanity and to choose an option that can work well "for" and "with" the community.

References

Dayaratme R (2016) Creating places through participatory design: psychological techniques to understand people's conceptions. Journal of House and the Built Environment:719–741

Duong T (2017) Ngam "Ngoi lang bich hoa" dep me man tai Quang Nam. Retrieved from dep. com.vn: https://dep.com.vn/ngam-ngoi-lang-bich-hoa-dep-me-man-tai-quang-nam/

European Commission (2012) Community participation. European Commission, Brussels

Le DA (2016) Why is different? Community development project in Tam Ky, Quang Nam. Quang Nam, Tam Ky

Lynch KA (1960) The image of the city. MIT Press, Boston

Nguyen HN (1999) Self-help housing of the urban poor in the Hanoi ancient quarter. IHS, Rotterdam

Nguyen HN (2005) Resettlement for the ancient quarter in Hanoi. Architecture and urbanisation in Vietnam. University of Seoul, Seoul

Nguyen HN (2016) Project ACCA in Binh Dong community, Ward 3, Tan An, Long An. Architecture by the community. Ho Chi Minh City, Vietnam Association of Architects

To K (2018) Public spaces in livable and human cities. Urban Plan:76–83

Chapter 25
Fifty Years of Inclusive Transport Building Design Research

John Harding

Abstract Inclusivity is a core commitment of the New Urban Agenda of UN Habitat (adopted in 2016). Moreover, the Rail Sustainable Development Principles call for railway developments that are customer-driven, putting rail in reach of people, providing an end-to-end journey, being an employer of choice, reducing our environmental impact, being carbon smart, supporting the economy, optimising the railway, and being transparent (RSSB, Rail Sustainable Development Principles. RSSB, London, 2016). However, research suggests there is still a long way to go before "we live and work in an inclusive world" (Clarkson J, Coleman R., Appl Ergonomics 46:235–247, 2015). To consider this problem, this review searched for on-line library sources employing keywords including transport, inclusivity, circulation, severance, level of service, pedestrian movement within journal articles, conference papers, theses, books, and government papers. The review commences with a review of older views and goes on to critically assess potential innovations in contemporary scholarly literature. Earlier research discusses general factors that influence station design include context, location, platform, train length, depth of construction, geological, engineering, and property constraints, and passenger demand (Harding J., Investigating the built environment: survey of inclusive design attitudes within London's Tube Stations. The Departments of Engineering and Architecture Cambridge/The University of Cambridge, Cambridge, UK, 2011). This critical review considers the scholarly literature concerning inclusive design issues in transport buildings. It explores (i) gaps in canonical pedestrian movement theory affecting inclusivity, particularly vertical severance (VS), (ii) the size, shape and selection of circulation elements affecting Level of Service (LOS), and (iii) key measures of crowdedness and inclusivity. Potential innovations identified include, new 'designerly ways of knowing' about lack of inclusivity, and, the use of computer simulation using Agent Based Modelling (ABM) to probe how different

J. Harding (✉)
WSP, London, UK

University of Cambridge, Cambridge, UK
e-mail: john.harding@wsp.com

circulation arrangements affects the movement of agents over time. Anticipated benefits include a strengthened 'transport chain', less VS, enhanced empathy, and improved user experience and safety. More research is needed in this field, particularly owing to the significant cost and time in developing urban railway projects. This review identifies key research questions that require further investigation. It argues that more interdisciplinary design research will result in safer and more inclusive stations that contribute more to society. It is hoped that this review contributes to this slowly growing body of knowledge of inclusivity within the field of transport.

Keywords Inclusive design · Service design · Transport · Congestion · Severance · Agent based modelling

Abbreviations

ABM   Agent Based Modelling
LOS   Level of Service
PRM   Persons with Restricted Mobility
SD   Service Design
VS   Vertical Severance

25.1 Introduction: First Generation Wheelchair Accessible Transport Buildings

Many passengers experience a broken 'transport chain' (From Exclusion to Inclusion by the Disability Rights Task Force (DRTF), 1999 Cited in Bichard 2014: 90), including those who travel with bags, prams or heavy luggage, or are elderly, frail or have medical issues (GLA 2010). An 'end to end journey' (RSSB 2016) is practically impossible for some groups of people who wish to use public transport in London (Harding 2011). Only ten London Underground (LU) stations provide an unbroken, step-free journey from street to train (GLA 2010), 61 out of 270 stations provide step-free entry from street to platform level (ibid), and three stations are accessible from street to train in the three busiest employment centres in London (Harding et al. 2016). Recent research findings are that boarding or alighting a train is impossible for some groups of people owing to gaps at the train-platform interface (Atkins 2005). Findings are: (i) staff are unavailable to help with a portable ramp, (ii) wheelchair users and assistants attempt to board or alight unaided, (iii) the ramp gradient is too steep, (iv) the platform vertical and horizontal gaps exceed the minimum distance (v) gaps are visible even when the most modern trains are employed, (vi) mobile ramps are forbidden at LU stations owing to a low LOS and frequent train service (GLA 2010). However, the Disability Discrimination Act (1995) required wheelchair access in public buildings and soon after, the first generation of underground transport buildings had wheelchair access from street to platform for the first time (Harding 2011). A possible reason for this delay is that earlier research claimed that 'it was not essential' to make underground stations accessible for wheelchair users (Goldsmith 1976: 401 item 77,200). In his defence, Goldsmith, an influential researcher and architect who suffered from polio and used a walking stick, claimed the costs of lifts (US$10 m for lifts on the BART in San Francisco, and US$44–60 m for lifts on the Washington DC metro built in 1971) were excessive (1976: 60 para. 1411). He may have changed his mind as later editions of his book exclude these comments (cited in Harding 2011). On the other hand, his views were influential owing to his books formed the basis of the British Standard Code of Practice on Access for the Disabled to Buildings (CP96) in 1967, later revised to BS5810 (1979), that developed into Part M of the Building Regulations in 1987 (Coleman et al. 2003: 5). Whilst the Equality Act (2010) has a wider remit to promote inclusivity across gender, age, disability, and sexual orientation, surveys of passenger experiences show the small size, minimal quantity and poor location of lifts limit the benefits to few users (Harding et al. 2016).

25.1.1 Pedestrian Movement Theory

Another explanation for the poor uptake of lifts in stations is that current protocols and pedestrian movement calculations disregard the contribution lifts could make in moving people vertically (Network-Rail 2011). The methodology to validate designs for station concourses (Network-Rail 2011: 12) and for pavement areas (Atkins 2010) uses canonical theory (Fruin 1971). Fruin's methodology allows building designers to determine the sizes and shapes and arrangement of spaces, concourses and pavements, airport terminals, bus interchanges and train stations (ibid: 77–78). Level of Service (LOS) provides a measure of six congestion levels; a low LOS (E-F) describes a highly congested density level where it is difficult to change direction; in contrast, a high LOS (A-C) describes a free-flowing space where a rapid change of direction is possible (Fruin 1971: 71). However, a low LOS creates concerns for disabled pedestrians including, (i) they may delay other passengers, (ii) that other passengers may be unaware of their impairments, and (iii) their slower pace makes them more likely to be pushed or tripped in busy, congested, fast-moving spaces (ibid 177–178). To improve the disabled pedestrian's comfort and confidence he suggests increasing the LOS in congested circulation areas (ibid 177–178). Fruin's (1971) theory contains two main problems for inclusivity, (i) he does not develop his idea to offer a higher LOS for disabled people (ibid 177–178), possibly owing to the thought that increasing LOS is only possible by increasing the size of the transport building and that would increase destruction of historic built environments that he was against (ibid: 2–11), (ii) in transferring and adapting evaluation methods from highway design theory to pedestrian movement theory in his PhD thesis (Highway Capacity Manual, Highway Research Board, Special Report 87, Washington, D.C. (1965) cited in Fruin 1971: 71), he misses a critical difference that whilst a vehicle easily moves up or down a hill, pedestrians face many difficulties moving vertically in train stations (GLA 2010). Furthermore, current design guidance does not suggest increasing LOS in stations (Network-Rail 2011: 12) or within the urban realm (Atkins 2010) to improve inclusivity.

25.1.2   A 'Wicked' Problem

Designers need to know how many passengers would need to use a lift to ensure they specify the correct size and quantity of lifts (Al-Sharif and Al-Adem 2014). The risk is, without knowing how many passengers would like to use a lift, designers may reuse past assumptions and prototypes owing to the lack of research (Harding 2011), time pressures, and the Einstellung effect, or 'design fixation' (Crilly 2015). Harding (2018) claims knowing how many people require help with inclusivity is a 'wicked' problem that cannot be solved by traditional scientific and engineering methods (Cross 2007). Next, we will review research that could addresses these concerns.

25.2   Researching Empathy

A possible explanation for the low take-up of inclusivity in transport buildings is that designers, clients and project managers who produce and control design all have differing experiences according to their age and gender (Harding 2011). The problem, as Warburton (2003: 255–256) claims, is that mostly male young designers design primarily for themselves and other young people who they perceive as sexy. Similarly, younger transport planners design 'systems for the able-bodied, not for those who were frail. There was a desire for a gentler, more comfortable environment.' (Marsden et al. 2008: 5). To investigate these claims, Harding (2011) quantitative study uses a questionnaire survey and a five-point Likert scale to study 'Tube' user experiences. All 47 respondents (34 men, 13 female) were frequent commuters, and influential participants who were employed to design and build several major new underground stations in London. The respondents were selected to reflect observed demographic composition of the organisation and the data were collected during a 4-week period in 2010. None of the respondents claimed to have a disability. Findings are that a gentle journey affords clear announcements, low noise, lack of fear of being lost or splitting from a group, few changes of levels and easy orientation. Older men in influential positions were generally satisfied with their experience. Women and younger men have a poorer experience of security, confidence and comfort. Whilst all groups had confident experiences, their journeys were not gentle. Causes for concern are 17% of men experienced crime compared to 7% of women, whereas 76% women had a higher fear of crime, compared to 32% of men. Moreover, women talked to more strangers (46% compared to 20% of men) which suggests that men might be safer if they had greater awareness of crime and talked to strangers (Harding 2011: 74). Interestingly, women experienced more anti-social behaviour (61%) then men (50%). That suggests anti-social behaviour might be less predictable and difficult to avoid (Harding 2011: 74). Moreover, three gay male respondents completed survey forms. The data indicates no increased fear of crime or experience of crime for gay men compared to straight men. Thus, if both gay and straight men adopt female crime avoidance strategies by being warier both groups could experience less crime. Possible explanations for why males felt more secure were owing to, i) their physical strength, ability and confidence to deal with dangerous situations (Harding 2011: 53), (ii) male respondents have the best experience of travelling on the Tube, (iii) males and engineers dominate the design of underground stations, (iv) males were more reluctant to learn about inclusivity (Harding 2011: 83). Features that appear harmful, such as long, narrow or twisting corridors, may expose everyone to greater harm or fear of crime. However, men experienced less fear (17% Q13) and more crime (17% Q18). In contrast, women had more fear (53% Q13) and experienced less crime than men (7% Q18). Additional differences are women are around a third weaker than men, owing to their smaller physique and childbirth (Bassey 1997: 289–297). Consequently, specialists such as psychologist Huppert claim that 'older women should therefore be a priority in inclusive design' (2003: 35). However, safety guidelines disadvantage women who find it burdensome to carry a child and a folded pram and bags on an escalator (GLA 2010). On the other hand, there is a risk and irony that men who may wish to develop designs that suit themselves, endanger themselves and others by creating less safe circulation spaces within stations (Harding 2011). Limitations of this research are that the respondents were unrepresentative of the public because most were professionals, working and held degrees. Nobody declared a disability. Harding (2011) suggests this could be owing to disabled people having trouble travelling, gaining and keeping a job, and keeping silent about their disability (Payling 2003: 395 cited in Harding 2011). Research suggests a negative reply to this question is common in surveys (ibid). Additionally, a known weakness of using positivistic questionnaires that use inductive and deductive logics of enquiry is that they remove important details and differences to produce simplistic explanations (Stainton-Rogers 2006: 81). In summary, this study suggests that comfort, security, gentleness, confidence experiences could become new proxies to compare inclusivity within demographic groups. Table 25.1 below, compares these proxies with different demographic groups. Therefore, this could be a new tool for designers or researchers to roughly compare

Table 25.1 New proxies to compare inclusivity to age and gender in transport (Harding 2011)

Proxies	Baseline	Young <25	Middle	Aging >55	Women	Men
Comfort	☺	☹	☹	☺	☺	☺
Security	☺	☹	☺	☺	☺	☺
Gentleness	☹	☹	☹	☹	☹	☹
Confidence	☺	☺	☺	☺	☺	☺

Good Experience= ☺ Average= ☺ Bad Experience= ☹

the different experiences of inclusivity to their own group (Warburton 2003; Marsden et al. 2008) that could promote reflexivity in practitioners who are working in uncertain, unstable, unique areas with conflicting requirements and values (Schon cited in Cross 2007: 3). Further research could probe using this tool to compare different circulation arrangements in praxis.

25.2.1 Understanding User Experience

The influence behind this study was to investigate inclusivity concerns in the context of a busy, urban underground station using mobile methods by leaving "… the design office and becoming- if briefly-immersed in the lives, environments, attitudes, experiences and dreams of the future users" (Battarbee 2004). Participant observation studies are frequently used in the fields of anthropology and sociology using methods developed by Malinowski (cited in Buzard 1997). This is the first time this methodology, is applied to investigating passengers within a crowded urban underground railway context (Harding et al. 2016). Their qualitative study investigates how people in the 'rush hour' circulate in a low LOS, typically shallow station, which is in a busy urban location with one lift (Harding et al. 2016). Research questions are, in what ways do we find train passengers suggestible as they move through congested underground train stations, and how do passengers protect themselves against suggestions that do not help them survive or be included within the design? (Harding et al. 2016). This method allows a researcher to collect video data, using a chest mounted camera, whilst moving within the station with other passengers. The video data was transferred from the camera onto a computer and analysed approximately a month after the data was collected. The good quality of the recording allowed the researcher to recollect events and analyse results in a detached way. This exploratory study stores and preserves video files as an innovation Crichton and Childs (2005) that clips audio recordings to preserve original participants' voices and roughly codes audio files using Excel to store large files. Harding's modified approach saves research time by reducing the amount of transcription, preserves the original materials until explication, and allows a 'thick' or wordy description of the event. This immersive method allows a researcher to move spontaneously within a busy station during the evening peak-hour commute to observe passenger experiences directly (Harding et al. 2016). It focuses our attention upon the qualitative aspects of the 'passenger journey' and identifies, from the user viewpoint surprising design or behaviour issues. These include glare from bright lights shining into the eyes of passengers, noise from announcements and the quietness of crowds waiting patiently for their next train. Findings highlight the difficulty to reach and find a lift when it is small and poorly located at the end of a long and congested route at platform level. In contrast, the lift location at concourse level has a higher LOS and is visible adjacent to busy escalators, however, there is lack of space in the lift to accommodate all the passengers who wish to use it. Other negative factors result from confusion, congestion, glaring lights, noisy announcements and warnings. A typical island platform configuration often results in significant queuing when trains are insufficiently frequent or too congested to board. Findings from indirect methods using a questionnaire (Harding 2011) exclude such details. Other advantages are preserving the video data allows other researchers and participants to make their own interpretation. It reports on subtle details of 'lived experiences' that may otherwise be lost, and provides a voice for the 'silenced' which may improve the autonomy, survivability and perhaps their inclusion in a next generation of station designs. The analysis focuses upon how to improve mobility in the station from a user's perspective. The use of auto-ethnography is discussed as part of a broader methodological debate about how to explore universal design issues from a user's perspective, and in the context of empathetic design (Harding et al. 2016). The video data and analysis is useful to either a researcher, or a designer, not living in London, or in a country without a railway, who wishes to experience a journey in a busy train station and may be unable to travel and gain an insightful experience in a 'real' station. Difficulties with this research method are no significant data should be taken without consent (Oates 2006) and findings are un-generalisable owing to the participant observer had no impairment, is male and middle aged that risks unconscious bias.

25.2.2 Safer Stations

Train stations are the riskiest place for accidents on the entire rail network in the UK and cost the industry approximately £90 m per year (RSSB 2015). Slips, trips and falls are common accidents. A 'fatality weighted index' (FWI) quantifies all the, albeit few, fatalities and the frequent minor and major injuries into approximately and occur most often on stairs (10), platforms (8), concourses (4), escalators (4) and other (2). There are approximately 30 FWI per year (excluding suicides). The data shows almost 50% (14) of FWIs occur on stairs and escalators (refer to Chart 39 RSSB 2015). However, findings from interviews with station managers and travelers are that to minimise slips trips and falls, the design of vertical circulation is considered less important compared to selection of flooring materials, waiting rooms, lighting; signage, cleaning and housekeeping (Victoria et al. 2014: 39 Table 8). In contrast, worldwide research papers show that elderly people, children, women with high heels and inebriated people as the most at risk from slips, trips and falls on escalators. (Greenberg and Sherman 2005). Unfortunately, mixed methods studies from Hong Kong (Chi et al. 2006) and USA (O'Neil et al. 2008), where escalators are common, do not explain why these groups are most at risk; nor do they suggest measures that could prevent harm. Rubenstein (2006) identifies in a meta-analysis study the causes of falls in elderly adults, describes the interaction of environmental, medical and age factors that cause the falls (ibid ii38). Slips, trips and falls should be considered a significant issue in train station design because '. . . . [unintentional] injuries are the fifth leading cause of death in older adults (after cardiovascular disease, cancer, stroke and pulmonary disorders), and falls constitute two- thirds of these deaths' (ibid ii37). It is recommended that further design research probes factors including circulation and LOS that could result in safer stations.

25.2.3   Reduce Vertical Severance (VS)

To address the risks and issues of moving vertically, Harding (2013) synthesises earlier observations and literature (Harding 2011) by defining this phenomenon as Vertical Severance. The "... separation from ground level to the platform that creates spatial mobility and socio-economic concerns for individuals. VS results in less diversity and more exclusivity within transport modes and the cities they serve." (Harding 2013: 13). For new VS free transport buildings, Harding (2013) recommends the following solutions: "....1) Accept Maynard's claim that well-designed lifts are beneficial to almost all individuals (2007), consequently develop designs with more and faster lifts. 2) Consider other property types that solved VS, for example, Heathrow Terminal 5 [that provides many large lifts to serve passengers carrying luggage... 6) Consider [either] omitting or provid[ing] fewer spatially inefficient escalators to deeper stations, or where passenger exit and entrance numbers are relatively low, provide more space for lifts and evacuation stairs. Note that escalators cost more to build, maintain and consume far more energy than lifts. ... In summary, such changes require a paradigm shift to provide VS-free designs ..." (Harding 2013: 12). Some scholars argue that vertically separating the pedestrian from the vehicle is beneficial to pedestrian comfort and safety (Fruin 1971: 183–196); others prefer active streets and pavements and oppose vertical segregation (Hillier 2004: 45). Other scholars argue severance or the 'wrong side of the tracks' phenomenon barely exists (Mitchell and Lee 2014) – they tentatively argue that socio-economic divisions in neighbourhoods on opposite sides of the Clyde River valley in Glasgow are explained by their steep banks. Further research could probe how different circulation choices could satisfy either 10% or 25% of overall passenger numbers who may wish to use a lift at stations (Harding 2013: 11), and identify actionable insights for LOS, inclusivity and VS theory and praxis.

25.2.4   New Disability Discourses

No discussion of inclusivity would be complete, without a brief review of recent discourses in disability studies that developed earlier medical and social models of disability to current interactional models (Riddle 2013: 33–35). The Medical Model claims the "impaired body must be restored, adapted and cured" (Scullion (2009) quoted in Gomez et al. 2014: 272). In contrast, the Social Model of Disability claims society's actions and inactions cause a person's disability. The trouble with Social Model theory is that owing to its insistence that society causes disability it ignores the possibility that advances in medical technology developed within the medical and technological field may remove the impairment (e.g. glasses, wearable technology, or prosthetics) (Corker and French 1999). Similarly, Watson and Woods (2005: 104) argue that wheelchair technology has an emancipatory impact upon the lives of wheelchair users and the use of technology is often neglected as an aid for social justice. The weakness in the social model argument is that ". . . the horse before the disability studies carriage is often politics, not science", (Vehmas 2008: 21 Quoted in Riddle 2013:2028). Recent inclusive design (Boys 2014) and interactional theory focuses upon removing the impairment from both the built environment and the body (Riddle 2013). Interactional theory also expands the discourse to more complex socio-political contexts (not just disability) including feminist, racial, gender, ethnicity and sexual topics (Stainton 2000); non-disabled concerns (Slack 1999: 23) and challenges us to consider questions about sufficiency; and medical-socio-material-economic-political challenges (Slack 1999). Socio-material-environmental thinking found in recent research takes a more holistic view (Bichard 2014). Consequently, many philosophers and bioethicists support this interactional approach (Riddle 2013: 23). Furthermore, design paradigms used in the built environment have different aims, for example, Universal Design's (UD) aim is to provide access to a wide range of users, whilst Inclusive Design's (ID) aim is to offer everyone access (Martens 2018: 122). However, it is unknown how these unsettled discourses have impacted inclusivity in praxis. Further research could probe how subtle differences between rival UD and ID aims and social model and interactional logics may impact the circulation design within buildings in praxis, and consequently impact inclusivity.

25.2.5   'Next, Next Generation' Design Research Methods

First generation design research methods that used 'systematic, rational 'scientific' methods sought to optimise design, however these approaches did not solve 'wicked' design problems (Cross 2007: 1). 'Next, next generation' methods are 'more relevant to architecture and planning rather than engineering and industrial design' (Horst Rittel (1973) cited in Cross 2007: 1). These include an idea from business studies called 'satisficing' that aims to develop satisfactory and appropriate solutions and not 'optimising' solutions (Simon 1979). 'Service design' (SD) research by the design and research company IDEO strengthens the 'passenger journey' by developing knowledge of each activity or 'customer touch points' as the passenger obtains information, plans the journey, travels to the station, enters the station, buys tickets, waits, boards the train, travels on the train, alights, and continues the journey (Bhavnani and Sosa 2008). However, integrating this concept is untested in recent underground train station design research (Harding 2018). Further design research could probe how new 'designerly' methods (Cross 2007) could address the aforementioned 'wicked' design problems of inclusivity by incorporating SD and the passenger journey concepts (Bhavnani and Sosa 2008).

25.2.6   New Tools

New research inquiry tools that could aid inclusivity research and praxis include, (i) 'Bit Kit' is a tool to assist visually impaired users with navigating in buildings (McIntyre and Hanson 2014), Wayfindr is an assistive navigation tool useful for visually impaired people to navigate by themselves in unfamiliar buildings including transport buildings (Giannoumis et al. 2018), (iii) Space Syntax is a tool that evaluates connectivity of streets within urban areas (Hillier 2004) but does not model congestion, (iv) However, Legion is an ABM-based pedestrian modelling trusted by many transit authorities to interpret and validate train station design (Network-Rail 2011: 53). Recent ABM research recommends further 'integration of simulation results with accessibility requirements for persons with restricted mobility [PRM] . . .for all pedestrian simulation modelling to ensure an equitable assessment of transport interchanges' (Clifford et al. 2016: 16). However, restricting studies to PRM passengers appears to generate problems of exclusivity that this paper argues against. Instead this paper argues new ABM research should probe the 'wicked' problem of not knowing how many people require help with inclusivity (Harding 2013), how rival circulation choices impact agents' behaviour (Harding 2018).

25.3   Conclusion

Findings from this review suggest that researching inclusivity is a 'wicked' problem and that 'design thinking' and 'next, next generation' methods (Horst Rittel (1973) cited in Cross 2007: 1) and Service Design (Bhavnani and Sosa 2008) could advance 'satisficing' solutions (Simon 1979) for inclusivity. This review also highlights the concern that VS is a problem that can be seen in transport buildings (Harding 2013) and urban areas (Mitchell and Lee 2014). The impact of congestion upon inclusivity in transport buildings and urban areas appear under researched suggesting that canonical pedestrian movement (Fruin 1971) and VS (Harding 2013) require further research to satisfy inclusivity needs in society. These questions are addressed in Harding (2019).

References

Al-Sharif L, Al-Adem MD (2014) The current practice of lift traffic design using calculation and simulation. Build Serv Res Technol 35:438–445
Atkins (2005) Significant steps. Department for Transport, London
Atkins (2010) Pedestrian comfort level guidance. In: London TfL (ed). London: TfL
Bassey E (1997) Physical capabilities, exercise and aging. Rev Clin Gerontol 7:289–297

Battarbee K (2004) Co-experience: understanding user experience in social interaction. University of Art and Design/Finland University of Helsinki, Helsinki, pp 237–245

Bhavnani R, Sosa M (2008) IDEO: service design. Insead, Paris

Bichard J-A (2014) Extending architectural affordance: the case of the publicly accessible toilet. The Bartlett School of Architecture/University College London, London

Boys J (2014) Doing disability differently: an alternative handbook on architecture, dis/ability and designing for everyday life. Routledge, London/New York

Buzard J (1997) Mass-observation, modernism, and auto-ethnography. Soc Psychol 4(3):93–122

Chi CF, Chang TC, Tsou CL (2006) In-depth investigation of escalator riding accidents in heavy capacity MRT stations. Accid Anal Prev 38:662–670

Clarkson J, Coleman R (2015) History of inclusive design in the UK. Appl Ergon 46(Pt B):235–247

Clifford P, Melville E and Nightingale S (2016) Pedestrian modelling for persons with restricted mobility at transport interchanges. European Transport Conference., 1–17

Coleman R, Lebbon C, Clarkson J et al (2003) Introduction-from margins to mainstream. In: Clarkson JP, Coleman R, Keates S, Lebbon C (eds) Inclusive design: design for the whole population. Springer, London

Corker M, French S (1999) Reclaiming discourse in disability studies. In: Corker M, French S (eds) Disability discourse. McGraw-Hill Education, Buckingham, pp 1–11

Crichton S, Childs E (2005) Clipping and coding audio files: a research method to enable participant voice. Int J Qual Methods 4:1–9

Crilly N (2015) Fixation and creativity in concept development: the attitudes and practices of expert designers. Des Stud 38:54–91

Cross N (2007) Forty years of design research. Des Stud 28(1):1–4

Fruin JJ (1971) Pedestrian planning and design. Metropolitan Association of Urban Designers and Environmental Planners, New York

Giannoumis GA, Ferati M, Pandya U et al (2018) Usability of indoor network navigation solutions for persons with visual impairments. In: Langdon PLJ, Heylighen A, Dong H (eds) Breaking down barriers CWUAAT. Springer, Cham, p 2018

GLA (2010) Accessibility of the transport network. Greater London Authority, London

Goldsmith S (1976) Designing for the disabled. RIBA, London

Gomez JL, Langdon PM, Bichard JA et al (2014) Designing accessible workplaces for visually impaired people. In: Langdon PM, JL AH, Dong H (eds) Inclusive designing – joining usability, accessibility, and inclusion. Springer, Cambridge, pp 269–279

Greenberg DT, Sherman SC (2005) Escalator injuries. J Emerg Med 28:75–76

Harding J (2011) Investigating the built environment: survey of inclusive design attitudes within London's tube stations. The Departments of Engineering and Architecture Cambridge/The University of Cambridge, Cambridge

Harding J (2013) Experiencing mobility in underground transport systems. In: LTA-UITP Singapore international transport congress and exhibition (SITCE 2013). LTA, Singapore, pp 1–15

Harding J (2018) Agent-based modelling could remove an ethical barrier to researching inclusivity in crowded places. In: Langdon P, Lazar J, Helylighen A, Dong H (eds) 9th Cambridge workshop on universal access and assistive technology. Fitzwilliam College, University of Cambridge, Cambridge, pp 33–42

Harding J (2019) Using agent-based modelling to probe inclusive transport building design in practice. In: Proceedings of the Institution of Civil Engineers – Urban Design and Planning. https://doi.org/10.1680/jurdp.18.00028

Harding J, Luck R, Dalton NS (2016) Journeys in the City: Empathising with the users of transport buildings. In: Emmitt S, Adeyeye K (eds) International conference on integrated design. Building our future. University of Bath, Bath, pp 324–335

Hillier B (2004) Can streets be made safe? Urban Des Int 9:31

Huppert F (2003) Designing for older users. In: Clarkson J, Coleman R, Keates S, Lebbon C (eds) Inclusive design: design for the whole population. Springer, New York

Marsden G, Jopson A, Cattan M, et al. (2008) Older people and transport: integrating transport planning tools and user needs. Sparc?

Martens K (2018) Ageing, impairments and travel: priority setting for an inclusive transport system. Transp Policy 63:122–130

McIntyre LJ, Hanson VL (2014) Buildings and users with visual impairment: uncovering factors for accessibility using BIT-kit. In: Kurniawan S, Richards J (eds) Assets '14. New York, proceedings of the 16th international ACM SIGACCESS conference on computers & accessibility. ACM, New York, pp 59–66

Mitchell R, Lee D (2014) Is there really a "wrong side of the tracks" in urban areas and does it matter for spatial analysis? Ann Assoc Am Geogr 104:432–443

Network-Rail (2011) Station capacity assessment guidance. In: DfT (ed). London

O'Neil J, Steele GK, Huisingh C et al (2008) Escalator-related injuries among older adults in the United States, 1991–2005. Accid Anal Prev 40:527–533

Oates J (2006) Ethical frameworks for research with human participants. In: Potter S (ed) Doing postgraduate research. Open University in association with SAGE Publications, Milton Keynes

Payling J (2003) The sense of independence. In: Clarkson JP, Coleman R, Keates S, Lebbon C (eds) Inclusive design: design for the whole population. Springer, London

Riddle CA (2013) The ontology of impairment: rethinking how we define disability. In: Wappett M, Ardnt K (eds) Emerging perspectives on disability studies. Palgrave Macmillan, New York, pp 23–40

RSSB (2015) Annual safety performance report 2014/15 a reference guide to safety trends on GB railways. RSSB, London

RSSB (2016) Rail sustainable development principles. RSSB, London

Rubenstein LZ (2006) Falls in older people: epidemiology, risk factors and strategies for prevention. Age Ageing 35:ii37–ii41

Simon HA (1979) Rational decision making in business organizations. Am Econ Rev 69:493–513

Slack S (1999) I am more than my wheels. In: Corker M, French S (eds) Disability discourse. McGraw-Hill Education, Buckingham, pp 28–37

Stainton T (2000) Review: disability discourse. Int Soc Work 43(2):265–266

Stainton-Rogers W (2006) Logics of enquiry. In: Potter S (ed) Doing postgraduate research. Milton Keynes: Open University in association with SAGE Publications, London, pp 73–91

Vehmas S (2008) Philosophy and science: the Axis of evil in disability studies. J Med Ethics 34

Victoria K, Patrick W, Brendan R et al (2014) Managing the risks of slips, trips and falls for the ageing rail passenger population – final report (COF-HSW-02). In: RSSB (ed) . Loughborough University; University of Nottingham, Nottingham

Warburton N (2003) Everyday inclusive design. In: Clarkson J, Coleman R, Keates S, Lebbon C (eds) ID for the whole population. Springer, New York

Watson N, Woods B (2005) No wheelchairs beyond this point: a historical examination of wheelchair access in the twentieth century in Britain and America. Soc Policy Soc 4:97–105

Part VIII
Energy

How can highly energy efficient buildings be realized, which are supplied with renewable energy sources.

Chapter 26
The Total Cost of Living in Relation to Energy Efficiency Upgrades in the Dutch, Multi-Residential Building Stock

Thaleia Konstantinou, Tim de Jonge, Leo Oorschot, Sabira El Messlaki, Clarine van Oel, and Thijs Asselbergs

Abstract Decarbonizing the housing stock is one of the largest challenges in the built environment today, and is getting attention not only from policymakers, but also from social housing corporations, financial and tenants' organisations. In line with the international Paris-Climate-Change-Conference 2015, Dutch cities and housing associations have embraced this challenge with the ambitions to become carbon neutral in 2050. To reach such ambitious goals, both the rate and depth of renovation need to increase significantly. In the Netherlands, the Energy Agreement for Sustainable Growth, indicates that 300.000 dwellings have to be renovated annually, in accordance with the Energy Performance of Buildings Directive adopted by the European Union, to improve the Dutch building stock towards energy neutrality. Several technical solutions to eliminate the energy demand in dwelling have been developed and tested. Nevertheless, the intake rate of deep retrofitting is low. Currently, most improvements in residential buildings consist of basic maintenance and shallow renovation, but broader or deeper energy renovation measures are required. Despite more recent developments, there are still significant barriers related to financing, lack of information, and user acceptance. Complex technical characteristics are not always taken into account by tenants; the focus is usually on the ease of use, comfort and living expenses.

To this end, the present study sets of to investigate the relationship between energy efficiency upgrade measures and cost of living. Focusing on the post-war, multi-family social housing in the Netherlands, a framework of refurbishment measures that affect the energy efficiency were identified, and their performance was simulated. The variations refer to the façade design, thermal envelope upgrade, winter-garden addition and reviewable energy. The energy efficiency indicator is the

T. Konstantinou (✉) · L. Oorschot · S. El Messlaki · C. van Oel · T. Asselbergs
Beyond the Current Research Group, Faculty of Architecture and the Built Environment, Delft University of Technology, Delft, The Netherlands
e-mail: t.konstantinou@tudelft.nl

T. de Jonge
Winket Bouwkostenadviesbureau, Roosendaal, Netherlands

© Springer Nature Switzerland AG 2020
R. Roggema, A. Roggema (eds.), *Smart and Sustainable Cities and Buildings*,
https://doi.org/10.1007/978-3-030-37635-2_26

energy cost reduction, as well as the carbon footprint of the energy use. Furthermore, the rental price adjustment was estimated, taking into account the refurbishment investment and the operation cost of the renovated dwellings. All tested combination of variables resulted in significant energy savings, up to 70%, while energy generation was proven to be cost-effective, as it has a considerable positive effect on the energy use and the energy cost, without increasing the rental price.

The results aim at supporting the decision-making discussion between the stakeholders, primarily housing associations and tenants. The relation between the energy consumption and rental price for the different options identifies the effect of design variation and demonstrated the attractive solutions that the tenants are more likely to accept, taking into account the overall cost of living and sustainability benefits.

Keywords Energy efficiency · Renovation · Cost of living

26.1 Introduction

Decarbonising the housing stock is one of the largest challenges in the built environment today, which is getting the attention not only from policymakers but also from social housing associations and other institutional real estate owners, financial organisations and users. Several studies (BPIE 2011, 2013; Crawford et al. 2014; IEAAnnex56 2012) have reported that huge potential for energy savings, improved health and comfort of the occupants', elimination of fuel poverty, and job creation lay in the technical upgrade of the existing buildings stock. In line with the international Paris-Climate-Change-Conference 2015, Dutch cities and housing associations have embraced this challenge with the ambitions to become carbon neutral in 2050.

To reach such ambitions, both the rate and depth of renovation need to significantly increase (Artola et al. 2016; BPIE 2011). In the Netherlands, the Energy Agreement for Sustainable Growth, indicates that 300.000 dwellings have to be renovated annually, in accordance with the Energy Performance of Buildings Directive adopted by the European Union, to improve the Dutch building stock to energy neutrality (DIRECTIVE 2010/31/ EU). Moreover, in the Netherlands, the housing associations have the ambition to achieve a carbon-neutral building stock by 2050 (AEDES 2017). A number of technical solutions to eliminate the energy demand in dwelling have been developed and tested. Those solution target different levels of energy efficiency, ranging from a small upgrade of the energy label, most commonly up to label B, to achieving zero-energy demand.

Nevertheless, the intake rate of deep retrofitting is low. Currently, most improvements in residential buildings consist of basic maintenance and shallow renovation, but broader or deeper energy renovation measures are required (Filippidou et al. 2016). Despite more recent developments, there are still significant barriers related to financing, lack of information, and user acceptance (Matschoss et al. 2013). The residents of the dwellings care less about the technical characteristics of a dwelling, but more about the use, comfort and living expenses.

To this end, the present study sets of to investigate the relationship between energy efficiency upgrade measures and cost of living. Focusing on the multi-family social housing in the Netherlands, a framework of refurbishment measures that affect the energy efficiency were identifies and their performance was calculated. The energy efficiency indicator is the energy cost, as well as the energy use. Furthermore, the rental price adjustment was estimated, taking into account the refurbishment investment and the exploitation cost of the renovated dwellings. The comparison of the energy use and rental price for the different options demonstrated the most attractive solutions that the tenants are more likely to accept, taking into account the overall cost of living and sustainability benefits. The results aim at supporting the decision-making discussion between the stakeholders, primarily housing associations and tenants.

26.2   Methodology

To provide insights into the study's question on the relation between energy saving renovation and cost of living, the evaluation of the refurbishment options is based on Key Performance Indicators (KPI's). The key performance indicator is a measurable value that demonstrates how effectively a system, in this case, the refurbished buildings, performs. The KPI's used in this study- as concluded out of focus groups with stakeholders, such as residents and housing associations- are the energy use and its resulting cost, the rent price, because it reflects the refurbishment costs as it will be explained in sect. 2.3, and the total cost of living, as the sum of energy cost and rent. The sustainability of the solutions is indicated by the energy demand since the same heating system, and fuel is applied to all options. Hence, the energy demand and CO_2 emissions are proportional.

The steps to quantify the KPIs are hence related to the strategic organisation of the refurbishment measures, for starters, and then quantifying their effect on energy use, cost and rent price. The investigation is based on applying and refurbishment strategies on a case-study building. The specifics of the building were taken into account for the design and assumptions considered for the energy and cost calculations. The study focuses on low-rise, multi-family, walk-up apartments, as they present considerable challenges for their energy upgrade. Currently, there are still 799.956 apartments of all types from the period 1906–1965 in The Netherlands, 400.000 apartments of which are located in the four major cities. The building shown in Fig. 26.1 was selected as a case study to apply the refurbishment options, as being a typical example of the post-war period (Platform31 2013).

Fig. 26.1 Case study building: Camera Obscura, Overvecht Utrecht, 2016

26.2.1   Define the Alternatives and the Combinations: General Transformation Framework

In order to be able to evaluate the solution, the alternative refurbishment measures need to be defined. The measures are defined per category and per function, creating a "General Transformation Framework". The parameters taken into account for the framework development came out of research the existing tenement building types of the inter-war and post-war period and their special characteristics and projects (Oorschot et al. 2018).

Moreover, analysis of realised refurbishment project and interviews with architects and housing association helped to define the state-of-the-art. In the scope of the present study, the measures discussed refer to a cluster of technical interventions that can be employed to improve the energy efficiency of the apartments. Additional socio-cultural interventions related to the functional and cultural heritage qualities are possible to be applied, but outside the present paper's scope.

As they are not likely to be applied individually, they have been combined into integrated solutions, before they can be evaluated regarding energy demand and cost. The alternative measures were defined based on analysis of current refurbishment practice, literature review and discussions with stakeholders. The aspects considered that have an impact on the energy use of the building are the following, as presented in below (Table 26.1):

26.2.2   Energy Demand Calculation and Indicators

For the refurbishment options to be evaluated and for the total cost of living to be calculated, the energy performance of the case-study building is estimated. Firstly, the energy use for both building and user-related sources is calculated using dynamic thermal performance simulation. Then, the energy use is simulated after the proposed, combined solutions have been applied. The software used for the thermal

Table 26.1 Overview of the alternative refurbishment solutions proposed

Aspects	Alternative		Description
Façade design	Existing		Existing façade design. Sill height 1 m. window-to-wall ratio 80%, operable 30%
	Half open		Half open facade with operable opaque ventilation openings, with respect of the most characteristic heritage elements. Window-to-wall ratio 60%, operable 0%
	Open		Open facade with glass from floor to floor, with respect of the most characteristic heritage elements. Window- to-wall ratio 100%, operable 50%
Thermal properties upgrade	Level B	Basic upgrade	Facade U = 0,20 W/m2K
			Roof U = 0,20 W/m2K
			Floor U = 0,28 W/m2K
			Windows double glazing / U = 1,2 W/m2K
	Level A	Advanced, towards ZEB standards	Facade U = 0,20 W/m2K
			Roof U = 0,15W/m2K
			Floor U = 0,25 W/m2K
			Windows triple glazing / U = 0,8 W/m2K
Extension	Winter-garden		Extension with a glass covered balcony.
			External wall: 100% glass Single.Open 80% at 24oC. Shading intern drapes
			Interior partition: Double glazing, 100%. Open 80%. Min temp for Nat vent 24oC
	No extension		No additional construction.
Renewable energy generation	None		No PV panels nor solar collectors
	PV		Calculated per apartment, based on the overall available area for PV application. Efficiency 255Wp
	PV + solar collectors		Solar collectors are assumed to be placed on the balcony, on the south side, producing up to 330kWh/m2, which covers the energy demand for hot water

simulation is DesignBuilder, which was chosen as appropriate for the purpose of this study because it can generate a range of environmental performance data such as energy consumption and internal comfort data. The actual data for the building's size and construction were used, data for the location climate were input, and occupancy data were based on the building's function, classified as "Tailored rating", according to European Standards EN15603 (2008).

26.2.3   Inputs

For every energy consumption calculation, the way the building is constructed and operated needs to be specified, as input. When comparing current and new energy demand, an assumption is that the usage patterns will not change significantly. A nuclear family (four-person household, two parents and two children) will be considered, as it is the largest percentage in the demographics of the case study. The different inputs are summarised in Table 26.2.

26.2.4   Comfort, Energy Demand, Energy Cost and Carbon Footprint

The simulation resulted in the amount of energy in kWh a dwelling requires per year, including HVAC systems, domestic hot water and appliances. Moreover, the internal temperatures were checked to calibrate the dwelling function and comfort, existing and refurbished, and ensure that overheating is avoided. The energy costs are based on the prices indicated there, considering fixed amounts for the grid, as well as different prices for peak hours, the following costs were calculated for electricity 0,18/ kWh and gas 0,77/ m3, including tax (Eurostat 2016). Those prices are then implemented to the simulation results, for electricity and gas demand respectively.

26.2.5   Total Cost of Living Calculation Method

The refurbishment strategies are evaluated regarding the effect the investment has on the rent price. To this end, a Life Cycle Costing (LCC) was performed. The increase in the rent price was based on the assumption that for sustainable housing to be financially feasible, all investments must be covered by the exploitation period rent income. Firstly, the investment costs of major renovations were determined without

Table 26.2 Energy simulation inputs

Parameter	Inputs
Location	Netherlands
Orientation	Depending on the specific building
Geometry and zones	Every room as a different zone, depending on activity (bedroom, living room etc.)
Occupancy	Based on zones function, for a four-person household
Openings	Layout: Building design. WWR between 60-100%
Heating/ DHW	Gas boiler, efficiency 80%
Ventilation	Natural inlet through windows/ mechanical outlet through bathroom and kitchen.

considering specific energy-saving measures. The investment costs have been defined according to the Dutch standard NEN 2699 (NEN 2017) as: the value in use of the existing building + the construction costs of the renovation + the additional costs such as fees, connection costs and taxes. The construction costs of all renovation measures have been estimated by EcoQuaestor (2014) cost database. As a result, the cost level of the budgets is consistent with Dutch building practice. The rent of the apartments after renovation but without specific energy-saving measures was determined by the "Appropriate allocation" scheme under the 2015 Housing Act. Subsequently, the investment costs of specific energy-saving measures are added to the initial investment.

The investment for both scenarios, with and without energy efficiency measures, is then included in a cash flow survey of operating costs and benefits according to the life-cycle costing (LCC) model of the NEN 2699 standard. The cost of maintenance, management costs and other property expenses are included. On the revenue side of the balance sheet, the present value of rental income was added, for an exploitation period, assuming 30 years is the exploitation period for an apartment in the social housing sector. In the renovation scenario, the extra investment costs of the specific measures were included in the cash flow analysis. The present value of the rental income was adjusted to close the balance. The increase in monthly rent was then calculated as the difference required to balance the cost and income in the LLC.

It needs to be clarified that this method can result in differences in the rent price for the same combinations of energy efficiency measures combinations. The reason for this discrepancy is that the rent after renovation, which also depends in other parameters, such as the additional number of rooms, or the construction of additional dwellings, which are not within the scope of the current study.

26.3 The Resulting Cost of Living for the Different Aspects

This section presents the effect of each aspect, as defined in Table 26.1, on the KPI's energy demand, energy cost, CO2, rent price and total cost of living. Not all KPI's are discussed in every case, depending on the significance of the effect. The numbers presented in the figures are based on averages values for the combination of measures that include the respective variations. These averages are the reason why the total cost of ownership is not always the sum of the average energy cost and the average rent in the following figures.

26.3.1 Façade Design

There were three different options for the façade design. Those options differ in the window- to-wall ratio (WWR), layout and operation. The design of the façade is important for how the building is perceived, and our proposals came out of the

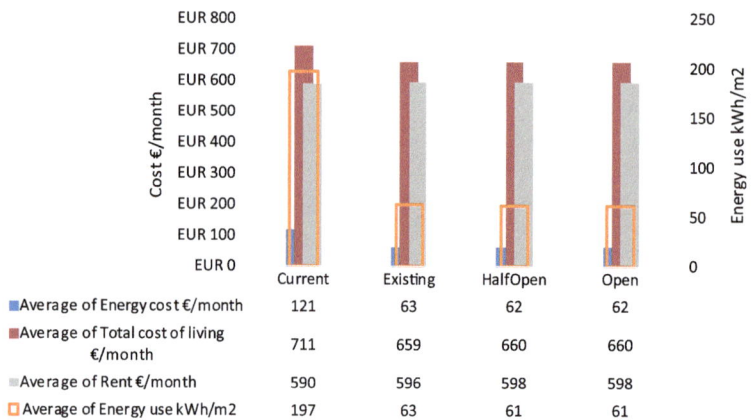

	Current	Existing	HalfOpen	Open
■Average of Energy cost €/month	121	63	62	62
■Average of Total cost of living €/month	711	659	660	660
Average of Rent €/month	590	596	598	598
□Average of Energy use kWh/m2	197	63	61	61

Fig. 26.2 Comparison of the Façade design variations and the current building, in terms of energy cost, rent, the total cost of living and energy use

analysis of the building characteristics and discussions with architects and housing associations.

Comparing the performance of the three façade designs, however, we can see that energy demand and, hence the energy cost, does not differ significantly, as shown in Fig. 26.2. This similarity can be explained by the thermal properties of the different options, which are all upgraded to high thermal resistance. It is also the reason why there is a 50% reduction in the energy costs and 68% reduction in the energy demand, compared with the current building. Moreover, the WWR is all three variations are relatively high, ranging between 60% and 100%. Therefore, the heat losses from the glazing, as well as solar heat gains are similar, resulting in similar energy use in the refurbished apartments. The choice of high WWR is consistent with heritage values of the existing building design.

Finally, the investment for the new façade, and the resulting rent increase is also similar, with the option of preserving the existing façade layout being marginally more economical. Nevertheless, the total cost of living is lower by 7%.

26.3.2 Thermal Properties Upgrade

The building envelope is upgraded with the application of insulation on the façade and roof, as well as replacement of the windows. The basic upgrade (B) is the minimum required by the regulations in the Netherlands, while the second option (A) is going towards zero energy standards. The main difference between the two options is the glazing and the higher thermal resistance of the roof. As can be seen in Fig. 26.3, the difference in the energy demand between the two variations is 5%, which is marginal. The marginal difference can be interpreted by the already good thermal performance of the basic upgrade. However, the investment for the more

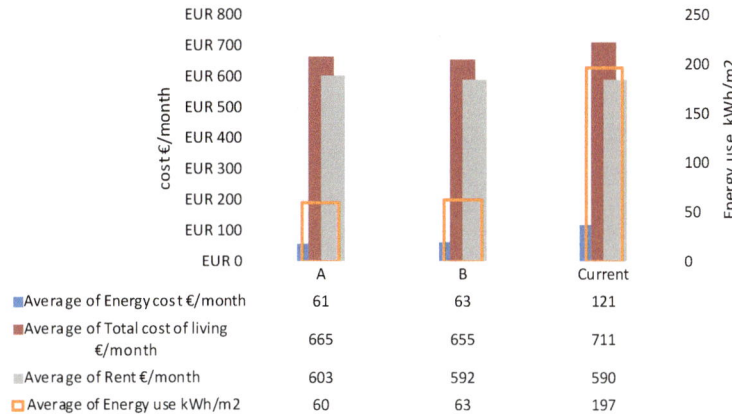

	A	B	Current
■ Average of Energy cost €/month	61	63	121
■ Average of Total cost of living €/month	665	655	711
▨ Average of Rent €/month	603	592	590
☐ Average of Energy use kWh/m2	60	63	197

Fig. 26.3 Comparison of the thermal properties upgrade options and the current building, in terms of energy

advanced upgrade has resulted in a rent increase greater than the energy cost savings. In this sense, the cost-effectiveness of the basic upgrade is better. It needs to be noted, that in both cases the saving to the current energy use is significant, as already mentioned.

26.3.3   Extension

The option to extend the living space is beneficial for improving the living conditions and functionality of the dwellings. Such examples range from the cladding of existing balconies to new construction. For the present study, the option considered included an additional construction, with mostly glazed external wall, having as a reference the project Tour Bois-le-Prêtre by Druot, Lacaton & Vassal. The new living space is not conditioned. Hence, the interior partition, previously external wall, featured insulated windows. Both interior and exterior windows are operable.

Figure 26.4 presents an overview of the KPI's with and without the extension construction, in relation with the thermal envelope upgrade. One of the first conclusions is that this investment does not affect the rent increase, as the average rent is the same. However, the energy use is higher in the dwellings with the winter garden. The higher energy use can be explained by the additional living spaces, which are not conditioned. The total cost of living in all cases is lower than in the current building.

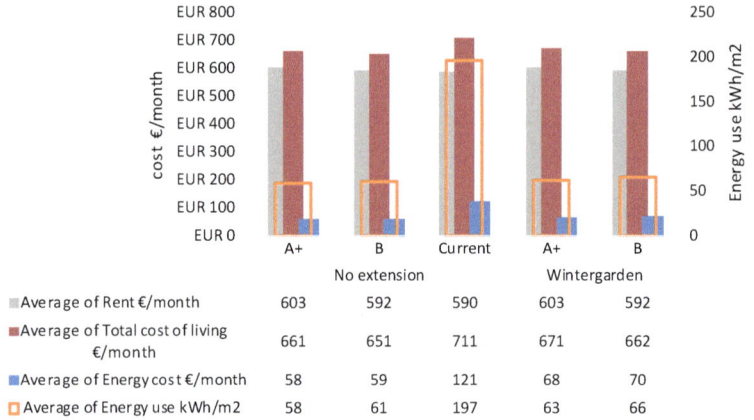

		A+	B	Current	A+	B
		No extension			Wintergarden	
■	Average of Rent €/month	603	592	590	603	592
■	Average of Total cost of living €/month	661	651	711	671	662
■	Average of Energy cost €/month	58	59	121	68	70
☐	Average of Energy use kWh/m2	58	61	197	63	66

Fig. 26.4 Comparison of the winter garden extension in relation to the thermal properties upgrades, in terms of energy cost, rent, the total cost of living and energy use

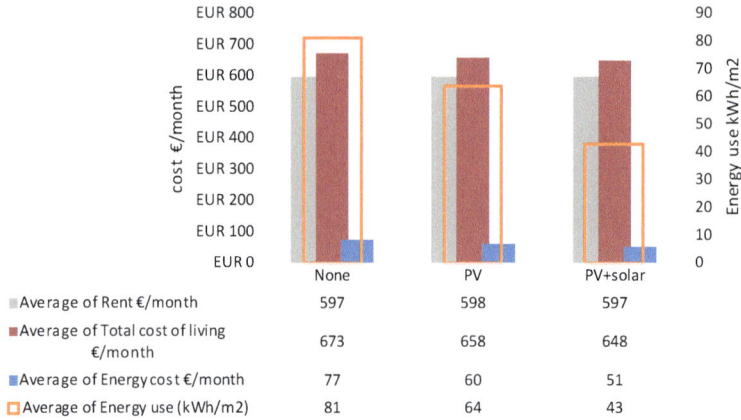

		None	PV	PV+solar
■	Average of Rent €/month	597	598	597
■	Average of Total cost of living €/month	673	658	648
■	Average of Energy cost €/month	77	60	51
☐	Average of Energy use (kWh/m2)	81	64	43

Fig. 26.5 Comparison of the thermal properties upgrade options and the current building, in terms of energy cost, rent, the total cost of living and energy use

26.3.4 Renewable Energy

Energy generation is a necessary step in the ambition to achieve energy neutrality on building level, and it is also a common consideration in energy efficiency upgrades. As shown in the results in Fig. 26.5, the application of renewable energy production technology can cut almost in.

half the energy use and 1/3 the energy cost. The rent, on the other hand, is not affected by the initial investment.

26.4   Discussion and Conclusion

The current paper described a methodology to combine the cost and the savings of energy efficiency upgrades in dwellings' refurbishment and identify the effect of design variation. Based on the aspects evaluated, the following main conclusions can be drawn.

- All tested combination of variables resulted in significant energy savings, up to 70%, due to the proposed the thermal envelope upgrade
- The variations in the façade design, given similar thermal properties, have a marginal effect on the energy demand
- The construction of a winter garden is possible without an increase in the rent
- Energy generation through the use of PV and solar collectors is cost-effective, as it has a considerable positive effect on the energy use and the energy cost, without increasing the rental price.

It is important to note that the savings on energy costs are greater than the capital burden of the energy-saving measures discussed in the current study. As a result, the total living cost to decrease in all cases. This conclusion is important to support the implementation of energy efficient measures; if the whole exploitation period is considered, the refurbishment is financially feasible, without burdening the household expenditure.

One of the main objectives of the study was not only to identify the effect the different parameters would have but also to inform the current practice in the context of energy efficiency upgrades of multi-residential buildings. To this end, the variations studied were selected based on commonly realized upgrades and focus groups with architects and users, and not in the interest of highlighting the effect on energy and cost. Thus, even if some of the variations result in non- significant differences for the KPI's, they are still valuable result to support decision making and provide options in the refurbishment strategy design.

The method presented in this paper was based on the energy efficient refurbishment measures and the specific KPI's. Other measures that may not be as cost-effective but do have additional environmental or living quality benefits, which can also increase the property value. These measures cannot be identified with the research method followed, which focused on energy efficiency. The conclusions on energy efficiency upgrades need to be considered both by the designers and other stakeholders, most importantly the occupants who will benefit of the reduced energy use, but also will need to pay the possible increase in the rent.

Acknowledgements The present article is the result of research being done by the Beyond the Current research group from May 1, 2016, to May 1, 2018. This work is part of the research programme Research through Design, project number 14569, which is (partly) financed by the Netherlands Organization for Scientific Research (NWO). The authors would like to thank all the involved parties for their collaboration and contribution to the project.

References

AEDES (2017) Aan de slag met de Routekaart CO2 -neutraal 2050 [Let's begin with the Roadmap CO2-neutral 2050]. Retrieved from https://ww.aedes.nl/artikelen/bouwen-en-energie/energie-en-duurzaamheid/vernieuwingsagenda/aan-de-slag-met-de-routekaart-co2-neutraal-2050.html

Artola I, Rademaekers K, Williams R, Yearwood J (2016) Boosting building renovation: what potential and value for Europe? Retrieved from http://trinomics.eu/project/building-renovation/

BPIE (2011) Europe's buildings under the microscope. Building Performance institute \Europe, Brussels

BPIE (2013) Boosting building renovation: an overview of good practices. Building Performance institute Europe, Brussels

Crawford K, Johnson C, Davies F, Joo S, Bell S (2014) Demolition or refurbishment of social housing? a review of the evidence. Retrieved from London: http://www.engineering.ucl.ac.uk/engineering-exchange/files/2014/10/Report-Refurbishment-Demolition-Social-Housing.pdf

DIRECTIVE (2010/31/EU) on the energy performance of building. THE EUROPEAN PARLIAMENT AND OF THE COUNCIL, Brussels

EcoQuaestor (2014) EcoQuaestor: de aanpak van Bouwprojecteconomie [EcoQuaestor: the approach to building project economy]. Retrieved 2017, from EcoQuaestor http://www.ecoquaestor.nl/

EN15603 (2008) Energy performance of buildings – Overall energy use and definition of energy ratings. European Commitee for Standardization (CEN), Brussel

Eurostat (2016) "Electricity and gas prices, second half of year, 2013–15" – Statistics Explained Retrieved from http://ec.europa.eu/eurostat/statistics-explained/index.php/File:Electricity_and_gas_prices,_second_half_of_year,_2013%E2%80%9315_(EUR_per_kWh)_YB16.png

Filippidou F, Nieboer N, Visscher H (2016) Energy efficiency measures implemented in the Dutch non-profit housing sector. Energ Buildings 132:107–116. https://doi.org/10.1016/j.enbuild.2016.05.095

IEAAnnex56 (2012) Cost effective energy and carbon emissions optimization in building renovation. Retrieved from http://www.iea-annex56.org/

Matschoss K, Atanasiu B, Kranzl L, Heiskanen E (2013) Energy renovations of EU multifamily buildings: do current policies target the real problems? Paper presented at the Rethink, renew, restart. eceee 2013 Summer Study. http://proceedings.eceee.org/visabstrakt.php?event=3&doc=5B-235-13

NEN (2017) Investerings- en exploitatiekosten van onroerende zaken - Begripsomschrijvingen en indeling (NEN 2699) [investment and operating costs of property -terminology and classification] in. Nederlands Normalisatie-instituut, Delft

Oorschot L, Spoormans L, El Messlaki S, Konstantinou T, de Jonge T, van Oel C, Asselbergs T, Gruis V, de Jonge W (2018) Flagships of the Dutch welfare state in transformation: a transformation framework for balancing sustainability and cultural values in energy-efficient renovation of postwar walk-up apartment buildings. Sustainability 10(7):2562. https://doi.org/10.3390/su10072562

Platform31 (2013) DOCUMENTATIE SYSTEEMWONINGEN '50 -'75. Retrieved from http://www.bouwhulp.nl/systeembouwwoningen%2D%2D-sev.html

Chapter 27
Analysis of the Energy-Saving in the Conference Center Atrium

Yamin Jamin Guan, Yimin Sun, and K. Xia

Abstract This paper investigates the influence of the Conference Center atrium design on energy consumption from an integrated perspective. Computer simulation techniques were used to assess the effects of the Stratification Cooling (SC) and passive energy-saving. The simulation results indicate that the effectively designed air supply system can perform two major functions: separation and utilization of natural air to help reduce the refrigeration zone and cool the upper zone. The simulation results of a case study show that the composite design method can make the air conditioning operation in the atrium of the hotel more energy efficient, and maintain a comfortable ambience in summer season. This is recommended.

Keywords Energy-saving · CFD · Conference Center atrium · Natural air ventilation

27.1 Introduction

What is a Conference Center atrium? Atriums are always located in the major axes of buildings. The modern building atrium has two attributes. It has indoor space specifics, (it is a major part of the whole building, a hub; public activities include arrival and distribution, multiple-function rooms), and outdoor specifics (multiple

Y. J. Guan
School of Architecture, South China University of Technology, Guangzhou, China

Y. Sun (✉)
State Key Laboratory of Subtropical Building Science, Guangdong Provincial Academy of Building Research Institute, Guangzhou, China

K. Xia
Guangdong Provincial Academy of Building Research Institute, Guangzhou, China

© Springer Nature Switzerland AG 2020
R. Roggema, A. Roggema (eds.), *Smart and Sustainable Cities and Buildings*,
https://doi.org/10.1007/978-3-030-37635-2_27

Fig. 27.1 Atrium photo

stories and natural scenery). A perfect atrium design should embody two kinds of functions: to express a theme entirely and to have unique characteristics.How about its shape? An atrium tends to be a tremendous space, which can create a natural atmosphere (Fig. 27.1).

The differences between atriums and other interior spaces can be clarified: a courtyard has no roof, a departure lounge does not have specific outdoor features, a shopping mall's long shape space has no scenery function, and the appreciating function is lacking. The Conference Center atrium is the place which requires more landscaping, to improve the image of the Conference Center and to attract more customers. Given the demands of the Conference Center atrium, as compared with other functioning buildings, it is open every day and emphasizes landscaping. Therefore, we should think about a different method for air distribution.

The disadvantage of this space is its huge energy consumption, especially in summer. However, it is based on the theory of stratified air conditioning, which states that only the lower space of the atrium needs to be cooled. Researches have pointed out that in summer the load of refrigeration (stratification) compared with refrigeration (total room) is about 2:3 (Shi 2011; Li 2011). Moreover, due to the large size of the atrium (eg. 40 m in diameter $\times$ 20 m high) nearly 25,000 m3, cooling load is still a great burden to the Conference Center's operators, because energy saving is inadequate. There are many researchers who are concerned about different functions of an atrium beyond a Conference Center, and most of them consider that the transition seasons can only use buoyancy driven natural ventilation. Some even try to find out if adding a chimney above the atrium can promote the chimney effect (Hussain and Oosthuizen 2013; Liu 2009). Using the atrium itself as an approximation of a tall chimney to help energy savings and integrate active with passive energy-saving methods, both of which are used in the hottest summer day to deal with largest consumption, is the topic of this research paper, the first of its kind. This paper will begin with reviewing some theoretical research to see whether it works well in real situations.

27.2   Background Investigation 1

Stratification Cooling (SC) is based on the theory of Thermal Stratification (TS). Some researchers (Torrance 1979; Rees et al. 2001) use two models to demonstrate that there are TS gravitational waves indoors. These two models are laminar flow and turbulent flow.It shows that through some methods (such as using correlation flow), there will be a few strata temperature zones indoor (horizontal layers).

The goal of SC is to cause TS, for which only the lower zone needs to be cooled. This paper will try to explain the reduction of the impact of the heat shift from the upper zone to the lower zone to save energy. A simple prototype is assumed to simulate the situation of using a glass tube filled with hot and cold water, and a glass lens to separate the hot water above from the cold water beneath. This was used to simulate the separate correlating air masses in the atrium (Fig. 27.2).

By looking at Fig. 27.2, the upper hot water will transfer its heat to the lower cold water. After a long time, the temperature of these two parts will become the same. If we continue to cool the cold water, it also cools the upper part. We use this water model to simulate the situations of SC indoors, beyond some steady loads (energy consumption such as building envelope load, internal fever, fresh air load, etc.).

In the atrium, there is convection loss and radiant heat transfer (by Gao 2007). The principles of the model about water separated by glass in the tube and the air curtain in the atrium are almost the same.

Fig. 27.2 Prototype tube of separate cold and hot water

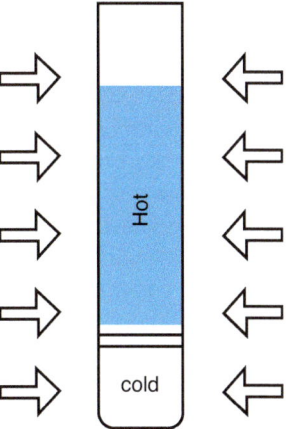

This shows that an important factor of Energy-Saving of SC is the main reason to affect the ratio of time periods between cooling the upper zone and the lower zone. It further indicates that we must consider how to cool the upper zone with less energy, and how to reduce its negative impact on the whole load in the atrium.

27.3   Background Investigation 2

The theory of the Chimney Effect is based on different air pressures between outdoors and indoors, due to the cold air infiltration into the hot air. The air on the top is usually warmer than the air below indoors. With the Neutral pressure plane, the outside air is pulled into the atrium from the bottom, then rises to the top and ejects into the sky. We can define the pressure as in Eq. (27.1).

$$\Delta P = 3460 \left(\frac{1}{T_0} - \frac{1}{T_S} \right) h \qquad (27.1)$$

where T_0 = indoor temperature at the top; T_S = out-door temperature, h = height from the neutral plane of atrium; and ΔP = pressure difference values. (Zhu et al. 2008). e.g. indoor 36.2°C for T0, outdoor 31.2°C for Ts, 11 meter high atrium for h/2), the ΔP is 83.72 Pa.

Can the chimney effect help to cool the upper zone? According to published research by Gao (Gao 2007), the effect is limited and little. His study is based on a CFD (Computational Fluid Dynamics) simulated model, Up-supply Down-return (common modern cooling system), combined with the natural ventilation. His conclusion is as following:

The difference between the two methods is that the temperature decreases by 0.1°C in the lower zone while the upper zone drop is more than 3°C. This makes us confused, why is the chimney effect invalid?

Investigating the conditions of his model, apparently the author determined that the height factor can make no difference. His method has no natural air autocycle method and a lot of heat air shifting into the refrigerated area. The upper zone has many loads, including the envelope glass wall which has been warmed by the outside environment that is producing high thermal radiation. In the upper zone, the flow of air is contrary to the former, and the hot air does not eject.

From the above, we have every reason to say that this method fails to effectively promote the discharge of the heat of the upper air. A method that takes advantage of the natural ventilation method to reduce energy consumption is needed. The air flow should be organized.

27.4   Methods

In this research, a case study was selected to demonstrate the feasibility of cutting down the summer consumption. The experiments are taken both by measurement and simulation that include dates of Temperature and Velocity. And performed by

HD and Hobo, or Airpack (a CFD software). Three sides of the atrium are surrounded by floors, except the north side. It is a greenhouse style atrium in a typical subtropical seaside Conference Center in Shenzhen city, whose north side is enveloped by a glass wall, on which the sun shines directly and which is affected by the surrounding environments (Fig. 27.1).

27.4.1   Improving Method One

Firstly, we have to figure out whether the original cooling system is suitable or not. Here are the data which are processed by survey of instruments. The former air distribution is shown in Fig. 27.3. The experiment is conducted at hottest summer day, (2018/06/17).

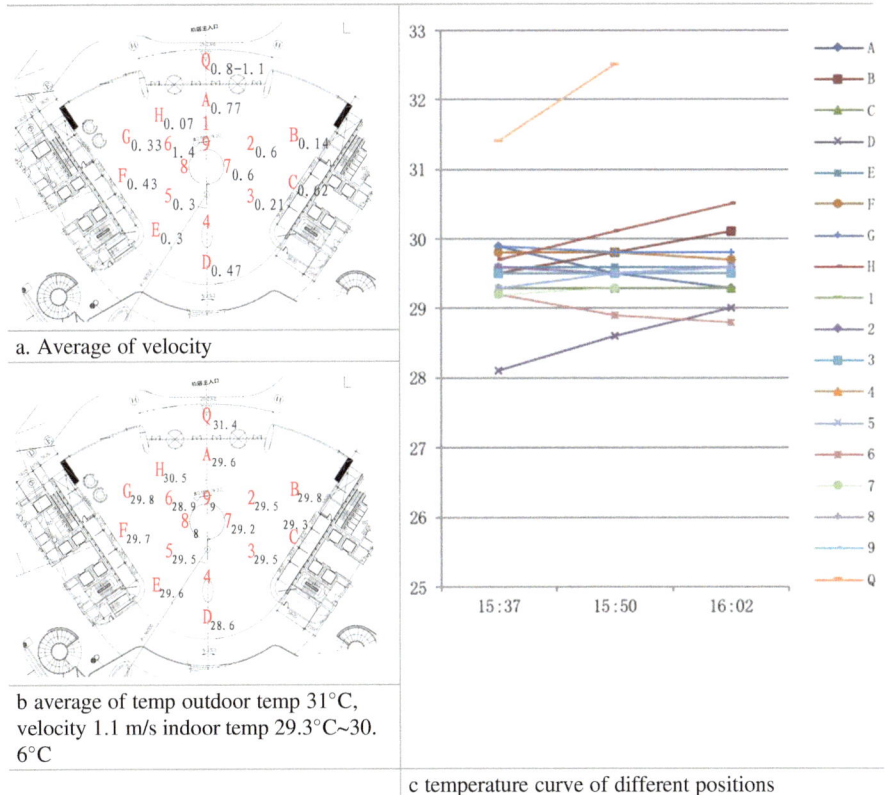

a. Average of velocity

b average of temp outdoor temp 31°C, velocity 1.1 m/s indoor temp 29.3°C~30.6°C

c temperature curve of different positions

As shown in Fig. 27.6, point 6 has the largest velocity, which is 1.4 m/s. This indicates that almost each ray of air flow converges at this point, while it also has the lowest temperature which is 28.9°C. Δ°C = 1.2°C; ΔV = 0.21~0. 6 m/s. The temperature and velocity are varied in the atrium and that indicates there is not

Fig. 27.3 Middle-supply and down-return (this method is being used mostly currently). *Photo of air distributor*

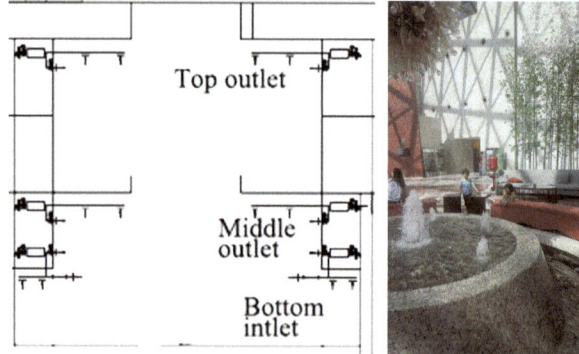

Top outlet

Middle outlet

Bottom intlet

HOBO   HD

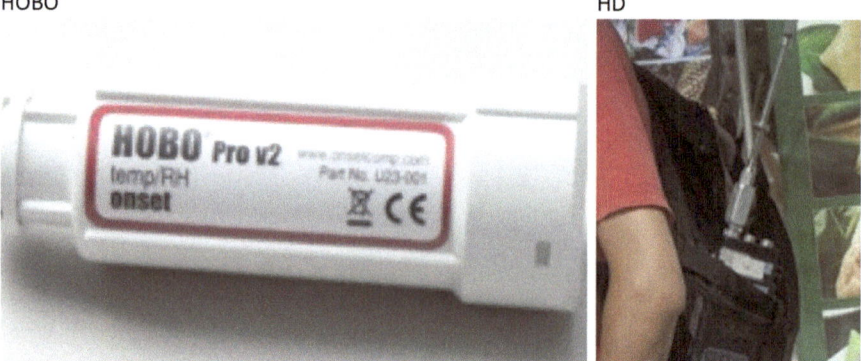

Fig. 27.4 Equipment of HOBO and HD

quite enough comfort in the atrium. Meanwhile, the energy waste deserves more concern.

If the Conference Center atrium uses different air distribution systems, such as an air supply column tube, it can greatly reduce the cooling load. Therefore, we have more than one choice (now cooling air is delivered into the middle from two bound walls). By setting the return inlet at 2.5 m high, it can improve the capacity of energy savings. Compared1 to the traditional middle supply mode, the dwindle range of cooling was shown in the shadow (Fig. 27.4).

This is integrated design, and it can either succeed or fail based on the satisfaction, usefulness, and aesthetic. So, next we analyzed the site function in the atrium to determine whether it is possible to use the integrated air supply system.

The atrium has a variety of functions, and yet is difficult to be clearly separated into different parts. The Conference Center has four functions as mentioned before. The first three functions (Fig. 27.5) are related to areas, except the hub function of human traffic flow, in which it is possible to use an upward air supply system and the air supply column in the rest areas.

Yellow represents sea view corridor; while blue represents a path of evacuation, and white represents an active area for the hub. Pink represents the reception area,

Fig. 27.5 Reference of Air
supply column type

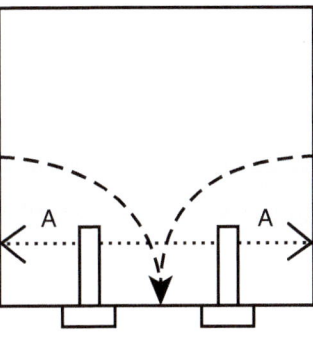

■Landscape display
function—bamboo,
pools, waterfalls,
potted plants
■Center transfer
function--reception
■Public distribution
and join together—
leisure area, coffee bar

Fig. 27.6 Atrium function

which is separated into two sides to ensure that customers can see the sea directly and clearly. By doing so, the designer wants to design spectacular scenery in the atrium.

The arrangement is as following: AHU equipment is set up in the basement right beneath the atrium center. Refrigeration air is supplied by six air columns, and then returned through six vents which are set on both side walls and one vent at the center fountain, then gathered to go back to AHU. Two electric windows are situated in the middle of north glass wall, to control the direction of the air. Two scuttles are placed on the roof, and one fan in the center to enhance the air flow. According to Fig. 27.5, air supply outlets can be uniformly placed (Figs. 27.6 and 27.7).

The outlets can be decorated in many methods to make them look beautiful, such as using vine climbing around the air supply column and then beside the bamboos, which are now set in the atrium (Fig. 27.8), and set the vent above the water or beside the potted plants (Fig. 27.8), or hide it under the vitreous table with shade loving plants. The flower can be seen through the vitreous table.

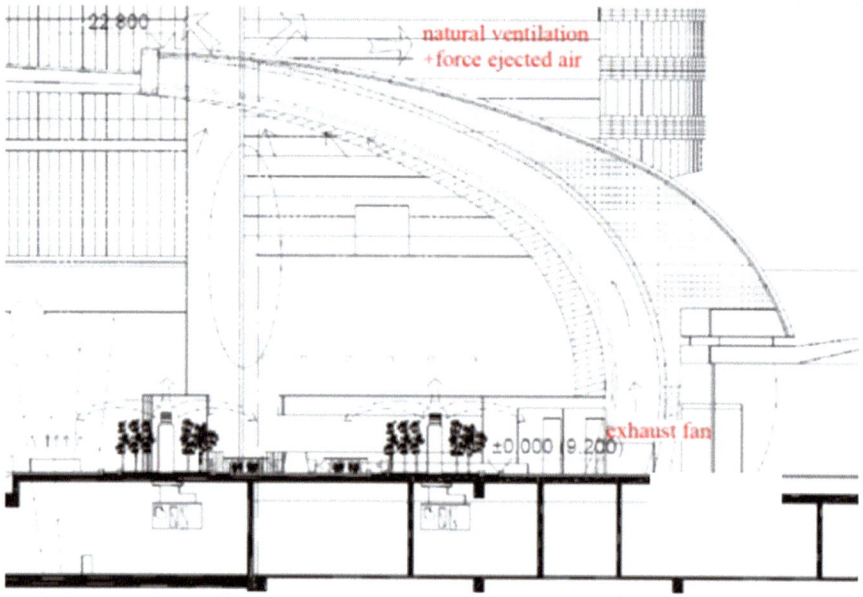

Fig. 27.7 Upward air supply system combine with fans

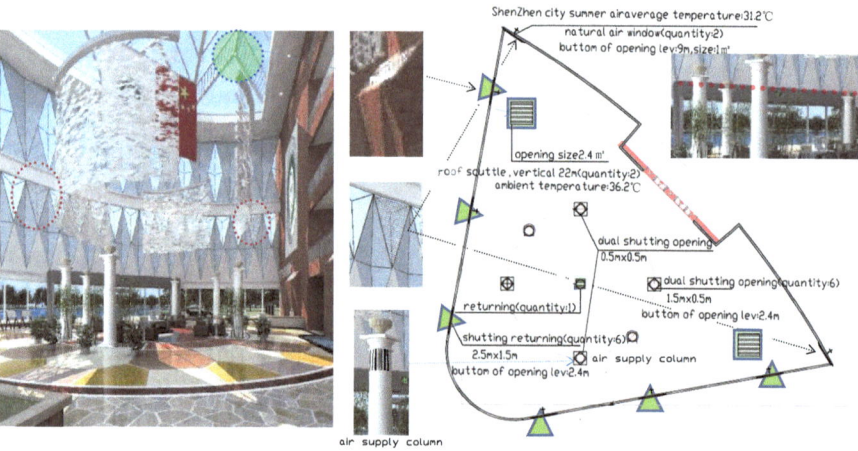

Fig. 27.8 Integrated air supply system combine with natural vent

Air supply columns seldom be used in Conference Center atrium, because of its ugly appearance, but when properly decorated and carefully arranged, the utilities can be merged into the scenery, and make the system more aesthetically acceptable. They can share the appearance (Fig. 27.8).

Next, we will discuss this to take one more step further to improve the energy saving by using the nature air cycle, based on the Method one.

27.4.2   Method Two

Our goal is to improve the chimney effect, so as to bring out the extract heat in the atrium. The negative influence should be avoided or decreased. To facilitate the model establishment, some adjustments have been added. The simulation of air-flow is performed by Airpack, which is a CFD program.

27.4.3   Research Mode (Table 27.1)

Table 27.1 Building database

	Horizon	Vertical	Long
Room	45 m	22 m	45 m
	Size	Quantity	Heat conductivity(outside temp)
Envelop			
Roof			78
Glass Wall-Front			50°C
Door			30.9°C
Floor			Ambient
Man and equipment			
Man	1.5 m	32	180 W
Lamp	0.5x0.5x1m	25	34 W
Copier	0.5x0.5x1m	2	85 W
Printer	0.5x0.5x0.5 m	2	160 W
Computer	0.5x0.5x0.5 m	4	65 W

27.4.4 Routing (Fig. 27.9)

There are two kinds of analogy shown in the Table 27.2. One is Natural-B method, which can be used in the transition seasons, and the other Down-C is used in hot summer.

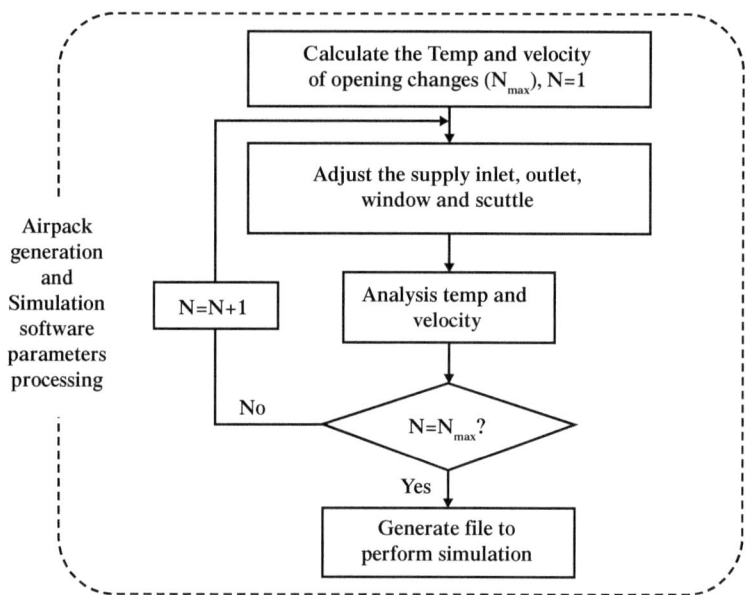

Fig. 27.9 Generation and simulation program flow chart

Table 27.2 The properties of Inlet and Outlet

	Position	Size	Parameters	Temperature°C	Vertical (m)	Quantity
Natural-B	Opening					
	Wall-F	4.6x2.4 m	Am-	31.2	10.5	2
	Vent					
	Roof	3.4x3.4 m	1.5 m/s	36.2	22	2
	Exhaust fan	1 x1m	0.5 m/s	Am-	22	1
Down-C	Opening					
	Colu-Fr	1.4x1m	5 m/s	19	2.7	2
	Colu-mi	1.4x1m	4 m/s	19	2.7	2
	Colu-Ba	1.4x1m	2 m/s	19	2.7	2
	C-M-in	1.4x1m	3 m/s	19	2.7	2
	Colu-Fr	0.7 x1m	3 m/s	19	2.7	2
	Door-a-c	3 x0.5m	3 m/s	25	2.7	1
	Vent					
	Wall-w &e	2.5 x1m	-50 pa	26	3	6
	Floor	1 x1m	-50 pa	26	1	1

27.4.5   Opening Size and Position

27.4.6   Simulation Process and Result

Five air inlet locations representing different situations: total heat gain (Off-A), natural flow with only middle window and top outlet (Nat-B), green house type atrium with air conditions (Down-C), Down-C with exhausting fan(Precedent) and integrated middle natural air inlet and artifact air conditions (All-D) are considered in the atrium, which reflect the cooling effects. The air flow organization is as shown in stimulated results.

Figure 27.10 shows the lower zone (0~2.5 m) $t_e^d = 32$~35°C, middle zone (2.5~9 m) $t_e^m = 36$~54°C, upper zone (9~20 m) $t_e^u = 54$~62°C.

Figure 27.11 shows $t_e^d = 30$°C, $t_e^m = 30$°C, $t_e^u = 36$~40°C. According to Fig. 27.11, the outside natural air flow goes inside, after that, most of the flow descends and is sucked through the roof outlet. The Velocity magnitude of enter air is 1.5 m/s; The speed of exhale air is 0.7 m/s.

Figure 27.12 shows $t_e^d = 22$~28°C, $t_e^m = 35$~40°C, $t_e^u = 49$°C.

Figure 27.13 shows $t_e^d = 22$~29°C, $t_e^m = 35$~39°C, $t_e^u = 45$°C.

Figure 27.14 shows $t_e^d = 21.5$~28°C, $t_e^m = 28$°C, $t_e^u = 33$~40°C (Figs. 27.15 and 27.16).

With proper arrangement, the negative effect can be decreased. Furthermore, air distributed uniformly, as opposed to use middle-supply and down-return system, makes the air temperature converge, and improves thermal comfort (Figs. 27.17 and 27.18).

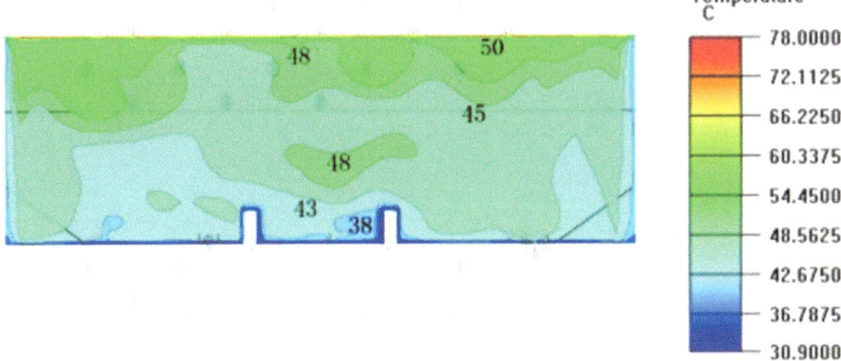

Fig. 27.10 Off-A

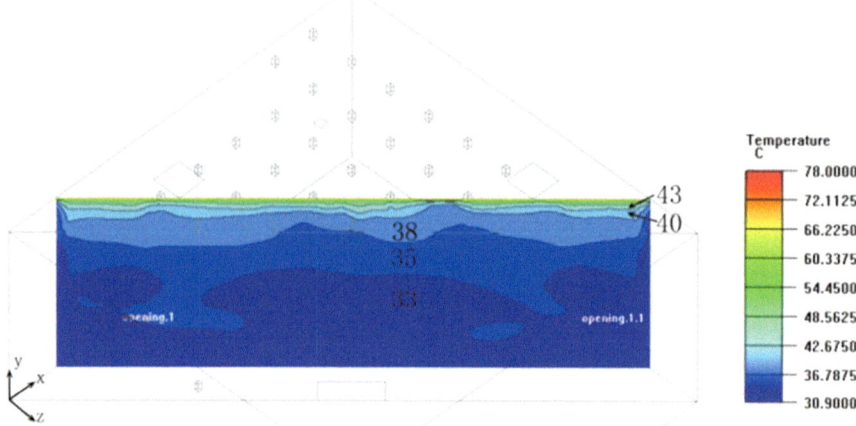

Fig. 27.11 Nat-B

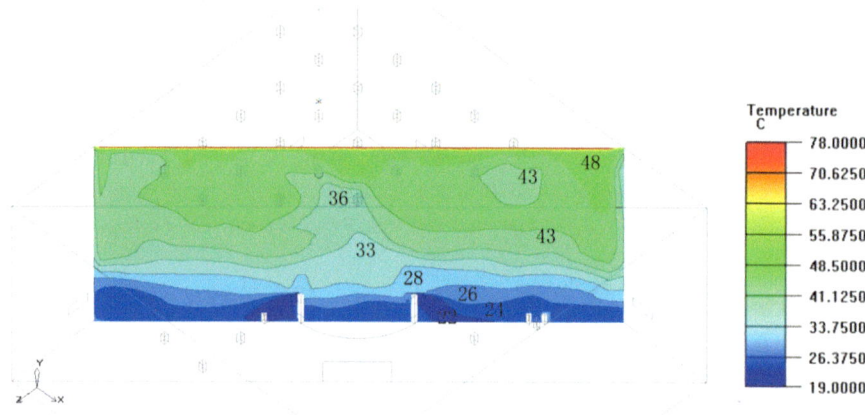

Fig. 27.12 Down-C

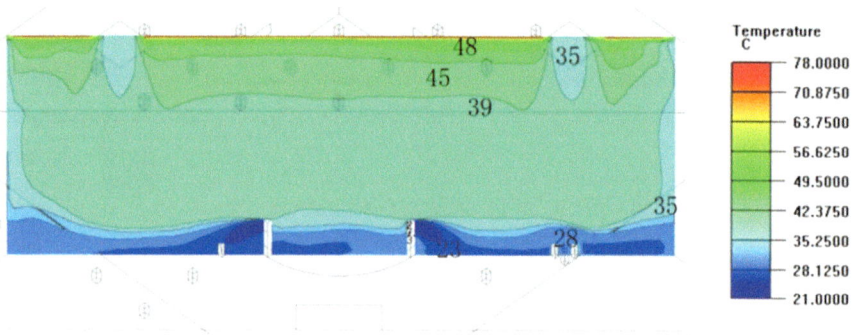

Fig. 27.13 Precedent

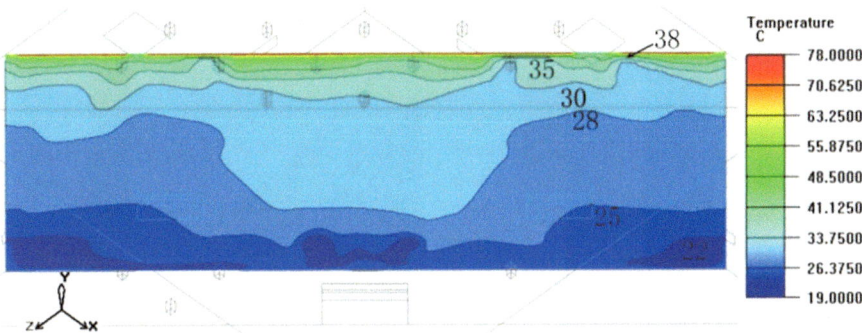

Fig. 27.14 All-D

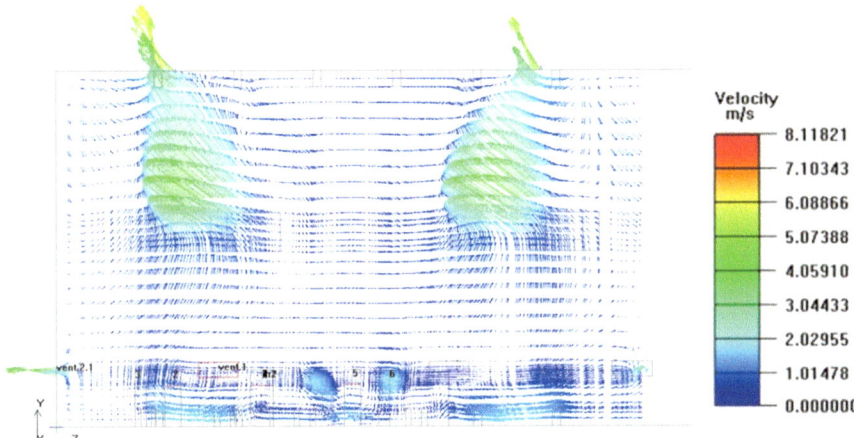

Fig. 27.15 Vertical velocity of method All-D

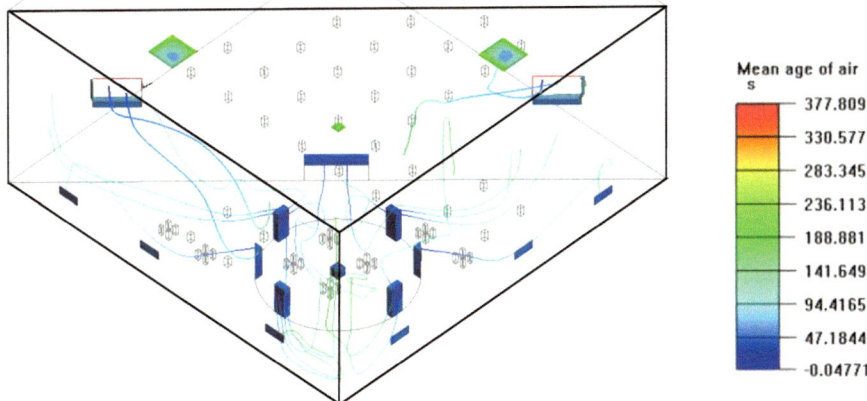

Fig. 27.16 Mean age air

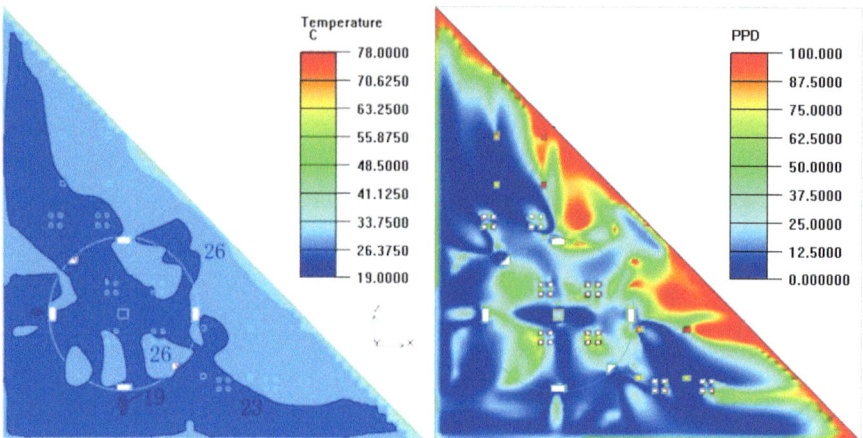

Fig. 27.17 Temperature (left) & PPD (right) of 1.5 height

Fig. 27.18 Temperature curve of five methods

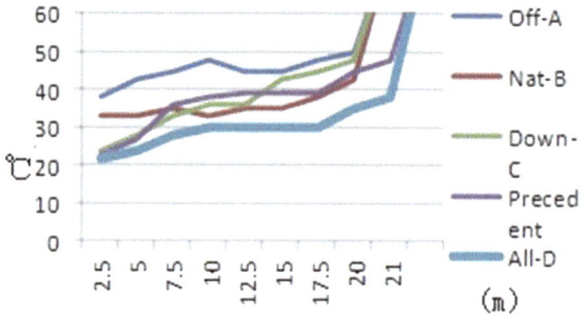

27.5 Results Analysis

The simulation results of total heat gain without cooling facility (Off-A) indicate that there is a temperate ladder in the atrium, therefore, we use the natural air ventilate (Nat-B), under the pressure of chimney effect, 2~5°C (middle-zone), 13°C (upper-zone). This kind of middle opening method barely influences the lower zone, but cooling the upper zone is a big deal. When (T0 - Ts) = 5°C, even without addition facility, the air rising is still significant.

27.5.1 The Effect of Mechanical Facility

There is a lot of research using the exhaust fan to eject the hot top air outside, aiming to decrease the cooling load. According to Fig. 27.14, partly refrigerated air in the lower zone has been sucked to scuttle directly 3°C, 4°C. The result is similar to

Gao's experiment. The height factor has not been fully considered, the only difference between use of middle-supply, and down-return system and supply column, is that the air temperature becomes more converged.

27.5.2   The Effect of the Integrated Method

Comparing the conditions of (ALL-D) and (Precedent) 0.5~1°C (down-zone), 7~11°C (mid), 12~5°C (up). Usually a Conference Center operation schedule is more than 14 h per day, some famous Conference Centers such as Sheraton are operational 24 h. Based on long term operation, reducing the upper zone total heat gain could tremendously cut down the whole load in the atrium. After the temperature becomes steady, we could increase the supply air temperate 1°C, with the associate of cooled upper zone, the lower zone can keep the comfortable environment. In long hot summer of Shenzhen, with more than 14 h continued working, the lower 1°C supply air can save-energy greatly. Moreover, it is a passive method.

The best air conditioning design in this case had been figured out. It is significant, because it is in summer. The simulation results prove the assumption of glass tube which was discussed before, even though there is no way to avoid the negative effect of natural air partly influencing the refrigerant zone, as shown in the velocity, and most of the hot air in the upper zone is ejected out. Another positive effect is that the natural air forces the hot air to follow the curved glass wall on the north side, which separates the heat from the outside as shown in Fig. 27.15. Certainly, that can eject the exhaust air outside at the top (which can be a combined design with fire served system) to extinguish the stagnate layer. As shown in Fig. 27.15, the method can make the upper zone (above 2 m high) form a continuous flow of air to utilize the chimney effect. That means the method further effectively reduces the upper zone temperature.

One thing which had to be noticed is that there is a misunderstanding in the SC system as to what should be used for the lateral supply down return organization (Fig. 27.19). However, this has led to a contradictory situation of using the chimney effect and natural ventilation. As seen in Fig. 27.20, by Lu (Lu 2008), he considers

Fig. 27.19 Usually
method. By Lu (Lu 2008)

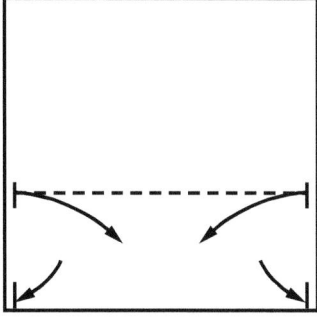

Fig. 27.20 Combine
method. By Lu (Lu 2008)

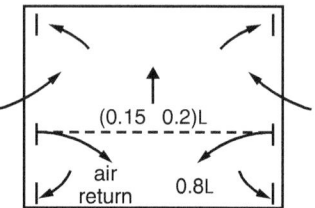

this will take advantages to fully use lower temperatures contained in extracting
waste gas which can help to cool the upper part. But we already know that there is an
air pressure coming from outdoor that will interrupt the flow organization because of
the middle supply system, forcing the part of cool air to rise up, and increase the
cooling range and refrigerant load. It is no wonder that this method fails to properly
use the natural ventilation if we use the column supply system instead, while setting
fresh air inlet higher than the cooling area. As a result, it will reduce the whole air
flow, and can also receive the SC effect to make sure just the lower area is cooled.
With the upper area is cooled by the chimney effects, moreover, it can save a lot of
energy.

27.6 Conclusions

As with the CFD stimulated case study, passive means combined with active means
have two main advantages: separation and utilization of natural air buoyancy.
Firstly, the cooling zone only covers the lower part, which, to be specific, covers
two meters height of human activity areas, as opposed to use of a nozzle set at 6 m or
even higher. It can reduce the range of refrigeration more efficiently, almost half in
this case. As is well known, lower height air with lower temperature returning to the
cooling below saves energy; Secondly, natural air performed energy-preservation is
cooling the upper zone.

Using these two methods, less investment and saving of energy can be achieved.
The integrated refrigeration system is not only beneficial for saving energy largely,
but also provides fresh air with comfortable thermal environment that is attractive
and easily accepted by people.

The further study will be a comparison of the energy consumption between the
original design and the improved method. Hopefully, we can see exactly how much
energy is saved using the latter method.

References

Gao J (2007) Research of the theory of thermal stratification in buildings and its applications. Harbin Institute of Technology, Harbin

Hussain S, Oosthuizen PH (2013) Numerical investigations of buoyancy-driven natural ventilation in a simple three-storey atrium building and thermal comfort evaluation. Appl Therm Eng

Li CC (2011) Ventilation and energy saving strategy for large space buildings. China Architecture & Building Press, Beijing

Liu PC, Lin HT, Chou JH (2009) Evaluation of buoyancy-driven ventilation in atrium buildings using computational fluid dynamics and reduced-scale air model. Build Environ 44(9):1970–1979

Lu YQ (2008) The practical heating and air conditioning design handbook. China Architecture & Building Press, Beijing

Rees SJ, McGuirk JJ, Haves P (2001) Numerical investigation of transient buoyant flow in a room with a displacement ventilation and chilled ceiling system. Int J Heat Mass Transf 44 (16):3067–3080

Shi LJ (2011) Study on load in upper and lower zones of stratified air conditioning Technology of Terminal. Chongqing University, Chongqing

Torrance KE (1979) Natural convection in thermally stratified enclosures with localized heating from below. J Fluid Mech 95:477–495

Zhu J, Huo R, Sun X (2008) Experiment study on the influence of side opening in the vertical shaft. Eng Mech 25(9):188

Chapter 28
Sharing Urban Renewable Energy Generation Systems as Private Energy Commons

Craig Burton, Seona Candy, and Behzad Rismanchi

Abstract This study tested a new methodology for simulating shared electricity generation among small groups of neighbours with Ostrom (1994) principles of common pool resource (CPR) (human behaviour-based) efficiencies. The approach does not anticipate exclusive off-grid communities, but instead, diverse energy users taking advantage of the averaging effects of aggregation, the social benefits of a CPR, and direct action on emissions. The study tested three groups of 5 adjacent- or same-building-neighbours for 3 months, to measure how electricity demand (import) is affected by an in-home display issuing nudges and sanctions by the group around a simulated (limited capacity) shared solar and battery system. A control group of 6 homes' energy data was obtained for the same period. All three groups reduced their energy demand with weak but significant correlation between stimulus and reduced energy demand, and one group significantly shifted demand toward available shared solar energy resources during the intervention.

Keywords Solar power · Commons theory · Microgrid · Behaviour change · Energy sharing

28.1 Introduction

The present trajectory for the Energy Transition in Australia in the absence of substantial demand management for electricity will see a doubling of demand to 2050 (CSIRO and ENA 2015) and the likely scale of renewables will be twice the nameplate capacity of dispatchable generators (AEMO 2013) in order to compensate

C. Burton (✉) · S. Candy
Faculty of Architecture, Building and Planning, University of Melbourne, Melbourne, Australia
e-mail: cburton2@student.unimelb.edu.au; seona.candy@helsinki.fi

B. Rismanchi
Faculty of Engineering, University of Melbourne, Melbourne, Australia
e-mail: behzad.rismanchi@unimelb.edu.au

© Springer Nature Switzerland AG 2020
R. Roggema, A. Roggema (eds.), *Smart and Sustainable Cities and Buildings*,
https://doi.org/10.1007/978-3-030-37635-2_28

407

for variability. That is, four times the generation power of 2018 carbon-based generators may be needed in 2050. The transition to very low carbon has already been estimated to cost AUD1,140 billion (CSIRO and ENA 2015), a large component of this cost being met by private solar owners.

Accordingly, some form of demand management is often assumed to occur in popular "zero carbon" plans (for example Ison and Lyons 2013) to the extent that electricity demand is expected to be *halved*. Usually, this is proposed to occur through efficiency gains. However, a turnaround in demand of this magnitude would be unprecedented (Kelly 2010). This electricity demand history is already moderated by long running government programmes for energy efficiency, behaviour change, domestic solar stimulus, electricity price increases, and other effects that should suppress demand. These are delivered via energy saving information (DELWP 2014), but also efficient device labelling (DEE 2018), efficient product rebates, solar tariffs (DELWP 2016), and others. The motivations for these approaches include that energy consumers can change their habits. However, the literature increasingly does not support this.

28.2   Demand Management

Demand management is a mature field which has rested heavily on Bandura's Theory of Planned Behavior (TPB) (Bandura et al. 1977). The fundamental premise of this approach is that consumers are rational and when served with information and incentives, they will choose to consume differently. Studies of studies (Delmas et al. 2013) find that the largest effects of very many forms of demand management signals based on TPB seem to have a final impact on actual electricity demand of only 10%, usually less, and with many studies having identified methodological problems (also Abrahamse et al. 2005). That is, interventions ranging from rewards, to goals, penalties, pecuniary incentives, home audits, and others do not have large effects on actual demand, and often, the removal of the intervention results in demand returning to where it was. Most recently, a campaign for low-income families in several cohorts across Australia (n = 18,889) with respondents self-reporting changed energy practices (30–80% reported changes) after efficiency interventions were able to reduce their demand by only 2–12% (Russell-Bennett et al. 2017). Consumers are informed, they make changes to their routines, but the returns in avoided energy demand are small.

The challenge of demand management likely has at least four impediments: campaigns typically deal with the consumer in insolation, in the absence of their social environment or even family environment (for example Lowe et al. 2015); electricity is offered as limitless, cheap and highly reliable (Abbott 2001); electricity use is very difficult to disaggregate from social practices (Shove and Walker 2014); and, there are no consequences for high demand - only higher electricity bills.

There are of course some notable exceptions to these. High-touch interventions such as personal in-home audits perform well compared to others (Delmas et al.

2013). High-tech interventions such as in-home displays (IHDs) have been observed to bring about a 20% reduction in electricity demand (Gans et al. 2013) and finally, the subject of this study, social approaches to sustainability (EcoTeams, energy communities), have established long running energy conservation behaviours that even improve in time (Hargreaves et al. 2008). Reframing within social structures already seems successful for demand management in water. Prior to the Victorian Millennium drought (1999–2009) water was felt to be inexpensive, unlimited and reliable (Allon and Sofoulis 2006, p. 49) - as electricity is presently understood by consumers. Melbournians discuss water now as a limited and valuable resource after a large utility-led demand programme (Liubinas and Harrison 2012). Can such a re-framing be achieved for electricity? This may certainly be harder to bring about because the visible impact of the drought has no equivalent for energy, and even the very likely climate change effects behind the large bushfires of 2009 were attributed to ocean currents (Australian Bureau of Meteorology 2014) and not human-forced climate change. Simulating a re-framed electricity supply may cause greater conservation, but the re-framing itself needs to be compelling.

28.3   Commons

This study proposed another way to potentially re-frame the electricity supply as a means to conservation - that electricity should be presented as a Common Pool Resource (CPR) which is managed by a commons system. Traditionally, CPRs have been natural resources which have limited productivity and have to be managed by those who appropriate from them. Not managed properly, the resource is overused and it collapses. Elinor Ostrom's work *Rules, Games and Common Pool Resources* (1994) spells out eight design principles for commons which have since been rigorously tested in hundreds of studies (Cox et al. 2010). It is also promising that rural and isolated distributed renewable energy systems can be successfully shared and that the naturally limited performance of the systems can be accommodated in behaviour changes (for example see Gardiner 2017). Similarly, commons systems successfully self-manage shared pasture, fisheries, and forests. Can there be urban energy commons?

28.3.1   Commons Theory

Ostrom's principles are derived from observations of (very broadly): borders, appropriation and maintenance, rule-making, monitoring, sanctions, conflict resolution, government interference, and scale. The definition of a managed commons is still not broadly understood - Wikipedia is not a managed commons: it is certainly managed and has its own system of governance and rules, but anyone can consume a Wikipedia article without impacting other users (it is not rivalrous, and there are no sanctions for misusing it). This makes Wikipedia a *public good*. Less certain are club goods or toll goods (Bollier 2014) such as a collective solar and battery system that is owned in shares. This is not a commons according to Ostrom's principles, because access to the toll good is artificially limited (by access limited to owners) and non-rivalrous (until there is congestion) (Ostrom et al. 1994, p. 7). Instead the energy commons (the *non-stationarity*) was defined in this research as the collective energy sharing agreement itself. To test this, the applicability of commons principles for sharing energy, principles 4 and 5, *monitors* and *graduated sanctions* (respectively) were chosen for testing, because these principles were most often absent from public goods and toll goods management.

Electricity use, at present, occurs without any kind of signals about limits, let alone reciprocity. That is, one can use a large amount of electricity and, apart from an eventual large bill, there are no consequences. Since electricity costs no more than fifty cents a kWh, it may cost fifty dollars to run a large air conditioner in summer. In reality, running a large air conditioner on the hottest days of the year can cost the utility *AUD1,500* (Wood et al. 2014, p. 9) to deliver the electricity *to one air conditioner*. This is because in fact, electricity *is* rivalrous at the extremes and the cost of peak demand infrastructure (to support the grid for the top 10 demand days a year) exceeded 45 billion dollars over the last 10 years (Hill 2014). This signal, however, is completely hidden from residential consumers and this high cost is spread across all electricity users, whether they use air conditioning or not.

The energy source to share in the proposed commons is a fictitious shared solar and battery system that a group of adjacent neighbours will operate and use. This system could be placed on the larger roofs, and batteries could be installed on some of the other houses or where there is space. They are privately connected together (on a private easement) or there is some arrangement with a retailer to allow the exchange of electricity to happen among them via the distribution system. The homes (or in fact flats in the same building) are adjacent because a real private easement would require this, but also there is likely social cohesion among such homes which is identified as valuable for CPRs (Bollier 2014). Avoiding opt-in for energy sharing (instead recruiting with a spatial constraint) should reduce some self-selection effects in energy collectives (Bauwens and Eyre 2017).

Thus, the methodological approach considers the impediments listed above and attempts to meet or avoid them: the intervention targets the group as the subject, not the individual home - it is asserted here that this will take advantage of group effects. Second, the electricity supply is re-framed to a limited and valuable resource; the intervention does not rely on signals that target disaggregated behaviours; and, a social sanctions approach is proposed (as part of a commons system) so that there are social outcomes for using electricity at the extremes. The apparatus is an in-home display (IHD), a form of energy use feedback that has rendered promising results for self-management of electricity demand.

28.3.2 In Home Display

In order to signal to home occupants that they are draining a rivalrous, limited electrical supply, an in-home display was built for this study. The device is a Raspberry pi microcomputer with a 7-inch touchscreen in a 3D printed case. The computer has a 3G cell modem and a Radiohead low power packet radio receiver. An accompanying smart meter reader was created that counts the LED flashes on two brands of smart meters and radios the count to the IHD in the home. Twenty of each device (IHD and reader) were made. This setup is similar to the proprietary Watt's Clever IHD system (SmartUser 2018) except that the study system is a networked colour touchscreen display with Linux operating system.

Fifteen homes took part in the study, recruited via "champions" who live among them and were reached via personal and professional networks. Five homes at one site were in a CBD apartment building. Two other sites were five homes each, freestanding 1920–1930's era renovated wooden dwellings. About half the free-standing homes had solar panels, but the study disregarded these "real" systems and simulated a larger, shared system with battery modelled for each group. The modelling was performed with a numeric solver, accepting the half-hourly aggregate demand of all participants in the group from 2017 along with previous solar records, energy costs, and equipment CAPEX costs. The model produced solar and battery sizing and modelled performance for each group. This was used to configure each IHD. The size of the modelled system was broken down into unit holdings propor-tionate to the household demand size of 2017, so that large energy consuming households would not be unduly penalised (and small consumers would not be unduly rewarded).

Software was written for the IHD and a central server. Each IHD used its 3G modem to reach the central server and report hourly on interactions the IHD experienced with the occupants. Each IHD had a reinforcement schedule programmed to deliver signals and interactive prompts. These are given in Table 28.1 below. To norm the system, government-provided energy tips (Sustain-ability Victoria 2014) were programmed to be delivered over 20 days along with energy use information to make sure the IHD had the impact observed of other IHDs elsewhere. Signals about the presence of the group and the limited availability of the system generation began after 20 days, and from 40 days, the system detected

Table 28.1 Signals, schedule and function

Signal and schedule	Form and purpose
Tips 1 to 10, every 2 days	Energy saving information
Sanctions 1 to 3, triggered for 20–40 days	Solicit for fines for an over-using household then report the fine
Rewards 1 to 3, triggered for 20–40 days	Solicit for rewards for an under-using household then report the award
Satisfaction question every 5 days	Prompt for satisfaction with the service to detect problems

Fig. 28.1 Negative and positive sanction triggered by home energy demand cause solicitations to the other group members, who may choose to fine or reward the household. An avatar, Rachel, reports the outcome to the affected household

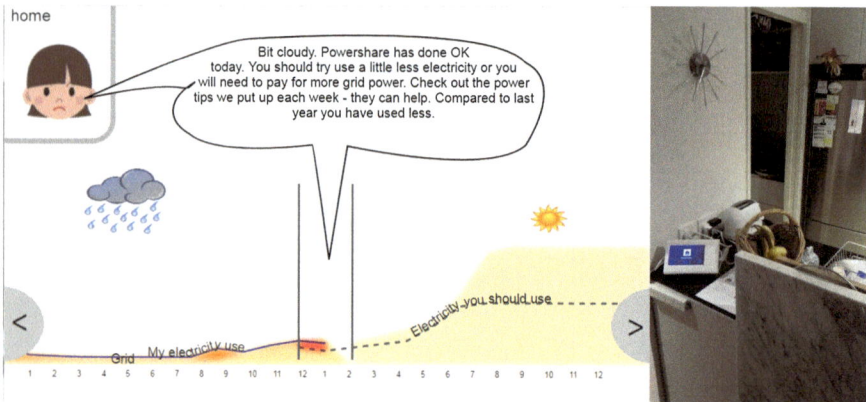

Figs. 28.2 IHD screen capture showing ten different features to reinforce conservation and shifting of demand loads toward available simulated solar and battery power. (1) A centre "goals" area of the graph shows immediate situation (at 1 pm) with (2) prediction at right (3) weather, (4) demand and battery/solar) and (5) performance today at left with (6) actual solar and (7) recent demand. Grid draw (8) is shown in red. Rachel an avatar, (9) emotes and gives (10) plain language versions of statistics The tabs with < and > are for navigating. Image of an IHD (right) in the kitchen of a participant home

exceptional energy demand events and solicited the group to give positive or negative feedback to a deviant home.

In addition to the timed prompts, pages of the system offered either statistics or commentary by a cartoon avatar, Rachel (Fig. 28.1). Together with graphs, these systems essentially offered the same information in three different modes. Figure 28.1 shows the pop-up prompts that were served on top of the graphs shown in Fig. 28.2.

A page of the system intended to reinforce a group goal, called the Paydown page, was removed close to deployment because the short run of the study meant that only 3 months of performance could not be effectively graphed against a payment duration of perhaps 8 years. Still, commentary about paying down the system and certain statistics about this were retained.

28.4 Result

About a 100,000 event records were collected for the three groups along with a selection of 6 other homes used as controls in the period 1 Jan 2018 to 14 August 2018. The IHD signals ran for 100 days to 14 August. There were 617 interactions with the IHD collected and counted. A 4th order polynomial regression was fitted to the control group energy data series, and this spline was subtracted from the study data series to adjust for seasonal, price, and other externalities. The period after 1 Jan 2018 but before May 2018 was observed for study effects which may have contributed a Hawthorne effect to the outcome. None was observed. That is, the effect of recruitment, collateral, and a smart meter reader before the IHD deployment did not have an observable effect on demand and all observed effects were more likely due to the IHD and its signals.

Analysis of these results sought to find a negative correlation between demand and the cumulative signals with an increasingly negative trend toward the end of the run, to reflect increasing effects of more group- and socially-oriented signals on consumption of electricity. As an example, Fig. 28.3 shows group 1 cumulative responses and electricity demand. This weak but statistically significant negative correlation is present for all three groups.

Analysis determined the correlation between demand and the hourly availability of simulated solar and battery (called Past Distributed Renewable Energy - PDRE). This was determined as an indicator of load-shifting behaviour change. Figure 28.4 above shows group 3 weekly correlation between demand and available simulated solar and battery. The overall Spearman Rho correlation for energy demand and solar availability within the intervention period for this group was $r = 0.12$ ($P = 0.0001$). For groups 1 and 2 there was no significant correlation for time of day demand and solar-availability in the IHD period.

In the 100 days of the intervention in fact no sanctions messages were executed since no sanction events were triggered. Group messages were present from 20 days

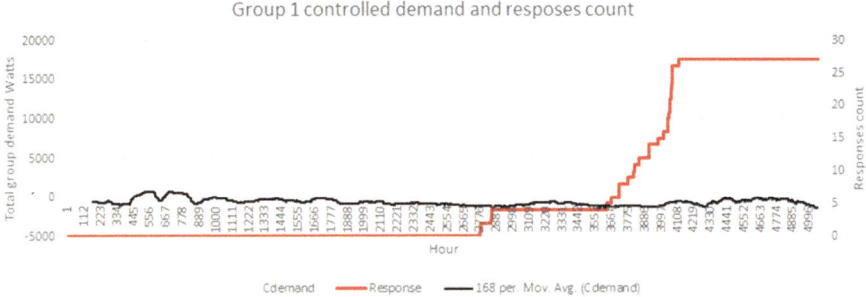

Fig. 28.3 Group 1 responses to stimuli presented on the IHD are accumulated in red. This is plotted against moving average of a week of demand data (which have had control group effects subtracted) in black. Correlation between demand and responses in the IHD period is $r = -0.156$ ($P = 0.000$)

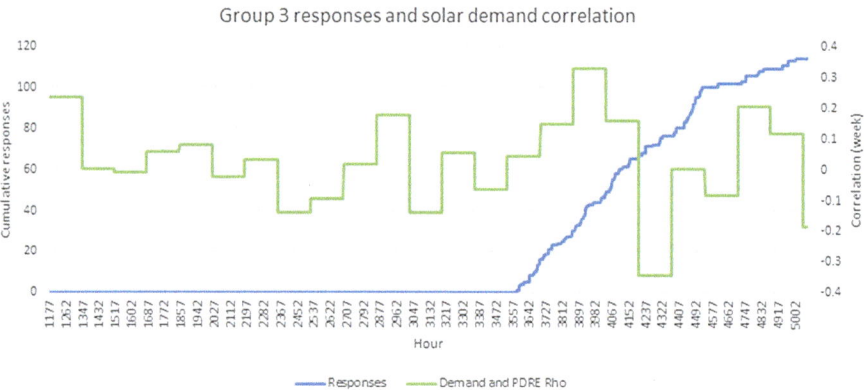

Fig. 28.4 Group 3 Weekly Spearman's Rho correlation between PDRE and total group demand (green). Responses from the group are shown cumulatively (blue). The graph starts at hour 1177 which is when solar accounting began. From hour 2420 to the end, r = 0.12 (P = 0.0001)

on per the schedule and are attributed to the increasing trends in declining energy use and demand-solar correlation.

28.5   Discussion and Conclusions

This study deployed a specially designed in-home display (IHD) that simulated a large, shared solar and battery systems among 3 groups of 5 adjacent homes. The signals delivered by the study in-home display simulated a commons arrangement by purporting that the shared system provided limited, rivalrous, unreliable electricity. This, together with prompts to positively or negatively message group members who under- or over-used electricity (respectively) attempted to reproduce Ostrom's (1992) fourth and fifth commons principles: monitors and graduated sanctions (respectively) are identified as critical to actual demand moderation.

This approach adds to the field by testing a more black-box approach to home energy use, in contrast with Bandura's Theory of Planned Behaviour (Bandura et al. 1977) which sought to modify rational behaviour in energy consumption of individual occupants. This study exploited the more modern community-as-consumer (Mackenzie-Mohr and Smith 1999) and simulated a large shared solar and battery system.

The study found statistically significant support for behaviour change due to the IHD and signals. Due to the scale and short time frames of the study, outcomes for this cohort are perhaps limited in their applicability, but they do indicate a promising methodology for directly testing the potential of an energy commons without the need for infrastructure or utility cooperation. In future, it may also be possible to force the sanctions events, since none were triggered by consumption extremes.

IHDs are a promising way to provide signals - a large Northern Ireland study found 20% demand reduction (Gans et al. 2013) but these results were not as strong in Australia before the Victorian smart meter rollout (McKerracher and Torriti 2013). There may be a new opportunity to pair IHDs with Victorian smart meters, but also to go further and use the IHD, effectively a new messaging device, as a way to coordinate groups of homes as this study has demonstrated. The value of commons theory in this approach is promising, because previous demand management studies did not reframe electricity as different from highly available, unlimited, and highly reliable. Renewable supply will not match these qualities, and instead requires a kind of compromise between demand behaviour, available power, and conservation. The greater agency of people in groups is already important for environmental conservation more broadly (Bandura 2000) and the very difficult challenge of changing social practices around energy should also be a social effort. Shared energy can exploit these opportunities, and commons theory provides a very rich history of stable, non-market, self-governing social efforts around critical limited resources.

This study does not propose urban off-grid communities but it may be possible for a private microgrid to be legally off grid (islanded) for electricity exchanges. The grid utility provides considerable value but it does not provide effective signals to moderate demand and the GHG emissions from electricity are very great. Instead, the modelling in this study showed a majority of electricity needed by a group of homes can be obtained efficiently (both in terms of energy and cost) from a shared solar system, but also in concert with a demand management programme. This approach should be explored further for its potential to bring about substantial shifting- and reduction- of energy demand if renewable generation is to plausibly meet our future demand.

Acknowledgement This research was funded along with scholarship RP5005 supported by the Low Carbon Living CRC.

References

Abbott M (2001) Is the security of electricity supply a public good? Electr J 14(7):31–33

Abrahamse W, Steg L, Vlek C, Rothengatter T (2005) A review of intervention studies aimed at household energy conservation. J Environ Psychol 25(3):273–291

AEMO (2013) 100 PER CENT RENEWABLES STUDY – MODELLING OUTCOMES. Australian

Allon F, Sofoulis Z (2006) Everyday water: cultures in transition. Aust Geogr 37(1):45–55

Australian Bureau of Meteorology (2014) What is El Niño and what might it mean for Australia? Retrieved June 2, 2018, from http://www.bom.gov.au/climate/updates/articles/a008-el-nino-and-australia.shtml

Bandura A (2000) Exercise of human agency through collective efficacy. Curr Dir Psychol Sci 9 (3):75–78

Bandura A, Boone T, Reilly A, Sashkin M (1977) Social learning theory, vol 2. Prentice-Hall, SAGE Publications, Englewood Cliffs

Bauwens T, Eyre N (2017) Exploring the links between community-based governance and sustainable energy use: quantitative evidence from Flanders. Ecol Econ J Int Soc Ecol Econ 137:163–172

Bollier D (2014) Think like a commoner: a short introduction to the life of the commons. New Society Publishers, Gabriola Island

Cox M, Arnold G, Villamayor Tomás S (2010) A review of design principles for community- based natural resource management. Ecol Soc 15(4):38

CSIRO and ENA (2015) Electricity network transformation roadmap - Interim program report. Energy Networks Association. Retrieved from http://www.energynetworks.com.au/sites/default/files/roadmap_interim_report_final.pdf

DEE (2018) Independent review of the Greenhouse and Energy Minimum Standards (GEMS) Act 2012. Department of the Environment and Energy. Retrieved from http://www.energyrating.gov.au/sites/new.energyrating/files/documents/GEMS%20Review%20 -%20FINAL%20Discussion%20Paper_20180228%2B%2B.pdf

Delmas MA, Fischlein M, Asensio OI (2013) Information strategies and energy conservation behavior: a meta-analysis of experimental studies from 1975 to 2012. Energy Policy 61:729–739

DELWP (2014). Top 10 ways to save energy. Retrieved July 19, 2017, from https:// www.victorianenergysaver.vic.gov.au/more-ways-to-save/top-10-ways-to-save-energy

DELWP (2016) Standard feed-in tariff and transitional feed-in tariff – DELWP. Retrieved March 31, 2017, from http://delwp.vic.gov.au/energy/electricity/victorian-feed-in-tariff/closed- feed-in-tariff-schemes/standard-feed-in-tariff-and-transitional-feed-in-tariff

Energy Market Operator. Retrieved from https://www.environment.gov.au/system/files/resources/d67797b7-d563-427f-84eb-c3bb69e34073/files/100-percent-renewables-study-modelling- outcomes-report.pdf

Gans W, Alberini A, Longo A (2013) Smart meter devices and the effect of feedback on residential electricity consumption: evidence from a natural experiment in Northern Ireland. Energy Econ 36:729–743

Gardiner K (2017) The small Scottish isle leading the world in electricity. Retrieved September 29, 2017, from http://www.bbc.com/future/story/20170329-the-extraordinary-electricity-of-the-scottish-island-of-eigg

Hargreaves T, Nye M, Burgess J (2008) Social experiments in sustainable consumption: an evidence-based approach with potential for engaging low-income communities. Local Environ 13(8):743–758

Hill J (2014) Power corrupts. Retrieved August 30, 2018, from https://www.themonthly.com.au/issue/2014/july/1404136800/jess-hill/power-corrupts

Ison N, Lyons M (2013) Homegrown power plan summary. GetUp and SolarCitizens

Kelly MJ (2010) Future energy needs and engineering reality. Retrieved from http://www.nscj.co.uk/JECM/issues/JECM-1-3.pdf

Liubinas A, Harrison P (2012) Saving a scarce resource: a case study of behavioural change. In Academy of Marketing Conference

Lowe B, Lynch D, Lowe J (2015) Reducing household water consumption: a social marketing approach. J Mark Manag 31(3–4):378–408

Mackenzie-Mohr D, Smith WFSB-A (1999) Introduction to community-based social marekting. Gabriola Island New Society Publishers, Gabriola Island

McKerracher C, Torriti J (2013) Energy consumption feedback in perspective: integrating Australian data to meta-analyses on in-home displays. Energ Effic 6(2):387–405

Ostrom E, Gardner R, Walker J, Walker J (1994) Rules, games, and common-pool resources. University of Michigan Press, Ann Arbor

Russell-Bennett R, Bedggood R, Glavas C, Swinton T, McAndrew R, O'Mahony C, Pervan F, Willand, N. (2017). Power shift project one: driving change – identifying what caused low-income consumers to change behaviour, final report. Queensland University of Technology and Swinburne University of Technology

Shove E, Walker G (2014) What is energy for? Social practice and energy demand. Theory Cult Soc 31(5):41–58

SmartUser (2018) Watts clever energy monitor – EW4500 – smartuser. Retrieved July 23, 2018, from http://www.smartuser.com.au/product/watts-clever-energy-monitor-ew4500/

Sustainability Victoria (2014) Household Energy Action Guide. Victorian State Government. Retrieved from http://www.sustainability.vic.gov.au/-/media/SV/Publications/You-and-your-home/Save-energy/Households-Energy-Action-Guide.pdf

Wood T, Carter L, Harrison C (2014) Fair pricing for power. Grattan Institute Melbourne, Australia

Chapter 29
Identifying Bottlenecks in the Photovoltaic Systems Innovation Ecosystem – An Initial Study

Kristian Widén and Charlotta Winkler

Abstract Solar energy is likely to play a major role in future renewable energy systems. One important part in this is the integration of photovoltaic (PV) systems into the built environment. Earlier studies show that the institutional framework plays a major role in achieving a broad implementation of PV systems. It has, however, also shown that the value network of PV systems needs to be understood and developed further. In that respect, earlier research on innovation diffusion into the built environment shows the necessity of involving and understanding key stakeholders. Stakeholder analysis may help in identifying key stakeholders but fail in assessing the stakeholders' role in the value network as it does not, for example, take into account the relational effects. Innovation Eco Systems is an approach that has the potential to do this, as it addresses the alignment structure of the partners needed for the value proposition to occur.

The aim of this initial study is to address the use of the innovation ecosystem as a way of assessing implementation of PV systems in the built environment. Two structured workshops with two key stakeholder categories, Clients and Suppliers, were held to identify the main barriers for a broader implementation of PV systems into the built environment in Sweden. The main results show that the earlier studies were right in that the institutional framework is a major factor, but also that the value network is important and that the problems in the value network are perceived somewhat different between the two categories. This suggests that it will be necessary to address the value network from the perspective of the actors by applying an innovation ecosystem analysis. It also helped in identifying other important stakeholders in the value network that will be needed to include in the future studies. To

K. Widén (✉)
School of Business, Engineering and Science, Halmstad University, Halmstad, Sweden
e-mail: kristian.widen@hh.se

C. Winkler
School of Business, Engineering and Science, Halmstad University, Halmstad, Sweden

WSP Environmental, Stockholm, Sweden
e-mail: charlotta.winkler@wsp.com

© Springer Nature Switzerland AG 2020
R. Roggema, A. Roggema (eds.), *Smart and Sustainable Cities and Buildings*,
https://doi.org/10.1007/978-3-030-37635-2_29

419

summarize, the findings of this initial study suggest that innovation ecosystem will address a more comprehensive picture on the implementation of PV systems in the built environment. However, to be able to identify bottlenecks and subsequent solutions to these bottlenecks further studies of the complete innovation ecosystem, with its stakeholders, is necessary. The ongoing project is currently carrying out these studies in a Swedish context.

Keywords PV · Implementation · Innovation · Ecosystems · Bottlenecks · Stakeholders

29.1 Introduction

Addressing the Agenda 2030 and in the 17 Global sustainability goals, renewable energy solutions have been identified as one important contributor. Photovoltaic electricity is an important element in renewable energy systems. The pace of implementation differs in different countries. Germany has approximately 8% photovoltaic electricity in the electricity mix, while Sweden only has about 0.2%, although the level of solar radiation is about the same. Achieving 100% renewable electricity of the electricity mix demands a larger share of photovoltaic electricity and a significant increase in the implementation rate of photovoltaic systems in the built environment. The Swedish Energy Agency states in its strategy for photovoltaic electricity that the potential is between 5 and 10% of the electricity mix. They also state that it will be necessary to reach these levels if Sweden is to reach its target on 100% renewable production of electricity by 2040 (Energimyndigheten 2016). The same strategy identifies regulations and public policies, as well as bureaucratic administration, as the main bottlenecks to an increased implementation of photovoltaic systems. A report from the Swedish Energy Agency maps the innovation system for the energy sector as a whole, to identify overarching strengths and weaknesses through, for example, drivers and hinders for development and implementation. The result is mainly targeting public bodies, decision and policy makers. With specific regards to photovoltaic electricity production though, it is noted that there is an increased need of understanding the value chain and that the supply chain network is underdeveloped (Energimyndigheten 2014). As a large part of the photovoltaic production of electricity will need to be implemented in the built environment, it relates to the construction value chain, and the supply chain network needs to be understood and developed in that context.

Developing, diffusing, and implementing innovations in the construction value chain has its specific challenges. A short-term perspective often characterizes the construction value chain and fragmented processes (see for example Barlow 2000; Widén and Hansson 2007). It is also characterized by innovations being implemented in projects (Winch 1998) and knowledge spreading between construction projects (Senaratne and Sexton 2008), both of which are hampered by the short-term perspective. From earlier research, it is known that the rate of innovation diffusion is depending on the involvement of key stakeholders already during the development stage of the innovation process (Widén et al. 2014). It is also known that different parts of the construction value chain experience different issues as main barriers to innovation implementation (Chen et al. 2018). In order to understand and to enable development of the construction value chain related to the implementation of the photovoltaic electricity production system, there is a need for an approach that takes the project-based nature of construction, the different perspectives of different actors, and how it affects the value chain into consideration.

The research presented in this paper is an initial step in assessing PV related construction value chains using the approach innovation ecosystem. The aim with this initial study is to address the use of innovation ecosystem as a way of assessing implementation of PV systems in the built environment. The aim of the ongoing research is to identify bottlenecks in the value chain hindering or slowing the pace of PV systems implementation in the built environment.

29.2   Innovation Ecosystem

The traditional way of looking at innovation is that an inventor, for example in the context of construction a supplier or a contractor, generates value for a client by supplying/selling a product or a service within a project (Winch 1998). Research has shown that, in reality, quite often there is a need for other stakeholders to change their behavior, or in some instances even innovate as well, for the client to realize the value (Adner and Kapoor 2010). An innovation ecosystem is a way to connect or relate the different stakeholders that have the potential to influence the possibility to create value from innovations from an actor's perspective (compare with Adner 2017; Moore 1998). This system does not follow the traditional linear process of value creation and many of the stakeholders lie outside of the traditional value chain (Iansiti 2004). It takes its starting point in value creation and analyzes what stakeholders need to interact and in what way they need to interact in order to realize the value (Adner 2017). In this case the value of PV produced electricity in the built environment. Innovation ecosystems can be said to complement the understanding between stakeholder analysis and technological innovation systems, by relating the effects of the actions of stakeholders to the result for a single stakeholder. The stakeholders in the system generate value for the client as a system of interdependent companies, rather than as individual companies. In other words, innovation ecosystems enable the understanding of value creation for the client that individual companies would not be able to do (Adner 2006).

Innovation ecosystems can be studied or seen as either "ecosystem as structure" or ecosystem as affiliation" (Walrave et al. 2017). The former is more focused than the latter. In this research, we have opted for "ecosystem as structure" thereby following that the ecosystem's value proposition is defining the innovation ecosystem (Walrave et al. 2017). This in turn means that the system boundaries are defined by those stakeholders and other elements that will have an effect on the realization of the ecosystem's value proposition (Adner 2017). It is important to note that the focus on realizing the ecosystem's value proposition it is from the perspective of the end user (Clarysse et al. 2014). Another important aspect to be aware of is that due to the interdependencies between the stakeholders, failure to contribute by one of them may have negative effect on the others (Brusoni and Prencipe 2013).

In the context of construction, an innovation ecosystem analysis is a tool for identifying stakeholders that are affected or can affect the implementation of an innovation in order to relate key stakeholders to the innovation process (compare with Widén et al. 2014). It is also important to analyze in what way stakeholders need to change/develop their behavior and what the incentive is to do so (compare with Adner and Kapoor 2010).

In summary, innovation ecosystems allow for the analysis of how different stakeholders, for example consultants, grid owners etc., need to interact, develop, change business models and more, as well as what incentives are necessary to limit the effects of bottlenecks and increase the implementation of PV systems. In other words, it provides a tool to understand how the value chain as a whole need to develop to ensure value creation through implementation of PV systems in the built environment.

29.3   Method

In order to investigate the innovation ecosystem, there are three features that need to be addressed:

(1) The innovation ecosystem need to be identified, what stakeholders are part of the system and their relations. Within the system, (2) the bottlenecks must be located and (3) it must be identified which ones are the easiest to address. As described earlier, stakeholders in the innovation ecosystem may be stakeholders not part of the traditional value chain. These are particularly important to identify.

We have opted for a structured workshop method (compare with Björkdahl and Holmén 2016) where the participants with strategic insight are given the task to identify and rate the key challenges, as well as rank them according to ease of addressing them for a particular area. In this case, the particular task was "what are the key challenges to increased implementation of PV systems in the built environment". The ecosystem's value proposition is improved value for property owners by implementation of PV systems. The aim of the workshop was to identify the three features describe above. The process of the workshops follows a pre-defined structure (Björkdahl and Holmén 2016):

- Identification
- Clustering
- Relationship
- Ranking

In the identification stage, the participants identify the five top challenges for the particular problem area, in this case, bottlenecks creating value for the client through implementation of PV Systems. An important part at this stage is to ensure that everyone understands each challenge. In the second stage, clustering, the initial identified challenges are clustered into problem clusters, and similar challenges are brought together. The problem areas are then assessed for relationship to any other problem cluster, for example, if one problem cluster leads or contribute to another problem cluster. The last stage is to rank the problem clusters according importance and manageability (Björkdahl and Holmén 2016).

In this initial part of the research, the clients/property owners and suppliers were targeted, as those two were identified as the two types of stakeholders that without question would be part of the innovation ecosystem. The plan for the research following these initial studies is to target the types of stockholders that are identified in these two workshops as well as in the subsequent workshops. The two types of stakeholders were represented by the Solar Energy Association of Sweden, for suppliers and Swedish Energy Agency's group for procurement of energy-efficient multi-family buildings, 'BeBo', for clients/property owners.

29.3.1 Solar Energy Association of Sweden

The Solar Energy Association of Sweden is the national organization with approximate 200 professional members representing Swedish industry, as well as Swedish research institutes, working with solar energy. The association works for utilization of solar energy (direct transformation from solar radiation to heat and electricity) and plays an important role in a sustainable Swedish energy system.

Activities within the association and the secretary board consist of communication with authorities in order to improve the conditions for solar energy in Sweden and assembling information material for the public and decision-makers. They are the owner of the annual Solar Energy Award (Solenergipriset) for an exemplary plant and an exceptional contribution to the development of solar energy in Sweden, organizes meetings for the solar industry in order to discuss common activities, collects branch statistics over installed systems.

29.3.2 Swedish Energy Agency's Group for Procurement of Energy-Efficient Multifamily Buildings, 'BeBo'

BeBo has been active since 1989 and is a network of property owners, with the Swedish Energy Agency as a financier for the coordination of the network. The main focus is to reduce the dependency on energy in the form of heat and electricity in multifamily houses, and thereby reduce the impact on the environment. BeBo's activities will lead to the introduction of energy- efficient systems and products through a combined procurement competence. The Swedish Energy Agency therefore contributes funding and expertise to BeBo, which in turn, goes on to the property owners by means of, among other things, demonstration projects carried out with the help of the members.

29.4   Results

The results of the workshop with the clients/property owners in BeBo is presented in Table 29.1. The factor that is considered to be the largest bottleneck is the public policy system. The second factor has to do with the problem of transferring electricity produced between property boundaries and the connection point, and the third correlates to technical development. Of all the bottlenecks identified, needs profile/dimensioning is the considered easiest to manage, with connection point on second and building regulations on third

As can be seen in Table 29.2, the client/property owners related almost all bottlenecks to their profitability. The only two bottlenecks that were not related to profitability were public procurement and the construction process. Two bottlenecks were understood to have relations to more than one other bottleneck. Public system affects both connection point and profitability, while Lack of knowledge affects public procurement, profitability and needs profile/dimensioning.

The result from the workshop with the supplier representatives is presented in Table 29.3. The main bottleneck they identified was the lack of a stable subsidies system. The second was lack of knowledge and the third has to do with profitability. Lack of knowledge is considered as the one easiest to manage and profitability the second. Subsidies, tax system, and building regulations were §joint third.

Table 29.1 Bottlenecks according to clients/property owners

Factor	Need	Need - weighted	Manage-ability	Manage-ability - Weighted
Policy system	12	**3,00**	0	0,00
Connection point	8	**2,00**	6	**1,50**
Building regulations	5	**1,25**	5	**1,25**
Procurement	3	0,75	3	0,75
Construction process	3	0,75	4	1,00
Technical development	6	**1,50**	2	0,50
Profitability	3	0,75	2	0,50
Lack of knowledge	2	0,50	1	0,25
Needs profile/ dimensioning	0	0,00	10	**2,50**
Investment	0	0,00	3	0,75

Table 29.2 Identified relations by clients/property owners

Policy system	Connection point	Building regulations	Public Procurement	Construction process	Technical development	Profitability	Lack of knowledge	Needs profile/ dimensioning	Investment
Policy system	x								
Connection point						x			
Building regulations						x			
Public procurement									x
Construction process								x	
Technical development						x			
Profitability									
Lack of knowledge			x			x		x	
Needs profile/ dimensioning						x			
Investment						x			

Table 29.3 Bottlenecks according to the suppliers

Factor	Need	Need - weighted	Manage-ability	Manage-ability-Weighted
Subsidies	16	**2,29**	8	**1,14**
Tax system	7	1,00	8	**1,14**
Power grid	5	0,71	7	1,00
Politicians	4	0,57	0	0,00
Building regulations	7	1,00	8	**1,14**
Public procurement regulation	5	0,71	0	0,00
Construction process	1	0,14	0	0,00
Lack of knowledge	13	**1,86**	24	**3,43**
Profitability	10	**1,43**	13	**1,86**
Development recourses	0	0,00	2	0,29

The suppliers group identified more diverse relations than the client/property owners group (see Table 29.4). There are three bottlenecks that are affected by three or more other bottlenecks: (1) public procurement, affected by politicians, construction process, and lack of knowledge, (2) construction process, affected by power grid, building regulations, public procurement, and lack of knowledge, and (3) their profitability, affected by subsidies, tax system, power grid, and public procurement. There are two bottlenecks that affect six other bottlenecks; (1) politicians, affecting subsidies, tax system, power grid, building regulations, public procurement, and development resources, and (2) lack of knowledge, which affects tax system, power grid, politicians building regulations, public procurement, and construction process.

29.5   Analysis

Both groups identified 10 problem clusters. Quite a few of the bottlenecks are the same but they did not always use exactly the same words. Both groups verified the earlier research that identified the importance of supportive regulations and public policies. There are also several other identified bottlenecks that relate both to the traditional construction value chain, but also point towards stakeholders not traditionally being part of that vale chain. The main difference between the two groups has mainly to do with how they understand the relation between the different bottlenecks. The client/property owner group related almost all bottlenecks to their ability to profit from the investment. This is understandable, as that is their main goal with their business. One of the participants in this workshop actually said that it was interesting to see how their perspective on bottlenecks actually changed from the initial part of the workshop to the end. Initially, everyone was talking about how the institutional framework was the major bottleneck, but in the end, it was actually profit, or rather their inability to profit, from the implementation that was the main

Table 29.4 Identified relations by the suppliers

	Subsidies	Tax system	Power grid	Politicians	Building regulation	Public procurement	Construction process	Lack of knowledge	Profitability	Development recources
Subsidies									x	
Tax system									x	
Power grid							x		x	
Politicians	x	x	x		x	x				x
Building regulations							x			
Public procurement							x		x	
Construction process						x				
Lack of knowledge		x	x	x	x	x	x			
Profitability										x
Development recourses										

bottleneck. The supplier group had, at least to some extent, more focus on what bottlenecks affected others, whereas politicians and lack of knowledge came through as the most important ones. It is understandable that the two groups have different perspectives on this, as their roles in both the value chain and in the innovation ecosystem are different.

In regard to which stakeholders should be included in the PV innovation ecosystem, the identified bottlenecks suggest that apart from the traditional construction value chain stakeholders, power grid owners and potentially politicians/policy makers should be included. Both groups have identified the potential to use existing power grids to transfer electricity between properties. This has both to do with regulation and a lack of interest from the power grid owners to find solutions to this issue. Both groups also identified a number of different factors being bottlenecks that have to do with political decisions suggesting they be included, on the other hand that could be considered as boundary conditions. Both groups, quite naturally, mentioned the construction process and related to that, the stakeholders of the construction process as stakeholders of the PV innovation ecosystem.

29.6   Conclusion

To conclude, earlier research has been verified, and both groups identified the need for a supportive institutional framework. It is also possible to conclude that there is a need to develop the understanding of the (extended) value chain. More importantly, the findings of this initial study suggested that using the innovation ecosystem framework has the potential of getting a more detailed understanding. For example, just from these two stakeholders' perspectives, one stakeholder, the power grid owners, was identified, that would not have been identified through a traditional value chain or supply chain analysis.

Although quite similar, there are also some clear differences between the two groups in both of the identified bottlenecks, but more in the perception of the relations between the bottlenecks. However, it is too early to draw a conclusion on the extent of the similarities and differences, as well as the effects of these. There is a need to validate the findings from the workshops quantitatively. It is also too early to say anything about the actual innovation ecosystem. Therefore, workshops with other identified stakeholders need to be carried out, and those in turn need to be validated.

To summarize, the findings of this initial study suggest that innovation ecosystem will address a more comprehensive picture on the implementation of PV systems in the built environment. However, to be able to identify bottlenecks and subsequent solutions to these bottlenecks, further study of the complete innovation ecosystem, with its stakeholders, is necessary. The ongoing project is currently carrying out these studies in a Swedish context.

References

Adner R (2006) Match your innovation strategy to your innovation ecosystem. Harv Bus Rev 84:98–107

Adner R (2017) Ecosystem as structure: an actionable construct for strategy. J Manag 43:39–58

Adner R, Kapoor R (2010) Value creation in innovation ecosystems: how the structure of technological interdependence affects firm performance in new technology generations. Strateg Manag J 31:306–333

Barlow J (2000) Innovation and learning in complex offshore construction projects. Res Policy 29:973–989

Björkdahl J, Holmén M (2016) Innovation audits by means of formulating problems. R&D Manag 46:842–856

Brusoni S, Prencipe A (2013) The organization of innovation in ecosystems: problem framing, problem solving, and patterns of coupling. Collaboration and Competition in Business Ecosystems

Chen L, Manley K, Lewis J, Helfer F, Widen K (2018) Procurement and governance choices for collaborative infrastructure projects. J Constr Eng Manag 144:04018071

Clarysse B, Wright M, Bruneel J, Mahajan A (2014) Creating value in ecosystems: crossing the chasm between knowledge and business ecosystems. Res Policy 43:1164–1176

Energimyndigheten S (2016) Förslag till strategi för ökad användning av solel

Energimyndiheten S (2014) Teknologiska innovationssystem inom energiområdet, En raktisk vägledning till identifiering av systemsvagheter som motiverar särskilda politiska åtaganden. Stockholm

Iansiti IMALR 2004. The keystone advantage: what the new dynamics of business ecosystems mean for strategy innovation, and sustainability. Harward Business School Press

Moore JF (1998) The rise of a new corporate form. Wash Q 21:167–181

Senaratne S, Sexton M (2008) Managing construction project change: a knowledge management perspective. Constr Manag Econ 26:1303–1311

Walrave B, Talmar M, Podoynitsyna KS, Romme AGL, Verbong GPJ (2017) A multi-level perspective on innovation ecosystems for path-breaking innovation. Technological Forecasting and Social Change

Widén K, Hansson B (2007) Diffusion characteristics of private sector financed innovation in Sweden. Constr Manag Econ 25:467–475

Widén K, Olander S, Atkin B (2014) Links between successful innovation diffusion and stakeholder engagement. J Manag Eng 30:04014018

Winch G (1998) Zephyrs of creative destruction: understanding the management of innovation in construction. Build Res Inf 26:268–279

Chapter 30
A User-Led Approach to Smart Campus Design at a University of Technology

Alfred B. Ngowi and Bankole O. Awuzie

Abstract A South African university of technology (UoT) has embarked on a transformative journey towards achieving a SMART campus status. Due to the plurality of perspectives concerning the meaning of the SMART campus concept among end-users and other stakeholders, and for the avoidance of pitfalls associated with end-user apathy towards new innovations, a user-led approach (participatory design) was adopted. This study provides a detailed narrative on the utility of the participatory design process in the design of the SMART campus initiative at the UoT. Relevant user categories within the UoT were identified, and a pre-determined number of individuals purposively recruited from each of these categories. A workshop was convened to elicit their opinions concerning the SMART campus concept. Discussions at the session were recorded and subsequently transcribed. Data emerging from the discussions were analyzed thematically relying on a coterie of pre-set themes. Findings from the study highlight the benefits accruable from the proper usage of the participatory design approach. The adoption of this approach enabled among other things, a consensus on an institution-wide/institution-specific definition of the SMART campus concept, an identification of the state-of-the-art of the institution's ICT platform, users' perception of the performance of extant ICT platforms, their expectations, as well as aspects of the SMART campus they believe should be prioritized during implementation. The study highlights the utility of the participatory design approach in the design and implementation of innovative solutions like the SMART campus initiative in organizations.

Keywords SMART campus · South Africa · Participatory design · User-centrism

A. B. Ngowi · B. O. Awuzie (✉)
Central University of Technology, Bloemfontein, South Africa
e-mail: angowi@cut.ac.za; bawuzie@cut.ac.za

© Springer Nature Switzerland AG 2020
R. Roggema, A. Roggema (eds.), *Smart and Sustainable Cities and Buildings*,
https://doi.org/10.1007/978-3-030-37635-2_30

431

30.1 Introduction

Successive governments have expressed commitment to Sustainable development and SMART City/Campus initiatives (Happaerts et al. 2010). This is evident in the number of national governments that are signatories to the sustainable development goals (SDG) framework and its precursor, the millennium development goals (MDGs) (Waage et al. 2015). Presently, society is witnessing an increase in the adoption of sustainable development tenets by organizations.

The criticality of data and its subsequent deployment towards enabling new approaches to sustainable consumption cannot be overemphasized (Mellouli et al. 2014). Yet, the management of the unprecedented inflow of data due to the advent of the internet makes the need for a veritable platform for data analytics, interpretation and utilization, imperative. This has given rise to the concept of smartness. SMART cities are salient facets of the smartness concept which have evolved to cater for this aspiration, albeit from a city-wide perspective. Therein, information and communication technologies are deployed to collect and manage data related to patterns of consumption within cities (Angelidou 2015). The processed data is used in making futuristic projections whilst allowing for effective decision-making processes and development of interventions to cater to the resolution of the present challenges confronting society (Celino and Kotoulas 2013).

The time-honoured position of universities as microcosms of society effectuates the need for them to play pivotal roles in the transformation of society towards smartness (Leal Filho 2011). Such roles proceed beyond the normal boundaries of knowledge creation for SMART cities. Societal expectations dictate that these institutions should, relying on multi-, inter- and transdisciplinary skillsets domiciled within, constitute themselves into living laboratories for enabling the SMART city scenario. The SMART Campus initiative has taken off across several universities in response to these expectations. Universities in South Africa are not left out from this broad-based initiative as a cursory look at the websites and associated marketing paraphernalia of these universities indicate their aspirations to transform their approach to operations, pedagogy, and research through the adoption of internet-enabled communication and digital technologies. These efforts are not solely aimed at process optimization and efficiency savings but also at transforming learner-experiences within these institutions (Malatji 2017). In short, it has become a value proposition to attract students and staff. Although transition efforts are yet on-going and have been alluded to enhance user-experience in these institutions, there is little evidence to show that the views of the users are being incorporated during the conceptualization and design stages of these initiatives. Rather, scholars bemoan the uni-directional nature of SMART city implementation programmes (Rha et al. 2016).

The SAUoT is an institution where the systemic transition to a SMART campus is at an embryonic stage. The desire to make a success of this initiative in this institution necessitates that the programme's proponents/designers elicit and incorporate the views of users at this stage. This study seeks to report on the engagement process adopted in eliciting such worldviews- a juxtaposition of the user-centered systems design and participatory design approach, and the benefits accruing from this methodology.

The rest of the paper is structured into: a review of the theoretical perspectives of the study's underlying concepts, a description of the case study's institutional context, a narrative on the research methodology adopted, a report on the reflections of the authors on the outcomes of the user-engagement brainstorming session, and a conclusion.

30.2   Theoretical Perspective

30.2.1   SMART Campus – A Review

A SMART Campus is a campus that is an efficient, safe, sustainable, responsive and enjoyable place to live and work, underpinned and enhanced by digital/internet-based technologies. Its evolution has been linked to the need to foster a new paradigm in higher education due to the overt reliance on the information and communication technologies (Rha et al. 2016). Considered as a microcosm of a SMART city, a SMART campus can be described as an environment which aligns the aspirations of the "university as a city" and stronger connections across and outside the campus. As universities embrace the concept of a "SMART campus", three elements need to be incorporated (Davies 2015): First is the concept of the university as a city in its own right, a collection of people, amenities and assets which respond to, and are shaped by, the values, expectations and shifting demands of its "citizens". Secondly, the idea of connectedness, which includes some of the more operational and transactional capabilities that come with the idea of a SMART campus. The notion of a SMART campus encapsulates sensor-based SMART parking and new ways to use digital lighting to make campus facilities more accessible, safer and more sustainable energy-oriented technologies to totally reshape the spaces for learning and interaction and to broker new and more nuanced relationships between students; these relationships extend to the wider communities of alumni, business and community partners within which a university is embedded. And thirdly, it requires investments in infrastructure and services upon which the previously mentioned concepts rely. Suffice to say that a SMART campus seeks to align the aspirations of "university as city" and stronger connections across and outside the campus with the necessary investments in requisite technology assets and capability.

Abuarqoub et al. (2017) observe that significant investments were being made by higher education institutions in the transition towards smartness due to a desire to optimize service delivery to members of the university community. Furthermore, they opined that universities with SMART campuses will fare better than those without, in terms of cost and time savings, protection of the environment, effective monitoring of attendance of staff and students as well as effective space planning and utilization efforts. In furtherance to these, the adoption of implementation of

SMART campus projects in universities will enable the collection of critical data about operational facets thereby enabling optimal decision-making.

Examples of the SMART campus initiative implementation as chronicled in the literature include: the development of an anytime-anywhere learning within a SMART campus environment (Hirsch and Ng 2011), SMART parking (Bandara et al. 2016), frameworks for modelling movements on a SMART campus (Fan and Stewart 2014), development of platforms for energy management and optimization on campuses (Barbato et al. 2016,), dynamic timetabling systems (Campuzano et al. 2014), as well as the use of apps for on-campus navigation and information dissemination purposes (Dong et al. 2016). These examples go to prove utility of SMART Campus in engendering better learning environments as well as effective and efficient resource (mostly energy, time and space) allocation and usage. In a nutshell, it can be adduced that the SMART city and SMART campus initiatives are both aimed at enhancing the standards of livability of cities whilst enhancing the productivity levels of its citizens through participatory decision making and governance based on the availability of credible data.

30.2.2 A Case for User-Centricity in SMART Campus Design Through Participatory Design

The failure of strategy and policies in organizations and society has been attributed to the inability of policymakers and strategists to incorporate the viewpoints of not just the implementers but also the expected beneficiaries of the policy/strategy (Hupe et al. 2014). A cursory look at implementation-related studies indicates a tendency among strategists/policymakers to believe that they have an integral understanding of the needs of the expected beneficiaries (Ritter et al. 2014). Yet, this assumption has culminated in significant instances of beneficiary apathy and opposition to the policy/strategy programmes hence leading to underwhelming performance.

Most conventional methods for new product design have been described as uni-directional. In this instance, the developers source information from users about their preferences, design, develop and deliver systems which have been configured to those initially elicited preferences and then gauge the levels of usability afterwards. This is the case with user-centered design. But this has often resulted in failed delivery of products and solutions especially within the technological domain where rapidly changing features remain the norm (Ritter et al. 2014).

According to Muller (2003), participatory design (PD) refers to the procedural framework which chronicles the incorporation of potential end-users as critical participants in the activities which culminate in the design and development of software and hardware computer-based solutions. This framework enhances the designer's ability to among other things, achieve a clarification of design goals, needs formulation (co-design of problematic areas) between the designer and the beneficiaries (Simonsen and Hertzum 2012). PD makes a salient contribution towards enhancing mutual learning and co-production of relevant knowledge between designers and design beneficiaries, hence bridging the hitherto existing disconnect. This refers to the dual principles of reciprocity and mutuality as unique selling points of the PD approach during systems design. Summarily, the integration of PD and user-centric design holds significant potentials in the case of SMART Campus design.

The involvement of users in the design, development and subsequent implementation of the SMART campus initiative is deemed imperative for different reasons. Of significance is the need to ensure the adoption of democratic and collegial platforms during the design, development and implementation of the initiative. Blomberg and Henderson (1990) reiterate the significance of user involvement in systems design based on their expert knowledge of the work environment and the interaction of previously designed systems on workplace quality. Following from the foregoing, the adoption of such platforms encourages the strengthening of resource weak communities (Bjögvinsson et al. 2012) within the university community who often feel compelled to adopt and utilize new technological advancements in the execution of assigned roles. Such feelings inadvertently affect productivity, both at individual and organizational basis.

Having dwelt substantially on the advantages of users and user-perspectives in the design and development of actionable systems and programmes in organizations, this study will proceed to provide a narrative, in subsequent parts, on how user-centrism and PD were utilized during the design of a SMART Campus initiative at SAUoT.

30.2.3 Description of Case Study Context

The catch phrase of SAUoT Vision 2020 is "social and technological innovation". Furthermore, "innovation" is one of the five SAUoT values. Innovation is an improved product, process or service that benefits society in a timely and, sometimes, transformational manner. It is a team activity at the intersection of different fields, bringing together diverse ideas, abilities and/or methods to result in the creation of value (Patil et al. 2015). Innovation creates societal value (through an existing or new product, process or service) and one interesting aspect is internal innovation whereby an organization tweaks its processes to bring about efficiencies.

In the quest for digital transformation, one entity at SAUoT has taken several steps ranging from digital scholarship aimed mainly at enhancing student experience by placing all learning materials at a learning management system that is accessible anytime, anywhere; followed by comprehensive digital strategy aimed at bringing about operational integration and optimization through digital workflows and storage of information. However, these initiatives have been constrained by lack of their uptake by other entities across the campus. Recently, a project on digital transformation across the campus has been launched and it is expected that what has worked so far will be taken up by all entities.

Therefore, to ensure that all initiatives towards digital transformation are integrated, a SAUoT strategy for SMART campus has been proposed. The SAUoT campus is an ideal environment and 'vehicle' through which to research, develop and evaluate a diversity of SMART Campus, and potentially SMART City concepts, being embedded within an existing city desirous of making similar transition towards SMART City status in the near future, and encircled by major facilities, including a psychiatric hospital, police station, high court, and churches. In addition to the obvious benefits of a more efficient, connected and responsive campus, such a campus would also enhance and reinforce the University's reputation as a progressive academic institution and a university of technology. Such potential benefits and support are also likely to strengthen the case for further funding e.g. from the Department of Higher Education and Training (DHET), Technology and Innovation Agency (TIA); Department of Science and Technology (DST) and other funders who would like to be associated with efficient and technologically connected working environment.

However, before the complexity of challenges that developing a SMART Campus poses can be addressed, a formalization of needs as opposed to abilities is crucial. Also, it is necessary to identify what has already been done, both internally and externally. Indeed, an awareness of the later, would not only help to avoid simply replicating the work of other Universities, but enable a refocusing of efforts on identified areas of strength. It is also imperative that a 'people' oriented approach be advocated rather than 'hardware-centered' and avoid simply finding uses for new technologies and data, rather than focusing on the actual needs of those who use and service the campus.

30.3 Research Method

A brainstorming workshop session was convened by the lead author to elicit viewpoints from various stakeholder categories within the SAUoT concerning the design and implementation of the SMART campus initiative. Effort was made to identify and recruit discussants purposively from different stakeholder groups present in campus. This was considered necessary because of the need to provide these groups with the opportunity to participate in the development of a protocol for the institution's SMART campus transition. In total, 19 participants were recruited apart from the lead author who acted as facilitator and the co-author. Stakeholder groups from which these participants were drawn included: non-academic personnel from the Academic planning, Registry, Finance/Accounts, Procurement, Facilities, and, Information and Communication Technology (ICT) departments respectively. Also, in attendance were members of academic staff representing different disciplines and a select number of students from the Student Representative Council (SRC). In summary, the workshop featured a truly representative audience comprising of the internal stakeholders of the university community.

To break the ice, the facilitator had requested for researchers in the audience to make presentations on the utility and application of the SMART ideology according to their different specialisms. These presentations lasted for 10 min each on the average. PowerPoint presentations on themes such as SMART Buildings, SMART Energy, SMART Water, SMART Mobility and the Internet of Things (IoT) was carried out. In the aftermath of these presentations, questions around salient issues were posed by the facilitator to achieve the objective of the workshop – the development of a common ontology among different stakeholder categories concerning a SMART Campus and identification of priority areas where the incorporation of SMART features were deemed imminent.

Questions posed to the audience during the deliberations were centered on the following thematic areas:

1. A context-specific definition of the SMART Campus,
2. Stakeholders' expectations of a SMART Campus environment;
3. An appraisal of the state-of-art SMART infrastructure at SAUoT.
4. The level of integration and optimization of available SMART Infrastructure on Campus,
5. A SWOT analysis concerning the transition towards a SMART Campus environment, and;
6. Finally, a consensus on the priority implementation areas.

Discussants were requested to write down their answers on a notepad once a question was posed. A round of discussion ensued upon receipt of the notepads and the facilitator tried to achieve a consensus among participants on key issues concerning that question. This process lasted for 3 hours with breaks in-between.

The authors thematically analyzed the texts provided by the participants during the workshop. A comprehensive document outlining the details of the workshop was compiled by the authors and subsequently shared with the participants later. At this point, these participants were availed with a one-week window to either express their reservations on the information provided or make clarifications where necessary concerning the emerging implementation objectives included in the document. At the end of this period, all participants agreed that the content of the compilation was indeed, a valid reflection of their contributions and stance on the SMART Campus initiative.

Based on the foregoing, a set of strategic goals were formulated. These goals include:

1. Leveraging digital platforms to integrate and optimize operational activities,
2. Create user-friendly platforms for interaction with key stakeholders,
3. Optimize all utilities-energy, water, space-through reliance on Internet of Things (IoT),
4. Provide a secure and safe campus using SMART technologies
5. Leveraging digital platforms to enhance student life-both academic and social, and;
6. Optimize the functions of all SAUoT buildings using intelligent devices.

Accordingly, the development of these strategic goals informed the design of a user-led SMART Campus architecture at the institution. Yet, this paper only seeks to report on the utility of the processes applied towards enabling a consensus on these strategic goals.

30.4 Reflections on the Utility of the Process

In this section, the authors reflect on their experiences during the brainstorming workshop session as well as the utility of the process therein in engendering the development of a common ontology concerning SMART Campus and its design and implementation.

This study adopts the standpoint of Muller (2003) on the place of workshops in PD. In that study, workshops are referred to as providing a platform for bringing together, parties with divergent views, enabling robust exchange of ideas and information on a set of goals. A salient outcome from such platform will usually include a shared commitment towards enunciated goals by the parties. This was the essence of the brainstorming workshop convened by the first author. Studies into implementation failures in Higher Education institutions have identified stakeholder apathy, because of the top down approach to policy/strategy design, development and implementation, as a major hindrance. In the case of the SMART Campus agenda at SAUoT, the authors sought to ensure that not only did the design remain congruent to the ideals and expectations of a significant cross-section of the university community but also to ensure that its implementation was driven by this population.

The need to create shared knowledge among this population was acknowledged and served as a guiding principle during the identification of discussants. An understanding of the disparity of knowledge concerning the SMART Environment episteme led to the selection of researchers engaged with research in this area to make simplistic presentations to the audience prior to the commencement of brainstorming session proper. Also, the facilitator sought to allay the fears of the participants concerning the centrality and overt dependence of the SMART Environment episteme on technology. This was achieved through an attempt to develop a context-specific definition of SMART Campus from the perspective of the discussants. This was necessitated by the notion that the transition towards smartness will engender a loss of jobs at the university through the transfer of hitherto information technology-based competencies to external solution providers as was the case in successful SMART campus exemplars. Besides arriving at a context-specific definition of the SMART campus, the centrality of human engagement in the plan was reiterated to the cohort.

In furtherance to this, discussions availed discussants with the opportunity to critique the present level of smartness at the university and project the potentials of a SMART-enabled future therein. Regarding the former, discussants identified the various advances made concerning smartness at the university in their respective disciplines/departments and enumerated the available systems which could be leveraged upon to drive the initiative. Also, the cohort identified the challenges with the ability of available SMART infrastructure to boost the transformation. Significant among these challenges was poor integration of systems and platforms which resulted in a prevalence of information and knowledge silos. In envisioning the future, the discussants listed their expectations from the transition to SMART environments, especially as it concerned improved productivity in the workplace, security and usability of available networks. This encouraged consensus building on prioritization of different milestones in the design and implementation plan.

The brainstorming workshop boosted a multi-disciplinary user-led design of the SMART campus through the provision of an enabling platform for not just eliciting information about user preferences and expectations (user-centrism) but also allowing users to participate in the design of the SMART Campus initiative at the user. The advantages witnessed therein are like those professed elsewhere- see Muller (2003), Dalsgaard (2012), Kübler et al. (2014). Discussants were also apportioned roles in the implementation plan which evolved out of that workshop. These roles have subsequently culminated into the identification of capstone projects at the university which are meant to serve as harbingers of the SMART Campus. These projects include the SMART Farm, SMART Building, SMART Access/ Mobility, etc. Yet, the absence of external stakeholders during the workshop was considered a major limitation of the session. The decision of the authors to confine participation to only internal stakeholders was described as faulty by discussants as the university's role in societal transformation implied the need for external stakeholders to be identified as users. This is not new as Bjögvinsson et al. (2012) in their study, describe the tendency of PD facilitators to focus only on projects with identifiable users as a drawback of the process. They assert the need to move away from actual use to envisioned use during this process.

30.5   Conclusion

This study set out to understudy and report on the utility of a user-centric participatory design methodology in the design of a SMART campus at an SAUoT. The study was considered imperative as cases of user apathy towards Smart technologies on university campuses are increasingly being reported. Critical among the reasons deduced for this apathy was the non-incorporation of user perspectives during the design process. To avoid such issues at SAUoT-a university that is at the initial stages of transforming into a SMART campus, the authors attempted to explore the utility of a PD approach to eliciting their views and shaping the decision-making process for implementation with such views. Findings from the study indicate that whereas user-centric PD approaches served as a veritable platform for achieving consensus among various stakeholders, it also had its pitfalls. Despite some of the pitfalls elucidated in the study, the use of the user- centric PD in product design activities, such as the case in the SAUoT scenario, holds positive implications for stemming burgeoning levels of user apathy for digital solutions associated with the SMART campus initiative.

References

Abuarqoub A, Abusaimeh H, Hammoudeh M, Uliyan D, Abuhashim M, Murad S, Al-Jarrah M, Al-Fayez F (2017) A survey on internet of things enabled SMART campus applications. In: Proceedings of international conference on future networks and distributed systems. IFNDS, Cambridge

Angelidou M (2015) SMART cities: a conjuncture of four forces. Cities 47:95–106

Bandara H, Jayalath J, Rodrigo A, Bandaranayake A, Maraikar Z, Ragel R (2016) SMART campus phase one: SMART parking sensor network. In: Manufacturing and industrial engineering symposium: innovative applications for industry, MIES, 2016

Barbato A, Bolchini C, Geronazzo A, Quintarelli E, Palamarciuc A, Pitì A, Rottondi C, Verticale G (2016) Energy optimization and management of demand response interactions in a SMART campus. Energies 9:398

Björgvinsson E, Ehn P, Hillgren PA (2012) Design things and design thinking: contemporary participatory design challenges. Des Issues 28:101–116

Blomberg JL, Henderson A (1990) Reflections on participatory design: lessons from the trillium experience. In: Proceedings of the SIGCHI conference on human factors in computing systems, 1990. ACM, New York, pp 353–360

Campuzano F, Doumanis I, Smith S, Botia JA (2014) Intelligent environments simulations, towards a SMART campus. In: 2nd International Workshop on SMART University, 2014

Celino I, Kotoulas S (2013) SMART cities [Guest editors' introduction]. IEEE Internet Comput 17:8–11

Dalsgaard P (2012) Participatory Design in Large-scale Public Projects: challenges and opportunities. Des Issues 28(3):34–47

Davies B (2015) Internet of everything-powering the SMART campus and the SMART city: Geelong's transformation to a SMART city [Online]. Deakin University IBM CISCO, Geelong. Available: https://www.bhert.com/events/2015-06-08/SMART-Cities-Round-Table-Report-June-2015.pdf. Accessed 17 May 2017

Dong X, Kong X, Zhang F, Chen Z, Kang J (2016) OnCampus: a mobile platform towards a SMART campus. Springerplus 5:974

Fan J, Stewart K (2014) An ontology-based framework for modeling movement on a SMART campus. In: Analysis of movement data, GIScience workshop, Vienna, Austria, 2014

Happaerts S, Van Den Brande K, Bruyninckx H (2010) Governance for sustainable development at the inter-subnational level: the case of the network of regional governments for sustainable development (nrg4SD). Reg Fed Stud 20:127–149

Hirsch B, Ng JW (2011) Education beyond the cloud: anytime-anywhere learning in aSMART campus environment. In: International conference for internet technology and secured transactions (ICITST), IEEE, pp 718–723

Hupe P, Nangia M, Hill M (2014) Studying implementation beyond deficit analysis: the top-down view reconsidered. Public Pol Admin 29:145–163

Kübler A, Holz EM, Riccio A, Zickler C, Kaufmann T, Kleih SC, Staiger-Sälzer P, Desideri L, Hoogerwerf E-J, Mattia D (2014) The user-centered design as novel perspective for evaluating the usability of BCI-controlled applications. PLoS One 9:e112392

Leal Filho W (2011) About the role of universities and their contribution to sustainable development. High Educ Pol 24:427–438

Malatji EM (2017) The development of a SMART campus-African universities point of view. In: 8th international renewable energy congress (IREC), pp 1–5

Mellouli S, Luna-Reyes LF, Zhang J (2014) SMART government, citizen participation and open data. Inf Polity 19:1–4

Muller MJ (2003) Participatory design: the third space in HCI. Human-Comput Interact Dev Proc 4235:165–185

Patil L, Dutta D, Bement A Jr (2015) Educate to innovate: factors that influence innovation: based on input from innovators and stakeholders. National Academies Press, Washington, DC

Rha J-Y, Lee J-M, Li H-Y, Jo E-B (2016) From a literature review to a conceptual framework, issues and challenges for SMART Campus. J Digital Converg 14:19–31

Ritter FE, Baxter GD, Churchill EF (2014) Foundations for designing user-centered systems. Springer, London. https://doi.org/10.1007/978-1-4471-5134-0_2

Simonsen J, Hertzum M (2012) Sustained participatory design: extending the iterative approach. Des Issues 28:10–21

Waage J, Yap C, Bell S, Levy C, Mace G, Pegram T, Unterhalter E, Dasandi N, Hudson D, Kock R (2015) Governing the UN sustainable development goals: interactions, infrastructures, and institutions. Lancet Glob Health 3:251–252

Part IX
Comfort

What are ways to increase the outdoor and indoor quality of buildings. This theme discusses thermal comfort and visual performance.

Chapter 31
Outdoor Comfort in Metro Manila: Mitigating Thermal Stress in Typical Urban Blocks by Design

Juanito Alipio A. de la Rosa

Abstract Fast-developing Metro Manila has urban microclimatic conditions that continue to veer away from comfort, primarily due to urban warming caused by the built environment. The hot and humid background climate, projections on global warming, and the urban heat island effect further aggravate the situation, thus causing more people to seek comfort in air-conditioned environments. With real estate developers and local government units building at an unprecedented rate, thermal stress is expected to worsen, unless interventions are put in place.

This paper examines workable solutions to bring comfort back to outdoor spaces and extend the range of their use, particularly in Metro Manila's central business districts. With the aid of RayMan and ENVI-met as computer simulation tools, the impact of microclimatic strategies on the Physiologically Equivalent Temperature (PET) is assessed. PET is the thermal index used for its appropriateness in gauging outdoor comfort. The four selected interventions are the following: vegetation, shading, water features, and high-albedo materials. Their combinations are likewise explored.

Simulating an idealized model of the urban block on the warmest day and hour, it was found that as a single intervention, vegetation (trees and grass) is the most effective, reducing average PET by up to 2.4 °C. As for the combined interventions, the use of vegetation, shading and water proved to be the most effective design strategy, decreasing average PET by up to 7.2% or 2.8 °C, while providing, at the same time, better environmental quality.

Although the proposed interventions brought the PET values closer to the outdoor comfort range of 22–34 °C and improving from Strong Heat Stress to Moderate Heat Stress, a more drastic approach in coverage is needed to keep outdoor urban spaces resilient and sustainable amid soaring temperatures.

J. A. A. de la Rosa (✉)
Architectural Association School of Architecture, London, UK

Makati Development Corporation, Taguig, Philippines
e-mail: De-La-Rosa1@aalumni.org

© Springer Nature Switzerland AG 2020   445
R. Roggema, A. Roggema (eds.), *Smart and Sustainable Cities and Buildings*,
https://doi.org/10.1007/978-3-030-37635-2_31

Lastly, the practical approach and replicability of the interventions are envisioned to facilitate local government units and real estate developers into incorporating outdoor comfort in planning and design.

Keywords Outdoor thermal comfort · Hot and humid · Urban · Sustainability · EnviMET · Rayman · PET

31.1   Introduction

Comfort in the city, particularly in hot and humid outdoor environments of Metro Manila, is quite a challenge. On top of the background climate that is already getting warmer over the years, the city's morphology and human activities within tend to further aggravate warming of the urban environment. So is the case in Metro Manila's existing and emerging Central Business Districts (CBD), which are seen to alter the urban microclimates negatively.

Differences in temperature between central and suburban sites have been documented in Metro Manila, where temperatures could reach values up to 10–14 K higher than surrounding rural areas (Estoque and Maria 2000). Tiangco et al. (2008) also reveals that Metro Manila's Urban Heat Island (UHI) profile peaks at the Makati CBD, which indicates it as the hottest area (Fig. 31.1).

The UHI phenomenon is further compounded by the larger global warming effect. Global increase in air temperature was estimated at around 0.15 K per decade over the last 20 years (Hulme and Jenkins 1998). In the Philippines, the Manila Observatory reported a 1 K increase in air temperature over a 35-year period or equivalent to an increase of 3 K in 100 years (Estoque and Maria 2000).

More people continue to migrate towards the metropolis, and, given such conditions, more are likely to experience thermal stress in outdoor urban environments.

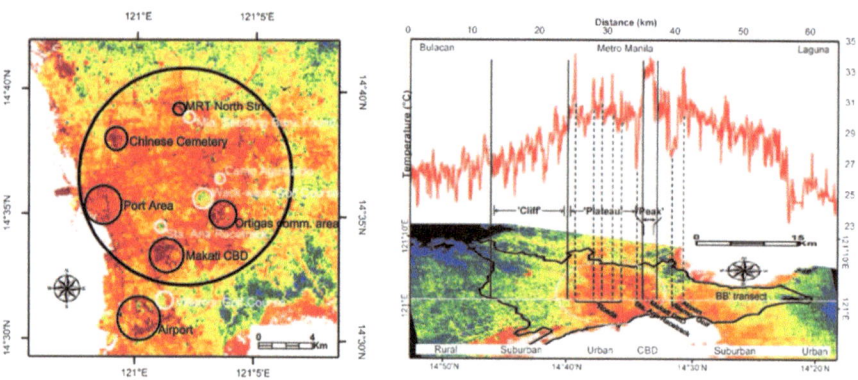

Fig. 31.1 Metro Manila urban heat island map with profile peaking at the Makati CBD. (Source: Tiangco et al. 2008)

Consequently, this leads to less use of public outdoor spaces and reliance on air-conditioned environments will continue to increase.

Despite the apparent thermal stress being experienced by millions of its citizens on a daily basis, very few studies are conducted regarding outdoor comfort in the Philippine setting. Local government units and real estate developers, who are both at the forefront of the country's fast-paced infrastructure development, could highly benefit from performance-based design in their planning.

This research examines interventions to improve urban microclimate conditions in a Philippine CBD setting, with the aim of mitigating thermal stress on the pedestrians. Thermal comfort is investigated using RayMan (Matzarakis et al. 2010), which estimates the Physiologically Equivalent Temperature (PET) – an appropriate thermal comfort index for outdoors. ENVI-met parametric simulations are carried out to assess the impact of vegetation, shading, water bodies, and high-albedo materials on both air temperature and mean radiant temperature (after Chatzidimitriou 2015). ENVI-met considers more variables in determining the mean radiant temperature, such as local wind speed and diffuse reflected radiation and longwave radiation of buildings and vegetation (Naboni et al. 2017). The meteorological output from ENVI-met is then used to calculate PET via RayMan.

31.1.1 Outdoor Thermal Comfort Index

This study uses PET as the thermal comfort index given its suitability for outdoor environments. Based on the Munich Energy-balance Model for Individuals, PET is defined as the air temperature, at which in a typical setting (without wind and solar radiation), the heat of the human body is balanced with the same core and skin temperature as under the complex outdoor conditions to be assessed (Hoppe 1999).

Matzarakis and Mayer (1996) developed the PET thermal index range, which is useful but not necessarily applicable to all settings due to climate specificities. In warm climates, for example, the mean radiant temperature can be twice as significant as the dry bulb temperature due to lighter clothing (Szokolay 2014). Matzarakis and Mayer's range remains applicable for Western and Middle European locations but for the Philippines, a more fitting range is used based on a study by Lin and Matzarakis (2008) for Taiwan's hot and humid climate.

Table 31.1 shows the adjusted range for Taiwan, with neutral temperatures at 26–30 °C, as opposed to the 18–23 °C used in Europe. The neutral temperatures are considered within comfortable range. However, it is also true that expectations for outdoor spaces are different, thus allowing for a wider range of tolerable temperatures below and above the neutral. Light cold stress (PET 22–26 °C) and light heat stress (30–34 °C) do not necessarily render the outdoors uncomfortable, especially with adaptive opportunities that could be available at the site. This research endeavors to study solutions that could bring the PET closer to the defined adaptive outdoor comfort range of 22–34 °C.

Table 31.1 PET range for Western/Middle Europe and Taiwan/Philippines

Thermal sensation	PET range Western/Middle Europe	PET range Taiwan/Metro Manila	Physiological stress
Very cold	<4 °C	<14 °C	Extreme cold stress
Cold	4–8 °C	14–18 °C	Strong cold stress
Cool	8–13 °C	18–22 °C	Moderate cold stress
Slightly cool	13–18 °C	22–26 °C	Light cold stress
Neutral	18–23 °C	26–30 °C	No thermal stress
Slightly warm	23–29 °C	30–34 °C	Light heat stress
Warm	29–35 °C	34–38 °C	Moderate heat stress
Hot	35–41 °C	38–42 °C	Strong heat stress
Very hot	>41 °C	>42 °C	Extreme heat stress

Source: after Matzarakis and Mayer (1996) and Lin and Matzarakis (2008)

31.1.2 Background Climate and Fieldwork

Located $14°36'$ North of the equator, Metro Manila has a tropical monsoon (Am) climate as per Köppen-Geiger classification. This is characterized by hot and humid conditions all year long with a narrow diurnal temperature change of 5–6 °C. The monthly mean minimum temperature is 25 °C and mean maximum is 31 °C. The annual average temperature is 29 °C and relative humidity, 77%. Annual average wind speed is 3.5 m/s.

The warmest months in Metro Manila are from March to May. This period is also the meteorological summer (as opposed to astronomical summer) for the entire country, during which the schools are on holidays. With April being the hottest month, this study focuses on this period as regards analysis and proposed interventions.

Spot measurements during a fieldwork done in April at the Makati CBD, where the UHI profile peaked, are considered and used as RayMan input to determine the outdoor thermal comfort, or lack thereof. The hours of 9:00 to 18:00 are also chosen, representing the typical busy hours in the CBD and during which foot traffic is also highest.

Table 31.2 shows PET values reaching around 45 °C from morning until early afternoon, which reveals how much heat stress typical pedestrians endure outdoors. Worth noting is that 12:00 noon registered the highest Ta and PET. Being under the shade, however, results in PET values that are comparable with the Ta and within the defined outdoor comfort range. By 18:00, when there is little or no direct solar radiation, PET significantly improves. However, many people are already on the

Table 31.2 Makati CBD (Ayala Triangle area) PET calculation for selected hours using spot measurements and RayMan

Hour	Ta	RH	Wind	Tmrt	PET	PET (shade)
09:00	30.9 °C	67%	0.7 m/s	60.6 °C	**45.0 °C**	31.5 °C
12:00	33.6 °C	54%	1.2 m/s	61.2 °C	**45.8 °C**	33.3 °C
15:00	32.1 °C	52%	1.1 m/s	60.6 °C	**44.5 °C**	32.4 °C
18:00	30.2 °C	66%	1.9 m/s	21.9 °C	26.5 °C	26.5 °C

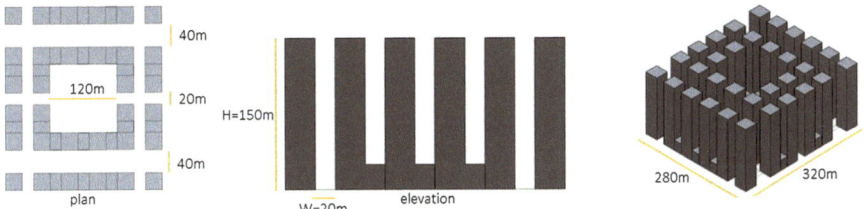

Fig. 31.2 Dimensions of the idealized block representative of Makati CBD

way out of the CBD and back to their respective homes at this time so the critical period is defined from morning until mid-afternoon with noontime peak.

31.2 Improving the Typical Urban Block

With more CBDs proliferating Metro Manila, taller buildings crowd the typical urban block. The remaining open spaces are the streets, parking lots, and some existing pocket parks. With the urban block as the area of interest for this study, an idealized model is defined for the purpose of parametric analysis. The model is intended to represent a typical scenario, which can be applicable to other CBDs in Metro Manila.

Since the 1960s, taller buildings are being erected across the metropolitan capital. For this study, the urban block model assumes 150-meter high buildings to represent the average height of the newer skyscrapers. The block measures 320 × 280 meters and the buildings are arranged around an open common space (Fig. 31.2).

The ENVI-met model is made within the model area of 95 × 95 × 35. The size of the grid cells are 5 m for both dx and dy and 10 m for dz, which are within the 0.5–10-meter recommended size for modeling (Bruse et al. 2014). The vertical grid has an equidistant generation. The simulation time is set for 12 h starting at 6:00 in the morning to cover the peak working hours in the CBD wherein thermal stress is likely. April 21 is set as the start date, which is within the typical warmest week in the Philippines. The following meteorological conditions are used (Table 31.3):

The base case run for the typical urban block generates results comparable to the fieldwork, having air temperatures at around 30 °C, Tmrt at approximately 60 °C,

Table 31.3 ENVI-met input data for meteorological conditions

Wind speed at 10 m height:	3 m/s
Wind direction:	45° (Northeast)
Roughness length at site:	0.01
Initial temperature of atmosphere:	302.87 K (29.72 °C)
Specific humidity at model top (2500 m):	7.0 g/kg
Relative humidity at 2 m:	73%

and PET also reaching close to 45 °C. With the base case set, proposed interventions are modeled and subjected to runs.

31.3 Proposed Interventions and Analysis

With the aim of mitigating heat stress at the typical urban block especially at peak summer conditions, the following variables are studied (after Chatzidimitriou 2015):

31.3.1 Vegetation

A typical tree is modeled in ENVI-met to mimic an ideal tree species for streets in the Philippines: the "dita" tree or alstonia scholaris. Bigger trees are used for major roads and courtyards while smaller trees are assumed for minor roads. A leaf area density (LAD) of 2 is used, which is typical of tropical trees. A total of 68 big trees and 72 small trees were added to the model's open spaces. Grass with 10 cm average density was also used, having a total coverage of 16,000 m².

31.3.2 Shading

The next parameter for testing is the effect of shading. Opaque canopies attached to the buildings along the pedestrian walkways are added to the base case model, with a height of 5 m from the ground. This follows the typical ground floor height of commercial and office building lobbies in the CBD.

31.3.3 Water

Water features in the ENVI-met model assume a pond depth of 0.60 m and fountain height of 2 m.

31.3.4 High-Albedo Materials

The pavement and roadways, which are of dark granite and asphalt respectively, are changed into higher-albedo surfaces in the model.

The individual and collective impacts of these interventions (Table 31.4) on the urban block are analyzed with respect to the base case. Results are compared in terms of Ta, Tmrt, and ultimately PET, using data for 12:00 PM being the hottest hour as per fieldwork and weather data.

Comparing the individual effects of each parameter on the base case, vegetation and shading seems to be the most promising, thus reaffirming the choice of these two adaptive measures in Lin's (2009) study on a Taiwan public space. The vegetation demonstrated the highest reduction in air temperature (1.4%), whereas shading reduced the Tmrt the most (up to 6.2%). Water did not do as good as expected, with only less than 1% reduction in Ta, Tmrt, and PET. The high-albedo materials

Table 31.4 ENVI-met input data for each of the parameters

1. Vegetation

	Tree species:	Alstonia scholaris
	Leaf Area Density:	2
	Tree height:	20 m, 15 m
	Tree width:	15 m, 9 m
	Foliage albedo	0.20
	Grass ave. density:	10 cm

2. Shading

	Structure:	Building canopy
	Canopy height:	5 m
	Albedo:	0.20

3. Water bodies

	Type:	Pond, fountain
	Water depth:	0.60 m
	Fountain height:	
		2 m

4. High-albedo surfaces

	Roadway base case:	0.12
	Roadway albedo – Light new concrete:	0.40
	Pavement base case:	0.30
	Pavement – Medium coloured stone:	0.40

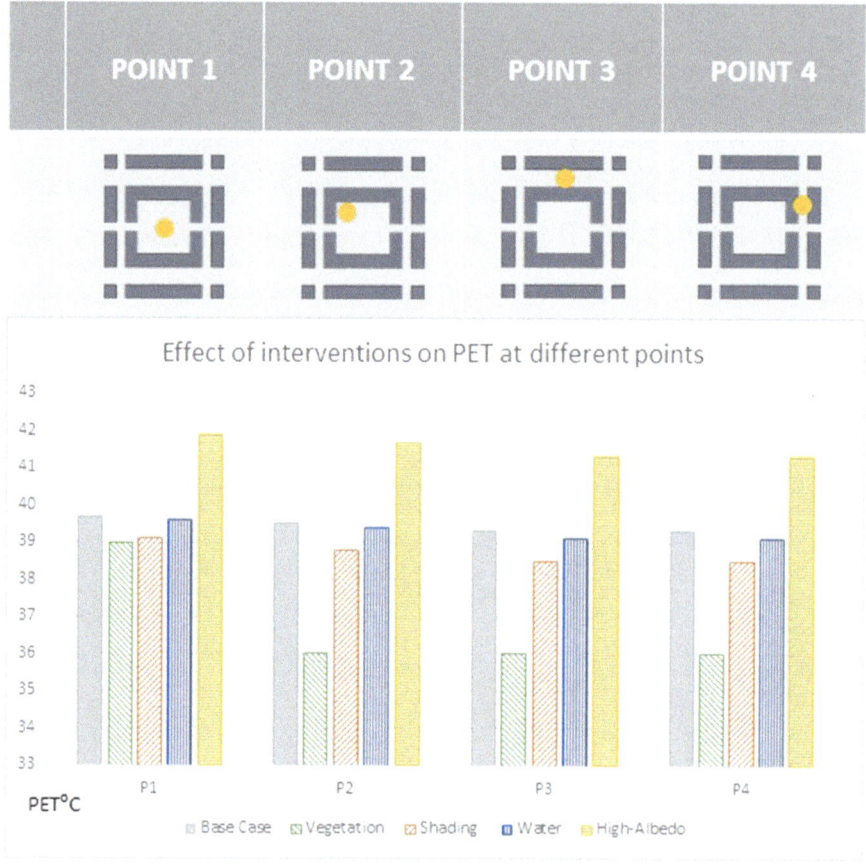

Fig. 31.3 Impact of the individual parameters on PET at the four identified locations within the block model

did reduce Ta more than shading and water, but it worsened the PET values due to higher Tmrt.

To further illustrate the parametric runs spatially, four points are identified in the block model (Fig. 31.3) to see the impact of location on thermal stress, aside from just employing the interventions.

Observing the four points (P1, P2, P3, P4), it can be inferred that the use of vegetation showed the highest potential in mitigating thermal stress, followed by shading. Areas closer to the buildings also generated better PET values than the area in the middle of the open space. This can be attributed to the shading provided by the buildings. In prioritizing improvement measures based on impact and according to simulation results, the following order of importance may be established: (1) Vegetation (2) Shading (3) Water features.

31.4   Optimization

Further analysis is carried out to assess the effects of combining the microclimatic measures, and examine if further improvement on outdoor conditions can be achieved. The following combinations are then simulated:

- Vegetation + Shading
- Vegetation + Shading + Water Features
- Vegetation + Shading + Water Features + High-albedo materials

Figure 31.4 shows the effects of the combined measures tested on air temperature. Results show that the combination of all four interventions managed to bring down Ta by up to 0.6 K.

For PET, however, the combination of first three elements namely vegetation, shading, and water, is more effective (Fig. 31.5) than all four combined. PET was reduced by up to 3.5 K, such as in P2. These PET values, now within the region of 35 to 37 °C, are much closer to the comfort range identified and have improved in scale from Strong Heat Stress to Moderate Heat Stress. Moreover, it is notable from the study that high-albedo materials have a negative impact on PET, even if they reduced the Ta.

In terms of average differences from the base case using data from the four points identified in the block, Table 31.5 summarizes the findings of the study, showing the potential of the interventions in improving outdoor comfort. As a single approach, use of vegetation alone lowered the PET by an average of 2.4 K. Combining vegetation, shading, and water reduced PET by an average of 7.2% or 2.8 K. This

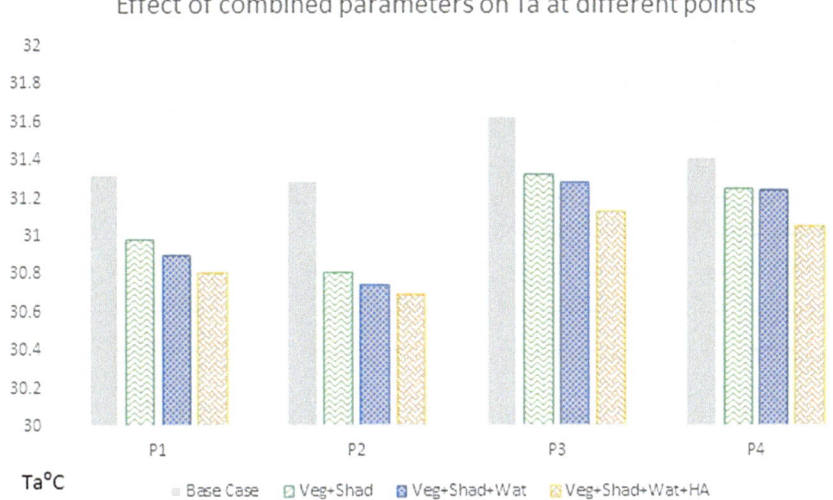

Fig. 31.4 Impact of combining the different parameters on air temperature (Ta) at four points within the block

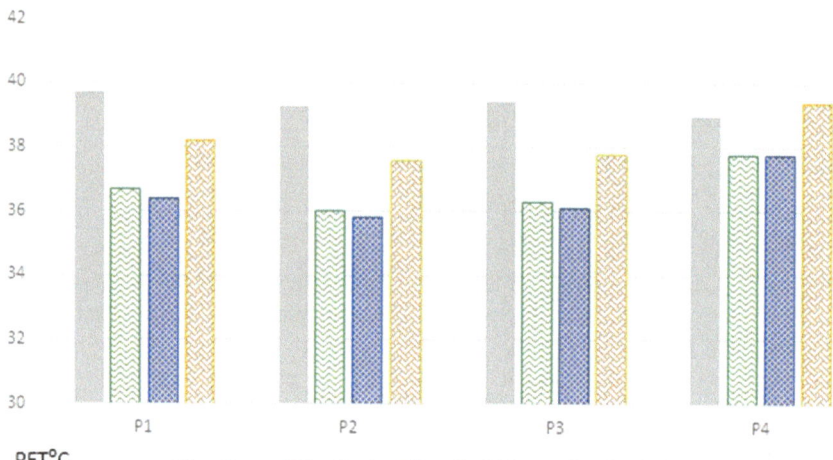

Fig. 31.5 Impact of combining the different parameters on PET at four points within the block

Table 31.5 PET average differences of intervention/s from base case based on ENVI-met run at 12:00 PM of April 21 (hottest day and hour)

Intervention/s:	PET	
	Ave. diff.	Ave. Δ%
1. Vegetation only	−2.4 °C	−6.2%
2. Shading only	−0.7 °C	−1.7%
3. Water only	−0.1 °C	−0.3%
4. High-albedo materials only	+2.2 °C	+5.5%
5. Vegetation + shading	−2.7 °C	−6.7%
6. Vegetation + shading + water	−2.8 °C	−7.2%
7. Vegetation + shading + water + high albedo	−1.1 °C	−2.8%

becomes a useful tool in decision-making for microclimatic improvements in urban areas, especially in hot and humid Philippine setting. These interventions, together with adaptation and perceived control by the occupants, strengthen the possibility of reducing thermal stress.

31.5   Conclusion

In Metro Manila's CBDs, people do experience thermal stress as confirmed by the fieldwork and simulations done for this research. As more CBDs emerge and more people move towards the city, the situation is likely to aggravate. Real estate developers and local government units, who are at the forefront of the country's physical development, continue the building spree with little guidance from performance-based design research.

The study proposes improvements in typical urban blocks within the CBD, with the aim of reducing thermal stress. Gauged by PET, which is deemed appropriate for outdoor comfort, interventions are studied singly and in combinations. An adjusted comfort range for PET is used, with 22 to 34 °C as the target region.

The four interventions studied are vegetation, shading, water features, and high-albedo materials. In reducing PET, only the first three have been effective based on the simulations. The most significant improvement is brought about by vegetation, which reduced PET by up to 3 K in extreme conditions and 2.4 K on average or 6.2% improvement. The use of shading followed, with an average PET reduction of 0.7 K. By combining both vegetation and opaque shading, PET is reduced by up 2.7 K on average.

Best performance is by the combination of vegetation, shading, and water features, which reduces the base case PET by 3.5 K in extreme conditions and 2.8 K on average or equivalent to 7.2% improvement. In simulating the extreme condition using data from the hottest day and hour of the year, this combined solution brought the PET values closest to the target comfort range.

Despite the overall reduction in PET (from Strong to Moderate Heat Stress), a more drastic approach is needed for the PET values to be comfortably within the range of thermal neutrality. For vegetation, even more trees and larger grass coverage are recommended. Increased coverage of opaque shading should also be in place. As temperatures continue to soar, these practical interventions will keep the open spaces in city blocks resilient and sustainable.

In Metro Manila where CBDs are mostly managed by private estates and local government units, budgets are usually very limited so this study could help the entities managing the estates to decide on which intervention to use based on budget, timeline, and progressive interventions. For instance, using vegetation alone would yield similar PET improvement as using the combination of vegetation, shading and water features. But in terms of cost or overall environmental quality, the differences are more pronounced. On the side of the government, integration of appropriate interventions discussed could form part of local ordinances and guidelines, which could be replicable in other existing and emerging CBDs across several cities in the metropolis.

References

Bruse M et al (2014) ENVI-met (Version 4.0). Available at: http://www.envi-met.info/public/files/previews/ENVI metV4setup.exe. Accessed 18 May 2015

Chatzidimitriou A (2015) Microclimate studies in open urban spaces. In Lecture, Architectural Association School of Architecture

Estoque MA, Maria MVS (2000) Climate change due to urbanization of metro Manila. Technical reports. Manila Observatory, Manila

Hoppe P (1999) The physiological equivalent temperature – a universal index for the biometeoro-
logical assessment of the thermal environment. Int J Biometeorol 43(2):71–75

Hulme M, Jenkins G (1998) Climate change scenarios for the United Kingdom. Scientific report.
UK Climate Impacts Programme, Climatic Research Unit, Norwich

Lin TP (2009) Thermal perception, adaptation and attendance in a public square in hot and humid
regions. Build Environ 44:2017–2026

Lin TP, Matzarakis A (2008) Tourism climate and thermal comfort in Sun Moon Lake, Taiwan. Int
J Biometeorol 52:281–290

Matzarakis A, Mayer H (1996) Another kind of environmental stress: thermal stress. WHO
Collaborating Centre for Air Quality Management and Air Pollution Control. Newsletters
18:7–10

Matzarakis A, Rutz F, Mayer H (2010) Modelling radiation fluxes in simple and complex environments: basics of the RayMan model. Int J Biometeorol 54:131–139

Naboni E et al (2017) An overview of simulation tools for predicting the mean radiant temperature
in an outdoor space. In: Proceedings CISBAT 2017 international conference

Szokolay S (2014) Introduction to architectural science. The basis of sustainable design, 3rd edn.
Routledge, London

Tiangco M et al (2008) ASTER-based study of the night-time urban heat island effect in Metro
Manila. Int J Remote Sens 29(10):2799–2818. Taylor & Francis

Chapter 32
Markov Logic Network-Based Group Activity Recognition in Smart Buildings

Hao Chen and Tae Wan Kim

Abstract The rise of the smart building has promoted various pervasive computing technologies to be used in the intelligent building system. People tend to be in groups inside a building, working together, taking classes, having meetings, etc. Being able to recognize such group activities will be key to making a functionalized building an activity-aware smart system. In that way, the system can adjust the surrounding atmosphere automatically according to the detected group activity and adapt to the needs of individuals or groups. Exiting works on group activity recognition (GAR) mainly focus on computer vision through surveillance hardware, which suffers from privacy and illumination problems. We decided to use the smartphone to identify GAR, considering it's pervasive and ubiquitous properties as well as various built-in sensors such as an accelerometer, gyroscope, microphone, etc. In this paper, we first conduct an extensive literature review relating to GAR in smart buildings. Our goal is to recognize fine-grained group activity by utilizing coarse-grained smartphone sensors.

Keywords Group activity recognition · Smart building · Markov logic networks · Sensors

32.1 Introduction

Smart building aims at integrating intelligence into the building environment to provide occupants with attractive services, reliable safety systems, and user-centered control information (Goldstein and Talon 2015). The building collects information from occupants to recognize their status, such as activity, location, and heating/lighting preference, to refine settings that enhance the occupant's satisfaction

H. Chen · T. W. Kim (✉)
Department of Architecture and Urban Design, Incheon National University, Incheon, Republic of Korea
e-mail: taewkim@inu.ac.kr

© Springer Nature Switzerland AG 2020
R. Roggema, A. Roggema (eds.), *Smart and Sustainable Cities and Buildings*,
https://doi.org/10.1007/978-3-030-37635-2_32

(Fruchter et al. 2016). More recently, however, the attention has shifted from the understanding of individual occupants' status to group status, comprised of multiple occupants, as nearly 70% of the time we spend is within groups to achieve goals, acquire emotional needs, and improve social circles (Moussaïd et al. 2010). Understanding group activity in a building has more meaningful prospects in achieving smart building.

Group Activity Recognition (GAR) in the building has many meaningful applications. Firstly, the building designers are able to explore the relationship between space usage by the group and corresponding group activity (Kim 2013), and group preference in choosing the space for specific group activity (Kim et al. 2018), to optimize building space design. Secondly, building management systems can provide an accurate response for occupants' group activity, for example, meeting activity needs a bright lighting, and comfortable temperature. Multiple discussion activity, on the other hand, might need cooler temperatures, considering the heating emission of the group. Also, GAR in the building may be related to evacuation research. When the building is in an emergency situation, people tend to stay in a group with friends or family for a sense of safety during evacuation (Mawson 2005). Group activity and group location will provide valuable information for rescues.

Our objective is to recognize group activity recognition using smartphones. We selected smartphones by considering its pervasive and ubiquitous characteristics, as well as scalability and popularity among modern people. we proposed the Markov Logic Network (MLN) based group activity, and first built up an ontology that encoded the relationship between group activity and its context information: individual activity, location, role information, sound activity. We mainly focused on two group activities: taking class and going to a meeting, but it can also be extended to more group activity. The generated ontology is then converted to the equivalent first order logic, which will be used to generate probabilistic ontology through the novel utilization of MLN. The probabilistic ontology serves as a knowledge base that offers consistent knowledge and enables uncertainty modeling, contextual modeling, and GAT in a unified framework. We validated the proposed methods in an academic building, and the results reveal the effectiveness of our method.

32.2   Related Work

Depending on the use of sensors in data collection, GAR can be categorized as vision based and sensor-based GAR. Vision based GAR has been in-depth explored in the computer vision domain (Murino et al. 2017). Although this approach can provide a global perspective towards a group, it may violate occupants' privacy and cause a lack of scalability to different scenes. Sensor-based GAR, monitoring the occupant's states using unobtrusive sensors, like a smartphone, ultrasonic, or wearable sensor, provides a ubiquitous and nonintrusive way to monitor occupants' states. Sensor based GAR can be re-classified as data-based GAR and knowledge-based GAR (Liang and Cao 2015):

Data-based GAR normally recognizes group activity in machine learning per-spective by extracting discriminative features from the collected dataset to detect group activity with uncertainty. Group activity can be characterized with different features. Firstly, group members' activity can provide important cues for their collective activity. Zhang et al. (2006) proposed a two-layer Hidden Markov Model (HMM) to detect individual activity and use it to inference group activity, similarly. Gordon et al. (2013) collected sensor data from cafe cups holding by group members, the individual activity recognized for each café cup are fused together with the group classifier to detect group activity like meetings, presentations and coffee breaks. Secondly, interaction among people in the process of group activity can be used to differentiate group activity. Wang et al. (2011) collected data from acceler-ation, audio and RFID, to model interaction among people and feed the extracted features to two probabilistic models – Coupled HMM and FCRF to recognize multi-user activities. In addition to individual activity and interaction features, the role which a group member is performing also provides important cues for GAR (Forsyth 2018). For example, in the speech activities, there will be two speech related roles: questioner and talker. A group member plays the role of context to guide the detection of group activity as the member is evolving. Hirano and Maekawa (2013) used GMM to cluster group members' roles and extract the feature from clusters, the extracted feature then feeds to HMM to recognize group activities. The proposed model is robust against changes in the number of participants and roles, and they do not need finely labeled training data. Data-based GAR needs an efficient collection of data to accurately model the group activity which introduces an overhead problem. This approach also has a re-usability problem as we need to build a new group activity model for new group members and different group sizes.

Knowledge-based GAR, on the other hand, constructs a reusable contextual model that represents and defines the knowledge related to the group activity with location, time, and individual activity (Chen et al. 2012). This approach can also be categorized as mining-based, logic-based, and ontology-based GAR (Liang and Cao 2015). Gu et al. (2009) developed a mining-based knowledge pattern, "Emerging Patterns (EP)", to describe significant changes between different activities. The EP is mined from a training dataset that serves as the knowledge base for further use. Ijsselmuiden et al. (2014) deduced group situations from annotated person tracks, object information, and annotated information about gestures, body poses, and speech activity. The author utilizes situation graph trees and fuzzy metric temporal logic rules to build up a knowledge base, thus inferring the corresponding group activity. Also, the starting time and duration of group activity also performs an important role in characterizing group activity. Choi and Yong (2015) built up rule-based knowledge that encodes group activity with relevant duration time, location, and members' behavior. Loke and Abkenar (2017) proposed a knowledge-based group detection language named "GroupSense-L" that encodes context information like location, member's activity, member's sound level, as well as start time, to differentiate group activity. Although knowledge-based GAR has properties like scalability and reusability, it has limitations in performing reasoning over temporal and uncertain data, making the GAR a challenging task.

To address the above issues existing in both data-based and knowledge-based GAR, an activity model that integrates the advantages and eliminate the shortcoming of above two methods is in its urgent. Markov Logic Network (MLN) (Richardson and Domingos 2006) is one of the methods that satisfy the requirements. MLN is a probabilistic logic which applied the ideas of the Markov Network to inductive logic programming, thus enabling uncertain inference on knowledge base.

32.3   Proposed Method

Figure 32.1 shows an overview of our system. The framework is segmented into three layers: Data layer, Individual layer, and inference layer. We first collected raw data from the smartphone, which collected sensor data monitoring occupant's motion information, and iBeacon (received signal strength index) RSSI information that localized spatial position. The individual layer aims to recognize occupants' status at a certain time, the raw dataset needs to be pre-processed, and discriminative features that represent people's status, including activity, location, and speech

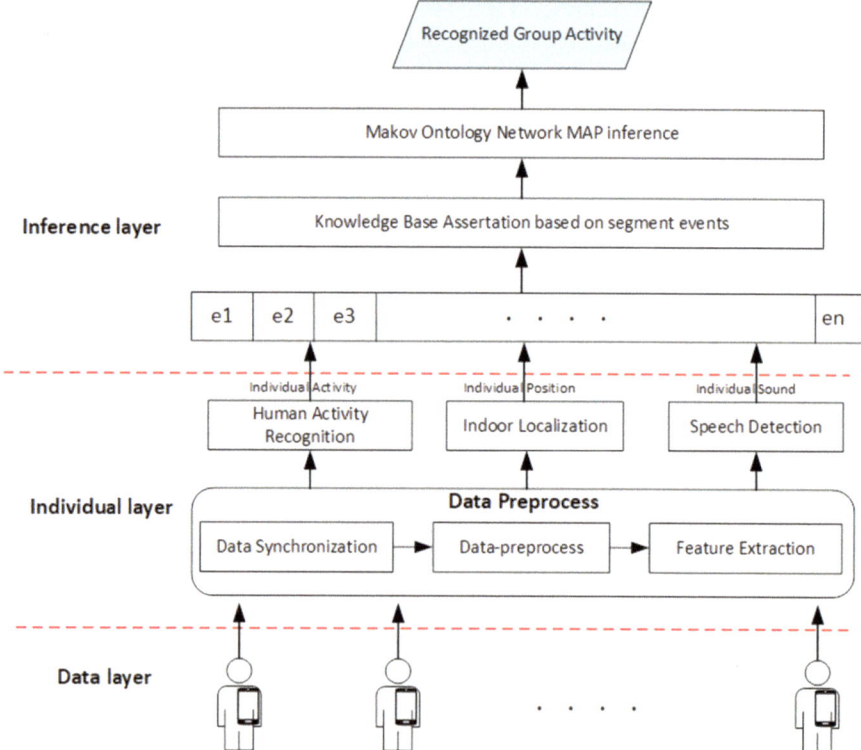

Fig. 32.1 Overall framework

recognition are then extracted, the processed data then feeds to three parts: human activity recognition, indoor localization, and speech recognition.

- Data Layer: In the data layer, we collect two types of data from people. Firstly, sensor data that monitors user's motion, orientation, rotation etc. In our work, we mainly collected accelerometer, gyroscope, magnetometer, and rotation sensor data from the smartphone. Secondly, we also collect RSSI from surrounding Beacon tags which are pre-installed in the building. This information will be used for indoor localization.
- Individual Layer: In this layer, we first synchronize group members' data to avoid device heterogeneity, and set up the same starting point, the data is then cleaned using resampling and butter low-pass filter techniques to remove outliers and smoothen raw data. The human activity recognition part is to recognize individual simple activity followed this work (Bulling et al. 2014), that is, in our work, sit, stand, and walk. Indoor localization parts localize occupant with the help of RSSI information and trilateral positioning method, we follow (Zhou et al. 2017) for its effectiveness and simple implementation, while speech detection followed by the threshold-based method by assuming that a person is considered speaking when the corresponding sound level is larger than 300db.
- Inference Layer: takes individual status (activity, location, sound) as input, based on the sound feature, we developed an algorithm to segment time series data into small event, which will feed into trained MLN knowledge-base to compute the inference group activity. More specifically, we first make an assertation of events and add them into ABox and make inference on the Markov Network.

32.3.1 Buildup Ontology

We use Web Ontology Language (OWL), a Description Logic (DL)-based markup language to formally define the semantics of group activity, individual activity and context factors. Specifically, DL is a formalism for knowledge representation that describes a given domain by defining relevant concepts (Terminological Box) and vocabularies of asserted individual and their relationship, also different operators and development of SWRL rules makes OWL 2 DL even more expressive and extensive in terms of knowledge representation. In this research, we proposed a Group Activity Ontology comprised of four main entity: Person, Activity, Location, Role, and the corresponding relationships among them, and all together defined the what, who, where, and how for each event. Here we provide an excerpt of our ontology shown in Fig. 32.2.

The basic idea is that complex group activity (CGA) can be decomposed into a hierarchal simple group activity (SGA), while SGA comes directly from ontology reasoning based on the input user's status information. We use "Action-Role-Action-Duration" schema to define SGA, for example, "StandTalkSit_long" can be represented as any group activity which has only one member that stands, and

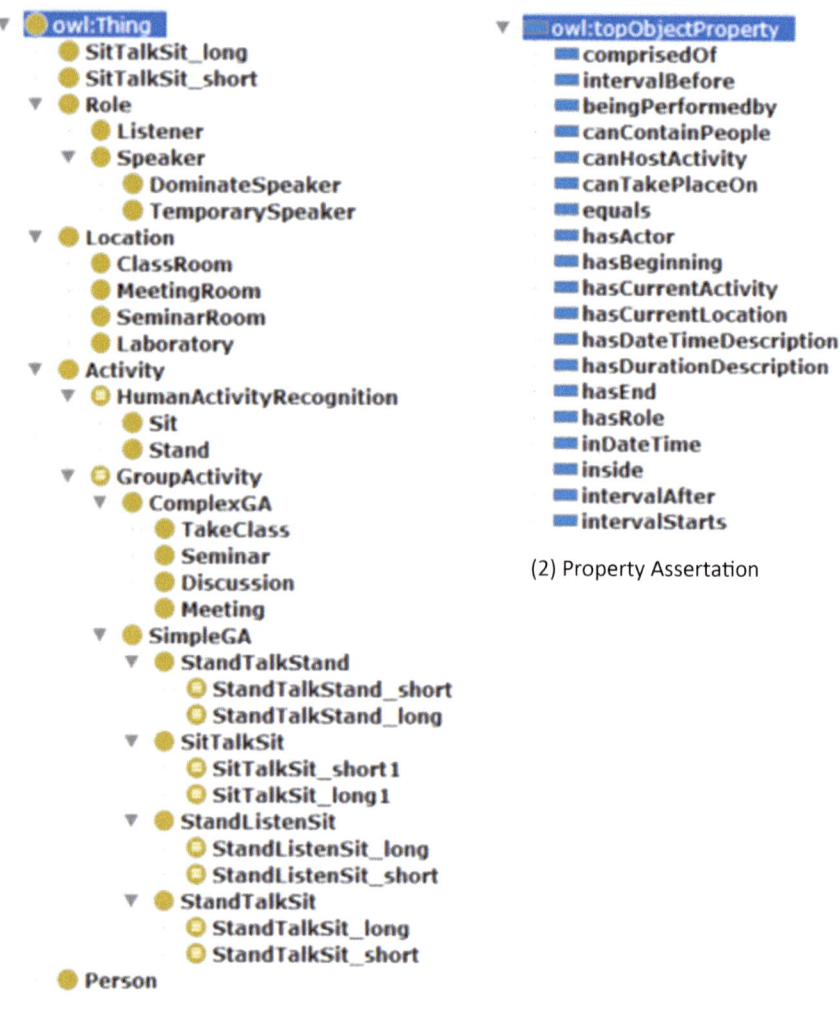

(1) Class Assertation

(2) Property Assertion

Fig. 32.2 Group activity ontology

performs DominateSpeaker, and has at least one member that sits and perform the listener role. In terms of role definition, we define explicit rules for representing role information:

Person(?p), hasRole(?p, ?r), Role(?r), hasSoundLevel(?p, ?sound), xsd:integer[>= 330](?sound), hasDuration(?p, ?du), xsd:float[>= 90.0f](?du) ->DominateSpeaker(?r)

Where the above rule defines that any person who has soundlevel larger than 330 db and consistently speaking for more than 90s are considered to perform "DominateSpeaker" Role.

32.3.2 *Markov Logic Network Modeling*

Markov Logic Network (MLN) is a powerful approach in representing uncertain knowledge, as it combines first-order logic (FOL) and Markov networks in the same representation. The main idea behind Markov logic is relaxing the hard rule with attached weight on the FOL, so a world that violate the FOL is not impossible, but less probable (Oliveira de 2009). Higher weight represents a bigger difference between the dataset and the rules. Using the weighted formulas, we are able to model probabilistic uncertainty on well-defined ontology knowledge base, in our words, combine the advantage of data-based and knowledge-based characteristics.

A MLN M is a finite set of pairs

$$(F_i, w_i), 1 \leq i \leq n$$

where F_1 is a formula in FOL and w_1 represents formula's weight. Together with a finite set off constants $C = \{c_1, \ldots, c_n\}$ it defines a Markov network M_L. A binary node on M_L is built based on each possible grounding of each atom in M, and also defines a log-linear probabilistic distribution on a ground Markov network:

$$P(x) = \frac{1}{Z} \exp \left(\sum_{i=1}^{F} w_i n_i(x) \right) \tag{32.1}$$

where F is the number of formulas in the MLN. $n_i(x)$ is the number of true groundings F_i in the world, w_i and is the weight of formulas.

32.4 Experiment Analysis

To evaluate our approach, we conducted an experiment in an academic building with different functional rooms, such as a classroom, meeting Room, Seminar Room, laboratory, etc. We deployed Beacon tag with 4 m intervals on the ceiling to avoid human effect when receiving signal. Samsung android devices were used to sample the accelerometer, gyroscope, and magnetometer at 50 Hz and record Beacon RSSI data at 1 Hz.

The smartphone was put into the right leg pocket for capturing motion of human center and 2 group activities were performed by 5 subjects (P1, S1-S4): (1) Take Class, (2) Meeting for 10 min each. The subjects were made up of 4 males and 1 female. The collected data from 5 subjects were synchronized, re-sampled to 50 Hz. For the human activity part, time and frequency domain features were extracted to recognize simple activity, while indoor localization parts takes RSSI data as input, and compute distance from an unknown point (current device position)

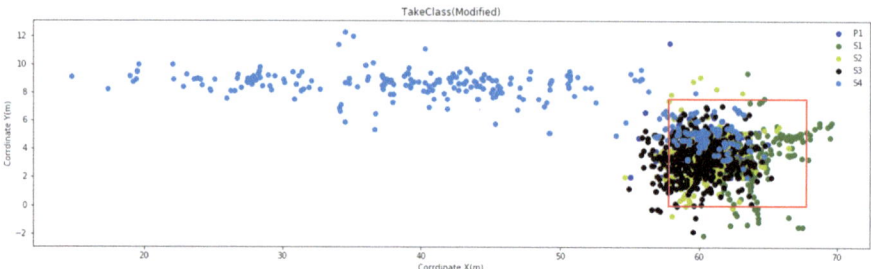

Fig. 32.3 Take class data distribution

to pre-installed Beacon coordinates using signal propagation equations (Zhou et al. 2017). Figure 32.3 shows the localization results for the take class activity:

Clearly, we can easily observe that teacher and student share similar motion characteristics, noticing that S4 seems strange compared to other trends. This is because S4 is a late student, who joined class activity after 5 min. we designed this scene to make the real-world situation more realistic. Moreover, P1 dominated the whole speaking activity, while S3 served the lowest sound level.

Acknowledgements This research was supported by Basic Science Research Program through the National Research Foundation of Korea (KNRF) (2018025981).

References

Bulling A, Blanke U, Schiele B (2014) A tutorial on human activity recognition using body-worn inertial sensors. ACM Comput Surv 46:1–33

Chen L, Hoey J, Nugent CD, Cook DJ, Yu Z (2012) Sensor-based activity recognition. IEEE Trans Syst Man Cybern Part C Appl Rev 42:790–808

Choi JI, Yong HS (2015) Conceptual group activity recognition model for classroom environments. In: International conference on ICT convergence 2015: innovations toward the IoT, 5G, and Smart Media Era, ICTC 2015, pp 658–661

Forsyth DR (2018) Group dynamics. Cengage Learning, Boston

Fruchter R, Cristina F, Rodriguez G, Law KH (2016) Space-mate: computational modeling for building and occupant cooperative sustainable performance

Goldstein N, Talon C (2015) Smart offices: how intelligent building solutions are changing the occupant experience. In: INTEL Sponsored White Paper

Gordon D, Hanne JH, Berchtold M, Shirehjini AAN, Beigl M (2013) Towards collaborative group activity recognition using mobile devices. Mobile Netw Appl 18:326–340

Gu T, Wu Z, Wang L, Tao X, Lu J (2009) Mining emerging patterns for recognizing activities of multiple users in pervasive computing. In: Proceedings of the 6th annual international conference on mobile and ubiquitous systems: computing, networking and services, pp 1–10

Hirano T, Maekawa T (2013) A hybrid unsupervised/supervised model for group activity recognition. In: Proceedings of the 2013 international symposium on wearable computers, ACM, pp 21–24

IJsselmuiden J, Münch D, Grosselfinger AK, Arens M, Stiefelhagen R (2014) Automatic understanding of group behavior using fuzzy temporal logic. J Ambient Intell Smart Environ 6:623–649

Kim TW (2013) Predicting space utilization of buildings through integrated and automated analysis of user activities and spaces. CIFE, Stanford

Kim TW, Cha S, Kim Y (2018) Space choice, rejection and satisfaction in university campus. Indoor Built Environ 27:233–243

Liang G, Cao J (2015) Social context-aware middleware: a survey. Pervasive Mob Comput 17:207–219

Loke SW, Abkenar AB (2017) Assigning group activity semantics to multi-device mobile sensor data. KI – Künstliche Intelligenz 31:349–355

Mawson AR (2005) Understanding mass panic and other collective responses to threat and disaster. Psychiatry Interpersonal Biol Process 68:95–113

Moussaïd M, Perozo N, Garnier S, Helbing D, Theraulaz G (2010) The walking behaviour of pedestrian social groups and its impact on crowd dynamics. (Chirico, G., Ed.). PLoS One 5: e10047

Murino V, Cristani M, Shah S, Savarese S (2017) Group and crowd behavior for computer vision. Academic Press, Oxford

Oliveira de P (2009) Probabilistic reasoning in the semantic web using markov logic. Master's thesis, University of Coimbra

Richardson M, Domingos P (2006) Markov logic networks. Mach Learn 62:107–136

Wang L, Gu T, Tao X, Chen H, Lu J (2011) Recognizing multi-user activities using wearable sensors in a smart home. Pervasive Mob Comput 7:287–298

Zhang D, Gatica-Perez D, Bengio S, Mccowan I (2006) Modeling individual and group actions in meetings with layered HMMs. IEEE Trans Multimedia 8:509–520

Zhou C, Yuan J, Liu H, Qiu J (2017) Bluetooth indoor positioning based on RSSI and Kalman filter. Wirel Pers Commun 96:4115–4130

Chapter 33
Impacts of Highly Reflective Building Façade on the Thermal and Visual Environment of an Office Building in Singapore

Jianxiu Wen, Nyuk Hien Wong, Marcel Ignatius, and Xinzhu Chen

Abstract The study examines the impacts of glare caused by highly reflective building façade material on the thermal and visual performance of surrounding development under tropical climate context. A 9-story commercial building with a curved stainless-steel façade was investigated, as it produces discomfort glare to the surrounding buildings due to direct exposure to the sunlight. A 6-story nearby office has been affected by the reflective sunlight from the stainless- steel façade, and thus it was selected for the case study.

Nine measurement points were selected and evenly distributed at level 2, 4, and 5 of the affected office building. Vertical and horizontal alignments were maintained for all measurement points at respective locations in the building. Sensors were put in strategic locations to record air temperature (T_i), glass (T_g) and wall surface temperatures (T_w), and indoor illuminance (I_{in}). Parameters such as weather condition, horizontal location, vertical location, and orientation were evaluated based on the data collected from the on-site measurement.

Data analysis has been carried out in diverse aspects. Results indicate that weather condition plays a vital role in both indoor thermal and visual environment. The average T_i, T_g, T_w, and I_{in} are higher on a sunny day that has high incoming solar radiation. Specifically, on a sunny day, the increase in average T_i, T_g and T_w is more than 1 °C compared to the rainy day. The difference of the maximum T_i between a sunny day and a rainy day can reach 3.26 °C, and the maximum I_{in} on a sunny day is 610.64 lux higher than that on a rainy day. These temperatures and illuminance differences are deemed substantial in the working environment.

Additionally, horizontal location and orientation also affect indoor thermal and visual performance of the affected building. Data evaluation concludes that points

J. Wen · N. H. Wong (✉) · M. Ignatius
School of Design and Environment, National University of Singapore, Singapore, Singapore
e-mail: bdgwnh@nus.edu.sg

X. Chen
Graduate School of Design, Harvard University, Cambridge, MA, USA

© Springer Nature Switzerland AG 2020
R. Roggema, A. Roggema (eds.), *Smart and Sustainable Cities and Buildings*,
https://doi.org/10.1007/978-3-030-37635-2_33

467

directly facing and nearer to the affecting reflective façade indicate worse thermal and visual performance during the affecting period which is 16:00 to 18:00 in this case study. When the stainless-steel façade was exposed to the sun, reflective sunlight generated and a sharp increase in T_i, T_g and T_w can be observed at the measurement points which are directly facing to the reflective façade. This phenomenon is more evident at the point nearest to the reflective façade than other points.

Furthermore, vertical location is another significant factor. Results show that temperatures and illuminance at the higher level will normally be higher than that at the lower level. However, the external heat source such as the reflective sunlight, in this case, will lead to an abnormal increase in temperatures and illuminance during the affecting period. This changing trend is more obvious for points at the lower level. During the affecting period from 16:00 to 18:00, the average T_i at level 2 is 1.66 °C higher than that at level 4.

Keywords Reflective building façade · Reflected sunlight · Thermal performance · Visual performance

33.1 Introduction

The trend of using highly reflective building materials such as glass or mental claddings has been increasing in the building industry in Singapore. However, the highly reflective building façade materials can generate reflective sunlight which may affect both thermal and visual environment of surrounding developments (Danks et al. 2016). This phenomenon is more urgent in Singapore than other regions due to the strong solar radiation of tropical climate.

Previous studies have been conducted to examine the relationship between the building façade and reflected sunlight. Wong (2016) developed a ray tracing model to investigate the impact of solar reflection from building façade on the surrounding environment. Brzezicki (2012) evaluated the effects of glare reflection on the surrounding environment by comparing differently shaped glass facades. Yang (2013) conducted one parametric study using Rhino and Grasshopper to simulate and compare the reflected daylight from different building envelopes. Suk (2017) found that there are strong correlations between human visual discomfort and excessive sunlight reflections from building envelopes. However, few researchers have studied the impact of highly reflective building façade on the indoor thermal and visual environment of surrounding developments. Hence, the study to investigate how highly reflective façade will affect the thermal and visual performance of the surrounding building is necessary and meaningful.

The research objectives are: to study the impact of reflected sunlight on the indoor visual and thermal environment of the surrounding building; and to analyse the impacts of design parameters (vertical location, horizontal location, and orientation) and weather parameters on the behaviour of reflected sunlight.

33.2   Field Measurement

In this case study, the affecting building is a 9-story commercial building developed with a curved façade which features mental claddings and glazed curtain wall as shown in Fig. 33.1. Due to the reflective property of the building façade material, complaints were received from a 6-story office building nearby. Field measurement was conducted in the affected office building during the period of 1st April 2018 to 4th May 2018. The distance between the affecting building and the affected building is 46 meters. The affected office building is air-conditioned from 7 am to 7 pm during weekdays.

Based on the feedbacks received from building occupants and information collected during the site visit, nine measuring locations were selected in the affected building for the field measurement. As demonstrated in Fig. 33.2, these nine points were evenly distributed in level 2, 4 and 5, and the vertical alignment was maintained for all measurement points at respective locations in the building.

Fig. 33.1   Affecting building with a curved stainless-steel façade

Fig. 33.2   Location and alignment of measuring points at the affected building

The interior surface temperatures of the concrete wall and the window glass were measured by using HOBO thermocouple sensors and data loggers. Air temperature sensors and data loggers were installed on the internal glass surface to measure the indoor air temperature near the window area. Additionally, illuminance sensors and data loggers were placed on the internal glass surface to measure and record the illuminance level of every measuring point. All sensors were installed at the same horizontal level which was 1.6 m above the floor. The time interval for recording the data was set to be 1 min for all data loggers.

33.3   Results

To examine the impacts of reflected sunlight, measuring points with different horizontal locations, vertical locations, and orientation were compiled and compared with different weather conditions. Building thermal performance was studied by analyzing data collected from indoor air temperature, glass and wall surface temperatures sensors. Meanwhile, building visual performance was investigated by looking at the changes on indoor illuminance level.

33.3.1   Impact of Weather Conditions

Data collected on the 25th and 27th of April were compared as these days displayed contrasting weather trends, which assists evaluation of the weather impact. As shown in Fig. 33.3, the 25th of April was a sunny day with high incoming solar

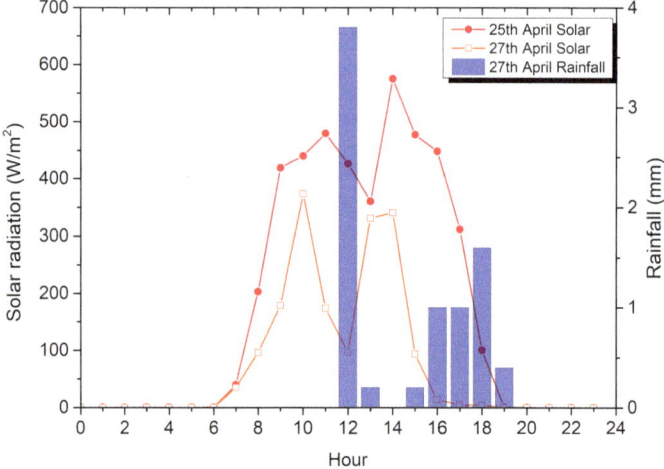

Fig. 33.3 Solar radiation and rainfall on the 25th and 27th of April

radiation during the daytime and no rain, whereas 27th of April illustrated low solar radiation with rain from 13:00 to 20:00.

33.3.1.1 Building Visual Performance

Point S-2-3 was chosen to examine the impact of different weather conditions. Table 33.1 describes the average and maximum illuminance values at point S-2-3 from 7:00 to 19:00 on the 25th and 27th of April. It is viewed that the difference of maximum I_{in} level between selected days can reach to 610.64 lux, which is significant in the working environment. Data shown in Table 33.1 indicate that higher solar radiation generates higher I_{in} value, and consequently affects the visual performance of the affected building.

33.3.1.2 Building Thermal Performance

Table 33.2 indicates the average T_i, T_g, and T_w on both the 25th and 27th of April. It can be found that even it is an enclosed and air-conditioned environment, higher incoming solar radiation during the daytime can result in higher temperatures. On average, the ascent of temperatures is more than 1 °C when there is a sunny day. Looking at the maximum temperatures during the daytime on selected days which is shown in Table 33.3, T_g illustrates the largest temperature difference which is 5.75 °C between the 25th and 27th of April. Meanwhile, the maximum T_i difference

Table 33.1 Average and maximum illuminance during the daytime (7:00–19:00)

	Daytime average (lux)	Daytime maximum (lux)
25th of April	230.85	810.47
27th of April	71.96	199.83
Difference	150.89	610.64

Table 33.2 Average temperature during the whole day

	T_i (°C)	T_g (°C)	T_w (°C)
25th of April	25.00	26.11	25.44
27th of April	23.91	24.46	24.41
Difference	1.09	1.65	1.03

Table 33.3 Maximum temperature during the daytime (7:00–19:00)

	T_i (°C)	T_g (°C)	T_w (°C)
25th of April	28.42	32.18	28.10
27th of April	25.16	26.43	25.83
Difference	3.26	5.75	2.27

and T_w difference between selected days are 3.26 °C and 2.27 °C respectively. It can be concluded that higher incoming solar radiation can generate higher maximum temperature during the daytime period.

33.3.2   Impact of Horizontal Location and Orientation

The impact of horizontal location and façade orientation on the respective indoor thermal and visual environment is investigated by comparing the data collected at measuring points S-4-1, S-4-2, and S-4-3. As illustrated in Fig. 33.2, point S-4-1 was located on the façade that was not directly facing the affecting building, whereas S-4-2, and S-4-3 were parallel, facing the affecting facade. T_i, T_g, T_w, and I_{in} at these three points were evaluated and compared.

33.3.2.1   Building Visual Performance

To minimize the glare issue caused by stainless-steel façade of the affecting building, $3M^{TM}$ sun control window film has been installed on the windows of the affected building. Based on the tests, the sun control film can reduce the illuminance by 93%. As shown in Fig. 33.4, from 16:00 to 18:00 on the 25th of April, I_{in} at point S-4-2 and S-4-3 (directly facing to the reflective façade) increased sharply when the affecting façade was directly exposed to the sun. Such abnormal illuminance increase did not occur during 16:00 to 18:00 on the 27th of April. On the rainy day, the illuminance profile is more analogous to the profile of solar radiation which peaks at 10:00 and 14:00.

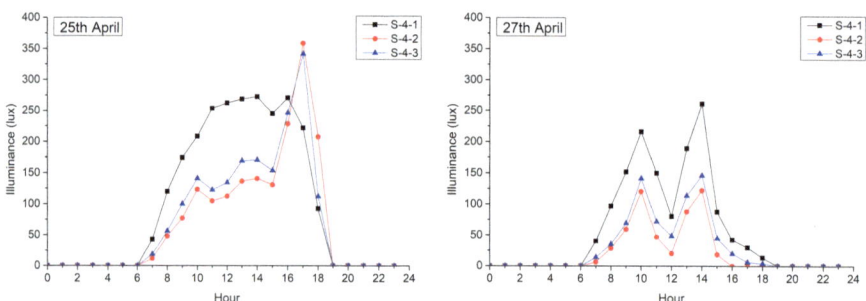

Fig. 33.4 Illuminance level at point S-4-1, S-4-2, and S-4-3 on the 25th and 27th of April

33.3.2.2 Building Thermal Performance

Air Temperature

Figure 33.5 displays the changing trend of T_i at selected points on the 25th and 27th of April. On the sunny day, from 9:00 to 16:00, point S-4-1 illustrates the highest T_i, followed by S-4-2, and S-4-3. During the period from 16:00 to 18:00, the affecting building façade was directly exposed to the sun when the sun is moved to the west side. This generated reflected sunlight entering the affecting measuring areas. It is observed that there is a sharp increase in T_i at point S-4-2 and S-4-3 during this affected period on the 25th of April. This phenomenon did not occur at point S-4-1 whose location was not directly facing to the stainless-steel façade. On the rainy day, the temperature increase did not occur during 16:00–18:00.

Glass Surface Temperature

The glass surface temperature at point S-4-1, S-4-2, and S-4-3 were compared. T_g at these three points shares the same trend as T_i. As shown in Fig. 33.6, from 8:00 to 16:00, the T_g at point S-4-1 was considerably higher than that at point S-4-2 and S-4-3. It is noticed that the trend of T_g at point S-4-2 and S-4-3 experienced a significant

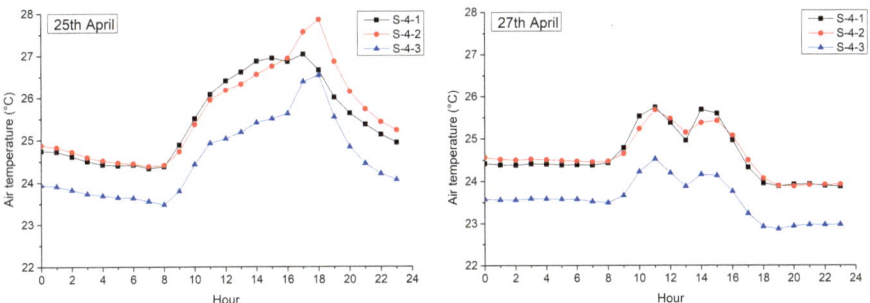

Fig. 33.5 T_i at point S-4-1, S-4-2, and S-4-3 on the 25th and 27th of April

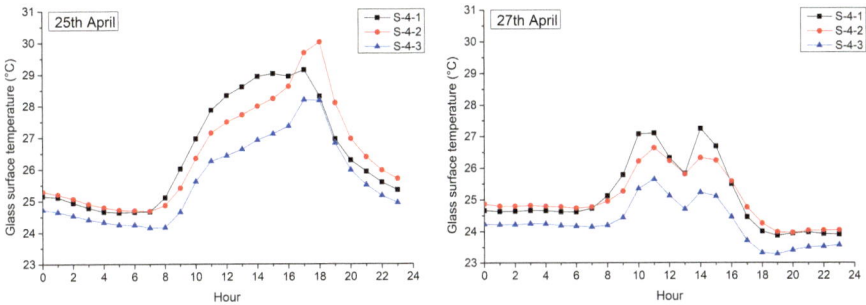

Fig. 33.6 T_g at point S-4-1, S-4-2, and S-4-3 on the 25th and 27th of April

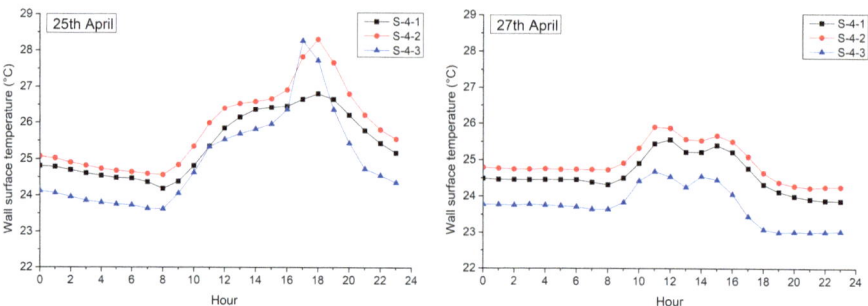

Fig. 33.7 T_w at point S-4-1, S-4-2, and S-4-3 on the 25th and 27th of April

growth from 16:00 to 18:00 on the 25th of April. Whereas on the rainy day, such phenomenon did not take place.

Wall Surface Temperature

The internal wall surface temperature at measuring points S-4-1, S-4-2, and S-4-3 were compared. As illustrated in Fig. 33.7, T_w at point S-4-2 ranks the highest and followed by point S-4-1, and S-4-3 from 0:00 to 16:00. While, during the affected period which is from 16:00 to 18:00, an obvious increase in T_w at both points S-4-2, and S-4-3 which are nearer to the affecting building can be viewed. The average wall surface temperature difference between point S-4-2 and S-4-1 is 0.81 °C during 16:00 to 18:00 on the 25th of April. Whereas on the rainy day, the three points share the same T_w changing trend.

33.3.3   *Impact of Vertical Location*

The impact of vertical location on the respective indoor thermal and visual environment was examined by comparing and analyzing the data collected from three measurement points which allocated at different levels (Level 2, 4, 5) while sharing the same horizontal location in the selected building, namely, S-2-3, S-4-3, and S-5-3.

33.3.3.1   **Building Visual Performance**

Illuminance value at point S-2-3, S-4-3, and S-5-3 were compared to evaluate the impact of vertical location on the visual performance of the affected building. As illustrated in Fig. 33.8, from 10:00 to 18:00 on the 25th of April, S-2-3 displays the highest I_{in} followed by S-4-3 and S-5-3. It is noticed that an anomalous ascent curve occurred at all three points from 16:00 to 18:00, and point S-2-3 performed the most

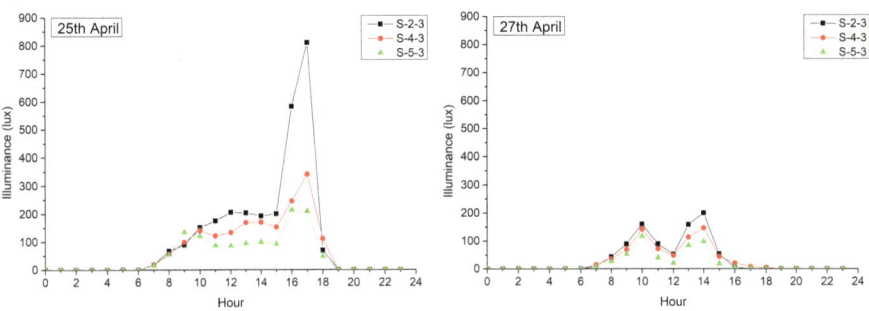

Fig. 33.8 Illuminance at point S-2-3, S-4-3, and S-5-3 on the 25th and 27th of April

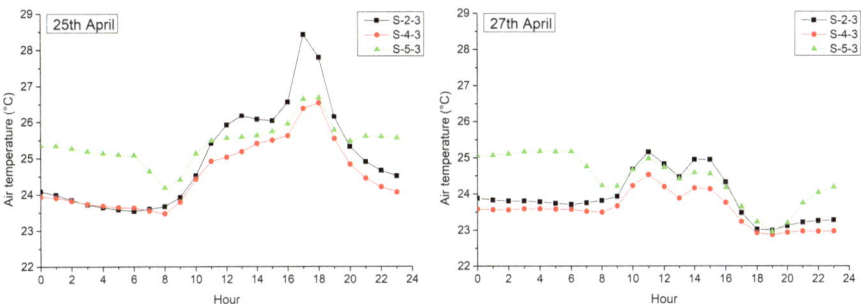

Fig. 33.9 T_i at point S-2-3, S-4-3, and S-5-3 on the 25th and 27th of April

considerable change. The maximum I_{in} at point S-2-3 reached to 810.47 lux during 17:00–18:00, which is 468.89 lux higher than the illuminance at point S-4-3. On the 27th of April, the differences among the three points were insignificant.

33.3.3.2   Building Thermal Performance

Air Temperature

As illustrated in Fig. 33.9, on the 25th of April, there is a sharp increase in T_i occurred in all three points from 16:00 to 18:00, during which the selected building was influenced by the reflective sunlight from the affecting building. This phenomenon is more obvious at point S-2-3, whose maximum T_i reaches to 28.42 °C during 17:00–18:00, followed by S-5-3, and S-4-3. During 16:00–18:00, the average temperature difference between S-2-3 and S-5-3 was 1.19 °C, and that between S-2-3 and S-4-3 was 1.48 °C. On the 27th of April, the temperature differences among the three points were insignificant during 16:00–18:00.

Glass Surface Temperature

As shown in Fig. 33.10, on the 25th of April, when the façade of the affecting building was directly exposed to the sun (16:00-18:00), a significant increase in T_g occurred at all three measurement points, point S-2-3 illustrates the most obvious change. The maximum T_g at point S-2-3 was 32.18 °C during 17:00–18:00. The average difference in T_g between S-2-3 and S-5-3 was 2.64 °C and that between S-2-3 and S-4-3 is 2.90 °C during 16:00–18:00. Whereas on the 27th of April, no such temperature increase can be observed during the affected period.

Wall Surface Temperature

Internal wall surface temperatures at point S-2-3, S-4-3, and S-5-3 were also studied and analysed. As shown in Fig. 33.11, from 0:00 to 16:00 and 19:00 to 23:00 on the 25th of April, point S-5-3 which located at relatively higher level displays the highest T_w, followed by S-2-3 and S-4-3. However, in the period from 16:00 to 18:00, it can be observed an abnormal growth in T_w at all three points, which is due

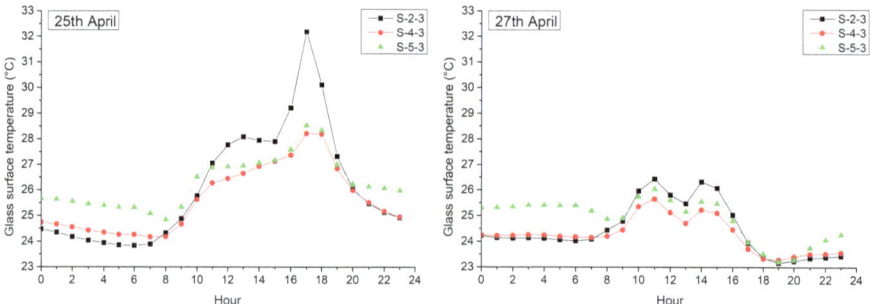

Fig. 33.10 T_g at point S-2-3, S-4-3, and S-5-3 on the 25th and 27th of April

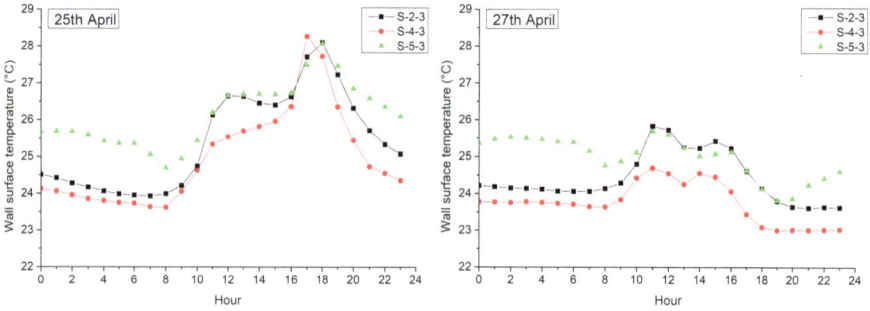

Fig. 33.11 T_w at point S-2-3, S-4-3, and S-5-3 on the 25th and 27th of April

to the reflective sunlight generated from the stainless-steel façade on the opposite side. On the 27th of April, point S-2-3 shows the highest T_w during the daytime, followed by point S-5-3 and S-4-3.

33.4 Conclusions

The internal and external heat sources are two main contributors to the increase of air temperature and surface temperatures. For the affected building, the internal heat sources which include occupants, lighting, equipment and human activities, are assumed to be constant throughout the measurement period. Thus, the external heat sources, mainly the reflected and diffused sunlight, are considered as the main reason that leads to the increase in temperatures and illuminance.

Weather condition is crucial in indoor thermal and visual environment. On a sunny day with high incoming solar radiation, the ascent of average indoor air temperature, glass and wall surface temperatures is more than 1 °C compared with a rainy day. By comparing a sunny and a rainy day, the difference of maximum indoor air temperature near the window area can reach 3.26 °C, and the difference of maximum illuminance value can be 610.64 lux, which are considerable in the working environment.

Horizontal location and orientation also play a vital role in the indoor thermal and visual environment. Points which orient directly facing and nearer to the affecting reflective façade indicate worse thermal and visual performance.

Furthermore, vertical location is another considerable element that will affect both the indoor thermal and visual environment. Data analysis provides us with a conclusion that temperatures and illuminance at the lower level will be more affected by the highly reflective facade. The external heat source such as the reflective glare, in this case, will lead to an abnormal increase in both indoor temperatures and illuminance level during the affecting period.

Parametric study through computer modeling will be conducted in the future to propose a guideline for building designers to mitigate the negative impacts from nearby building façade on the indoor visual and thermal environment.

Acknowledgement This project is funded by the Building and Construction Authority of Singapore (WBS: R-296-000-176-490).

References

Brzezicki M (2012) The influence of reflected solar glare caused by the glass cladding of a building: application of caustic curve analysis. Comput Aided Civil Infrastruct Eng 27(5):347–357
Danks R, Good J, Sinclair R (2016) Assessing reflected sunlight from building facades: a literature review and proposed criteria. Build Environ 103:193–202

Suk JY, Schiler M, Kensek K (2017) Reflectivity and specularity of building envelopes: how materiality in architecture affects human visual comfort. Archit Sci Rev 60(4):256–265

Wong JS (2016) A comprehensive ray tracing study on the impact of solar reflections from glass curtain walls. Environ Monit Assess 188(1):16

Yang X, Grobe L, Stephen W (2013) Simulation of reflected daylight from building envelopes. In: 13th conference of international building performance simulation association, pp 1–8

Chapter 34
A Field Survey on Thermal Comfort of Occupants and Cold Stress in CLT School Buildings

Timothy O. Adekunle

Abstract The goal of achieving a smart and sustainable built environment starts with the design, construction, and maintenance of an intelligent and sustainable occupied thermal environment. Such an environment must be designed and constructed to achieve thermal comfort and overall well-being of occupants. This paper presents a field investigation of occupants' comfort and cold stress in cross-laminated timber (CLT) school buildings during the cold seasons (fall and winter). The study was conducted from October to November 2017 for the fall season and from December 2017 to February 2018 for the winter season. The case study comprises of spaces constructed with structural timber products. The case study is a LEED certified school building. It has been identified as one of the first green school buildings in the Northeast region of the USA. The building explores HVAC systems, and it utilizes ground source heat pumps for heating and cooling. The research employed physical measurements of environmental variables such as temperature, relative humidity (RH), dew-point temperature, air velocity and CO_2 level in the selected spaces such as the administrative office, science, and art classrooms and the multi-purpose hall. The sensors were mounted on the internal walls at 1.1 m above the floor to measure the variables at every 60 min throughout the cold seasons. The study also calculated the Wet-Bulb Globe Temperature (WBGT) in the spaces to understand the average cold stress index within the thermal environment. The mean outdoor temperature was 11.3 °C in fall, and 0.5 °C in winter. In the fall season, the results showed that the mean indoor temperature was 21.2 °C. In the same season (fall), the mean RH was 50.7%, and the average dew-point temperature was 9.3 °C. In the winter, the average indoor temperature was 20.5 °C while the average RH was 23.9% and the mean dew-point temperature was −1.9 °C. The overall mean temperatures measured in the spaces during the cold seasons were within the comfort temperature thresholds (20.3 °C/23.9 °C) recommended by ASHRAE. In the fall, the mean RH was within the comfortable range (30–60%). The mean RH value was

T. O. Adekunle (✉)
Department of Architecture, College of Engineering, Technology, and Architecture (CETA),
University of Hartford, West Hartford, CT, USA
e-mail: adekunle@hartford.edu

© Springer Nature Switzerland AG 2020
R. Roggema, A. Roggema (eds.), *Smart and Sustainable Cities and Buildings*,
https://doi.org/10.1007/978-3-030-37635-2_34

below the comfortable range in the winter. The study recorded a higher mean temperature, RH and dew-point temperature in the office space than the classrooms and the main hall during the cold seasons. Lower cold stress indexes were also calculated in the multi-purpose hall than the classrooms and office space. The study revealed occupants are more likely to experience cold temperatures in the hall than the office space and classrooms. The difference in the floor level (the main hall is on the lower floor while the classrooms are on the upper floor), hours of occupation (more extended hours of occupation in the office space), and floor area may be the contributing factors to the lower temperatures measured in the hall than the other spaces. By applying the WBGT mathematical model, the research recommends the WBGT of 16.0 °C and 13.7 °C as the cold stress indexes in the building for the fall and winter seasons respectively. Finally, the study recommends a WBGT of 14.9 °C as the average cold stress index in the spaces evaluated in this paper.

Keywords Field study · Occupants' comfort · Cold stress · Cross-laminated timber (CLT) · School buildings · Cold seasons · WBGT

34.1   Introduction

This study aims to contribute to the body of knowledge by evaluating occupants' comfort and cold stress indexes in school buildings during the cold seasons. Recent studies have stated that prefabricated lightweight building materials including engineered timber products such as cross-laminated timber (CLT) are steadily considered as major building materials for various structures (Adekunle 2014). Existing research also explained such structures are susceptible to elevated temperatures in summer (Adekunle and Nikolopoulou 2014, 2016), and low temperatures in cold seasons (Adekunle 2014). Based on this premise, it is important to consider a study that examines occupants' comfort in cold seasons to understand the cold stress index and the temperatures at which the occupants in CLT school buildings will be subject to cold stress.

34.2   Literature Review

Different studies on occupants' comfort in school buildings in various locations across the globe are reviewed and widely discussed in the existing literature (Rupp et al. 2015). Occupants' comfort in school buildings (Teli et al. 2013; Adekunle 2018; Adekunle 2019) and the effect of the various age in rating the thermal environment in the buildings have been examined (Corgnati et al. 2007; Hussein and Rahman 2009; Mors et al. 2011; Teli et al. 2013). A few studies have evaluated occupants' comfort in school buildings with natural ventilation (Hussein and Rahman 2009; Hwang et al. 2009; Teli et al. 2013), and hot, humid climate (Hussein and Rahman 2009). Hussein and Rahman (2009) stated most of the survey participants (that is, 80%) accepted the thermal environment of the buildings. Also, the responses provided by the participants regarding the sensation are well above the specified ASHRAE 55 baseline (ASHRAE 2017). Hwang et al. 2009 highlighted the acceptable rate by the respondents have a broader range, and the comfort range has a smaller range when it was assessed using the adaptive comfort standard.

Mors et al. (2011) evaluated different environmental parameters in warm and cold seasons in school buildings. The study (Mors et al. 2011) showed a preference for lower temperatures. In another research conducted in a warm and fully humid climate with cold winters, Teli et al. (2013) explained the comfort temperature was at least 2.0 °C above the comfort temperature computed from applying the adaptive thermal comfort standard. The existing research showed that people are prone to high temperatures within the thermal environment (Teli et al. 2013; Adekunle 2018; Adekunle 2019). A study surveyed over 4000 students in Italy during cold and warm seasons in approximately 200 classrooms (Alfano et al. 2013). The investigation (Alfano et al. 2013) revealed the effectiveness of the PMV model in predicting occupants' comfort in the buildings. The existing research also found that variations in air velocity in the school environment can influence occupants' perception of air being more refreshing and cooler when compared with the thermal environment if the air velocity is kept constant (Wigo 2013). Corgnati et al. (2007) also explained that people accepted the thermal environment even when responses ranged from neutrality to warm on the evaluation scale. Pereira et al. (2014) maintained that the acceptable temperature for the respondents in school buildings exceeded the recommended comfort zone. Nevertheless, none of the studies highlighted above evaluated the cold stress index for people in school buildings during cold seasons.

Liang et al. (2012) explained that the building fabric energy regulation has a considerable effect on the thermal comfort of people in school buildings. Hwang et al. (2006) stated that relative humidity has a less substantial impact on the thermal sensation of people. Zhang et al. (2013) noted that the occupants in school buildings located in hot and humid climates have more tolerance to high temperatures and relative humidity when compared with the occupants in school buildings located in temperate climates. Also, people in non-naturally ventilated buildings are likely to take adaptive actions to regulate the thermal environment than the people in naturally ventilated buildings (Zhang et al. 2013). Zhang et al. (2013) explained further that people in the non-naturally ventilated spaces are more sensitive with a higher perception of the thermal environment than the people in naturally ventilated areas. Serghides et al. (2014) mentioned excessive cooling in school buildings that can lead to low temperatures in summer and excessive heating that can cause elevated temperatures in winter.

According to the National Institute for Occupational Safety and Health (NIOSH 2015), indoor monitoring of environmental parameters is important to identify the variables that influence the occupants' perception of comfort in the spaces. NIOSH (2015) explains that the occupants' perception of comfort is closely associated with the physiological adjustments, the heat transfer from the human body to the thermal environment, and body temperature. The study (NIOSH 2015) stressed further the heat transfer from the human body to the immediate environment is influenced by various environmental parameters such as relative humidity, temperature, air movement, clothing layers, and metabolic activities. ASHRAE Standard 55 (ASHRAE 2017) recommended the operative temperature of 20.3 °C (68.5 °F) to 23.9 °C (75 °F) in the cold season. The Standard specified that the comfort temperatures baseline within the thermal environment are determined by humidity, seasonal change, activity level, clothing insulation and other variables (ASHRAE 2017), and a range of RH from 30–60% is recommended for indoor environments to reduce and avoid mold growth (EPA 2012). Also, relative humidity is specified not to exceed 65% within the thermal environment to prevent or reduce microbial growth (ASHRAE 62.1, 2013). Based on the literature, indoor temperatures of 20.3/23.9 °C will be considered as the lower and upper thresholds for comfort temperatures in this paper.

USDOE (2018) explained heating and air-conditioning systems are often operated in different seasons to improve the thermal comfort of occupants which make buildings consume more energy with additional financial implications. In some situations, the cooling or heating systems may come on when the exterior temperatures rise or fall due to the provision of thermostats in the buildings (Nicol and Humphreys 2007; Nguyen et al. 2014). Based on the gap identified from the literature, this study evaluates occupants' comfort and the thermal behaviour of the spaces in CLT school buildings. Moreover, the literature review considered the recent and current studies on occupants' comfort in school buildings. None of the studies examined occupants' comfort and cold stress at the same time in school buildings built with CLT. As a result, this study evaluates occupants' comfort and cold stress in cross-laminated timber (CLT) school buildings during cold season by assessing the Wet Bulb Globe Temperature Heat (WBGT) threshold. The results on occupants' comfort in the school buildings are also compared with the results presented in the recent research to determine if the comfort temperatures exceed the recommended comfort temperature thresholds for buildings during cold seasons.

34.3   Research Method and Mathematical Expression

This study considered the physical measurements of environmental parameters (temperature, relative humidity, air velocity, dew-point temperature, and CO_2 level) during the cold seasons (fall and winter) as the primary research method for data collection. The study also considered the mathematical expression to compute the Wet Bulb Globe Temperature (WBGT) using the variables measured in the case study. Regarding the physical measurements in the fall season, the survey was conducted from October to November 2017. The winter survey covered the whole meteorological winter months (that is, from December 2017 to February 2018). The parameters were measured at 15-min intervals. The HOBO sensors were installed at a 1.1 m height above the floor level. For this paper, the environmental parameters recorded at 60-min intervals were extracted and analysed. Also, the WBGT which has been recently used to assess the heat stress in other usable spaces (NEHC 2007; OSHA 2016). The detailed information regarding the WBGT heat threshold has been presented in the literature (Stull 2011; Lemke and Kjellstrom 2012). Regarding the windows, the users of the case study can open and close windows without using additional energy (that is, manually operated). The users operate the windows manually for ventilation purposes while mechanical ventilation and heating systems are used depending on the thermal environment of the spaces, external conditions, and seasonal change requirements.

As discussed in the existing study, the internal WBGT can also be described as $WGBT_{ind}°C$. According to Eq. 34.1 (Lemke and Kjellstrom 2012), the $WBGT_{ind}$ is defined as the function of natural wet-bulb temperature ($T_{nwb}°C$) and black globe temperature ($T_g°C$). In the field study that considered physical measurements of parameters, measuring the globe temperature within the thermal environment could be challenging. Based on this idea, Lemke and Kjellstrom (2012) also validated an existing equation to compute the WBGT index. The validated equation (expressed as Eq. 34.2) has been applied to assess the thermal environment by evaluating the function of air velocity (Vm/s), psychrometric wet bulb temperature ($T_{pwp}°C$), and dry bulb temperature ($T_a°C$). The air temperature is similar to the radiant temperature as expressed in Eq. 34.2. Also, the combined effect of air temperature and low air speed within the thermal environment is known as the operative temperature. Based on this premise, the operative temperature is a function of air temperature, radiant temperature, and low air velocity recorded within the thermal environment. The operative temperature is also influenced by other variables such as the radiant factor.

$$WBGT_{ind} = 0.7T_{nwb} + 0.3T_g \qquad (34.1)$$

$$WBGT = 0.67T_{pwp} + 0.33T_a - 0.048 \log_{10} V (T_a - T_{pwp}) \qquad (34.2)$$

Stull (2011) proposed that the psychrometric wet bulb temperature can be computed. From the explanation, Stull (2011) explained in Eq. 34.3 that psychrometric wet bulb temperature ($T_{pwp}°C$) is similar to the wet bulb temperature – T_w. Equation 34.3 shows the arctangent (atan) evaluates the values that are the same as values expressed in radians. From Eq. 34.3, T_w is a combination of T_a (°C) and RH (%). For Eq. 34.3, a standard pressure value of 101.325 kPa (101,325 Pa) is considered for the computation. The equation will be explored to compute the WBGT index for cold stress.

$$\begin{aligned} T_w = & T_a \text{atan} \left[0.151977(RH\% + 8.313659)^{1/2} \right] \\ & + \text{atan}(T_a + RH\%) - \text{atan}(RH\% - 1.676331) \\ & + 0.00391838(RH\%)^{3/2} \text{atan}(0.023101 \times RH\%) - 4.686035 \end{aligned} \qquad (34.3)$$

34.4 The Case Study

The case study is a school building. It is built with engineered structural timber products such as cross-laminated timber (CLT) panels. The study building is sited on an inner-city park area of over 8 hectares located in the New England region (that is, Northeastern part) of the United States. The study building is designed and developed as a mixed-use project for various purposes. The case study consists of a high school, an inner-city farm zone, and a dedicated education and resource centre for environment and sustainable living. The development is one of the first projects in the US to use CLT as the principal structural building material for the construction. Regarding the floor area, the study building has a total area of about $1300m^2$ and comprises different usable spaces such as a hall that is being used for multipurpose events including art and dance performances, sports activities, and other communal and local events. The other spaces include offices, classrooms, including laboratories and study areas, as well as an art classroom. The hall has a double volume, and it is located on the ground floor with the administrative office. The classrooms, including the laboratories, the study, and the art classroom are situated on the upper level of the building.

Regarding the construction of the building, the tension surface and ceiling finishes are implemented with the black spruce CLT panels. The bearing and shear walls are designed and constructed using vertical CLT panels. About the U-values of the CLT walls, the values varied between about 0.13 W/m^2K and 0.20 W/m^2K based on the thickness of the CLT panels. The insulation (cellulose) used for the case study is one of the approved insulation materials recommended by the US Department of Energy to improve the performance of newly built or refurbished buildings (USDOE 2018). The case study is a recipient of numerous sustainable and green buildings awards and recognition. The building is also a recipient of the LEED-NC 2009 v3 based on the performance rating of 61% energy cost-saving and more than 25% below the ASHRAE Standard 90.1 baseline recommended for lighting power density in school buildings.

Environmental monitoring occurred in the main hall located on the ground floor, the administrative office on the ground floor, the classrooms, including the laboratories and the study areas, located on the upper level. In summer, the building is naturally ventilated but also provided with mechanical heating, ventilation, and air conditioning (HVAC) systems. The project also utilizes ground source heat pumps for heating and cooling. The case study meets the High-Performance Schools criteria set by the regulatory body in the state, and it is aiming to attain the LEED Platinum status. The case study features an exemplary earth-coupled energy system. Due to the period of the physical measurements, the systems were in operation. The HOBO sensors were placed in each of the spaces selected and approved by the school administrators to record and log temperature, relative humidity, air velocity, CO_2 level, and dew-point temperature at equal intervals during the cold seasons. The outdoor weather data recorded at a nearby weather station were used for the analysis. During the cold seasons, the case study was in use from 8 am to 5 pm and partially or not occupied from 6 pm to 7 am. Environmental monitoring considered in the spaces and the analysis of the data collected will be discussed in this study. In terms of the placement of the sensors, the sensors were placed in the following spaces: administrative office (south orientation), multipurpose hall (northeast orientation), science classrooms (southwest, west, and southeast orientations), and the art classroom (north orientation).

34.5   Analysis of Data

The analysis of the annual external weather data collected (historical) from the nearby weather station revealed that the climatic condition of the study location is fully humid, cold winter and cool summer (Dfc) based on the Koppen-Geiger climate classification (Kottek et al. 2006). The mean highest temperatures (monthly) varied from 1.7 °C in January to 28.3 °C in July. The analysis showed that the period of the field investigation is the coldest period of the year. The detailed information on the average temperatures (such as high, low, record high, record low) and mean precipitation is presented in Fig. 34.1.

For the outdoor weather data collected during the field investigation, the mean outdoor temperature ranged from 6.5 °C to 16.0 °C for the fall season and from −2.0 °C to 3.3 °C for the winter season (Table 34.1). The overall average outdoor temperature for the period of the field study is 4.8 °C. The average external dew-point temperature was 6.0 °C in the fall and − 5.0 °C in the winter. Regarding the average outdoor RH, it ranged from 45.8%–91.4% with an average RH of 69.3% in the fall and 66.0% in the winter. The average atmospheric pressure (hPa) at sea level for the duration of the field study is found to be 1019 hPa in the fall and ranged from 1017 hPa to 1022 hPa in the winter. The average outdoor WBGT indexes were 9.4 °C and − 0.7 °C in the fall and winter in that order. The summary of the essential

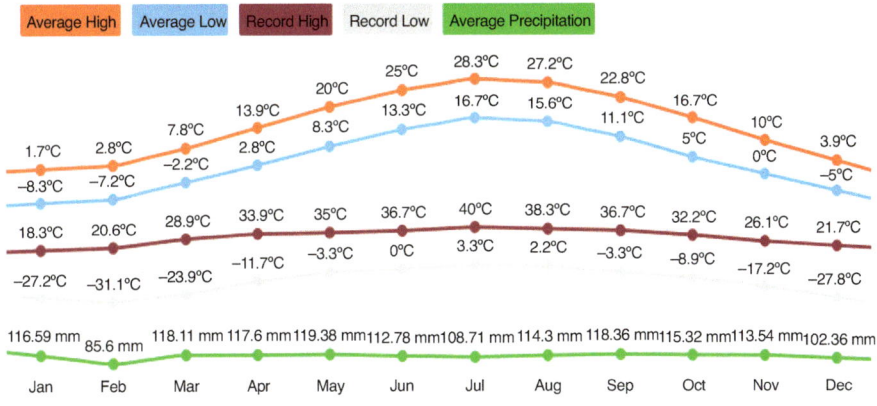

Fig. 34.1 The mean monthly external variables in the study location (historical data)

Table 34.1 Summary of the outdoor weather data for the fall and winter seasons

| | Fall season (Oct. – Nov. 2017) | | | | | | Winter season (Dec. 2017–Feb. 2018) | | | | | | | | |
| | October 2017 | | | November 2017 | | | December 2017 | | | January 2018 | | | February 2018 | | |
Variables	High	Mean	Low	High	Mean	Low	High	Mean	Low	High	Mean	Low	High	Mean	Low
Max. temp (°C)	27.0	22.0	22.0	22.0	17.0	11.0	15.0	12.0	90	120	7.0	3.0	180	11.0	6.0
Min. temp (°C)	140	8.0	20	4.0	-1.0	-6.0	-9.0	-11.0	-130	-100	-12.0	-17.0	-1.0	-6.0	-9.0
Average temp (°C)	21.2	16.0	11.5	11.6	6.5	15	3.5	0.3	-2.9	2.3	-2.0	-6.2	7.5	3.3	-0.6
Dew–point (°C)	22.0	11.0	-3.0	17.0	1.0	-180	13.0	-6.0	-22 0	11.0	-8.0	-25.0	120	-1.0	-19.0
RH(%)	91.4	72.1	52.9	84.9	66.4	47.8	78.2	62.0	45.8	78.0	64.0	50.0	906	720	53.4
Precipitation (mm)	1943.2	132.1	0	3607	43.2	0	2184	27.9	0	401.3	35.6	0	749.3	134.6	0
Wind speed (m/s)	102	5.6	10	10.1	5.6	12	9.7	5.6	16	10.2	6.0	1.8	9.4	5.1	0.8
Aim Atmospheric. press. (hPa)	1023	1019	1015	1023	1019	1015	1022	1017	1013	1026	1021	1016	1029	1022	1016
WBGT	20.6	14.0	86	10.7	4.8	-0.6	2.5	-1.1	-4.8	1.4	-3.2	-7.6	7.0	2.1	-2.4

features of the outdoor weather data from the location of the field survey is presented in Table 34.1. The analysis revealed that people in the outdoor environment of the case study might be subject to thermal discomfort such as freezing to slight cold stress at a temperature below the freezing point (0 °C) especially in the fall. In the winter, the occupants in the study location may also experience cold stress at about −13 °C and possibly strong cold stress at a temperature close to −27 °C. The investigation revealed that the stress index was below the slight cold stress range for some days during winter.

34.6 Results and Discussions

The overall average temperature of 21.2 °C was recorded in the spaces during the fall while the overall temperature of 20.5 °C was observed in winter. Concerning the average temperature in the spaces during office hours (8 am–5 pm), 20.6 °C was reported as the average temperature in the fall. In winter, the average temperature of 20.9 °C was recorded in similar spaces during office hours (Table 34.2). Likewise, mean temperatures of 19.9 °C and 20.0 °C were measured in the spaces during non-office hours (that is, from 6 pm–7 am) in the fall and winter respectively. The results showed the average temperatures recorded in the spaces during the office hours and non-office hours were slightly higher in the winter than the fall. The results showed heating of the spaces might be a contributing factor to higher average temperatures measured in the spaces during the office hours than the non-office hours. The results showed the mean temperatures did not exceed the threshold for comfort temperature (23.9 °C) in the spaces in the fall and winter. However, mean temperatures fall below the baseline for comfort temperature (20.3 °C) in the main hall in the fall especially during office hours as well as non-office hours. The results showed that occupants might experience thermal discomfort within the space in the fall.

The average dew-point temperatures of 9.3 °C and − 1.9 °C were observed in the spaces during the fall and winter correspondingly. The mean relative humidity was 50.7% in the fall. In winter, the average RH was 23.9%. Also, the average air velocity measured in the spaces was 0.1 m/s. At different seasons and periods considered in this study, higher temperatures were recorded in the office space than the classrooms and the main hall. The more extended hours of the occupation of the users in the office space and the frequent use of control (such as heating) may contribute to the higher temperatures recorded in the office space. Moreover, the orientation of the office space (south facing) and its location on the ground floor, which can help reduce various temperatures swing, may have also contributed to the higher temperatures recorded in the space than other spaces in the building. On the contrary, the double-volume, large floor area, as well as regular use of the entrance doors of the multipurpose hall, which are manually operated by people, may contribute to the lower temperatures reported in the main hall. For the classrooms, some parts of the spaces were used as laboratories. This factor may contribute to the

Table 34.2 Maximum, minimum and mean values of parameters recorded in the fall and winter

Spaces/parameters	Fall season (October–November 2017)									Winter season (December 2017–February 2018)								
	Max. temp (°C)	Mir. temp (°C)	Avg temp (°C)	Max. dew-point (°C)	Min. dew-point (°C)	Avg dew-point (°C)	Max. RH (%)	Min. RH (°C)	Avg RH (°C)	Max. temp (°C)	Min. temp (°C)	Avg temp (°C)	Max. dew-point (°C)	Min. dew-point (°C)	Avg dew-point (°C)	Max. RH (%)	Min. RH (%)	Avg RH (%)
Main hall (ground floor)	26.4	17.6	19.2	16.1	7.7	11.0	76.6	385	59.6	30.2	13.6	20.5	20.4	−15.8	−1.9	71.2	7.9	23.9
Classrooms (upper floor)	23.2	12.3	20.3	18.0	−7.3	6.7	78.5	190	42.9									
Office space (ground floor)	23.0	18.4	21.0	18.9	1.4	10.1	77.9	32.0	49.7									
Main hall (8am–5pm)	26.4	17.8	19.8	15.8	8.0	113	73.7	385	58.2	30.2	14.3	20.9	20.4	−15.8	−1.8	71.2	8.2	23.5
Classrooms (8am–5pm)	23.2	13.3	20.7	16.1	−7.3	6.6	71.6	190	41.6									
Office space (8am–5pm]	22.9	18.5	21.2	15.5	1.5	9.6	63.2	32.0	47.6									
Main hall (6pm–7am)	20.8	17.6	18.8	16.1	7.9	10.9	76.7	51.5	60.7	24.9	13.6	20.0	158	−15.0	−2.0	61.8	7.9	24.2
Classroom (6pm–7am)	22.9	12.4	20.1	18.0	−5.5	6.7	78.5	193	43.9									
Office space (6pm–7am)	23.0	18.4	20.9	18.9	1.4	104	77.9	32.0	50.9									

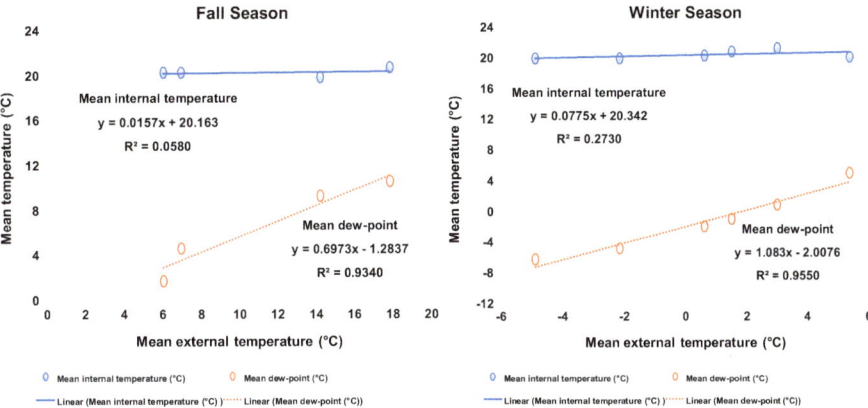

Fig. 34.2 The relationship between the mean internal and the external variables in the fall and winter

lower temperatures measured in the classrooms. Also, a higher CO_2 level was measured in the spaces during office hours than the non-office hours. The results also revealed higher dew-point temperatures and higher relative humidity were measured in the main hall than the classrooms and the main office in the fall and winter as well as the different periods (office hours and non-office hours). The maximum, minimum, and mean values of the parameters recorded during the field surveys in the fall, winter and different periods are presented in Table 34.2.

On the one hand, a relationship exists between the mean dew-point temperatures and the outdoor temperatures in the fall ($R^2 = 0.9344$) and winter ($R^2 = 0.9556$). On the other hand, no relationship exists between the mean indoor and the outdoor temperatures in the fall ($R^2 = 0.0581$). A relationship is also found between the parameters in winter ($R^2 = 0.2734$). The range of mean indoor temperatures and dew-point is slightly higher in winter than the fall (Fig. 34.2).

The WBGT values were calculated using Eq. 34.3. The mean air velocity of 0.1 m/s (measured) was used for the computation of the WBGT indexes. The results showed the mean WBGT values were higher in the office than the classrooms and the hall during the fall including office and non-office hours. In the winter, the WBGT index for the spaces was higher during office hours than non-office hours. Across the spaces, higher mean WBGT values were computed in the fall than winter. The research revealed the study building users might be subject to slight cold stress in the spaces, especially in winter. Considering, the number of hours above the ASHRAE comfort temperature thresholds for the cold season, the results showed a substantial amount of hours above 20.3 °C/23.9 °C in the spaces during the different seasons and periods. The investigation revealed that the building users might be subject to discomfort in the cold seasons. By relating the mean WBGT values in this research during the fall (16.0 °C) and winter (13.7 °C) with the recent study on heat stress in the CLT school buildings (Adekunle 2018), lower WBGT indexes were computed in this paper than the value (19.8 °C) calculated in existing research.

Table 34.3 Comparison of the average values measured with the ASHRAE comfort temperature thresholds

Spaces/ parameters	Fall season (Oct. – Nov. 2017)				Winter season (Dec 2017 – Feb. 2018)			
	Avg temp (°C)	Avg RH (%)	Avg WBGT (°C)	No of hours above 20.3/ 23.9°C	Avg temp (°C)	Avg RH (%)	Avg WBGT (°C)	No of hours above 20.3/ 23.9°C
Main hall (ground floor)	19.2	59.6	16.0	861/ 6	20.5	23.9	13.7	1162/ 3
Classrooms (upper floor)	20.3	42.9	15.4					
Office space (ground floor)	21.0	49.7	16.7					
Main hall (8am–5pm)	19.8	58.2	16.4	404/0	20.9	23.5	13.9	564/1.5
Classrooms (8am–5pm)	20.7	41.6	15.6					
Office space (8am–5pm)	21.2	47.6	16.6					
Main hall (6pm–7am)	18.8	60.7	15.7	457/370	20.0	24.2	13.3	598/1.5
Classrooms (6pm–7am)	20.1	43.9	15.3					
Office space (8am–7am)	20.9	50.9	16.7					

Note – The total hours in the fall – 1464 h. The total hours in the winter – 2160 h.

Table 34.3 summarizes the mean values for the temperature, RH, WBGT and number of hours above the comfort temperature thresholds.

Strong correlations exist between the mean WBGT and RH within the spaces in the fall and winter (Fig. 34.3). The investigation revealed higher humidity values at warm temperatures have a considerable effect on WBGT while higher air velocity at warm temperatures does not have a substantial impact on WBGT.

On the one hand, relationships are found between the mean indoor temperature, the mean WBGT and the mean dew-point in winter. On the other hand, low R^2 values were reported between the variables in the fall (Fig. 34.4). Seasonal changes may influence variations in the occupants' adaptation to the thermal environment, which can contribute to the extent to which they use the control to regulate the environment. A combination of variables (temperature, air speed, CO_2 level, RH, dew-point temperature) could influence the cold stress values within the thermal environment. The research showed an increase in the WBGT has an impact on the temperature and dew-point in the winter rather than in the fall. As a result, the WBGT can be applied to determine people's vulnerability to cold stress in CLT school buildings. This paper also highlighted that the WBGT could be explored for assessing cold stress in different thermal environments.

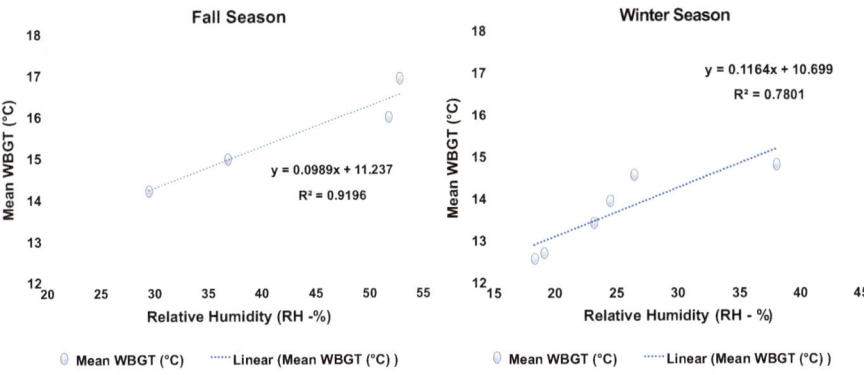

Fig. 34.3 The relationship between the relative humidity and the mean WBGT in the fall and winter

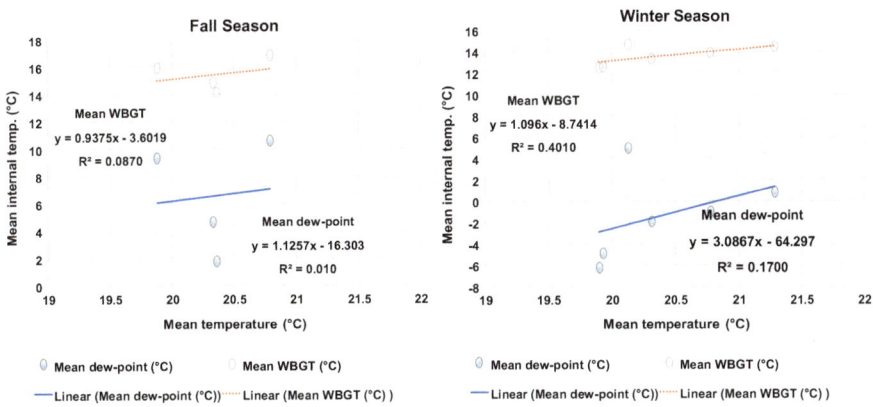

Fig. 34.4 The relationship between the mean WBGT, dew-point and internal temperatures

34.7 Conclusions

The study assessed the occupants' comfort and cold stress in CLT school buildings. The mean temperatures in the office space, main hall, and classrooms varied from 18.8 °C to 21.2 °C in the fall. The values ranged from 20.0 °C to 20.9 °C in winter. The overall mean temperature of 21.2 °C was reported in the fall while a mean temperature of 20.5 °C was recorded in winter. The mean temperatures in the cold seasons (fall and winter) were within the ASHRAE comfort temperature thresholds (20.3 °C/23.9 °C) for people in cold seasons. The mean RH in the fall was within the comfortable range (30–60%). However, it was below the range in winter. The office space was found to be the warmest space due to the longer hours of occupation and frequent use of control (heating) to regulate the thermal environment. The hall is the coolest space due to various contributing factors such as its location on the ground floor, double-volume of the space, large floor area which may require additional energy for heating, and frequent use of the large entrance doors which may contribute to constant heat loss within the space during the cold seasons. The opening of the entrance doors should be regulated to reduce heat loss in cold season. The introduction of a transitional space between the entrance doors and the main hall may also reduce heat loss in cold seasons, but the approach may contribute to summertime overheating in the space.

Regarding office hours (8 am–5 pm), the mean temperatures were 20.6 °C and 20.9 °C in the fall and winter respectively. The mean temperature during the non-office hours was 19.9 °C in the fall, and it was 20.0 °C in winter. The temperature exceeded the recommended comfort temperature lower threshold (20.3 °C) for about 59% of the time in the fall, and it exceeded the limit for approximately 54% of the time in winter. During office hours, the temperatures exceeded the comfort temperature lower threshold for 28% and 26% of the time in the fall and winter respectively. The temperature did not exceed the comfort temperature's upper threshold (23.9 °C) for more than 1% of the time at any season or duration. The lower external temperatures reported during the cold seasons may be a contributing factor to this finding.

The results showed a higher mean temperature, RH and dew-point temperature in the office space than the classrooms and the main hall during the cold seasons. The users are more likely to experience cold temperatures in the hall than the office and classrooms. By considering the WBGT mathematical model, lower cold stress indexes were calculated in the hall than the classrooms and office space in cold seasons. The average WBGT ranged from 15.3 °C to 20.0 °C in the fall while the average WBGT varied from 13.3 °C to 13.9 °C in winter. Overall, the research proposes a WBGT of 14.9 °C as possible cold stress index for the vulnerable people in CLT school buildings during the cold seasons. The study showed higher RH have a considerable impact on the WBGT.

Acknowledgments The author would like to thank the appropriate authority including the school officers that allowed the investigation to be carried out in the case study. The author appreciates the staff of the case study for their support throughout the field investigations.

References

Adekunle TO (2014) Thermal performance of low-carbon prefabricated timber housing in the UK. PhD Thesis, University of Kent, UK

Adekunle TO (2018) Thermal comfort and heat stress in cross-laminated timber (CLT) school buildings during occupied and occupied periods in summer, In Proceedings of 10th Windsor Conference, UK. April 12–15

Adekunle TO (2019) Occupants' comfort and stress indices in a structural timber school building in the northeast US in different seasons. Building Research & Information. https://doi.org/10.1080/09613218.2019.1662714

Adekunle T, Nikolopoulou M (2014) Post-occupancy and indoor monitoring surveys to investigate the potential of summertime overheating in UK prefabricated timber houses, In Proceedings of 8th Windsor Conference, UK. April 10–13

Adekunle TO, Nikolopoulou M (2016) Thermal comfort, summertime temperatures and overheating in prefabricated timber housing. Build Environ 103:21–35

Alfano d'Ambrosio FR, Ianniello E, Pallela BI (2013) PMV-PPD and acceptability in naturally ventilated schools. Build Environ 67:128–137

ANSI/ASHRAE Standard 62.1 (2013) Ventilation for acceptable indoor air quality. American Society of Heating, Refrigerating and Air-conditioning Eng, Atlanta

ASHRAE Standard 55 (2017) Thermal environmental conditions for human occupancy. American Society of Heating, Refrigeration and Air-conditioning Engineers, Atlanta

Corgnati SP, Filippi M, Viazzo S (2007) Perception of the thermal environment in high school and university classrooms: subjective preferences and thermal comfort. Build Environ 42 (2):951–959

EPA (2012) A brief guide to mold, moisture, and your home. Environmental Protection Agency (EPA), Washington, USA. https://www.epa.gov/sites/production/files/2016-10/documents/moldguide12.pdf

Hussein I, Rahman MHA (2009) Field study on thermal comfort in Malaysia. Eur J Sci Res 37 (1):134–152

Hwang R-L, Lin T-P, Kuo N-J (2006) Field experiments on thermal comfort in campus classrooms in Taiwan. Energ Buildings 38(1):53–62

Hwang R-L, Lin T-P, Chen C-P, Kuo N-J (2009) Investigating the adaptive model of thermal comfort for naturally ventilated school buildings in Taiwan. Int J Biometeorol 53(2):189–200

Kottek M, Grieser J, Beck C, Rudolf B, Rubel F (2006) World map of the Koppen-Geiger climate classification updated. Meteorol Z 15(3):259–263

Lemke B, Kjellstrom T (2012) Calculating workplace WBGT from meteorological data: a tool for climate change assessment. Ind Health 50:267–278

Liang H-H, Lin T-P, Hwang R-L (2012) Linking occupants' thermal perception and building thermal performance in naturally ventilated school buildings. Appl Energy 94:355–363

Mors S-t, Hensen JLM, Loomans MGLC, Boerstra AC (2011) Adaptive thermal comfort in primary school classrooms: creating and validating PMV-based comfort charts. Build Environ 46 (12):2454–2461

NEHC (2007) Prevention and treatment of heat and cold stress injuries, Navy Environmental Health Centre (NEHC), USA

Nguyen JL, Schwartz J, Dockery DW (2014) The relationship between indoor and outdoor temperature, apparent temperature, relative humidity, and absolute humidity. Indoor Air 24 (1):103–112

Nicol F, Humphreys M (2007) Maximum temperatures in European office buildings to avoid heat discomfort. Sol Energy 81:295–304

NIOSH (2015) Indoor environmental quality, The National Institute for Occupational Safety and Health (NIOSH). https://www.cdc.gov/niosh/topics/indoorenv/temperature.html

OSHA (2016) Minnesota department of labor and industry. Occupational Safety and Health Division, Minnesota. www.dli.mn.gov

Pereira DL, Raimondo D, Corgnati SP, da Silva MG (2014) Assessment of indoor air quality and thermal comfort in Portuguese secondary classrooms: methodology and results. Build Environ 81:69–80

Rupp RF, Vasquez NG, Lamberts R (2015) A review of human comfort in the built environment. Energ Buildings 105:178–205

Serghides DK, Chatzinikola CK, Katafygiotou MC (2014) Comparative studies of the occupants' behaviour in a university building during winter and summer time. Int J Sustain Energy 34 (8):528–551

Stull R (2011) Wet-bulb temperature from relative humidity and air temperature. J Appl Meteorol Climatol 50:2267–2269

Teli D, James PAB, Jentsch MF (2013) Thermal comfort in naturally ventilated primary school classrooms. Buil Res Inf 41(3):301–316

The United States Department of Energy (USDOE) (2018) Types of insulation, USDOE. Saver https://www.energy.gov/energysaver/weatherize/insulation/types-insulation

Wigo H (2013) Effect of intermittent air velocity on thermal and draught perception – a field study in a school environment. Int J Vent 12:249–255

Zhang Y, Chen H, Meng Q (2013) Thermal comfort in buildings with split air-conditioners in hot-humid area of China. Build Environ 64:213–224

Part X
Green Building

The way sustainable and green buildings can be designed, so they do not only perform in a sustainable way, but are also of high architectural quality.

Chapter 35
Towards Self-Reliant Development: Inhabitant Housing Capacity Gap of Rural Inhabitants on Mt. Elgon

Michaël Willem Maria Smits

Abstract Rural communities in developing countries show a socially inclusive, resilient, and self-reliant model for their housing, despite the lack of individual capacities. However, due to scarce opportunities, many people move to the cities, often returning to challenging living conditions. As a result, both urban and rural inhabitants struggle to reach the desired living standards and well-being. This article explores general capacities of rural inhabitants in Kenya and identifies what shortages prevent inhabitant well-being within their housing. Outcomes of the interviews performed on two hundred families (four communities) evaluate whether the different communities still build housing by themselves, if they would like to continue this 'self-reliant model', or would prefer professionals ro realize their housing. The conclusion indicates that inhabitants would prefer to build housing by themselves and exposes why these communities change to 'external', housing solutions. Housing alternatives which lie within their capacities, play a crucial role in sustaining the communities' self-reliance in relation to their housing.

Keywords Self-reliance · Inhabitant capacities · Inhabitant-led development

35.1 Introduction

Existing informal rural (vernacular) architecture offers a flexible model based on locally available (renewable) materials and building methods. This model often evolved over centuries, passed down to every new generation. Due to the nature and character of the vernacular archetype, extensive maintenance is often needed. Even though the maintenance is considered inconvenient, the continuality allows the

M. W. M. Smits (✉)
Faculty of Architecture and the Built Environment, Delft University of Technology, Delft, The Netherlands
e-mail: m.w.m.smits@tudelft.nl

© Springer Nature Switzerland AG 2020
R. Roggema, A. Roggema (eds.), *Smart and Sustainable Cities and Buildings*,
https://doi.org/10.1007/978-3-030-37635-2_35

497

community to practice its execution. This makes them highly resilient towards change (Nel and Binns 2000). The circular sustainable model is still widely used among many rural African communities. Rural communities have been trying to improve the living quality, however the change introduced industrialized materials and construction methodologies. In practice, this means, despite that durability and maintenance have improved, the process created significant external dependency in material, construction and skills (labor). Causing unsustainable, non-circular and climatic undesirable solutions. What is equally important, it diminishes the community's self-reliance[1] towards their built environment (Smits 2014). The reasons are manifold for rural inhabitants to improve the quality of the existing vernacular housing. In most cases they are restricted to only use local, natural materials and traditional construction methods.

In an effort to change the existing housing model, they now often use materials and techniques that lay outside the inhabitants' knowledge sphere. If these communities are to continue the self-reliant housing model, they need a way to upgrade it (extend durability, lower maintenance) without damaging its qualities: self-building practice, climatic orientation, and renewable materials. To sustain both self-reliance and house qualities it is vital to evaluate inhabitant capacities[2] and use them in decision-making (Smits 2017). However, inhabitant capacities often seem to contradict the ones necessary to build their desired house. Inhabitants are aware of the housing they would like, however, lack the capacities (materials, knowledge, skills and finance) to build the house by themselves. Therefore, this study investigates the conditions in which inhabitants are living in right now and how they would prefer to live. The rural area of Mt. Elgon proves a representative study area[3] in which communities with various levels of 'capacities' can be found. This explores general capacities of rural inhabitants and helps identifying what shortages are preventing them from improving their housing.

For this purpose, over two hundred families participated in a survey conducted in February 2017. Due to the sensitive context of the survey a questionnaire was combined within an interview. Here the interviewer had the opportunity to answer any questions and explain the interviewee's privacy rights (informed form). To have a representative sample of the Mt. Elgon area, four communities with different levels of income, housing and ownership were targeted. To have a representative sample per community (around 50% of the population) communities were targeted of approximately 100–120 households. Moreover, did one male and one female researcher investigate every community, sampling 25 females and 25 males from various households.

This article focusses on the type of housing most of the rural inhabitants of Mt. Elgon live in right now, the housing type they desire, and what their capacities are. This will help expose the discrepancies between the capacities inhabitants have

---

[1]Self-reliance: to which extend a person or family can rely on their own capacities.

[2]Capacities: collection of all available resources, tools, knowledge and skills.

[3]Representative study area: the level of 'development' in the area is representative for many others.

and those they desire. This article will elaborate the executed study on Mt. Elgon in four steps. Firstly, explaining the context of Mt. Elgon and relevance of the targeted communities. Secondly, describing the methodology and consecutive execution of the study in February 2017. Thirdly, elaborating the most important outcomes of the study. Fourthly, describing the conclusions and restrictions to the study. This article will conclude the importance of assessing and incorporating inhabitant capacities towards their housing, which the author coined: "capacity informed decision-making". As this article will show, the communities in the Mt. Elgon area have a shared notion of their desired housing. This shared image is studied in detail, including: housing size and materialization. However, almost half of all the participants of the presented study estimate that they won't be able to afford desired housing, resulting in a large part of the population remaining in challenging conditions: 75% of the studied communities live in mud-based houses. Indicating a need for alternative housing solution(s) for a large part of the community.

35.2   Mt. Elgon Area & Targeted Communities

Over 70% of the built housing worldwide, is built informally and often by the inhabitants themselves (UN-Habitat 2013). South Asia and Sub-Sahara Africa will have one of the most significant shifts from rural to urban in the upcoming decades (UN-Habitat 2015). This shift has posed a great threat to the wellbeing of vulnerable families in the past and predict huge problems ahead. In Africa, projections are that over half of the urban population (61.7%) lives in slums and by 2050, Africa's urban dwellers are projected to have increased from 400 million to 1.2 billion (UN-Habitat 2015; United Nations 2012). It is needless to say that it will be vital to understand one of the largest contributors to this urbanization, namely: rural-urban migration (Tacoli et al. 2014). Therefore, this article studies the current living situation of rural inhabitants on Sub-Sahara Africa as it contributes to the fastest urbanizing area on the continent. With 20–25% of the countries' population urbanizing in the next 20–30 years (World Bank 2016) Kenya proves to be representative case.

In particular West Kenya has a large number of growing cities Kisumu, Eldoret, and Nakuru (World Bank 2016), also called the 'western hub'. In the left image of Fig. 35.1 this urbanization is projected. Here, Mt. Elgon is one of the rural areas that potentially hold rural-urban migrants.

In this area four communities were sought to analyze their current and desired housing. Considering the available resources for this study, a total of 200 inhabitants could be interviewed. Based on this scope several criteria were chosen to identify the communities: firstly, to have a representative sample, at least 40–50% (Thompson 2012) of each population had to be included in the study. Therefore, four communities of around 100–120 families were sought in Mt. Elgon area. Secondly, to prevent a subjective representation, communities with variable levels of income were selected (only selecting poor communities would support the claim that

Fig. 35.1 (Left to right): urbanization index (World Bank 2016), location of Mt. Elgon in Western Kenya and location of selected communities. (Source: Google Maps)

capacities do not meet the desired housing). The communities on the North-eastern slope of Mt. Elgon have varied levels of income (areal employers include: Mt. Elgon Orchards, ADC Japata and ADC Suam), good schooling, and healthcare. Thirdly, a mixture of housing quality had to be identified. It was crucial to show that the mismatch between capacities and desired housing is found amongst different levels of income and quality of housing.

Four researchers from Nairobi University and a local social worker deliberated with the village elder and areal chief for suitable communities in the Northeast area of Mount Elgon. Here, 12 rural communities were evaluated according to previously mentioned criteria. The considered communities were (estimated inhabitants): Chepchoina (70), Cherubai (200), Habitat (94), Japata (90), Kaisheber (150), Kaptega (50), Koronga (550), Nabeki (420), Njoro (300), Sokomoko (100), Vamia (150) and Wangu (30). Finally, four communities in proximity to each other were selected and grouped:

1. No/low income, doesn't own plot, mainly renting/self-build houses
2. Low/regular income, doesn't own plot, mainly/self-build renting houses
3. Low/regular income, owns plot, mainly self-build houses
4. Regular/high income, owns plot, mainly commercially build house.

The Japata settlement near Kaptega river was selected as group 1 (Fig. 35.2: red marker). This community of approximately 70 households, was allowed to temporarily settle themselves as farm workers and since independence (1963) have been living there. They do not own the plot they live on, are not allowed to build permanently, and have low/non-existent incomes.

Chepchoina village was selected as group 2 (Fig. 35.2: green marker). This community of approximately 110 households lives around the Chepchoina village market. The plot is privately owned; most of its residents rent a house in this area. The families have a mixed income and often combine small business with farming,

Fig. 35.2 Map of the selected communities on Mt. Elgon. (Source: Google Maps)

generating a low/regular income. This community has its own marketplace and bus stop, which influences landownership.

Vamia was selected as group 3 (Fig. 35.2: blue marker) consisting of approximately 120 households. The plot belongs to the inhabitants and they mainly have a regular income combining a commercial position with farming their lands. The Habitat community was selected as group 4 (Fig. 35.2: orange marker), consisting of 94 households owning their plots. The majority works fulltime for a commercial farm and have a regular/high income.

With the four communities selected, the next section will elaborate on the methodology used to interview the communities and consecutive questionnaire.

35.3 Survey, Mixed Methodology: Interview & Questionnaire

Studying inhabitant capacities in relation to their housing, involves both quantitative and qualitative aspects. Quantitative capacities consider measurable aspects such as: income, size of family, ages, etc. Where, qualitative capacities consider why and how they live at the moment, moreover, understanding their housing preferences. For this purpose, a mixed method was used, where both questionnaire and structured interview are performed in a survey framework (Creswell 2013; Fowler 2013). The questionnaire was used to register quantifiable answers, closed questions and later on to compare the 200 outcomes. The structured interview was used to address open questions and help to understand motives. A structured interview is chosen to ensure that the interviews follow the exact same procedure. The questionnaire supports the structured interview to ask the same questions in the same order amongst all 200 participants of the survey.

35.3.1   Interview Context

The survey was performed in a vulnerable environment where many of the partic-ipants have difficulties to sustain a living (below international poverty line: $1.90 p. p.p.d.). Moreover, many participants live in a traditional house and traditional relation between man and woman. Therefore, it was essential to take preliminary precautions. As the community elder, chief, and local social workers were already involved, they were also aware of the survey and informed the communities. To get a balanced perspective all households, 100 surveys were conducted with women and 100 surveys with men. To prevent social/cultural dilemmas two female and two male researchers were hired. Sophie E. Kibuywa, is head of a local organization (Desece: development education services for community empowerment) and has decades of experiences in conducting local researches. She recruited the researchers and instructed them for the survey. Evaluating the experiences within the four selected researchers Pauline was appointed as team leader (she was the most experienced) (Fig. 35.3).

During the survey, the researchers were staying separately (men/women) in the middle of the targeted communities. Two communities were next to their place of residence and two communities were in a short travelling distance (max. 5 min on motorbike). There was an office arranged at the local hospital where they were able to work.

35.3.2   Interview Instruction & Guide

To prevent any inconsistencies in executing the surveys, a questionnaire instruction sheet was prepared for the researchers. The instruction explains step by step how the survey should be performed and what the points of attention are. It starts by explaining the context, in which the survey is positioned, gives the objective and aim, continues by introducing the composition (targeted age and such), and explains

Fig. 35.3 (left-right): B. Sawenja, K. Hamphrey, S. Kibuywa, P. Nabalayo and A. Nyangugu. (Source: Author)

the practicalities of the questionnaire: location, recruitment of the participants, picture/audio recording, venue, breaks and ethical issues. Ensuring that the surveys were taken in a safe environment, with the participants of appropriate age and gender and not invading the privacy of the participant. The interview guide has a similar purpose to the instruction. However, it gives the exact questions that need to be addressed during the interview. The guide was written according to the advised structure of an interview guide by: Qualitative Research Methods (Hennink et al. 2010). It starts by introducing the research purpose and explains the attached consent-form (see attachment). The researchers are asked to read the consent form and answer any questions of the participant. When all of them are answered the interview can be conducted. The questions are divided in three sections: general information, questions about current house, and questions about the desired house.

– Section 1: The general information questions are closed quantitative questions that are relatively easy and comfortable to answer. Questions are meant to evaluate: family size, occupations, ages, financial capacities, and landownership. The answers will help understanding the extent to which these capacities enabled the current and desired type of housing.
– Section 2: The questions on current habitation aim to understand the house people live in. The questions emphasize ownership, amount of structures, house size, in/outdoor functions, used materials, self-build practice, help by community members, satisfaction, maintenance, and the reasons for not realizing desired housing.
– Section 3: Questions in this section focus on the participants' desired housing. The closing questions in this section emphasize if they would be able to afford[4] the desired house based on their existing capacities. Moreover, if they would prefer to build the house by themselves, supported by their community.

The questionnaire was available via Google-sheet, accessible by smartphone (all researchers had one). All interviewers had a printed version of the questionnaire, interview guide and instruction with them.

35.3.3 Pilot & Adjustments

On the 30th of January 2017, the first pilot was run amongst the researchers. Here the team was requested to test the survey (using the printed English questionnaire, making audio recording and pictures) amongst each other. The team concluded that there was a necessity of translating the questionnaire to Swahili as it was too difficult to do this simultaneously during the interview. The cross-cultural survey guidelines of Mohler et al. (2010), also called The Team Translation Model Procedures (TRAPD) provide with an appropriate team translation model that suited the

---

[4]Afford: to what extend the capacities enable or disable a realization.

Fig. 35.4 (top to bottom): samples of the Japata ADC & Habitat community. (Source: Author)

requirements of this study. The group of researchers was divided in two teams and separately made their translation. In the review session they compared their translations, discussed the differences and made a concept translation. The results were reviewed by Sophie E. Kibuywa and returned to the team. They had a second adjustment session where they debated the review and made a final translation.

35.3.4 Executing the Survey

The survey started with one research team in the Japata ADC and one in the Habitat community. Every time locating one household that had a mother present and another that had a father present. According to the set target, every team conducted between 8 and 10 interviews per day. The researchers used the physical print to write down the answers of the participants and their phones to make the audio recordings. After each survey, the researchers took a picture of the participants. Afterwards they were given 1 kg of sugar per household as compensation. At the end of the week the researchers used 3 days to digitalize the 100 answer sheets and upload the pictures and audio recordings (Fig. 35.4).

On February 16th, the survey continued in the Vamia and Chepchoina communities, following the same procedures as the Japata ADC and Habitat community. The research teams were able to finish the second round of 100 surveys by February 24th (Fig. 35.5).

35.4 Outcomes Survey

In the following sections the outcomes of the survey are compared between the four communities. In each consecutive part of the questionnaire the most important findings are shown and explained.

Fig. 35.5 (top to bottom): samples of the Vamia & the Chepchoina community (Source: Author)

Table 35.1 Shared income, Income stability & family size

5. How much is your shared income?

	Japata	Chepchoina
<1000	0,0%	4,7%
1000–2499	9,1%	4,7%
2500–4999	20,5%	9,3%
5000–7499	40,9%	16,3%
7500–9999	13,6%	16,3%
10,000–24,999	15,9%	37,2%
25,000–49,999	0,0%	9,3%
50,000–99,999	0,0%	2,3%
100,000–500,000	0,0%	0,0%

6. Is this stable or does it fluctuate?

	Japata	Chepchoina	Vamia	Habitat
Stable	18,4%	28,9%	19,5%	64,2%
Fluctuates	81,6%	71,1%	80,5%	35,8%

8. Number of children

	Japata	Chepchoina	Vamia	Habitat
0–3 years	42,9%	69,4%	36,6%	32,1%
4–7 years	38,8%	22,4%	43,9%	58,5%
8–11 years	18,4%	8,2%	17,1%	3,8%
12+ years	0,0%	0,0%	2,4%	5,7%

35.4.1 General Information Questions

Table 35.1 projects shared income, income stability and the family size, between the four communities. Although the Habitat and Chepchoina community have a higher average income, the majority of inhabitants (>50%) earns up to 25,000 KsH (roughly $250) per month. Considering that the majority of the community has between 0 and 7 children this leaves the households with $4 per person per day (2-person household), $1,6 in a five-person household and worst-case $0,8 in a nine-person household. With income fluctuating in at least 70% of the households in three

Table 35.2 Questions on: farmland, ownership and the contribution to livelihood

12. Do you have farmland (shamba)?				
	Japata	Chepchoina	Vamia	Habitat
Yes	38,8%	54,2%	80,5%	94,3%
No	61,2%	45,8%	19,5%	5,7%

13. Do you own this farmland?				
	Japata	Chepchoina	Vamia	Habitat
Yes	0,0%	34,7%	70,7%	84,9%
No, companyland	79,6%	0,0%	0,0%	1,9%
Unknown	20,4%	22,4%	17,1%	1,9%
Familyland	0,0%	12,2%	9,8%	7,5%
No	0,0%	24,5%	0,0%	3,8%
Rented	0,0%	6,1%	2,4%	0,0%

14. Does it generate income?				
	Japata	Chepchoina	Vamia	Habitat
Yes: grow crops for family	36,7%	28,6%	63,4%	66,0%
Yes: for family and selling	2,0%	18,4%	14,6%	18,9%
Unknown	59,2%	22,4%	22,0%	1,9%
No	2,0%	30,6%	0,0%	13,2%

out of four communities, questions arise if the families are able to sustain basic life necessities (as they are far under the international poverty line: $1,90). It is important to state that Japata has a considerably lower average income.

Table 35.2 shows that although most households did not state that they are farmers (<15%) three out of four community has a majority that has a farmland (>50%), which contributes to their daily livelihood. Current capacities in the communities show that some of the households have been able to secure a stable and substantial income. However, the vast majority of the households have a daily budget below the poverty line and the income in most cases fluctuates often. It makes the households highly vulnerable and indicates that making means to an end is difficult. In relation to their built environment that in most cases the financial capacities for materials and labor are marginal. The next section reflects on how these capacities relate to current habitation.

35.4.2 Questions on Current Housing

Ownership in the communities differentiates substantially: see Fig. 35.6. The government owns the land on which the Japata community lives, inhabitants are mainly workers of the Japata ADC farm. Japata has an almost equal ownership and renting division. However, as they do not own the land it is questionable to what extent they are allowed to live there. Chepchoina has almost solely renting residents (>95%) and

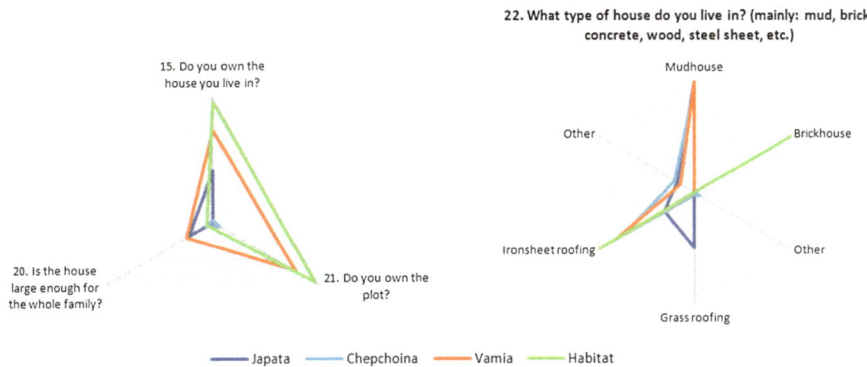

Fig. 35.6 Left: house ownership and Land ownership, Size suitability; right: type of current house

therefore the majority has no land rights. The opposite is true for the neighboring Vamia. Here, the majority (>70%) owns both land and house. Despite the differences in all three mentioned communities, the vast majority (>90%) of their households live in mud-based houses. This occurs despite the fact that Chepchoina and Vamia on average have a much higher income then Japata. Even renting does not seem to enable households with an average higher income to live in an 'improved house'. Which can be explained by two factors: availability of brick houses and fluctuations in income. The latter explained by the 70–80% of households in these communities have seasonal/unstable jobs. The Habitat community stands quite the contrary to the other three communities. Here, the land is individually owned, however, via a collective. Considering the height and the stability of their income they are the only community who could afford a brick house.

However, it seems that in none of the communities the current capacities have offered sufficient living space for the whole family (Fig. 35.6). With the majority of the households having between 0 and 7 children in a house between 5,7 and 13,7 square meters this problem can be explained.

In the case of Japata and Vamia the majority of the materials (75–100%) are not bought but collected. The only costs involved are to cover transportation. Table 35.3 shows the large amount of natural materials used in constructing houses, which makes the materials affordable, especially amongst the communities with a low income.

Looking at the self-built practice (Fig. 35.7) especially in Japata and Vamia this influences the maintainability of the house. The opposite happens in the Habitat community where more than 90% is not able to maintain the house by themselves. A more worrying trend seems to be the ability to afford maintenance in case income becomes low or stops altogether. The Japata community actually has the most positive score in this section. Here, over 65% of the households think they will be able to pay for the maintenance on the house, due to the availability of materials.

Although the capacities and living situations differ strongly, they all seem to result in an opinion of dissatisfaction on the house (Fig. 35.8). The Habitat

Table 35.3 Material cost & availability

If yes: a. Did you have to pay for the materials or are there other ways of collecting/ acquire these materials				
	Japata	Chepchoina	Vamia	Habitat
Pay	14,8%	0,0%	0,0%	100,0%
Free	7,4%	14,3%	0,0%	0,0%
Collected	14,8%	4,8%	0,0%	0,0%
Collected and paid for transportation	63,0%	19,0%	100,0%	0,0%
N/A	0,0%	61,9%	0,0%	0,0%

25. Are those materials local natural resources (e.g. mud or straw) or Manufactured (e.g. cement, iron sheet)?

	Japata	Chepchoina	Vamia	Habitat
Natural	100,0%	78,3%	90,0%	0,0%
Industrial	0,0%	21,7%	10,0%	0,0%
Both	0,0%	0,0%	0,0%	100,0%

Fig. 35.7 Surface of current house

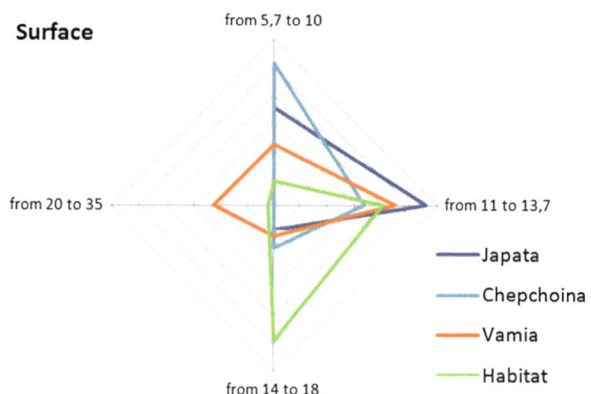

Fig. 35.8 Self-build practice, repair ability and affordance

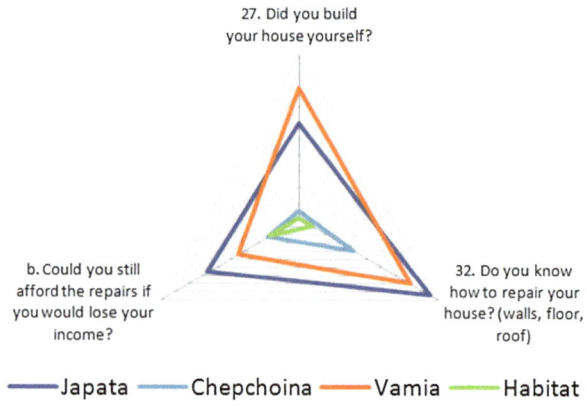

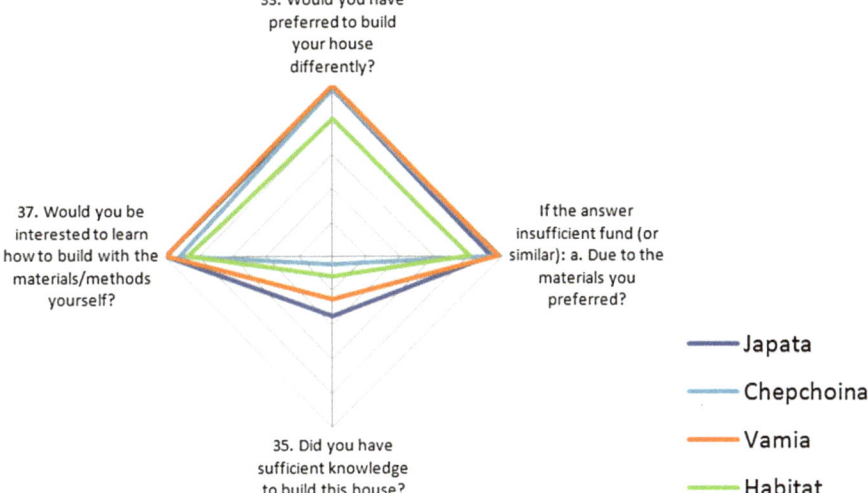

Fig. 35.9 House preference, sufficiency funds, building knowledge and willingness

community shows a little more content with the existing house, however, >80% still prefers to build the house differently. When asked why, the majority answered: due to the lack of funds, which most likely is linked, to the type of materials they would have preferred to build with (>80%). Moreover, when asked if they would know how to build this house by themselves, more than 64% of all respondents do not think they are able to do so (Fig. 35.9).

The last question on the existing house inquires if the inhabitants would be interested to learn how to build their desired house. What they most likely do not consider is the knowledge, skill level and training needed to build such a house. Building such a house would actually require extensive professional knowledge, skills and training, such as: mason, steelworker, carpenter, etc. Those types of trainings would either take many years in training or could be learned on the job. Indicating not only a problem in comprehending the needed requirements but also that there is a considerable knowledge, skill and training gap between the currently used and desired building technique. In the last section on the results this topic will be further explored.

35.4.3 Questions on Roofing

Based on a previously made observation in the area: the houses built with thatch in comparison to roofing sheets, seem to be cooler during the day and warmer during the evening. When it rains the roofing-sheets produce a lot of noise in comparison to the thatched roof. To better understand if the inhabitants had similar observations and how they reflected on material suitability, a short section was included in the

Fig. 35.10 (left to right): roofing sheet Chepchoina, thatched roof Vamia. (Source: author)

Fig. 35.11 Left: opinion about existing roofing sheet; Right: opinion about desired roofing sheet

survey. Figure 35.10 shows the results on the existing house (left image). With the majority of the communities having roofing sheets (Japata >32%, Chepchoina, Famia & Habitat 75–100%) they have sufficient experience to reflect on the effects of the roofing sheet (Fig. 35.11).

Results show that the majority of the households find the roofing sheet radiating heat when the sun is shining (>90%) and makes noise during rains (50–95%), confirming the initial observation made in the communities. Despite these disadvantages the majority still uses roofing sheets. Moreover, does the majority not know any cheaper alternatives (50–90%). The rest of the respondents do point out thatched roof as existing alternative. Respondents admit that those alternatives would react better to sun (50–90%) and rain (80–100%), indicating that there are no cheaper alternatives, however, they have better characteristics than roofing sheets. Which indicates a possible knowledge gap of alternative roofing solutions within the communities.

The same questions were asked after households stated their preferred type of house. Here, between 70 and 90% of the households (Fig. 35.10, right image) answered that they prefer using iron sheet roofing. When asked if the iron sheets made noise during rain or radiate heat when the sun is shining, the answers were quite the opposite to their current housing. Here, the majority of the households (rain noise: 40–70% & sun radiation: 55–70%) stated that the iron sheets do not have this effect. In the interviews, many households stated that the main reason there are no cheaper alternative is the difficulty they have to find grass locally. Due to this shortage people started to sell grass as a building product. The available 'free' grass has to come from such a distance that the transport costs are almost equal to buying roofing sheets. Moreover, in their opinion the grass roofing requires more maintenance and leaks more often. Other reasons for preferring roofing sheets are: fire resistance and insect-proof. It seems that these reasons influence their perspective on the disadvantages of the roofing sheet.

35.4.4 Questions on the Desired House

The questions in the third section of the questionnaire focused on desired housing. In the Japata and Chepchoina community respondents would all prefer to own both the house and land. Among all the communities 95–100% of all the households would prefer to own their house and the land they live on (Table 35.4).

When asked which materials they would prefer to build their desired house from (see Fig. 35.12) the majority chose bricks (45–75%) and iron sheets (70–95%). Most households state that the preferred materials are expensive (see Table 35.5).

Figure 35.13 shows that inhabitants prefer to build the house by themselves (75–95%) and if they can't or won't build the house themselves that they will need to hire labour (90–100%). Japata and Vamia think that their community would help most of them in building the house (>95%), which in Chepchoina (mainly renting) and Habitat (formed community) is quite the contrary. It could be argued that these communities are differently organised and therefore inhabitants are reluctant to help each other. This, in the Habitat community is strange considering the fact that they own the land communally. What is most worrying, is that three out of four communities will not be able to make house repairs when their income diminishes.

The willingness to learn how to build the desired house is very strong (Fig. 35.13) amongst all households: 95–100%. Indicating that self-build practice is preferred. Although in some communities there are doubts if community members would be

Table 35.4 Desired house/land ownership

39. Would you prefer to own or to rent the house?				
	Japata	Chepchoina	Vamia	Habitat
Own	98,0%	100,0%	100,0%	100,0%
Rent	2,0%	0,0%	0,0%	0,0%
44. Would you prefer to own or to rent a plot?				
	Japata	Chepchoina	Vamia	Habitat
Own	100,0%	97,9%	100,0%	100,0%
Rent	0,0%	2,1%	0,0%	0,0%

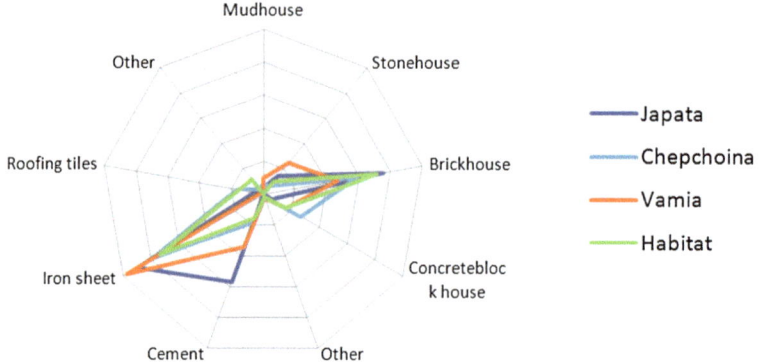

Fig. 35.12 Preferred materials, material costs

Table 35.5 Material costs

47. Are those materials expensive or cheap?				
	Japata	Chepchoina	Vamia	Habitat
Expensive	81,6%	63,0%	87,8%	56,6%
Cheap	18,4%	37,0%	12,2%	43,4%

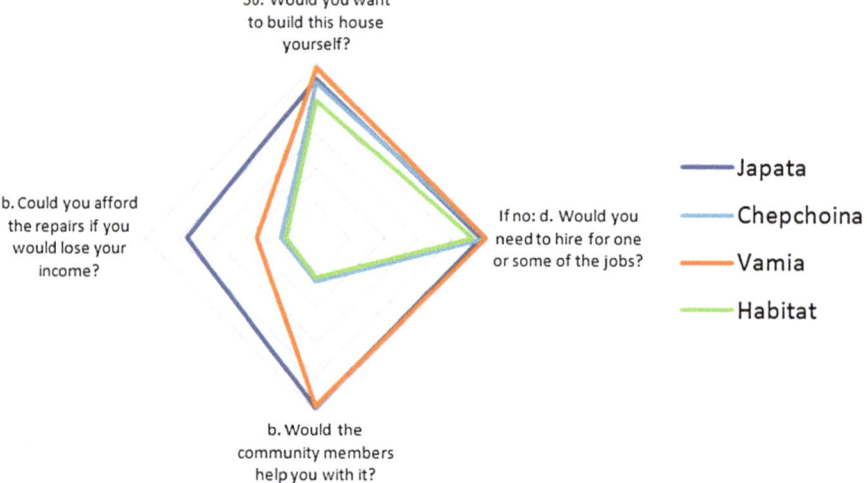

Fig. 35.13 Self-build preference, community help and affordance

willing to help build a new house. However, almost all households are willing to help (95–100%) a community member if they can learn how to build in return, indicating there is a strong willingness to learn by helping each other. What might be even more interesting is that again the vast majority of the households are willing to help constructing public building in order to learn how to build in an 'improved' way (Fig. 35.14).

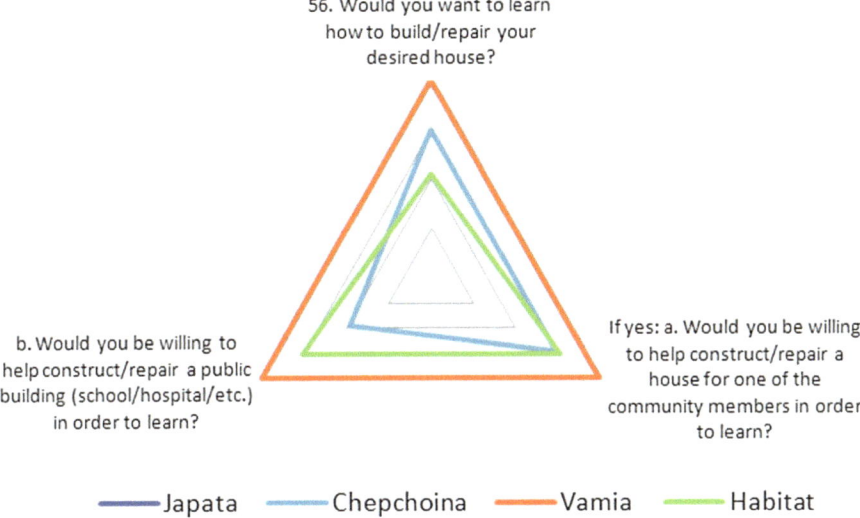

Fig. 35.14 Repair preference house, help of community members and willingness build community infrastructure

35.5 Conclusion

This study proves that the majority of the interviewed households are living in challenging housing conditions. These living conditions are in most cases in mud-based houses often too small for the entire family. Although the households living in these conditions have an idea on their desired housing, they lack the capacities to realize such housing. Landownership is an important restriction for households in achieving better housing. The Japata community lives on government land and is not allowed to build an improved house. The Chepchoina community mainly rents and is therefore very vulnerable to changes in income. The Habitat community has severely restricted land rights and is not allowed to make any extensions/additions. The Vamia community has the most households owning their land and house. With an acceptable and stable income, it is unclear why they were not able to realize desired housing.

The problem seems complex, however, it revolves around two elements: current capacities and the capacities needed to build the desired house. The majority of the interviewed households have more than sufficient capacities to build a house by themselves. This 'traditional' way of building is a shared practice within the family and their community. This practise suits all their capacities: local/natural/free materials, local/available tools, financial and required skills. However, it is clear that almost all households desire a different way of housing. Looking at what those preferences would require it is clear there is disparity between the capacities inhabitants have and those they need to realize improved housing. The lack of locally available alternatives in typology, material and building methodology, limit the scope in which the households consider alternative options. Additionally, is the inhabitants' understanding of materials and skills limited on what possible harm they inflict. Moreover, are possible alternatives difficult to articulate without a substantial knowledge base. This makes the formulation of a possible alternative 'desired' house by the inhabitants themselves difficult. Integrating their current capacities into alternative housing solutions will play a vital role to its success and implementation. As shown in this study, considering alternative solutions that do not meet the inhabitants' capacities is simply not viable. The study proves that there is a high willingness to build by oneself, help each other and help to build community infrastructure. This sense of community could be fundamental in advising rural communities how they can improve their living environment without losing their self-reliance. In a consecutive article the methodology developed for analysing inhabitant capacities will be explained. Furthermore, will show how these capacities can be used in what the author calls: capacity-based decision-making.

References

Creswell JW (2013) Research design: qualitative, quantitative, and mixed methods approaches. SAGE Publications. Retrieved from: https://books.google.nl/books?id=PViMtOnJ1LcC

Fowler FJ (2013) Survey research methods. SAGE Publications. Retrieved from: https://books. google.nl/books?id=CR-MAQAAQBAJ

Hennink M, Hutter I, Bailey A (2010) Qualitative research methods. SAGE Publications. Retrieved from: https://books.google.nl/books?id=rmJdyLc8YW4C

Mohler P, Dorer B, De Jong J, Hu M, Harkness J, Mohler PP (2010) Cross-cultural survey guidelines translation: overview. Retrieved from: http://ccsg.isr.umich.edu/images/PDFs/ CCSG_Translation_Chapters.pdf

Nel E, Binns T (2000) Rural self-reliance strategies in South Africa: community initiatives and external support in the former black homelands. J Rural Stud 16(3):367–377. https://doi.org/10. 1016/S0743-0167(00)00003-6

Smits M (2014) An architect's investigation into the self-reliance of a Sub-Saharan African community. In: Tomasz Jeleński EW-S, Juchnowicz S (eds) Tradition and heritage in the contemporary image of the city: monograph. Challenges and responses. Krakow, Wydawnictwo PK, pp 119–125. Retrieved from: https://books.google.nl/books?id=JLKcAQAACAAJ

Smits M (2017) Formulating a capability approach based model to sustain rural Sub-Saharan African inhabitant's self-reliance towards their built environment. Int J Sustain Dev Plan 12 (2):238–251. https://doi.org/10.2495/SDP-V12-N2-238-251

Tacoli C, Mcgranahan G, Satterthwaite D (2014) World migration report urbanization, rural–urban migration and urban poverty. Retrieved from: https://www.iom.int/sites/default/files/our_work/ ICP/MPR/WMR-2015-Background-Paper-CTacoli-GMcGranahan-DSatterthwaite.pdf

Thompson SK (2012) Sampling. Wiley. Retrieved from: https://books.google.nl/books?id=-sFtXLIdDiIC

UN-Habitat (2013) Global reports on human settlements. Retrieved October 24, 2017, from: http:// mirror.unhabitat.org/content.asp?catid=555&typeid=19&cid=12336

UN-Habitat (2015) Habitat III issue papers 22 – Informal Settlements. In: Habitat III issue papers: 22- informal settlements, p 9. Retrieved from: https://unhabitat.org/wp-content/uploads/2015/ 04/Habitat-III-Issue-Paper-22_Informal-Settlements.pdf

United Nations. (2012). State of the world's cities 2012–2013: prosperity of cities. United Nations Pubns. Retrieved from: http://mirror.unhabitat.org/pmss/listItemDetails.aspx?publicationID=3387

World Bank (2016) Republic of Kenya; Kenya urbanization review, 192. Retrieved from: http://documents.worldbank.org/curated/en/639231468043512906/pdf/AUS8099-WP-P148360-PUBLIC-KE-Urbanization-ACS.pdf

Chapter 36
Mainstreaming *Real* Sustainability in Architecture

Luke Middleton

Abstract The popularity of sustainability has grown significantly in the last decade with new buildings, cars, furniture and other consumer goods all carefully wording their *sustainable* marketing approach to appeal to the consumer. The reality, however, is that most of these products tend to over-sell their commitment to sustainability in their relevant fields. In the built environment, for example, this is due to many interconnected problems: "a failure to take issues of sustainability seriously enough, incoherent ESD [Ecologically Sustainable Development] policy, the predominantly aesthetic agenda of architecture and the assumption that certain technologies in themselves will deliver sustainability." (Willis, The limits of 'sustainable architecture'. Paper delivered at shaping the sustainable millennium, Queensland University of Technology, Brisbane, Australia, 2000). At EME Design, we believe that a shift in the approach and processes driving architecture and the built environment needs to occur in order to create sustainable and resilient buildings, communities, and cities.

Keywords Passive house · Sustainable · Architecture · Energy · Comfort · Resilience

36.1 Introduction

Why has 'sustainable' architecture become so popular in recent decades? And is it even understood completely? Unfortunately, most architecture firms use this as a marketing technique and *real* sustainability is not well understood by the general public. We are being lied to about what *real* sustainability is and people are led to accept less than optimal and bolt-on solutions that depend on high-performing and expensive technologies. As we will discuss later in this paper, we define *real*

L. Middleton (✉)
EME Design Pty. Ltd., Collingwood, Australia
e-mail: luke@emedesign.com.au

© Springer Nature Switzerland AG 2020   517
R. Roggema, A. Roggema (eds.), *Smart and Sustainable Cities and Buildings*,
https://doi.org/10.1007/978-3-030-37635-2_36

sustainability in architecture as; beautiful, educational, positively impacting (environmentally, energy, materials, etc.), comfortable and healthy, and with a multi-disciplinary approach to problem solving and design.

"Architecture is a very dangerous job. If a writer makes a bad book, eh, people don't read it. But if you make bad architecture, you impose ugliness on a place for a hundred years" (Luscombe 2011). More relevant today, we would argue that "ugliness" in this quote be replaced with something along the lines of *if you make bad architecture, you impose an energy intensive and fragile built environment on a place for a hundred years*. Renzo Piano, one of the many *starchitects* with limited commitment to sustainability really puts things in perspective for us here. Architecture has become an issue in aesthetics rather than something that should be concerned with the wellbeing of people and the built and natural environment.

Architecture has become too reliant on technologies and consultants to achieve an acceptable level of indoor comfort in what would be considered a poorly performing building. Silos have been created between professions limiting the dialogue of process and exchange that result in an in- depth understanding from all involved professions.

Architects used to play King when creating buildings from their wildest dreams and imaginations. But with more complexities in the profession evolving, consultants such as engineers (and more recently ESD consultants) have been brought in to ensure the architect doesn't have to be an expert in every field. They handballed their responsibilities to other experts, and while this was a great move forward in developing multiple professions and specialties, it generated silos that segregated professions which made total understanding and deep collaboration redundant. While there are some professionals and practices that take a multi-disciplinary approach to building design and construction, the majority are still dependent on the expertise of others, without fully comprehending what is going on.

We believe that a fundamental shift needs to happen in the education system in order to produce holistically sustainable buildings in our ever-changing built environment. There needs to be more focus on multi-disciplinary approaches to improve understanding and the flow of knowledge through projects and systems.

EME Design's approach to sustainability is one that involves and educates the client, builders, contractors, neighbors and the wider community in the entire process. This is what we call the Elastic Loop Process of the project which aims to ensure a more informed and holistically sustainable approach. Rather than a bolt-on, fake 'sustainable' building, the Elastic Loop generates results which create large energy and waste reductions and savings, as well as improved year-round comfort, and air quality – all while not sacrificing beauty. The Elastic Loop Process is by nature one that is more permeable and open to accepting positive influence. It is founded on a process which depends on dialogue and exchange of knowledge and learnings. The process embraces the challenge of harmonizing the pragmatic fundamentals (the physics of super-efficient buildings) with the poetics of space and light. It is interested in real outcomes and involves detailed post occupancy analysis to ensure the reality is measured against the rhetoric (something sadly missing in most architecture today, for a reason). Shifting the profession and industry will require a multi-facetted approach starting with universities, government and professional bodies.

36.2   Elastic Loop Process

Since 2000 the Elastic Loop process has been developed and adapted by EME Design. It consists of three main phases which allow a project to continually evolve and adapt over time to improve the overall benefits and gains. The process, by its nature, ensures that the system and approach continuously evolve and improve.

36.2.1   *Exploration/Gathering*

This first phase (like all other phases) can be revisited and re-viewed multiple times throughout the elastic loop process. It involves clients and other parties to engage in the process of gathering information to understand the local and broader context. This initial phase is very much focused on self-education where all parties become educated to shift their focus (and thus the project brief) into a holistic and sustainable approach. It is about mapping the potentials (energy, materials, climate, movement, beauty, etc.) to see how they can be exploited and (potentially) quadrupled in the manifestation phase (Roggema et al. 2009).

36.2.2   *Harmonizing/Aesthetic*

The information collected from the exploration/gathering phase then informs decisions made in the harmonizing/aesthetic phase to "form an "armature", the backbone of the design response. This is tested and must be found to have stability when challenged" (Middleton and Friedlander 2009). The client's perceived norms are challenged in this phase as the design progresses through the problem-solving period.

36.2.3   *Manifestation*

This phase brings in people from other fields – builders, tradespeople, to create a dialogue between clients, contractors, science, media, etc. This results in the final project, but previous stages can be constantly re-visited to evolve and improve the final and future manifestations.

36.2.4 Post-manifestation

After the physical manifestation of the project is completed, the Elastic Loop Process does not stop. Projects with a complete Elastic Loop approach are then monitored post-occupancy to ensure the building project is performing as expected or exceeding expectations. This collected data then allows future projects to share the obtained knowledge to continually improve and evolve building techniques, processes and outcomes.

36.3 Elastic Loop Process and Passive House

As our process and experience at EME Design has developed and improved over time, just like our approach to building projects, we have moved towards the Passive House approach. The Passive House approach was developed in Germany and is a voluntary method of certification which results in an extremely energy efficient home (or building) with high quality indoor air, improved health and year-round comfort (Passipedia 2018). Its five basic principles to achieve this are (Fig. 36.1):

1. Thermal Insulation
2. High Performance Windows and Doors
3. Mechanical Ventilation with Heat Recovery
4. Airtightness
5. Thermal Bridge-Free Design

Fig. 36.1 The 5 basic principles of Passivhaus (Passive House Institute), https://passipedia.org/basics

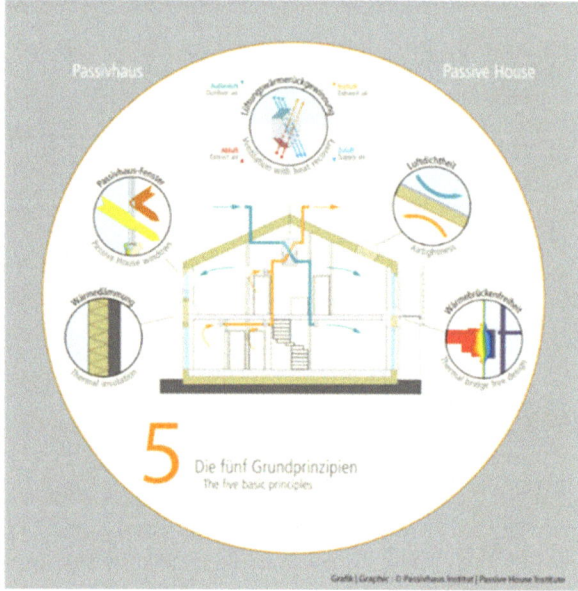

36.4 Key Principles of an Elastic Loop Process

36.4.1 Comfort

When people talk about a Passive House, the first thing that comes to mind is the energy efficiency. But there is so much more to a Passive House than this. They are extremely comfortable to live in with a radiation temperature asymmetry less than 5 °C. This means that two internal surfaces will never have a temperature difference of more than 5 °C in a Passive House which vastly improves the internal comfort levels. Draughts from glass surfaces do not occur as they do in most conventional homes, so no cold spots are created. Overheating occurs less frequently, too (maximum allowed for a certified Passive House is 10% of the time annually, but most are much less than this), reducing the need for air conditioning. Internal temperatures of Passive Houses have a very low diurnal range and are required to stay within 20–25 °C for increased comfort year-round.

The controlled stable temperature is only achieved with a rigorous approach which challenges the typical methods of design and construction. It results in an airtight building that removes uncomfortable draughts and while still providing constant filtered fresh air through the Mechanical Heat Recovery Ventilation unit at a controlled rate and temperature. With this improved quality of indoor air and thermal comfort, the health of the occupants is also significantly improved.

36.4.2 Health

Most typical constructions are prone to mold growth and unhealthy air quality due to high levels of VOC's (volatile organic compounds) and carbon dioxide from poor ventilation and badly designed building materials, furniture, and even toys (to name a few of the culprits!). Think about lead paint in homes or asbestos. How many years were we constructing with these materials until we realized they were bad for us? And how many homes around the world are still occupied by these toxins? Too many to count. Passive House, on the other hand, is a scientific approach which focuses on the health of the occupants and removes harmful toxins from the built environment – a focus where too many *sustainable* certification systems fall short.

36.4.3 Post-occupancy Monitoring

While the Passive House Institute does not require building monitoring to be conducted, we believe that it is an extremely important part of learning that contributes to the Elastic Loop process. From real data, we are able to analyze and compare predicted results of a building's performance with real results. We are able to use this gained knowledge on future projects and to expand and improve on systems and approaches. The post-occupancy monitoring works like an open source for people wanting to understand more about Passive Houses and their efficiency. It works as an educational tool that is continually growing and developing over time.

36.5   Passive House Case Study #1 – Passive Butterfly

The Passive Butterfly (Figs. 36.2 and 36.3) is a heritage renovation project that holistically upgrades a cold and draughty home into a twenty-first century sustainable and comfortable home. Passive House principles were applied to create an exemplary home that exceeds minimum building standards worldwide. Pioneering projects such as this don't eventuate without complete passion and collaboration from all parties. We were lucky to have an enlightened and educated client whose vision of a healthy and carbon positive world perfectly aligned with EME's ambitions. The client was also the perfect example of an actively engaged client involved in the process of dialogue and exchange enabling greater understanding and knowledge sharing between all.

The Elastic Loop process of the Passive Butterfly continues even today with the client actively involved with the Passive House Institution, and other sustainable organizations such as the ATA – Alternative Technology Association. Multiple papers and studies have been conducted on the performance of the Passive Butterfly (see Figs. 36.4 and 36.5), and all this information is openly shared as a resource to learn from.

Fig. 36.2 South-facing elevation of Passive Butterfly. (Photography by Amorfo)

Fig. 36.3 Rammed earth walls at Passive Butterfly. (Photography by Amorfo)

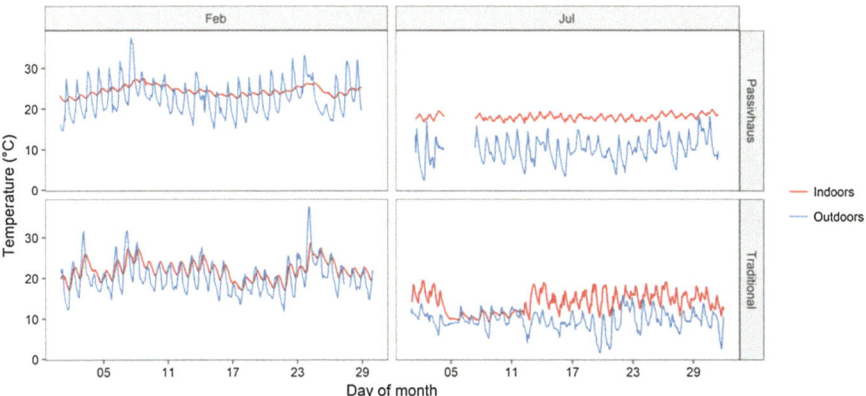

Fig. 36.4 Comparative temperature performance between Passive Butterfly (Passivhaus) and a typical Australian home (Traditional)

The above diagrams indicate performance testing conducted at the Passive Butterfly post- occupancy. Figure 36.6 shows how the CO_2 levels and relative humidity are controlled and stabilized through the Mechanical Ventilation Heat Recovery unit. As soon as the unit is switched off, CO_2 and relative humidity rise to levels that are uncomfortable and unhealthy to live in causing mold growth as well as increasing the risk of asthma. Once the unit is switched back on, a low volume of filtered fresh air is brought in to constantly provide 100% fresh air and healthy levels of CO_2 and relative humidity (Fig. 36.7).

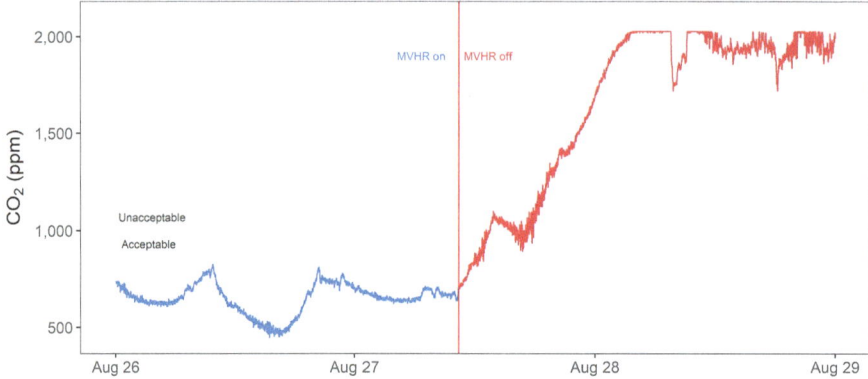

Fig. 36.5 CO_2 and relative humidity data from Passive Butterfly. Showing how the humidity increases significantly when Mechanical Ventilation Heat Recovery unit is switched off

Fig. 36.6 Asymmetrical gable, inspired by local heritage homes, design to capture winter northern sun. Multi-functional entry allows home to adapt over time

36.6   Passive Hybrid Case Study #2 – The 'MM-House'

This home was designed a hybrid, following both Australian passive solar design together with Passive House design principles (Fig. 36.8). It was pressure tested during construction achieving 0.6 ACH (∗), and is currently being monitored for comfort, performance, air quality. The home is a testing bed for other embedded design systems that EME have experimented previously. This project will be part of an on-going education and dissemination program actively pursued by EME.

Fig. 36.7 (**a**) Terraced Deck conceals 11,000 l of rainwater storage supplying irrigation, toilets and laundry. Rear garden includes permaculture productive garden; (**b**) Northern kitchen courtyard brings light and solar warmth – deciduous Japanese Maple provides shade in hotter months; (**c**) Highlight windows strategically placed to capture winter sun and provide cross flow ventilation in summer; (**d**) External blinds integrated into the western façade provide protection during summer. Northern high windows capture winter sun

This temperature monitoring graph of MM House shows a comfortable internal diurnal range, while external temperatures range 0.6–13 °C (Fig. 36.9).

Spikes in the internal temperature in the above graph are due to direct solar heat gain. This was a deliberate strategy to better monitor the direct solar gain on thermal mass.

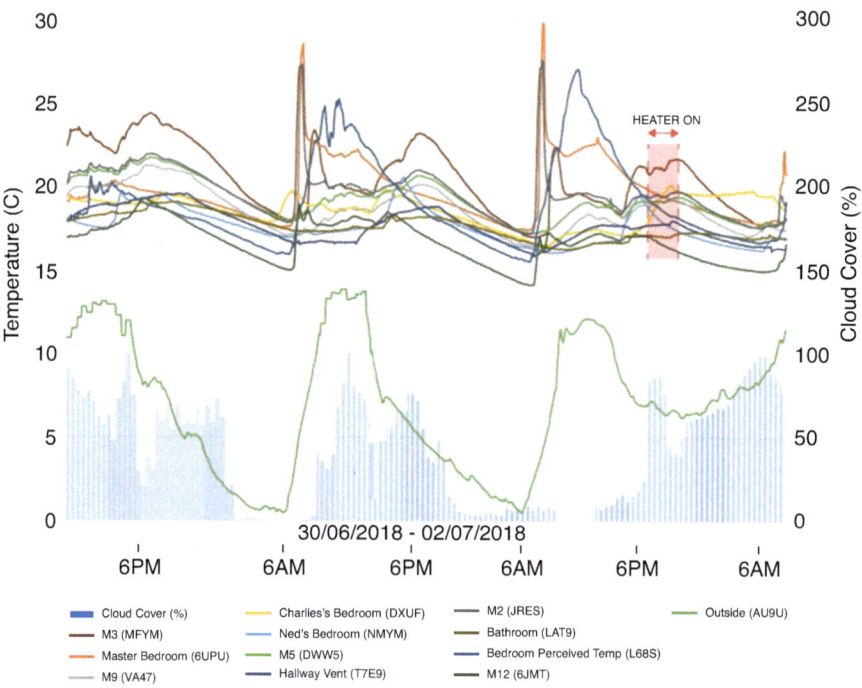

Fig. 36.8 MM House temperature monitoring. Minimal supplementary heating 3 h × 4Kw = 12kw over 3 days with very low ambient temperatures

36.7 Exponential Energy Saved

The Passive Butterfly, being the first Passive House project we embarked on, is the beginning of a new Elastic Loop period where new clients will be educated on the positive change they can make with a Passive House standard home. Two of our most recent projects – Passive Butterfly and MM House were both designed and built with passive house principles in mind. They're prototypes for us to learn from, and with continual performance monitoring, we're able to share with others how that knowledge approach and process can produce a building that is holistically sustainable (see Table 36.1).

36.8 Positive Changes

Steps have been taken gradually to improve policies and building standards both Australia-wide and internationally. The minimum green star rating, for example, was introduced in 2005 to require a minimum of 5 green stars for residential projects. The problem with this process, however, is that it looks at the "designed" "theoretical

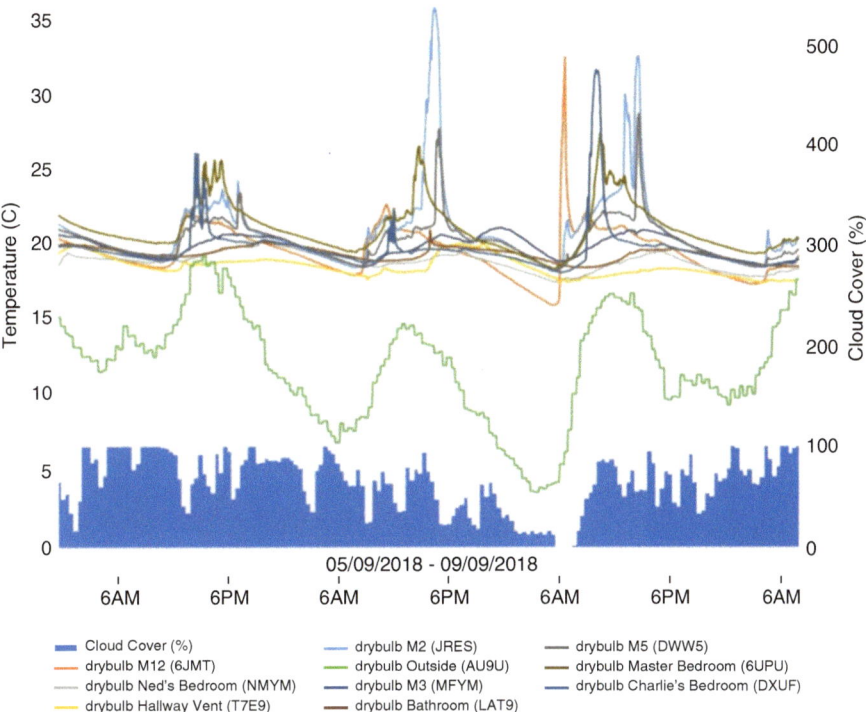

Fig. 36.9   MM House temperature monitoring (no heating)

Table 36.1   Energy saved

Project	Year	Energy generated	Energy consumed	Extra energy	Energy consumption per M^2
Passive Butterfly (143 m^2)	2017	7855	3947	3908	27.6
MM House (197 m^2)	2017	7785	4485	3300	22.7

Based on 15.7 kWh/household/day average (Ausgrid 2018)
[a]Energy shown in kWh/year
[b]Average electricity consumption = 5730 kWh/year

performance, rather than the performance during occupancy. The other concern with this system is there is limited understanding on how the calculations are made and what assumptions are made in terms of comfort levels (i.e. the tolerance internal temperature).

Other government initiatives, such as the Zero Net Carbon homes, for example, are happening not just in Victoria, but all-around Australia. The Zero Net Carbon homes initiative's objectives are to improve the supply of Zero Net Carbon homes, increase the consumer demand for Zero Net Carbon homes and pilot As-built

Verification (ABV) methodologies, standards and skills development through a collaborative approach with industry (Sustainability Victoria 2018).

While these initiatives are great, and more need to be executed, there still needs to be a greater shift in architectural education and approach to prevent aesthetic-obsessed buildings that depend on expensive bolt-on 'sustainable' solutions. No more sustainability as minimum standard imposed on BAU design solutions – a quantum shift in the benchmarks and the approach to design.

36.9 What Still Needs to Be Done

36.9.1 Education

The core issue is education. We need to engage others in the understanding and critical engagement of what *real* sustainability is and what a viable solution is (and isn't). There is a real issue in understanding and explaining to people working in industries outside of the built environment what a *real* sustainable building is. Unfortunately, big name architectural firms with the most media coverage are where they're getting the message from. This message is typically a non-holistic and bolted-on approach to sustainability that results in poor building performance, comfort and lifecycle. Butera (2005) sums up wonderfully how the built environment lies to us about *sustainable* buildings "Ferraris are beautiful cars, a perfect balance between advanced technology and beauty; but never they have been sold as ecological cars. The same should apply to fully glazed buildings: some of them are outstanding for beauty and for technological innovation; they are the Ferraris of modern architecture. But please, do not sell them as sustainable buildings."

36.9.2 Image

This point goes hand-in-hand with Education as it focuses on developing a *real* representation of what sustainable architecture is. The marketed image of a beautiful million-dollar home in the middle of a forest is not sustainable just because it's situated in a natural environment. And a glazed office tower is definitely not sustainable just because it has solar panels on its roof and uses rainwater for flushing the toilets.

In fact, "Anyone can build a net zero house by slapping enough solar panels on the roof; it can still be drafty, have too much glass, inappropriate orientation, ineffective shading and in efficient planning. Alternatively, anyone can build a sort of comfortable house by throwing enough radiant flooring and ground source heat pump air conditioning and other high-tech gizmo green at it. [. . .] comfort can also be achieved simply by design" (Alter 2014).

Another issue related to image is that *real* sustainability has also not yet reached a level of market appeal to attract the masses. It's *cool* to be *sustainable*, but as it's not understood from a holistic perspective, this '*cool sustainability*' is typically a bolt-on solution like putting solar panels on your roof and claiming you're now living a sustainable life.

36.9.3 Cost

There is also a misconception about the price of sustainability. People believe it comes at a cost that outweighs the benefits substantially. Typically, when using a bolt-on sustainability approach where everything is an afterthought, this might be the case. But when an approach like the Elastic Loop Process is utilized from the beginning of the project, overall lifecycle and running costs are significantly reduced. We also believe that the benefits of truly sustainable architecture can never be reduced to a price tag. How can you possibly put a price on comfortable, healthy, energy efficient and beautiful buildings? Especially when they're creating a positive impact and significantly reducing our footprint.

36.9.4 Approach

As discussed earlier, we believe that following the principles of the Elastic Loop process both enable and ensure that a shift in focus towards a resilient and sustainable built environment occurs. The Elastic Loop process, by its nature, is focused on education and knowledge sharing to create systems of positive feedback loops.

36.9.5 Policies

While we have seen positive steps taken towards incremental improvement in building performance, there is still so much more we can do to challenge the minimum requirements. Another problem is that these rating systems are also based on designed/theoretical, not the as-built performance.

The Passive House certification standard, for example, is completely voluntary, but we have seen a positive growth in popularity it has had in the last decade – especially in Australasia. Improvement in mandatory requirements are resisted by powerful interest groups thereby hampering dynamic and substantive policy change. We would like to see professionals as the leaders of change. Leading by example, setting new standards of architecture that are rigorously tested. The rhetoric has gone unquestioned too long.

We believe that it is vital that public and high-profile projects and projects that are published should be tested to ensure that the right message, knowledge and lessons are being communicated. After-all these are the projects that are effectively being held up as exemplars in their field, and more likely to be referenced by others, both professional and general public.

We feel industry professionals have an ethical responsibility to ensure they are communicating a genuine message. Implementing rules and regulations to ensure that the real-life outcomes of these high-profile projects, that gain more attention in the public eye, will have an exponential effect on lifting public knowledge and awareness. These regulations will also highlight the benefits of projects with holistically sustainable approaches (rather than bolt-on solutions) and will also increase understanding and awareness of what is sustainable and what isn't.

To combat the issue of *fake* sustainability or untested sustainable rhetoric, we suggest auditing of built projects to check their sustainability to compare the claims and the actual operation. All high-profile projects will be obligated to publish the real results during their operation. This measure will motivate positive change simultaneously at many levels. Genuinely sustainable projects will clearly shine above ones that have a superficial sustainable approach. The general public and industry peers will enjoy the benefit of better understanding what designs have real sustainability embedded in their design process. Developers and governments will be accountable for all the outcomes of their high-profile projects. The implementation of these measure will prevent design professionals from making misleading statements, thereby, motivating them to really practice what they preach.

High-profile projects to be audited would include:

1. Public projects (government, etc.)
2. High profile award-winning projects
3. Real sustainable projects
4. Projects published in mainstream and industry publications.

We suggest the following new policy to be implemented:

(a) All projects that fall within the criteria listed above would be required to be rigorously tested with details post-occupancy audits. All results of the audit must be published widely and extensively (to the same level as the publicity gained by the project prior to comprehensive testing). The publication of the results will be accompanied with a plain English commentary to ensure the interpretation of the results can be understood by the wider public as well as industry professionals.
(b) A register of these high-profile projects will provide historic bench-marking to enable comparisons to easily be made.

36.10 *Real* Sustainability

So, what is this talk of *real* sustainability, and what does it all mean? For us, we define *real* sustainability by the following criteria (in no particular order):

36.10.1 *Beautify and Delight*

Real sustainability gives beauty and brings delight. If a building is not beautiful, it will not be appreciated and there will be no desire to build it (or, much less desire than other aesthetic-driven architecture). As aesthetics is what preoccupies people most (typically) about architecture, we need to ensure that sustainable solutions are still beautiful ones – if not more beautiful than non- sustainable solutions.

36.10.2 *Educational*

Once education is re-calibrated to portray *real* sustainability to the general public, and through the teaching of architecture (and all other involved professions), only then can the image of sustainability and its benefits be properly represented and understood. Performance monitoring of built projects will also allow real data results to be shared and used as an educational resource and tool. It can also weed out those projects that claim to be sustainable but do not live up to the claim.

36.10.3 *Positive Impact*

This goes hand-in-hand with the core of sustainability. It is about making a positive impact on the environment by reducing energy consumption (or even better, making an energy contribution), using sustainable, recyclable and reusable materials, and being a generator for positive change.

36.10.4 *Multi-disciplinary Approach*

The Elastic Loop process is all about learning and sharing knowledge between disciplines to achieve a holistically sustainable project.

36.10.5   *Comfortable and Healthy and Resilient*

Comfort and health are key. A sustainable building provides both – with almost no inputs from bolt on technical/mechanical devices. A building that is robust and efficient at its core – providing shelter, comfort and a healthy environment to its occupants.

36.11   Conclusion

In our rapidly progressing society where a focus on sustainable solutions is key, new approaches to architecture are crucial in order to improve our built environments and reduce our footprints. Following the typical business as usual approach is not going to cut it anymore and a holistic approach such as the Elastic Loop is well needed to attack multi-disciplinary issues. Architecture can no longer only focus on the aesthetics, but instead needs to create a positive outcome and take a stance that focuses on sustaining our environment. "The requirement for architecture to contribute to social and environmental sustainability now charges architects with responsibilities that go beyond the limits of an autonomous brief." (Butera 2005). Our approach to sustainable architecture and communities relies on a significant shift in the way all professions, stakeholders, governments, and private developments undertake their processes. Only through the process of critical engagement and re-assessment of our typical approach to design and procurement, will we be able to achieve a quantum shift to rapidly improve our built environment's resilience.

References

Alter L (2014) The three most important things about passive houses are comfort, comfort and comfort. Treehugger. https://www.treehugger.com/green-architecture/three-most-important-things-about-passive-houses-are-comfort-comfort-and-comfort.html

Ausgrid (2018) Average electricity use. https://www.ausgrid.com.au/Industry/Innovation-and-research/Data-to-share/Average-electricity-use

Butera FM (2005) Glass architecture: is it sustainable? International conference 'passive and low energy cooling for the built environment', Santorini, Greece

Luscombe B (2011) 10 questions for Renzo Piano. TIME Magazine. http://content.time.com/time/magazine/article/0,9171,2079576,00.html

Middleton L, Friedlander M (2009) Elastic design – regeneration of education through architecture – case studies. Paper delivered for SASBE 2009, Sydney, Australia

Passipedia – The Passive House Resource (2018) Thermal comfort parameters. IPHA, Germany. https://passipedia.org/basics/building_physics_basics/thermal_comfort/thermal_comfort_parameters

Roggema R, Middleton L and van Den Dobbelsteen A (2009) Quadruple the potential: scaling the energy potential. PLEA2009 – 26th conference on passive and low energy architecture, Quebec City, Canada

Sustainability Victoria (2018) Zero net carbon home program – program overview, Melbourne, Australia

Willis AM (2000) The limits of 'sustainable architecture'. Paper delivered at shaping the sustainable millennium, Queensland University of Technology, Brisbane, Australia

Chapter 37
Green Buildings in Australia: Explaining the Difference of Drivers in Commercial and Residential Sector

Tayyab Ahmad, Ajibade A. Aibinu, and André Stephan

Abstract Green Building (GB) projects can positively affect society, the economy, and the environment. They are being adopted by different sectors of the building industry, including the office and residential sectors. As each building sector has a unique development environment, GBs in each sector are developed with particular drivers. Understanding the different drivers of GB development, the differences across sectors of building industry, and the reasons for those differences should give greater insight into how GB development might be promoted.

An investigation of the GB development-related drivers is conducted in Australian context with a focus on commercial-office and residential sectors. Interviews are conducted with Australia-based GB experts. These interviews are qualitatively analysed and explained. Overall, experts agreed that the commercial-office sector is highly driven towards GB development, while the residential sector lags behind in this regard. Some prominent GB drivers in case of commercial-office sector in Australia are found to be the high commercial value and marketability of these office spaces, regulatory requirements imposed by local authorities, health and well-being, energy efficiency, and the organizational Corporate Social Responsibility (CSR). In the residential sector, the main drivers are the sustainability awareness of the client and energy efficiency. Along with the drivers, this paper also presents the attributes of development environment in each sector, which are the core reasons for difference of GB development across the two sectors. These attributes discussed in this paper include the regional context, business case of GB development, risk driven motivations, and ownership structure, as well as organizational and individualistic thinking.

The understanding of GB development-related drivers across the residential and commercial sector can provide support to the policy frameworks involving these sectors. In the paper, only a comparison of residential and commercial-office sector is provided. Future research should also provide a comparison across additional

T. Ahmad (✉) · A. A. Aibinu · A. Stephan
Faculty of Architecture Building and Planning, University of Melbourne, Melbourne, Australia
e-mail: ahmadt@student.unimelb.edu.au

© Springer Nature Switzerland AG 2020
R. Roggema, A. Roggema (eds.), *Smart and Sustainable Cities and Buildings*,
https://doi.org/10.1007/978-3-030-37635-2_37

sectors of building industry, for instance retail, health care, hospitality, and the educational sector.

Keywords Development environment · Green building · Drivers · Commercial-office · Residential sector

37.1 Introduction

Different sectors of the building industry have unique project development environments. These development environments involve different stakeholders, stakeholder relationships and project expectations. The knowledge of the development environment of a building sector can be critical when different sectors are being compared. Such understanding can also help rationalize the difference in GB development across different sectors.

Sustainability is currently an important aspect of the building industry and a research related to the drivers for Green Building development can significantly contribute to our understanding of factors which strongly influence GB development. While acknowledging the value of GB driver- related research for understanding of sustainable development, it is important to note that the research on this topic needs to thoroughly explore the difference in drivers resulting from the respective development environments of different building sectors. Investigating development environment is particularly important as drivers for achieving a goal are highly dependent on the context in which the goal is being achieved.

The residential and commercial-office sector are significant and should be compared. The importance of considering these sectors is that they respond to the two most important routine matters of an inhabitant (that is living and working) and both these sectors have a significant environmental footprint. For instance, in Australia where the 94% of energy is generated from non-renewables, the residential sector accounted for 7.5%, while the commercial sector accounted for 5.6% of Australia's energy use in 2015–2016 (Ball et al. 2017). Both these sectors have considerably different development environments, and while the office sector in Australia has significantly embraced sustainability, the residential sector lags behind. An inquiry in this regard can have a significant role in theory and practice related to GB development. The inquiry presented in the paper highlights the drivers which contribute towards sustainability in these sectors. Further, this inquiry draws a comparison of the development environment of individual sectors, the intrinsic differences within their environment, and the drivers which suit the environment of each individual sector.

The aim of this paper is to investigate the underlying reasons for a difference in Green Building drive in the residential and commercial-office sectors in Australia.

37.2 Literature Review

Research has previously been conducted to discuss the drivers of GB development in Australia, at different levels of detail. Some prominent insights in this regard are by Wilson and Tagaza (2006) who discussed financial risks, construction risks, and regulatory environment in terms of GB-related drivers and barriers; and Liang and Wilkinson (2008), who investigated the role of social agenda in driving sustainable development of real estate in Melbourne, Australia. In another study by Bond (2011), a household survey was conducted to identify householders' lifestyle choices influencing energy use in homes and the people's motivation to conserve energy.

While the mentioned studies contribute to understanding the factors that drive GB development in the commercial and residential sectors in Australia, no previous study provided a comparison of the GB development in residential and commercial sector. Furthermore, the previous studies within Australian context are also limited in terms of investigating the attributes of development environment, which can affect GB development.

37.3 Methods

This paper reports some of the findings of semi-structured interviews conducted with GB experts in Australia. While interviews in Melbourne were conducted in-person, the interviews in other cities of Australia were conducted over the phone. The interview participants included design consultants, sustainability consultants, project managers, and developers. The interviews included a key question: *'What are the differences of GB development in Australia across the residential and commercial-office sectors?'* This paper reports the findings from 14 interviews. Some of the research findings related to the difference in a household's and an organization's decision-making for GBs are explained using the constructivist research paradigm.

The interview data is categorically classified as prevalent drivers of GB development and development environment attributes which result in a difference of GB drive across the residential and commercial sectors. It is recognized that even within the commercial sector, the office and retail projects have different GB drivers. Keeping this in consideration, this paper only reports the findings related to office projects in the commercial sector and the term 'commercial' is exclusively used for commercial-office projects.

The results section first presents the views of interview participants related to the development status of GB projects in residential and commercial sector. Subsequently, the drivers related to GB development are explained. Afterwards, the development environment attributes resulting in a difference of GB development across the two sectors are explained. Following the results section, the discussion section explains the interrelationships among different development environment attributes and different drivers of GB development.

37.4   Results

In general, it is agreed that in Australia, the GB development is highly focussed towards the commercial sector and residential projects have received relatively much less attention in this regard. As an interview participant (Int-13) said, *"residential sector as compared to commercial sector has taken longer to embrace sustainability."* The lack of GB development in the residential sector is owing to the lack of demand in this sector and the developers' perception of low demand. As an interview participant (Int-6) put it, *"from residential clients there is almost a non-existent request to develop sustainably."* While addressing this issue another interviewee (Int-11) stated, *"in the Australian residential sector, it is hard to identify the drivers which lead to Green Building development. There are some occasional residential projects which opt for sustainability, but overall the residential sector seriously lags behind the other sectors when it comes to sustainability. Although in Australia, there seems to be people who want to live in eco-efficient homes, the developers don't seem to listen to them and argue that there is no market demand for sustainable homes."*

As compared to residential sector, the commercial-office sector has aptly embraced the GB development practices, as according to an interview participant (Int-12), *"sustainability is not a novelty anymore for the commercial market [in Australia]. If your project is not sustainable anymore, then it's regarded as more uncommon these days."*

37.5   Drivers of Green Building Development

Some prominent drivers of GB development in the office sector include marketability, high commercial value, better working environment, energy efficiency, and regulatory requirements. Further, many business organizations need to occupy GBs, as it is part of their Corporate Social Responsibility (CSR). In general, CSR means that the corporations take responsibility of the social implications of their activities. While reflecting on this, an interview participant (Int-10) stated, *"marketing is the key factor when it comes to commercial-office buildings. The top tier buildings go for certifications with the green rating tools."*

While the regulatory requirements may also apply in the residential sector, in most cases these are minimal and do not act as a driver. The drivers prevalent in case of GB development in residential sector include the sustainability awareness of the owner, and energy efficiency. While stating client's awareness of sustainability as a driver, a participant (Int-3) stated, *"in case of Queensland, the drivers for sustainability in residential buildings are basically from the clients who understand the benefits of sustainable development. There are minimal compliance requirements for sustainability in residential buildings in the region."* Energy savings is also recognized by residential clients as a benefit of sustainability, as an interviewee (Int-8) stated, *"in residential buildings, people are mostly driven towards sustainability to help realize better living environment and savings from energy efficiency."* According to another interviewee (Int-6), *"for the clients of residential projects, energy is the major consideration. They want to get the value for their money"*

37.6 Development Environment Attributes Contributing to Variations in Green Building Development

Attributes related to the development environment can contribute to a lag of GB development in residential sector as compared to the commercial sector, and are discussed in this section. These attributes are often the regional priorities, business case of projects, stakeholders' ownership of projects, and difference of individualistic and organizational thinking.

37.6.1 Regional Context

In Australia, the building efficiency and sustainability requirements are governed by two separate regulatory systems as discussed by some interview participants (Int-13; Int-8; Int-9). Nationally, the building code of Australia imposes some requirements. In parallel the local councils impose special planning requirements for building projects within their regional jurisdictions. While there are plans for significant efficiency-related updates in the national building code in 2019, the national building codes lag behind the planning requirements which the inner-city councils in Australian metropolitans have enforced. A project merely meeting the national building codes, cannot qualify as a GB. Planning requirements for the local councils however, vary significantly. While some inner-city councils in Australian metropolitans have strongly embraced GB development in their regulations, the other councils have not considered sustainability in much details. According to an interviewee (Int-7), *"primarily, the drive for sustainability in commercial buildings is for town planning requirements. If the building size is greater than 5,000 m^2 then the Melbourne City Council requires an informal Green Star rating for the project. Although, this consideration varies from council to council, Melbourne City Council has the majority of office buildings in the region, so many office projects are bound to follow the sustainability regulations. Other than this, the clients also require sustainability in their projects for marketing reasons. However, the predominant driver is from the town planning requirements."*

Since a significant number of commercial-office buildings lies within the juris-dictions of inner city suburbs and a majority of residential development is within the jurisdictions of outer city suburbs, a difference can be observed in the GB development-related regulatory focus towards residential and commercial sector. According to an interviewee (Int-5), *"often the developments in countries [including Australia] are such that the commercial buildings are located in the city centres where the governments have much stronger control and have more ability to influence development. Residential sector on the other hand is spread out and often placed in different suburbs, therefore requiring more work from a government to be regulated in terms of sustainable development."*

Hence, a major reason for the variations in GB development across residential and commercial sector in Australia, is the non-uniformity of regulatory requirements across different regions. This is due to the geographical differences in the building stock, that is the commercial buildings majorly lying in the inner city and the residential buildings majorly lying in the suburbs.

37.6.2   Business Case of Green Building Development

The building industry is highly driven by decision-making frameworks in which economic benefits are often a high priority. For a developer, the economic benefits are paramount as an interview participant (Int-1) pointed out, *"a developer is not much concerned about the climate change issues, etc. To him [/her] what matters is that, whether developing a building as a Green Building would help him [/her] sell it better."* In a project, sustainability considerations like any other aspect of the project, are subjected to cost-benefit comparisons and are deemed feasible if the benefits are comparable with costs spent. As an interview participant (Int-1) pointed out, *"whether it is the residential or the commercial sector, the developer performs a cost-benefit analysis before undertaking a Green Building project."*

In the commercial sector, sustainability has become a widely accepted project attribute and is thought of as a 'must have' value in projects. Therefore, the sustainability aspect becomes a source of market competition for these projects. According to an interview participant (Int-13), *"green development has become an industry standard especially in the office sector. As a developer, you are interested in competitive development and sustainability is one of the areas to compete in."* In case of commercial buildings, a higher demand for sustainability exists from the side of the occupants and they are willing to pay extra for these projects, which acts as an incentive. Commenting on this aspect, an interview participant (Int-6) said, *"there is a lot of evidence of the [high] value of green commercial buildings and the [high] amount a tenant would pay to lease it. Such an evidence that the occupiers pay more for a sustainable house, does not exist in case of the residential sector."*

In the residential sector, however, the prevailing competition is to develop pro-jects at low costs. Owing to the mutually contradicting relationship of capital cost with sustainability, the competition for lower costs restricts the introduction of sustainability initiatives. According to an interviewee (Int-8*), "... the general trend of sustainable projects in Australia is quite bad. This is particularly so, in case of residential projects as in those it is all about saving the bottom dollar."* According to another GB professional (Int-6), *"from a residential point of view, in developing a Green Building the developer thinks that, is there a benefit to the customer in doing so? It is very difficult for a developer to embed anything in the development in case it has no clear benefits for the customer and this is because of the already very high costs of construction in Australia."* Regarding this issue another interviewee (Int-5) stated, *"... in residential sector, sustainability initiatives impact the cost of dwellings. Due to these reasons, there is reluctance in developing residential buildings as green buildings, particularly in the first home owner market. For the potential home owners, the challenge is to whether pay premium for a well-furnished kitchen and an extra bedroom, or to pay for the solar panels."*

A product can only become competitive in a particular aspect in case the added value is appreciated among the customers (Anderson and Narus 1998). In the commercial sector, sustainability is a value appreciated by tenants and owners, while affordability is a value appreciated by home owners in residential sector and consequently both these sectors compete in this regard. These findings can be compared with a Melbourne-based study (Liang and Wilkinson 2008). Although some respondents (n = 3) in the study were found to support Ecologically Sustainable Development (ESD) unconditionally without economic considerations, most (n = 13) of the study participants in support of ESD stated that in case of a drop-in profit margins on a project they would have to reconsider supporting ESD. These study findings highlight that beside some exceptions, in the majority of cases, the decision-making in projects is driven by economic considerations.

37.6.3 Risks Related to Non-Green Building Development

The business case of GB development is in-part driven by risk-related motivation. Since an office building incorporates sustainability features for attracting high profile tenants, the GB development can be considered as risk driven, as this helps the developer to avoid the risk of losing premium clients. In the residential sector, owing to the lack of GB demand, a developer may not confront the risk of losing tenants or potential buyers. According to an interviewee (Int-4), *"the clients for green office buildings are the government agencies or private investors. The clients in case of these projects go for sustainability because they want to improve their assets' value. They also do it for risk mitigation in terms of climate change. The people aware of it understand, that the tenants will pay more for a building which performs better. So, the key drivers for the sustainability in office buildings are about risk and money."* According to another interviewee (Int-4), *"once you start talking about legalities, risks etc. then people start to take notice and now that we have found a new language to talk with business people, people are beginning to understand that merely maintaining a building is not good enough, they have to step up and improve it and future proof it, so that they won't be sued by future occupants for under-performance. Once the conversation is in that direction, the building owners are much more likely to understand and work on it."*

Hence, in commercial-office building projects, GB developments are in-part driven by the developers who avoid the risk of losing good tenants, and by tenant clients who avoid the risk of low staff productivity and absenteeism. This is discussed by an interviewee (Int-14) stating, *"high-end clients want their staff to know that they are living in or using Green Star facilities."* According to another interviewee (Int-9), *". . . drivers [of green office buildings] are also the increased health and well-being of occupants and increased productivity."*

Similar risks do not prevail in the residential sector, as a significant number of home owners may not relate GBs with health and well-being. According to an interview participant (Int-6), *"housing customers are much driven towards economic benefits of sustainable development rather than social benefits. They don't tend to associate social aspects with sustainability. Health and well-being is not considered by customers as a part of sustainability."*

37.6.4 Ownership Structure

The ownership structure of a building can have a significant contribution to the motivation for developing a GB. The difference in prevalent ownership structure of the two types of development is a significant reason for the difference in adoption of sustainability in residential and commercial sector. For the speculative project developments and the GBs being developed to be sold, the developers have little interest in sustainability. A difference in sustainability motivation can be seen in two sectors, as in many cases, developers sell the housing units, yet keep the ownership of office buildings once developed.

> *"The commercial and residential projects have different types of clients. In commercial projects, the clients have different business reasons to get the formalized sustainability certifications in their projects. This can help them to charge better rents. It also matters in such projects, whether they are owner-occupied or developed to be sold. The developers who plan to sell the property, don't care a lot about sustainability. They are likely to get the formal certification and move on. However, if the developer is someone who is also going to occupy the building, he [/she] will make sure that the commissioning and tuning are performed well and the operational energy performance is, as expected . . . . In case of the town house developments, the developers move on once the projects are completed, so they have little interest in sustainability."* (Int-7)

> *"A reason for the limited drive of residential sector towards sustainability as compared to the commercial sector is owing to the interest of the developer. In case of residential projects, the developer sells the property to the individual buyers, and resultantly the relationship of developer and the final user is for a very short term. However, in case of commercial buildings, the developer mostly owns the building and leases it to different tenants. In such a situation, the developer has an emphasis to reduce the costs of energy and water, as it is in his [/her] interest to do so. In commercial projects, as the developers are*

mostly owning the building and leasing it to tenants, it is in their interest that the value is continuously created for such projects through the benefits offered by sustainable development." (Int-2)

"In case of residential projects, the developers who work on speculative projects are not genuinely interested in sustainability and they stick with the bare minimum requirements." (Int-3)

These interview findings corroborate with a Melbourne-based study (Liang and Wilkinson 2008), which through a survey of property developers, revealed that incorporation of ESD was most probable for projects undertaken by owner-occupier clients, and was least probable in case of 'speculative projects'.

37.6.5 Individualistic and Organizational Approach Towards Sustainability

There seems to be a difference of perception of the workplace and the living place in terms of environmental performance. According to an interview participant (Int-10), *"there is not a close correlation between working in a green building and living in a green building. While one may think that someone working in a green office, would end up buying a green home, it may be true for a small percentage, but not a reliable assumption. Our concepts of work and living are quite different."*

A client's wilfulness for sustainability in commercial and residential context can be explained by constructivist research paradigm. In short, constructivism is a theory about how people learn. It says that by experiencing things and reflecting on those experiences, people construct their own understanding and knowledge of the world (Bereiter 1994). This theoretical paradigm explains the change in aspirations and wilfulness for sustainability as the role of a person changes from a corporate to a residential household. Both the home and office have a significantly different socio-cultural context and accordingly the person learning from a particular context acts differently.

One of the reasons for an organization to partake in sustainability initiatives more than an individual, can be the accountability for actions. As far as the accountability of the environment-related actions is concerned, organizations may be more exposed to it than individuals. As an interviewee (Int-5) stated, *". . . . the people occupying these [green office] buildings also have a corporate responsibility and sustainability policy. If the company is ASX [Australian Securities Exchange] listed, it has to report for sustainability every year. So, the ball has been rolling for a long time, and the commercial-office building market is quite mature in Australia, though not much in Western Australia."*

At the organizational level, sustainability can become part of the corporate social responsibility, therefore motivating business firms to either develop, own, or lease GB projects. This leads to a high commercial value for GB office projects and therefore, even in the absence of government regulations there is still sufficient motivation for development of such projects. Therefore, business aspirations lead to commercial demand, which consequently leads to development of GB offices. Contrary to this, the ownership of a house is the choice of an individual rather than an organizational decision, which means that even though someone is aware of the need of sustainability, s/he may decide to opt for non-sustainable alternatives. As an interview participant (Int-10) reflected on this aspect, *"the commercial sector is more evidence driven to certify buildings with popular green rating tools, while the residential sector is more lifestyle driven."*

In an organizational and a person's approach towards GB development, awareness for sustainable development plays a very important role. As an interviewee (Int-13) mentioned, *"in office projects, the tenants over the years have become more sophisticated in their understanding of sustainability. They understand what sustainability means to their staff, and what it means to the operations of the building and the ongoing costs."*

The standard way of GB development prevalent in commercial sector, is majorly driven by unified organizational thinking, for instance within the developing, owning and leasing client organizations. Resulting from organizational thinking and understanding of sustainability, CSR policies are formulated which eventually lead to development, occupation and operation of the buildings in sustainable ways. While the clients in commercial sector have adopted standard approaches of perceiving sustainability, it is not the same case for residential projects. Regarding this, a design consultant (Int-4) mentioned in the interview, *"most of the clients who come to us for developing sustainable residential buildings, want their buildings to perform well. They want buildings which touch the ground lightly. These clients are well-educated and have a different understanding of the sustainability word. Since sustainability can be a very broad area, so for each client we go through the process of ascertaining what the sustainability means to them and what are the key elements they want. In this way, there is a back and forth process involved in which we educate each other."*

This points to the fact that commercial-office developments have standardized GB development, which is not the same case for residential projects. While green office buildings are inclined to get green certifications, owners of green residential buildings do not acknowledge the need for these certifications.

37.7 Discussion

Different factors resulting in varying levels of GB development in residential and commercial sector are well connected (as shown in Fig. 37.1). Further, these factors also explain the drivers for GB development in the two sectors.

Overall it can be stated that as a result of non-uniform regulations across different regions in Australia, regulatory requirements for commercial-office sector are stringent as compared to residential sector. Some aspects need to be collectively considered for understanding the difference of GB development among commercial and

Fig. 37.1 The interrelationships among the GB drivers and reasons of differences in Green Building development

residential sector. First, individuals as compared to groups (that is organizations) think and act differently, which is also realized in a study by Kugler et al. (2012). Secondly, residential buildings have a different ownership structure than office buildings, as houses are owned by individuals while office buildings are mostly owned by developers and leased by business organizations. As organizations think differently than individuals, they decide to own or rent workplaces which support their CSR, and result in higher productivity and well-being of their employees. Once the demand for GBs is established, this results in a better business case for GB office development. These findings are also corroborated by a study (Eichholtz et al. 2010) stating that there are both the economically tangible and intangible benefits related to certified commercial GBs.

It can be stated that majority of key stakeholders are directly benefitted from development and operation of a Green office building. The occupants are more productive and feel better in a GB; the business organization can fulfil its CSR and avoid the risk of high utility bills, low performing staff, and high staff absenteeism; and the developer in case owning the building, is paid higher rents. Once, the high market demand for commercial GBs is coupled with high regulatory requirements, the development of such projects becomes unavoidable.

However, in the residential sector, the focus of a typical house buyer is towards amenities. While an individual may like to have sustainable features in the building, s/he does not appreciate the house being unaffordable. As the sustainable features relate with higher initial investment, a typical house buyer prefers affordability over sustainability. Since an individual and not an organization, is the decision-maker in buying a house, the individual's personal choices prevail. Consequently, the business case for a green residential building becomes unfeasible for majority cases and the developer avoids partaking in such projects. Once the low market demand for residential GBs is coupled with low regulatory requirements, the development of such projects is seriously impeded. These findings are partially corroborated by a study (Pinkse and Dommisse 2009), which discussed that in the residential sector, the principle-agent problem prevails to some extent. This is because the energy efficiency benefits of a GB development are for the end-user and not the home builder.

37.8 Conclusion and Recommendations

This paper is part of a study aiming to develop a framework of factors affecting sustainability performance in GBs. The paper highlights the importance of development environment attributes in understanding the drivers of GB projects across different sectors. By using the development environment attributes in explaining the difference of GB development across residential and commercial sector in Australia, this paper establishes the importance of context-specific research for GB projects. Particularly, it highlights that the previous research studies related to GB drivers in particular regions, are limited in terms of rigour, because not only the GB drivers vary across different regions, they also vary across different sectors of the building industry. An important contribution of this paper for practice is that it demonstrates the need of context-specific initiatives to enable a trend of GB development.

This paper has established the differences in drivers among the residential and commercial sector and the reasons of those differences. It is argued that both these sectors are significantly different in their contexts and the drivers in one sector may not be applicable in the other sector. Although a significant gap exists among the sustainability drive of two sectors to date, this gap can be filled by increased regulatory focus towards residential sector. Some aspects which ease the GB development in commercial sector include the ownership structure and the business case. In case innovative ownership structures and business cases for residential projects are used in future with a particular consideration of increasing stakeholders' interest in green features, then it can also increase GB adoption. Furthermore, the relatively high drive in commercial sector for GBs can be expected to increase the adoption of GBs in residential sector as well. First, because of the high demand of GBs in office sector, construction industry has significantly matured in delivering such projects. This maturity will benefit the GB development in residential sector by providing adequate skillset, reliable technology, and relatively lower costs of green features. Second, GBs in office sector will also act as education hubs for building users. People will begin to notice the difference of GBs from traditional buildings, and they may like to imitate this in their houses to benefit from the privileges of a GB.

The study is limited in terms of comparison across different sectors as only the residential and commercial sector are the subject of investigation. Even within the commercial sector, office projects and retail projects often have a different development environment. Further research needs to be conducted along these lines to enable a broader understanding of the optimal practices of GB development.

References

Anderson JC, Narus JA (1998) Business marketing: understand what customers value. Harv Bus Rev 76:53–67

Ball A, Ahmad S, McCluskey C, Pham P, Pittman O, Starr A, Nowarowski D, Lamber N (2017) Australian Energy Update 2017. Department of the Environment and Energy

Bereiter C (1994) Constructivism, socio-culturalism, and Popper's world 3. Educ Res 23:21–23

Bond S (2011) Barriers and drivers to green buildings in Australia and New Zealand. J Prop Invest Financ 29:494–509

Eichholz P, Kok N, Quigley JM (2010) Doing well by doing good? Green office buildings. Am Econ Rev 100:2492–2509

Kugler T, Kausel EE, Kocher MG (2012) Are groups more rational than individuals? A review of interactive decision making in groups. Wiley Interdiscip Rev Cogn Sci 3:471–482

Liang S, Wilkinson SJ (2008) Is the social agenda driving sustainable property development in Melbourne, Australia? Prop Manag 26:331–343

Pinkse J, Dommisse M (2009) Overcoming barriers to sustainability: an explanation of residential builders' reluctance to adopt clean technologies. Bus Strateg Environ 18:515–527

Wilson JL, Tagaza E (2006) Green buildings in Australia: drivers and barriers. Aust J Struct Eng 7:57–63

Part XI
Construction

This theme discusses the implementation of circularity and cradle to cradle concepts in the construction industry.

Chapter 38
Sustainable Waste Management Practices During Construction Projects

Mandisi George, Eric Simpeh, and John Smallwood

Abstract Waste generation rates remain a major problem in South Africa, especially when compared to waste recovery rates. This is largely attributed to inadequate approaches to waste management within several industries; the construction industry is one of those. This research adds knowledge to addressing and achieving sustainable waste management within the construction industry. The study investigates prevailing waste management practices among design team members and construction firms in the City of Port Elizabeth with the aim of improving prevailing waste management practices and reducing the contribution of the construction industry to South Africa's existing waste problem. To achieve the aim and objectives of the study, a comprehensive survey of the waste management literature relating to the causes / sources of C&D waste, diversion methods, illegal dumping, and the causes of cost overruns forms part of the study. A quantitative approach was adopted, and questionnaires were distributed to construction professionals comprising construction project managers, construction managers, site managers, quantity surveyors, and architects within Port Elizabeth. A response rate of 44% was achieved. Some key findings include the lack of waste management policy and plans in construction firms, the prevalence of landfilling as the main means of disposal, poor waste and material handling affecting the cost of building projects, and the general positive attitude and behaviours of construction professionals towards illegal dumping and the environment. Sustainable waste management requires co-operation and commitment of all construction professionals from the

M. George · J. Smallwood (✉)
Department of Construction Management, Nelson Mandela University, Port Elizabeth, South Africa
e-mail: john.smallwood@mandela.ac.za

E. Simpeh
Department of Construction Management, Nelson Mandela University, Port Elizabeth, South Africa

Department of Construction Management and Quantity Surveying, Cape Peninsula University of Technology, Bellville, South Africa
e-mail: simpehe@cput.ac.za

© Springer Nature Switzerland AG 2020
R. Roggema, A. Roggema (eds.), *Smart and Sustainable Cities and Buildings*,
https://doi.org/10.1007/978-3-030-37635-2_38

design, planning, and management stages of a project onwards, and a top-down commitment from construction firms.

Keywords Construction & demolition · Cost · Diversion · Dumping · Waste

38.1 Introduction

Waste generation and diversion within the construction industry is a global predicament. Yet despite the plethora of research conducted and combined efforts in the form of legislation and fiscal measures, there is still a major challenge in diverting waste away from landfills (Ajayi et al. 2014). Empirical evidence indicates that there are high rates of waste generation, and low rates of waste diversion within the construction industry worldwide. For instance, in 2014, Hong Kong produced 57,547 tons of construction and demolition (C&D) waste per day and landfilled 3942 tons of C&D waste per day (Hong Kong EPD 2014). In the United Kingdom, C&D waste comprised the largest amount of waste generated in the region during the years 2010, 2011, and 2014, peaking at 120 million tons in 2014 (United Kingdom, Department for Environment Food and Rural Affairs 2017). The United States of America produced 534 million tons of C&D waste in 2014, with 166 million tons from buildings alone (USEPA 2016). 19 million tons were generated in Australia between 2008 and 2009, and up to 40% was landfilled (Hyder Consulting Pty Ltd 2011). According to Alam et al. (2013), 27% of the municipal solid waste landfilled in Canada is C&D waste, and it has been estimated that 9 tons of C&D waste is produced annually.

In the Republic of South Africa, the same global trend of high generation and low diversion of C&D waste by the construction industry is evident. South Africa produced close to 5 million tons of C&D waste, of which 84% was landfilled [South Africa Department of Environmental Affairs (SAEDA 2012)]. 43,000 m³ of builders' rubble is dumped and landfilled every month in the City of Cape Town. There is an estimated R1 to R1.4 million worth of sub-base material being landfilled every month (GMI Report 2015). In the most populated province in the RSA, 25% of the waste landfilled is C&D waste and C&D waste is a major contributor to the diminishing air space in Gauteng Province's landfills (Gauteng Department of Agriculture Conversation and Environment 2009). According to Ajayi et al. (2014), construction waste management has been ineffective in dealing with the waste enigma of the construction industry. Therefore, the problem that the study aims to address can be stated as: 'The construction industry has a high generation of C&D waste and has a trend of low diversion of C&D waste, including rampant illegal dumping and a resultant negative impact of poor material and waste handling on construction costs'. The purpose is pursued by analysing quantitative data obtained from design team members and construction firms in the City of Port Elizabeth, South Africa. The structure of this chapter summarises and presents brief discussions with regard to the extant literature relating to the causes / sources of C&D waste, diversion methods, illegal dumping and the causes of cost overruns due to C&D waste. This is followed by the methodological approach adopted for collecting and analysing the data. Thereafter, the findings from the descriptive analysis are presented and discussed. The final section presents the conclusions and recommendations of the study.

38.2 Literature Review

38.2.1 Root Causes of Waste in the Construction Industry

In the waste management literature, there are numerous studies focusing on the causes and sources of waste in an attempt to identify and limit those factors (e.g. Asmi et al. 2012; Adlan et al. 2015; Formoso et al. 2015; Damci et al. 2017). According to Damci et al. (2017), identifying the most critical causes of waste generation is the first key step in the successful implementation of waste management during construction projects. According to Asmi et al. (2011), the main causes of waste include: incorrect material storage; workers' mistakes; poor planning; leftover materials on-site; ordering errors; the effects of weather, and design changes which contribute to waste generation. Al-Hajj and Hamani (2011) identified the direct and indirect causes of C&D waste. The direct causes include workers lack of awareness, excessive off-cuts from poor design, rework, and contract variations. On the other hand, the indirect causes of waste include a lack of legal incentives and contractual incentives. Contractors have a tendency to neglect the environmental effects of waste and therefore do not focus on waste reduction. Lawal and Wahab (2011) postulate that C&D waste is produced from off-site and on-site operational activities. Off-site operational activities include prefabrication, project design, and manufacturing and transporting materials and components. Whereas the on-site operational activities are the physical construction process, which includes the substructure and superstructure of a building (Lawal and Wahab 2011). Asmi et al. (2012) also contend that the major contributors to waste include: poor site management and supervision; lack of experience; inadequate planning and scheduling; mistakes and errors in design, and mistakes during construction. According to Hewage et al. (2013), excessive waste can be generated due to errors, deficiencies, ambiguity, and unfair risk in contractual documents, which manifests in the form of rework required during the construction process.

38.2.2 Waste Management System

Ajayi et al. (2017) postulate that there are two main schools of thought regarding waste management. Firstly, the reduction of waste landfilled through recycling and re-using and, secondly, the prevention and minimisation of waste before and during the construction process. The first approach to waste is re-active given that waste is viewed as a necessary by-product of the construction process that cannot be eradicated. The second approach to waste is pro-active, and is more efficient (Ajayi et al. 2017). According to Crossin et al. (2016), waste management entails the collection, transporting, treatment, recovery, storage, and disposal of waste. The waste management activities require an integrated system, which achieves and maintains acceptable environment quality and promotes sustainable development. Ajayi et al. (2017) suggest a holistic effort of combining not only construction, but planning, design and the material procurement strategies required to reduce waste. Alam et al. (2013) poses a similar idea by proposing a far more comprehensive and integrated lifecycle-based C&D waste management framework that incorporates the 3Rs into planning, designing, construction, renovation and demolition. Furthermore, waste minimisation requires waste management planning and construction methods in the planning and design stage, effective management, awards and regulations (Esa et al. 2017). However, Sapuay (2016) contends that the construction industry's waste management is still hindered by managers who do not follow waste management methods clearly, which results in the industry being inefficient and not achieving sustainable waste management objectives.

38.2.3   Waste Diversion and Recycling

38.2.3.1   Landfilling

The construction industry has a reputation for excessive landfilling in different countries all over the world (Department of Sustainability, Environment, Water, Population and Communities 2011; Department of Environmental Affairs 2012; Alam et al. 2013; Environmental Protection Department 2014; Environmental Protection Agency 2016; Department for Environment Food and Rural Affairs 2017). Landfilling and ocean disposal of C&D waste has serious environmental and socio-economic implications, including the reduction of landfill space (Alam et al. 2013). Furthermore, according to Arslan et al. (2017), the construction industry will never reach zero waste-status and the current dominant disposing method of C&D waste i.e. landfilling is not sustainable.

38.2.3.2   Recycling and Re-Use

C&D waste can be recycled and re-used within the construction industry, thus reducing the amount of waste that ends up in landfills (Centeno et al. 2016). Recycling serves as a waste management and material reduction process and reduces the need for landfilling, the use of virgin material and, thus, has a net benefit for the environment (Crossin et al. 2016).

38.2.3.3   Illegal Dumping in the Construction Industry

According to Baum and Katz (2011), the causes of illegal dumping include: long transportation distances; high tipping fees; lack of enforcement measures, and lack of knowledge regarding recycling measures. Furthermore, Baum and Katz (2011) state that a further cause of illegal dumping is due to a lack of legal landfilling sites. However, Ichinose and Yamato (2011) discovered that the number of landfilling sites has a positive effect on the number of illegal dumping incidents, which contradicts Baum and Katz's assertion. The results were statistically insignificant. Ichinose and Yamato (2011) posit that intermediate waste management facilities can decrease the amount of illegal dumping incidents. Sufficient numbers of intermediate waste management facilities lower the cost of legal dumping. Furthermore, Ichinose and Yamato (2011) revealed that the number of landfilling sites have an ambiguous effect on the number of illegal dumping incidents. Hence, there is a positive relationship between the weight of the waste produced and the number of cases of illegal dumping.

38.2.4   The Influence of Material and Waste Handling on Construction Costs

According to Sagan and Sobotka (2016), the type and quantity of waste will affect the cost of its collection and may influence the project cost. The mean percentage that material waste contributes to project cost overruns is between 21–30% (Ameh and Daniel 2013). Therefore, there is a significant relationship between building material waste on construction-sites and cost overruns. The general improvement of waste levels on construction-sites can have a cost saving benefit and may enhance the construction industry's performance (Ameh and Daniel 2013). According to Saidu and Shakantu (2016), the percentage of material waste's contribution to project cost over-runs ranges from 1.96% to 8.01%.

38.3   Research Methodology

This study was undertaken by conducting a survey of the literature and conducting an empirical study. The study takes the quantitative approach by analysing and measuring the responses to structured questionnaires. The primary data for this study was acquired with the administration and distribution of structured questionnaires, comprising close-ended and open-ended questions. The respondents selected resided within the city of Port Elizabeth, and consisted of construction professionals from small, medium, and large firms. The selection criterion included construction experts who are registered with a professional body. To obtain valid and reliable data, the study required construction professionals with a reasonable number of years of experience, in a variety of construction projects. The respondents were from 15 different construction related firms and the technique of probability (simple random sampling) sampling was employed to select the research participants. 68 questionnaires were e-mailed, or hand delivered to construction professionals in their offices as well as on site. 30 / 68 questionnaires were returned and an overall response rate of 44.1% was achieved. It is worth noting that 47% (14 / 30) of the respondents were construction managers / construction project managers / site managers (CMs / CPMs / SMs), 30% (9 / 30) were quantity surveyors (QSs), and 23% (7 / 30) were architects. The questionnaire was divided into 6 sections to facilitate easy response. Section A related to general personal information, Section B explored respondents' opinions regarding factors influencing waste generated, Section C and D solicited information concerning waste management and waste handling and diversion respectively, Section E related to illegal dumping, and Section F examined the impact of waste on construction costs. The data analysis technique adopted for the study was the descriptive statistical method. Descriptive statistics were used to measure the central tendency such as mode, median, and mean, and the dispersion (standard deviation) of the data.

38.4 Research Findings

38.4.1 Section A: Personal Information

38.4.1.1 Qualification of Respondents

Table 38.1 indicates the formal qualifications of the respondents. It is notable that less than half of the respondents (11 / 30 = 36.7%), hold a National Diploma. Moreover, the minority of the respondents hold a Bachelor of Technology (5 / 30 = 16.7%), Bachelor of Science (8 / 30 = 26.7%), Bachelor of Science (Honours) (4 / 30 = 13.3%), and Master's degree (2 / 30 = 6.7%).

Table 38.1 Highest formal qualification of respondents

Qualification	No.	%
National Diploma	11	36.7
Bachelor of science	8	26.7
Bachelor of technology	5	16.7
Bachelor of science (honours)	4	13.3
Masters	2	6.7
Total	30	100.0

38.4.1.2   Occupation of Respondents

Table 38.2 indicates the occupations of the survey participants. CMs / CPMs / SMs constitute less than half (46.7%) of the respondents. 9 / 30 (30.0%) were QSs, and 7 / 30 (23.3%) were architects.

38.4.1.3   Number of Years Worked in Construction

Table 38.3 indicates the number of years worked in construction by the participants in terms of percentages. Notably, 30% of the participants have between $5 < x \leq 10$ years' work experience in construction. Moreover, 20% of the participants have between $1 < x \leq 5$, $10 < x \leq 15$, and $15 < x \leq 20$ years working in construction respectively. Only 10% of the participants have more than $20 <$ years of experience in the construction industry.

Table 38.2 Occupation of respondents

Occupation	No.	%
CM / CPM / SM	14	46.7
Quantity surveyor	9	30.0
Architect	7	23.3
Total	30	100.0

Table 38.3 Number of years respondents have worked in construction

Period of years	Response
$1 < x \leq 5$	20.0
$5 < x \leq 10$	30.0
$10 < x \leq 15$	20.0
$15 < x \leq 20$	20.0
$20 <$	10.0

Table 38.4 Respondents' age

Age	Response
$20 < x \leq 25$	6.7
$25 < x \leq 30$	33.3
$30 < x \leq 35$	16.7
$35 < x \leq 40$	20.0
$40 < x \leq 45$	10.0
$45 < x \leq 50$	6.7
$50 <$	6.7

38.4.1.4   Age of the Participants

Table 38.4 indicates the ages of the participants in terms of percentages. It is worth noting that 33.3% of the participants are between $25 < x \leq 30$ years old. Moreover, 20% of the participants are between the ages $35 < x \leq 40$ and 16.7% of the respondents are between $30 < x \leq 35$ years old. 10% of the respondents are between $40 < x \leq 45$ years old. Notably, 6.7% of the participants are between $20 < x \leq 25$, $45 < x \leq 50$, and $50 < x$ years old respectively.

38.4.2   Section B: Factors Influencing Waste Generation During Construction

Table 38.5 indicates the extent to which eight factors contribute to the quantity of waste on construction projects in terms of responses to a scale of 1 (minor) to 5 (major), and a mean score (MS) ranging between 1.00 to 5.00, the midpoint being 3.00. It is notable that 7 / 8 (87.5%) factors' MSs are > 3.00, which indicates that generally the waste causative factors contribute to waste on-site to more of a major, than a minor extent.

The highest waste causative factors, according to architects are rework (4.00) and poor site management (4.00), which indicates that the architects perceive the contribution to be between some extent to a near major / near major extent since the MSs are > 3.40 to ≤ 4.20. Inefficient design (3.85) and poor understanding of design drawings (3.57) are ranked third and fourth respectively, within the same MS range as rework and poor site management. Regarding CMs / CPMs / SMs, the highest MSs are relative to rework (4.33), and poor material handling on-site (4.33). The MSs indicates that CMs / CPMs / SMs perceive the contribution to be between a near

Table 38.5 Extent to which eight factors contribute to the quantity of waste on construction projects

Factor	Architects MS	Rank	CMs / CPMs / SMs MS	Rank	QSs MS	Rank	Overall MS	Rank
Rework	4.00	1	4.33	1	4.00	2	4.10	1
Poor material handling on-site	3.57	4	4.33	1	4.14	1	4.07	2
Poor site management	4.00	1	4.22	3	3.6	5	3.90	3
Poor planning on-site	3.43	6	4.22	3	3.92	3	3.87	4
Poor understanding of design drawings	3.57	4	3.89	6	3.69	4	3.70	5
Inefficient procurement practices	3.85	3	3.89	6	3.42	6	3.67	6
Inefficient design (e.g. design errors and omissions)	3.14	7	4.11	5	3.35	7	3.53	7
Contract variations	2.57	8	2.78	8	3.14	8	2.90	8

major to a major / major extent since the MSs are > 4.20 to ≤ 5.00. The highest ranked factors for QSs were poor material handling on-site (4.14), rework (4.00), and poor planning on-site (3.92). The MSs indicates that QSs perceive the contribution to be between some extent to a near major / near major extent since the MSs are > 3.40 to ≤ 4.20.

As indicated in Table 38.5, it is worth noting that rework is ranked first relative to all construction professional fields, with a MS of 4.14. Overall, poor handling of materials on-site is ranked second (4.07), and poor site management is ranked third (3.90). It is worth mentioning that poor planning on-site (3.87) is not too far behind poor site management (at fourth). Furthermore, 7/8 (88%) MSs > 3.40 to ≤4.20, which indicates that the contribution can be deemed to be between some extent to a near major / near major extent for the following factors: inefficient procurement practices; inefficient design (e.g. design errors and omissions); poor materials handling on-site; poor planning on-site; poor site management; poor understanding of design drawings, and rework. However, for contract variations, the perceived contribution is between a near minor extent to some extent / some extent since the MS is > 2.60 to ≤ 3.40.

38.4.3 Section C: Waste Management

38.4.3.1 Does Your Firm Have a Waste Management Policy?

Table 38.6 indicates whether participants have or do not have a waste management policy in place in terms of percentages. It is notable that five or more participants (≤ 16.7%) within each construction professional field reported not to have a waste management policy in place. With respect to architects, none of the respondents reported having a waste management policy within their respective firms. With respect to QSs, five (16.7%) reported that they do not have a waste management policy and three (10.0%) were unsure whether their firms have or do not have a waste management policy, only 3.3% reported having a waste management policy in place. Relative to the CMs / CPMs / SMs, nine (30.0%) reported having no waste management policy within their respective firms; on the other hand, five (16.7%) do have a waste management policy in place. Overall, it is notable that more than half of the respondents 20 / 30 (66.7%) reported having no waste management

Table 38.6 Number of respondents whose firm has a waste management policy

Response	Architects		CMs / CPMs / SMs		QSs		Overall	
	No.	(%)	No.	(%)	No.	(%)	No.	(%)
Yes	0	0.0	5	16.7	1	3.3	6	20.0
No	6	20.0	9	30.0	5	16.7	20	66.7
Unsure	1	3.3	0	0.0	3	10.0	4	13.3
Total	7	23.3	14	46.7	9	30.0	30	100.0

policy. This finding suggests that there is a lack of commitment to waste management, which contributes to the occurrence of waste during construction.

38.4.3.2   Does Your Firm Evolve a Waste Management Plan for Each Project?

Table 38.7 indicates whether participants evolved a waste management plan for each project, or not, in terms of percentages. It is notable that five or more participants ($\leq$ 16.7%) within each construction professional field reported that they do not evolve a waste management plan. With respect to architects, six (20.0%) reported that they do not prepare a waste management plan for each project, and one (3.3%) was unsure. With respect to QSs, five (16.7%) reported that they do not evolve a waste management plan, while two (6.7%) reported that they do not evolve a waste management plan. Moreover, two (6.7%) reported to be unsure. With respect to CMs / CPMs / SMs, 8 (26.7%) reported that they do not evolve a waste management plan for each project. However, six (20.0%), of CMs / CPMs / SMs responded that they do not evolve a waste management plan. Overall, 17 / 30 (56.7%) of the respondents reported that their firms do not have a waste management plan in place for each project.

38.4.3.3   To What Extent Do the Following Factors / Practices Reduce / Prevent the Amount of Waste Generated on Project Sites?

Table 38.8 indicates the extent to which nine factors / practices reduced or prevented the amount of waste generated on project sites, in terms of responses to a scale of 1 (minor) to 5 (major), and a MS ranging between 1.00 to 5.00, the midpoint being 3.00. It is notable that 8 / 9 (88.9%) of the participants MSs are > 3.00, which indicates that these factors reduced/ prevented waste generated on-site to a major extent, rather than to a minor extent.

Notably, amongst the architects, material re-use had the highest MS (4.57). This is followed by designing for waste reduction (4.29), and waste estimating (4.29). The MSs of the top three ranked factors indicate that the architects perceive the factors / practices to reduce / prevent the amount of waste generated between a near major extent to major / major extent since the MSs are > 4.20 to $\leq$ 5.00. With respect to

Table 38.7 Number of respondents reported to have a waste management plan

	Architects		CMs/CPMs/SMs		QSs		Overall	
Response	No.	(%)	No.	(%)	No.	(%)	No.	(%)
Yes	0	0.0	8	26.7	2	6.7	10	33.3
No	6	20.0	6	20.0	5	16.7	17	56.7
Unsure	1	3.3	0	0.0	2	6.7	3	10.0
Total	7	23.3	14	46.7	9	30.0	30	100.0

Table 38.8 Extent to which nine waste preventive factors reduce or prevent the occurrence of waste generated on-site

Factor	Architects		CMs / CPMs / SMs		QSs		Overall	
	MS	Rank	MS	Rank	MS	Rank	MS	Rank
Material reuse	4.57	1	4.14	1	3.67	5	4.10	1
Waste management training	3.57	7	3.71	2	4.11	1	3.80	2
Waste estimating	4.29	2	3.14	6	3.44	8	3.71	3
Waste management awareness	3.71	6	3.57	3	3.89	4	3.70	4
Waste management plan	3.86	4	3.43	4	4.00	3	3.70	4
Waste management policies	3.86	4	3.21	5	4.11	1	3.63	6
Designing for waste reduction	4.29	2	3.07	7	3.67	5	3.53	7
Efficient waste procurement system	3.43	8	3.07	7	3.56	7	3.30	8
BIM (building information modelling)	3.00	9	2.21	9	1.89	9	2.30	9

QSs, waste management policies (4.11), waste management training (4.11), and having a waste management plan (4.00) are the top ranked factors, which indicates that the QSs perceive the factors / practices to reduce / prevent the amount of waste generated between some extent to a near major / near major extent since the MSs are > 3.40 to ≤ 4.20. Material re-use had the highest MS (4.14) relative to CMs / CPMs / SMs. Waste management training (3.71), and waste management awareness (3.57) were ranked second and third respectively. This indicates that the factors / practices are perceived to reduce / prevent the amount of waste generated between some extent to a near major extent / near major extent, since the MSs are > 3.40 to ≤ 4.20.

Overall, material re-use (4.10) had the highest MS, followed by waste management training (3.80), and waste estimating (3.71). Having a waste management plan (3.70) and waste management awareness (3.70) are ranked fourth and fifth, respectively. Waste management policy (3.63) is ranked sixth overall. It is worth noting that BIM (Building Information Modelling) had the lowest MS (2.30), which indicates it is perceived to reduce / prevent the amount of waste generated between a minor to a near minor / near minor extent, since the MS is > 1.80 to ≤ 2.60. It is notable that 11 / 30 (36.7%) participants reported to be unsure of the extent to which BIM reduces the occurrence of waste generated on-site. Moreover, it is worth noting that 7 / 9 (77.8%) MSs are > 3.40 to ≤ 4.20, which indicates that the factors / practices are perceived to reduce / prevent the amount of waste generated between some extent to a near major extent / near major extent for the following factors: material reuse; waste management training; waste estimating; waste management awareness; having a waste management plan; waste management policies, and designing for waste reduction. However, efficient waste procurement system is perceived to reduce / prevent the amount of waste generated between a near minor extent to some extent / some extent since the MS is > 2.60 to ≤ 3.40.

38.4.4 Section D: Waste Handling and Diversion

38.4.4.1 What Is Your firm's Main Means of Waste Disposal?

Table 38.9 indicates the main means of waste disposal in terms of percentage responses. It is notable that more than half (57.1%) of the participants' main means of waste disposal is landfill disposal. Furthermore, 2 / 14 (14.3%) and 3 / 14 (21.4%) of the respondents' main means of waste disposal is avoidance/minimisation and re-use, respectively. Recycling is the main means of disposal for a minority of the participants 1 / 14 (7.1%).

38.4.4.2 To What Extent Do the Following Factors Contribute to Successfully Diverting Waste Away from Landfills?

Table 38.10 indicates the extent to which seven factors contribute to successfully diverting waste away from landfills in terms of responses to a scale of 1 (minor) to 5 (major), and a MS ranging between 1.00 to 5.00, the midpoint being 3.00. It is notable that all the MSs are > 3.00, which indicates that in general, the factors contribute more of a major extent than minor extent to successfully diverting waste away from landfills.

Table 38.9 Main means of waste disposal

	CMs/CPMs/SMs	
Means	No.	%
Avoidance/minimisation	2	14.3
Incineration	0	0.0
Landfill disposal	8	57.1
Re-use	3	21.4
Recycling	1	7.1
Other	0	0.0
Total	14	100.0

Table 38.10 Extent to which seven factors contribute to successfully diverting waste away from landfills

	Architects		CMs/ CPMs/SMs		QSs		Overall	
Factor	MS	Rank	MS	Rank	MS	Rank	MS	Rank
Recycling waste material	4.14	2	4.29	1	4.00	1	4.17	1
Re-use waste material	4.14	2	4.14	2	4.00	1	4.10	2
Waste management plan	3.57	5	3.86	3	3.56	4	3.70	3
Effective waste handling on-site	4.43	1	3.43	5	3.33	5	3.63	4
Government legislature and policies	3.43	6	3.50	4	3.67	3	3.53	5
Source segregation and sorting	4.14	2	3.36	6	3.22	6	3.50	6
Waste diversion objectives	3.43	6	3.07	7	3.22	6	3.20	7

With respect to architects, effective waste handling on-site (4.43) was ranked first. The MS of the top ranked factor indicates that the architects perceive the factor to successfully divert waste away from landfills between a near major extent to major / major extent since the MS is > 4.20 to ≤ 5.00. This was followed by re-use waste material (4.14), recycling waste material (4.14) and source segregation (4.14). Given that the MSs are > 3.40 to ≤ 4.20, the factors are perceived to successfully divert waste away from landfills between some extent to a near major / near major extent. With respect to CMs / CPMs / SMs, recycling waste material emerged as the top ranked factor with a MS of 4.29. The MS indicates that the CMs / CPMs / SMs perceive the factor to successfully divert waste away from landfills between a near major extent to major / major extent since the MS is > 4.20 to ≤ 5.00. This was followed by re-use waste material (4.14), and waste management plan (3.86), which indicates that the factors are perceived to successfully divert waste away from landfills between some extent to a near major / near major extent given that their MSs > 3.40 to ≤ 4.20. With respect to QSs, the top ranked factors are recycling waste material (4.00), and re-use waste material (4.00). The MSs indicates that the CMs / CPMs / SMs perceive the factors to successfully divert waste away from landfills between some extent to a near major / near major extent since the MSs are > 3.40 to ≤ 4.20.

Overall, 6 / 7 (85.7%) MSs are > 3.40 to ≤ 4.20 which indicates that the factors are perceived to successfully divert waste away from landfills between some extent to a near major / near major extent for the following factors: recycling waste material; re-use waste material; waste management plan; effective waste handling on-site; government legislature and policies, and source segregation and sorting. However, 1 / 7 (14.3%) MSs, namely relative to waste diversion objectives, is > 2.60 to ≤ 3.40, which indicates that the factor is perceived to successfully divert waste away from landfills between a near minor extent to some extent / some extent.

38.4.4.3 Does Your Firm Have a Method of Handling Waste on-Site?

Table 38.11 indicates whether the participants have or do not have a method of handling waste on-site. It is notable that more than half (64.3%) of the respondents reported having a method of handling waste on-site. On the other hand, 35.7% of the respondents reported having no method of handling waste on-site.

Table 38.11 Method of handling waste

Response	CMs/CPMs/SMs	
	No.	%
Yes	9	64.3
No	5	35.7
Unsure	0	0.0
Total	14	100.0

38.4.5 Section E: Illegal Dumping

38.4.5.1 Do You Know of a Contractor, or Have you Witnessed a Contractor Dumping Waste Illegally in the Last Twelve Months?

Table 38.12 indicates the number of respondents that know-of or witnessed a contractor illegally dumping waste during the last 12 months, in terms of percentages. It is notable that 19 / 30 (63.3%) of the respondents reported having not witnessed a contractor /did not know of a contractor that dumped waste illegally during the last 12 months. However, 9 / 30 (30.0%) reported to know or having witnessed a contractor illegally dumping waste during the last 12 months and 2 / 30 (6.7%) reported to be unsure. 6 (20.0%) of the architects and 7 (23.3%) of QSs responded no. Notably QSs and architects do not spend as much time on, or make as many trips to construction-sites as CMs / CPMs / SMs. With respect to CMs / CPMs / SMs, 8 / 14 (57.1%) reported knowing a contractor or having witnessed a contractor illegally dumping waste during the last 12 months.

38.4.5.2 Do You Know the Corrective Measures in Place by the Nelson Mandela Bay Municipality for Cases of Illegal Dumping of Construction Waste?

Table 38.13 indicates the number of respondents that are aware of the corrective measures put in place by the Nelson Mandela Bay Municipality (NMB) for cases of illegal dumping. It is notable that 17 / 30 (56.7%) of the respondents reported not being aware of the corrective measures put in place by the NMB municipality. The CMs / CPMs / SMs responded 'Yes' the most (20%). However, it is notable that

Table 38.12 Number of respondents that know of or witness a contractor illegally dumping waste in the last 12 months

Response	Architects		CMs/CPMs/SMs		QSs		Overall	
	No.	%	No.	%	No.	%	No.	%
Yes	0	0.0	8	26.7	1	3.3	9	30.0
No	6	20.0	6	20.0	7	23.3	19	63.3
Unsure	1	3.3	0	0.0	1	3.3	2	6.7

Table 38.13 Corrective measures put in place by the Nelson Mandela Bay municipality for cases of illegal dumping

Response	Architects		CMs/CPMs/SMs		QSs		Overall	
	No.	%	No.	%	No.	%	No.	%
Yes	0	0.0	6	20.0	1	3.3	7	23.3
No	6	20.0	5	16.7	6	20.0	17	56.7
Unsure	1	3.3	3	10.0	2	6.7	6	20.0

5 (16.7%) reported not being aware of the corrective measures put in place. It should also be noted that all the respondents responded that a fine is an appropriate corrective measure for preventing / reducing illegal dumping of waste during construction.

38.4.5.3   To What Extent Do You Agree / Disagree with the Following Statements Regarding Illegal Dumping?

Table 38.14 indicates the extent of agreement / disagreement with six statements regarding illegal dumping in terms of responses to a scale of 1 (strongly disagree) to 5 (strongly agree), and a MS ranging between 1.00 to 5.00, the midpoint being 3.00 (neutral). It is notable that 5 / 6 (83.3%) MSs are >3.00, which indicates that in general, the participants agree more than disagree with the statements.

The top ranked statement among architects is 'illegal dumping is detrimental to the environment' (4.86). This is followed by 'illegal dumping causes aesthetic damage' (4.71), and 'illegal dumping is a severe offence' (4.43). The degree of concurrence for the foregoing factors is considered to be between agree to strongly agree / strongly agree since the MSs are > 4.20 to ≤ 5.00. The top ranked statement among QSs is 'illegal dumping is detrimental to the environment' (4.78), followed by 'illegal dumping causes aesthetic damage' (4.67), and 'illegal dumping is a severe offence' (4.56). This indicates that the degree of concurrence can be deemed to be between agree to strongly agree / strongly agree, since the MSs are > 4.20 to ≤ 5.00. The top ranked statement among CMs / CPMs / SMs is 'illegal dumping is detrimental to the environment' (4.79). This is followed by 'illegal dumping causes aesthetic damage' (4.57), and 'illegal dumping is a severe offence' (4.34). This thus indicates that the degree of concurrence can be deemed to be between agree to strongly agree / strongly agree, since the MSs are > 4.20 to ≤ 5.00.

Table 38.14 Extent of agreement/disagreement with illegal dumping-related statements

Statement	Architects		CMs/ CPMs/SMs		QSs		Overall	
	MS	Rank	MS	Rank	MS	Rank	MS	Rank
Illegal dumping is detrimental to the environment	4.86	1	4.79	1	4.78	1	4.80	1
Illegal dumping causes aesthetic damage	4.71	2	4.57	2	4.67	2	4.63	2
Illegal dumping is a severe offence	4.43	3	4.34	3	4.56	3	4.43	3
The punishment for illegal dumping should be severe	4.42	4	4.00	5	4.44	4	4.23	4
Illegal dumping is a problem within the industry	3.86	5	4.36	4	4.22	5	4.20	5
Convenience is more important than the environment when disposing waste	1.86	6	1.86	6	2.22	6	1.97	6

Overall, 4 / 6 (83.3%) MSs are > 4.20 to ≤ 5.00, which indicates that the degree of concurrence can be deemed to be between agree to strongly agree / strongly agree for the following statements: 'illegal dumping is detrimental to the environment'; 'illegal dumping causes aesthetic damage'; 'illegal dumping is a severe offence', and 'the punishment for illegal dumping should be severe'. The degree of concurrence for illegal dumping is a problem within the industry can be considered to be between neutral to agree / agree since the MS is > 3.40 to ≤ 4.20. It is notable that the statement 'convenience is more important than the environment when disposing waste', had the lowest MS (1.97), which indicates that the degree of concurrence can be deemed to be between strongly disagree to disagree / disagree, since the MS is > 1.80 to ≤ 2.60.

38.4.6   Section F: Impact of Waste on Construction Costs

38.4.6.1   How Severe Are Your firm's Cost Overruns Due to Poor Material and Waste Handling On-Site?

Table 38.15 indicates the severity of cost overruns due to poor material and waste handling experienced by the respondents' firms in terms of percentage responses to a scale of 1 (no impact) to 5 (major impact), and a MS ranging between 1.00 to 5.00, the midpoint being 3.00. The MS is > 3.00, which indicates that in general, the severity of the respondents' firms' cost overruns is between less than an impact and an impact / impact, since the MS is > 2.60 to ≤ 3.40.

38.4.6.2   To What Extent Do the Following Sources of Waste Contribute to Cost Overruns?

Table 38.16 indicates the extent to which sources of waste contribute to cost overruns in terms of responses to a scale of 1 (minor) to 5 (major), and a MS ranging between 1.00 and 5.00, the midpoint being 3.00. It is notable that 6 / 7 (85.7%) MSs are > 3.00, which indicates that in general the sources of waste contribute more of a major than a minor extent to cost overruns.

With respect to architects, inefficient design (e.g. design errors and omission) (3.29) is the top ranked factor, together with poor material and waste handling on-site (3.29), which indicates that the factors contribute between a near minor

Table 38.15 Severity of cost overruns due to poor material and waste handling

Response (%)						
	No..Major					
Unsure	1	2	3	4	5	MS
0.0	7.1	7.1	50.0	28.6	7.1	3.21

Table 38.16 Extent to which seven waste causative factors contribute to cost overruns

	Architects		CMs/ CPMs/SMs		QSs		Overall	
Factor	MS	Rank	MS	Rank	MS	Rank	MS	Rank
Rework	3.14	3	4.21	1	4.00	3	3.90	1
Poor material and waste handling on-site	3.29	1	4.14	2	4.00	3	3.90	1
Poor planning on-site	3.14	3	3.93	3	4.33	1	3.87	3
Poor site management	3.00	5	3.93	3	4.11	2	3.77	4
Inefficient procurement practices	3.00	5	3.36	5	3.11	6	3.20	5
Inefficient design (e.g. design errors and omission	3.29	1	2.79	6	3.33	5	3.07	6
Contract variations	2.86	7	2.57	7	2.89	7	2.73	7

extent to some extent / some extent since the MSs are > 2.60 to ≤ 3.40. With respect to CMs / CPMs / SMs, rework (4.21) is the top ranked factor, which indicates that it contributes between a near major extent to major / major extent since the MS is > 4.20 to ≤ 5.00. With respect to QSs, poor planning on-site (4.33) is the top ranked factor, which indicates that the factor contributes between a near major extent to a major / major extent since the MS is > 4.20 to ≤ 5.00.

Overall, the top ranked factors are poor material and waste handling (3.90), rework (3.90), and poor planning on-site (3.87), which indicates that the factors contribute between some extent to a near major / near major extent since the MSs are > 3.40 to ≤ 4.20. Furthermore, it is worth noting that 4 / 7 (57.1%) MSs are > 3.40 to ≤ 4.20, which indicates that the factors contribute between some extent to a near major / near major extent: rework; poor material and waste handling on-site; poor planning on-site, and poor site management. Moreover 3 / 7 (42.9%) MSs are > 2.60 to ≤ 3.40, which indicates that the factors contribute between a near minor extent to some extent / some extent: inefficient procurement practices; inefficient design (e.g. design errors and omission), and contract variations.

38.5 Conclusion

38.5.1 Factors Influencing the Occurrence of C&D Waste and Improving Prevailing Waste Management Practices Within the Construction Industry

The descriptive statistics indicate that the main causes / sources of waste are rework, poor material and waste handling, poor site management, poor planning on-site, and poor understanding of design drawings. The analysis also revealed that to improve prevailing waste management practices within the construction industry, the following waste preventive measures can be adopted: material re-use; waste management training; waste estimating; waste management awareness, and having a waste management plan. With respect to waste management plan / policy, it should be noted that 66.7% of the respondents' firms have no waste management policy and 56.7% of the respondents' firms do no evolve a waste management plan for each project. For instance, architects do not view waste management as a priority, while CMs do not follow waste management practices and methods closely, and lack on-site realisation.

38.5.2 Dedicated Measures to Divert C&D Waste Away from Landfills

With respect to diversion of C&D waste, the study revealed that the main factors that contribute to diverting waste away from landfills are the following: recycling waste material; re-use waste material; waste management plan; effective waste handling on-site and government legislature and policies. However, there is very little diversion of waste amongst the respondents' firms and waste diversion means very little to the success of projects. For example, 57.1% of CMs / CPMs / SMs use landfill disposal as their main means of disposal and very few re-use, and recycle C&D waste. Because contractors are influenced by cost savings and the price for recycling is more expensive than regular methods of waste disposal. This is further exacerbated by the fact that barriers to recycling still exist, including a lack of recycling plants and a cultural reluctance towards re-use and recycling.

38.5.3 Effective Ways of Reducing Illegal Dumping of C&D Waste

The quantitative analysis revealed that construction professionals generally have positive attitudes towards illegal dumping and its effects on the environment. Measures that can be implemented to curb illegal dumping include immediate waste management facilities, waste minimisation, stronger penalties, and incentives for recycling, vehicle impounding, and manifest system.

38.5.4 The Impact of Ineffective Material and Waste Handling Methods on Construction Costs

The findings indicated that there is a relationship between building material waste on construction-sites and cost overruns. In addition, firms experience cost overruns due to poor handling and waste, to a very large extent, and the waste causative factors that contribute the most to cost overruns are the following: rework; poor material and waste handling on-site; poor planning on-site, and poor site management.

38.6 Recommendations

Waste generation rates remain a major problem in the South African construction industry, especially when compared to waste recovery rates. Thus, it is pivotal for the construction industry to play a role in ensuring environmental sustainability by tackling the waste enigma with an emphasis on C&D waste minimisation / reduction, and C&D waste diversion. To ensure sustainable waste management within the construction industry, the following are recommended:

- Design team members (architects, engineers, and QSs) must adopt a pro-active role in waste minimisation through the implementation of a waste efficient design that incorporates standardisation, dimensional co-ordination, modular design principles, and designing for deconstruction;
- Contractors should implement effective waste management practices on-site, that include waste segregation, material logistics management, maximising material re-use, effective material and waste handling, and contractual provisions for waste minimisation, and
- The construction industry professional bodies must play a role in creating sustainable waste management awareness and an emphasis on waste management training and education among construction firms, professionals and general workers.

References

Adlan MN, Aziz HA, Hassan SH, Johari I (2015) The causes of waste generated in Malaysian housing construction-sites using site observations and interviews. Int J Environ Waste Manag 15:295–308

Ajayi SO, Akinade OO, Alaka HA, Bello SA, Bilal M, Jaiyeoba BE, Kadiri KO, Owolabi HA, Oyedele LO (2017) Design for deconstruction (DfD): critical success factors for diverting end-of-life waste from landfills. Waste Manag 60:3–13

Ajayi SO, Akinade OO, Alaka HA, Bilal M, Owolabi HA, Oyedele LO (2014) Ineffectiveness of construction waste management strategies: knowledge gap analysis. In: Proceedings of the First International Conference of the CIB Middle East and North Africa Research Network (CIB-MENA 2014), 14–16 December 2014, pp 261–280

Alam MS, Eskicioglu C, Hewage K, Sadiq R, Yeheyis M (2013) An overview of construction and demolition waste management in Canada: a lifecycle analysis approach to sustainability. Clean Technol Environ Pol 15:81–91

Al-Hajj A, Hamani K (2011) Material waste in the UAE construction industry: Main causes and minimisation practices. Archit Eng Design Manag 7:221–235

Ameh JO, Daniel EI (2013) Professionals' views of material wastage on construction-sites, organization. Technol Manag Construct: Int J 5:747–757

Asmi A, Memon AH, Nagapan S, Rahman IA, Zin MR (2012) Identifying causes of construction waste – case of the central region of peninsula Malaysia. Int J Integr Eng 4:22–28

Asmi A, Nagapan S, Rahman IA (2011) A review of construction waste cause factors. In: Proceedings of Asian conference on real estate 3–5 October 2011, Thistle Hotel Johor Bahru, Malaysia

Arslan V, Kazaz A, Ulubeyli S (2017) Construction and demolition waste recycling plants revisited: management issues. Proced Eng 172:1190–1197

Hyder Consulting Pty Ltd (2011) Australia Department of Sustainability, Environment, Water, Population and Communities Construction and demolition waste status report. Melbourne: Australia Department of Sustainability, Environment, Water, Population and Communities

Baum H, Katz A (2011) A novel methodology to estimate the evolution of construction waste in construction-sites. Waste Manag 31:353–358

Centeno MG, Domínguez A, Domínguez AI, Ivanova S, Odriozol JA (2016) Recycling of construction and demolition waste generated by building infrastructure for the production of glassy materials. Ceram Int 42:15217–15223

Crossin E, Lockrey S, Nguyen H, Verghese K (2016) Recycling the construction and demolition waste in Vietnam: opportunities and challenges in practice. J Clean Prod 133:757–766

Damci A, Gurgun AP, Polat G, Turkoglu H (2017) Identification of root causes of construction and demolition (C&D) waste: the case of Turkey. Proced Eng 196:948–955

Esa MR, Halog A, Rigamonti L (2017) Strategies for minimising construction and demolition wastes in Malaysia. Resour Conserv Recycl 120:219–229

Formoso C, Koskela L, Rooke J (2015) A conceptual framework for the prescriptive causal analysis of construction waste. In: Proceedings of the 23rd Annual Conference of the International Group for Lean Construction, Perth, 29–31 July 2015, pp 454–461

Gauteng Department of Agriculture Conversation and Environment (2009) Gauteng provincial building and demolition waste guidelines

Hewage KN, Mendis D, Wrzesniewski J (2013) Reduction of construction wastes by improving construction contract management: a multinational evaluation. Waste Manag Res 31 (10):1062–1069

Ichinose D, Yamamoto M (2011) On the relationship between the provision of waste management service and illegal dumping. Resour Energy Econ 33:79–93

Lawal AF, Wahab AB (2011) An evaluation of waste control measures in construction industry in Nigeria. Afr J Environ Sci Technol 5(3):246–254

Sagan J, Sobotka A (2016) Cost-saving environmental activities on construction-site – cost efficiency of waste management: case study, Procedia engineering. World Multidisciplinary Civil Engineering-Architecture-Urban Planning Symposium 161:388–393

Saidu I, Shakantu W (2016) The contributions of construction material waste to project cost overruns in Abuja, Nigeria: review article. Acta Structilia: Journal for the Physical and Development Sciences 23:99–113

Sapuay SE (2016) Construction waste – potentials and constraints. Procedia Environmental Sciences, Waste Management for Resource Utilisation 35:714–722

South Africa Department of Environmental Affairs (2012) National waste information baseline report. Government Printer, Pretoria

United Kingdom Department for Environment Food and Rural Affairs (2017) Digest of waste and resource statistics – 2017 edition. UK, Her Majesty's Stationery Office

United States of America Environmental Protection Agency (2016) Advancing sustainable materials management: 2014 Fact Sheet, USA, US Government Printing Office

Chapter 39
Towards a Circular Economy in the Built Environment: An Integral Design Framework for Circular Building Components

Anne van Stijn and Vincent Gruis

Abstract The building sector consumes 40% of resources globally, produces 40% of global waste and 33% of emissions. The Circular Economy (CE) proposes an alternative to the current linear economy by decoupling economic growth from resource consumption. In a circular built environment, resource depletion and waste generation are minimised, and materials are circulated at their highest utility and value. Buildings consist of many components such as kitchens and facades which could be replaced by 'circular components' during maintenance and renovation, thus allowing a bottom-up implementation of the CE within the built environment. To support industry in the development of such components, an integral design framework is needed, including technical, supply chain and business specifications. Current frameworks do not suffice, as they are either fragmented (addressing only the business model or technical design) or are not developed for the building sector. In this paper, we present a framework to support industry in the integral design of circular components. The framework was developed in three stages. First, through a literature review, existing circular design frameworks were identified. Second, by combining and specifying these frameworks, the Circular Building Components Generator (CBC-Generator) was developed. Finally, the CBC-generator was tested in the development of an exemplary component: The Circular Kitchen (CIK). The CBC-Generator is a three-tiered design tool, consisting of a technical, industrial and business model generator. The generators are 'parameter based': consisting of a design template and option-matrix. By filling the templates through systematically 'mixing and matching' the options of each parameter, different variants for circular components can be formed. Employing the

A. van Stijn (✉) · V. Gruis
Department of Management in the Built Environment, Faculty of Architecture and the Built Environment, Delft University of Technology, Delft, The Netherlands

Amsterdam Institute for Advanced Metropolitan Solutions (AMS), Amsterdam, The Netherlands
e-mail: A.vanstijn@tudelft.nl; V.H.Gruis@tudelft.nl

© Springer Nature Switzerland AG 2020
R. Roggema, A. Roggema (eds.), *Smart and Sustainable Cities and Buildings*,
https://doi.org/10.1007/978-3-030-37635-2_39

CBC-generator, TU Delft, AMS-institute, housing associations and industry partners developed the CIK. The CIK consists of a docking station in which kitchen modules can be plugged in and out. The modules consist of a frame to which 'function modules' (appliances) and 'style packages' (front, countertop) can be easily attached, using click-on connections. The professional side of the business model applies a purchase with take-back model including circular KPI's and service subscriptions. A dealer offers extra kitchen modules and style packages to tenants through a variety of financial arrangements that motivate returning the product after use, including lease and sale-with-deposit options. After use, the kitchen producer and dealer 're-loop' the docking station, kitchen modules, parts and materials in a 'Return factory' and local 'Return streets'. The CBC-generator successfully supports the integral design of circular components: (1) it provides all the design parameters which should be considered; (2) it gives various design options per parameter; (3) the generator supports systematic synthesis of design options to a cohesive and comprehensive circular design. Further development can contribute to establishing causal links between 'parameter-options', and identification of 'the most circular variant' in terms of both environmental burden and Total Costs of Ownership/Use.

Keywords Circular economy · Design framework · Building components · Circular kitchen

39.1   Introduction

The building sector consumes 40% of natural resources globally, produces 40% of global waste and 33% of emissions (Ness and Xing 2017). The pursuit of sustainable buildings is dominated by a focus on carbon neutrality and green, often overlooking resource consumption and its contribution to greenhouse gas emissions and planetary degradation. Accordingly, this article seeks to highlight the importance of a resource-efficient built environment, which enables required functions to be delivered with less assets, and to put forward an approach toward this objective. In this regard, the transition to a Circular Economy (CE) in the built environment. The CE proposes an alternative model to the current linear *'take-make-use-dispose'* economic model and can be summarised in the following three principles (Ellen MacArthur Foundation 2013a). (1) preserving and enhancing natural capital by controlling finite stocks and balancing renewable resource flows; (2) optimizing resource yields by circulating products, components, and materials at their highest utility and value at all times in both technical and biological loops and (3) fostering system effectiveness by revealing and designing out negative externalities.

Looking at current practice in the building sector, the emphasis is still very much on how to deal with waste, or 'recycling', which is the outer loop of the CE. However, a main principle of CE is to first make optimal use of the inner loops such as 'maintain', 'reuse', and 'remanufacture', and thus to prevent waste. Looking at the built environment, maintenance and renovation of buildings is then by definition a way to make use of the inner loops of the CE. However, maintenance and renovation currently does not contribute to making the housing stock itself (e.g. components and material) more circular. Buildings consist of many components such as climate installations, kitchens, bathrooms and facades, which could be replaced by 'circular building components' during the natural maintenance and renovation moments. This would create an opportunity for a bottom-up implementation of circularity in the built environment without needing a complete overhaul of the system.

The theory of circularity in the built environment is still developing and circular design frameworks applicable for the built environment are a paucity. To facilitate development, and subsequently the implementation of circular components in the built environment, professionals (e.g., architects, manufacturers, contractors) would benefit from a specific framework which can support choices concerning for example material use, composition of the supply chain and financial engineering. Many researchers in the field agree that such a framework would require a systemic and integral approach to ensure the designed component is (used) circular along and beyond their life cycle (Bocken et al. 2016; Geldermans 2016; Mendoza et al. 2017; Mestre and Cooper 2017; Saidani et al. 2017). In an integral design a technical model, business model, and industrial model should be developed in cohesion with each other (Bocken et al. 2016). This paper presents a design framework to support the integral development of circular building components, including design of the following: (1) technical model (referring to aspects such as shape, construction and materialisation); (2) industrial model (referring to the composition of the supply chain); (3) business model (referring to aspects such as finance and economic feasibility).

39.2   Method

The framework has been developed in three stages:

The research stage, presented in Sect. 39.3.1 and 39.3.2, consisted of an analysis of existing circular design frameworks. The review considered peer-reviewed, conference as well as professional circular design frameworks. The articles were identified through Web of Science and Google Search engines using the following keywords: 'circular economy' and 'design' and 'framework', 'method' or 'tool'. From the search results, only articles that are concerned with a framework which supports the design of a circular technical, industrial and/or business model of an artefact and/or service and/or organisation, and (specifically) offers support in the synthesis of a design proposal were included. De Koeijer et al. (2017) summarise two main types of circular design tools and models: generative and evaluative, based on the tools' applicability in the front-end or back-end of the product development process, respectively (Bocken et al. 2014; Bovea and Pérez-Belis 2012; De Koeijer et al. 2017; Fitzgerald et al. 2005; Telenko et al. 2008). Both generative and evaluative design frameworks are important to fully support circular design. Yet, generative tools offer the initial support in synthesis of design variant(s) and, therefore, are the focus of this paper.

An aspect analysis was performed on the selected frameworks to identity gaps and employable elements for the development of the circular design framework. Three aspects were analysed: (I) scale on which the model can be applied. The CE can be applied at the macro (country, region, municipalities and urban agglomerates), meso (network, eco-parks and buildings) and micro (company, product and building components) level (Geng et al. 2012; Pomponi and Moncaster 2017). (II) The type of design support offered in the model, distinguishing: (1) guidelines, (2) design criteria, (3) checklists, (4) step-by-step guide, (5) design canvas, (6) architypes, (7) strategies, (8) parameters, (9) options and (10) examples. (III) The type of model for which the framework offers design support: business model, technical model or industrial model.

In the second stage, the identified employable elements from analysed design frameworks were adapted, built upon and specified using additional data from scientific, secondary literature and brainstorming to propose a design framework for supporting the integral development of circular building components (Sect. 39.3.3, 39.3.4, 39.3.5 and 39.3.6). In the third stage (Sect. 39.3.7, 39.3.8, 39.3.9 and 39.3.10), the proposed design framework was tested through application in the development of an exemplary building component: The Circular Kitchen. In Sect. 39.4, we reflect upon the resulting framework and the conclusions are summarised in Sect. 39.5.

39.3   Results

39.3.1   Literature Review

Using the selected keywords and the developed selection criteria, 30 of the found frameworks were regarded admissible. For the results of the aspect analysis, see Table 39.1.

Table 39.1 Results aspect analysis circular design frameworks

Source	Name framework	Scale			Type of support										Discipline		
		MA	ME	MI	1	2	3	4	5	6	7	8	9	10	TM	IM	BM
Achterberg et al. (2016)	The value hill			x				x					x				x
Antikainen and Valkokari (2016)	Framework for sustainable business model innovation			x					x			x			?	?	x
Bakker et al. (2014)	Products that last - framework			x						x	x			x	x		x
Bocken et al. (2016)	Circular design framework			x						x	x			x	x		x
Ellen MacArthur Foundation and IDEO (2017)	Circular design guide			x				x	x	?	?			x	x		x
Ellen MacArthur Foundation (2015)	ReSOLVE framework			x							x			x	x		x
Ellen MacArthur Foundation (2013b)	New framework on circular design			x								?		?		?	x
Evans and Bocken (2014)	Circular economy toolkit			x							x			x	x		x
Fischer and Achterberg (2016)	10 steps to create a circular business model			x				x				x			x		x
Forum for the Future and Novelis (n.d.)	Design for demand			x				x	x	x	?				x		x
Forum for the Future and Unilever (n.d.)	Circular business model toolkit			x						x				x	?		x
Geldermans (2016)	Circular building matrix and new-stepped strategy		x	x				x	x						x		
Gerritsen (2015)	Circular design checklist			x	x	x							x		x		
Gispen (n.d.)	Design framework			x		x	x					x	?		x	x	
Goldsworthy (2017)	Speedcycle			x						x		x			x		
Joustra et al. (2013)	Guided choices towards a circular business model			x				x			?	x	?	x	?		x

(continued)

Table 39.1 (continued)

Source	Name framework	Scale			Type of support										Discipline		
		MA	ME	MI	1	2	3	4	5	6	7	8	9	10	TM	IM	BM
Leising et.al (2018)	Collaboration tool for CE in the building sector		x		x			x							x	x	x
Lewandowski (2016)	Circular business model canvas			x					x			x					x
Mentink (2014)	Business cycle canvas			x					x			x				?	x
Mestre and Cooper (2017)	Multiple loop life-cycle design frame			x							x				x		
Mendoza et al. (2017) and Heyes et al. (2018)	BECE framework			x			x	x	x						x	x	x
Moreno et al. (2016)	Circular design framework			x	x					x	x	x			x		x
Nußholz (2017)	Circular strategies embedded in the business model canvas			x					x			x			x		x
Poppelaars (2014)	Design framework			x	x			x					x		x		
Van Dam et al. (2017)	Circular Pathfinder			x							x			x	x		
Scheepens et al. (2016)	Circular transition framework for business model innovation towards a CE	x	x						x			x		x	x	x	x
Sempels (2014)	Sustainable business model canvas			x					x			x					x
The Great Recovery and RSA (2013)	Design tools for a circular economy			x						x					x		x
van den Berg and Bakker (2015)	Circular design framework			x							x		x		x		
WIITHAA (n.d.)	Circulab board			x			x	x	x			x					x

39.3.2   Gaps and Employable Elements

From the analysis, several gaps are identified. Firstly, there is a paucity of frameworks which are developed for the 'meso' scale or buildings. The exceptions are the design frameworks developed by Geldermans (2016) and Leising et al. (2018). Secondly, although all of the frameworks recognise the need for a systemic and/or interdisciplinary approach, very few offer a convincing integrated framework; only Bakker et al. (2014), Bocken et al. (2016), Mendoza et al. (2017), Moreno et al. (2016) and Scheepens et al. (2016). Thirdly, there is a missing link between more comprehensive yet 'abstract' frameworks (e.g., Bakker et al. 2014; Bocken et al. 2016; Leising et al. 2018; Scheepens et al. 2016) and frameworks which offer (comprehensive) concrete design options. Finally, the business model and industrial model are often integrated as some parameters of these models overlap. Yet, in an integrated model, the industrial side of these parameters is insufficiently considered (e.g., mode of transport, and location of activities are omitted). Although none of the analysed frameworks provide complete integral design support for circular building components, they contain useful elements. First, the Circular Building Matrix by Geldermans (2016) contains a strong link between the meso and micro scale through the inventory matrix. It facilitates the unravelling of the building into elements of a system-tree, distinguishing the building, component, part and material level. It also facilitates the exploration of the circular potential of each element. Second, the majority of the analysed circular business model frameworks build on adapted versions of the business model canvas (Osterwalder and Pigneur 2010), providing an employable set of design parameters. Third, the design framework as developed by Bocken et al. (2016) contains the taxonomy of 'narrowing, slowing and closing loops', which offers a clear umbrella for circular design strategies. The integrated frameworks by Bakker et al. (2014) and Bocken et al. (2016) offer insight in how the technical, industrial and business model can be linked. Finally, the frameworks of Poppelaars (2014), Van den Berg and Bakker (2015), and Gerritsen (2015) provide valuable concrete design options. Together, these elements have been employed in the development of the circular design framework.

39.3.3   Development of the Circular Design Framework: The CBC-Generator

In this section, we present the design framework, building on elements of frameworks discussed in Sect. 39.3.1: 'the Circular Building Components (CBC) Generator'.

The CBC-Generator is a three-tiered design tool, consisting of a technical, industrial and business model generator, which together support the integral design of circular building components. The three generators are based on the developmental parameters of the technical, industrial and business model for a circular component. For each parameter, different possible design options are provided, which can serve as the 'building blocks' to create the design. For each generator, the parameters and options are listed in a matrix. The matrix is complemented with a template to support the synthesis of design options. The design framework should be easy to use, to offer a true support for practice. For designers, utilising a visual language runs parallel to the process of using words, when synthesising a design (Van Dooren et al. 2014). Hence, the generators were designed – and can be applied – using both textual and visual language. The tool is applicable in the different stages of a design process. A design process is characterised by an exponential information growth curve (Ullman 2010, p.19). A tool meant to support the synthesis of a first design idea should not require detailed information of the designer. To ensure the tool supports synthesis in different design stages, it includes three modi operandi.

39.3.4 Modi Operandi: From First Idea to Detailed Circular Design Proposal

The tool has three operational pathways: (1) 'ideate', (2) 'generate' and (3) 'refine', each supporting synthesis in a different stage of the design process. The modi operandi are organised in the design templates; each surpassing modus operandi requires the designer(s) to fill in more parts of the template and with a higher level of detail.

The first operational pathway, 'ideate', supports the development of first idea (s) for a circular building component design. The 'design template' is filled in by systematically 'mixing and matching' the design options of different parameters listed in the matrix. The outcome can be understood as a logical combination of technical, industrial and business model options which could be applied in a design (for an example see Sect. 39.3.8). The design team is free to start from the technical, industrial or business model generator, based on their preference. However, it is encouraged to always use the generators in parallel to achieve an integral circular design.

The second operational pathway, 'generate', supports the generation of circular building component concept designs. The combination of design options, as selected in the ideation stage, are applied as building blocks and translated to a concept design. Additional design options can be selected from the matrices. Parts of the template which are not yet relevant can be left blank, allowing a quick synthesis of design variants.

The third operational pathway, 'refine', supports the refinement of a circular building component design. The concept design is further detailed and refined to a comprehensive circular design proposal by completing and detailing all parts of the templates. The matrices can be consulted for additional options, and alternative options for parts of the design which were considered unfeasible or undesirable.

39.3.5   The Parameter-Option Matrices

The matrices include the parameters relevant for designing a circular technical, industrial and business model (see attachments B-D). The parameters were selected from the analysed design frameworks. For each parameter, various design options were specified.

The technical model matrix, employs elements (amongst all) from: Bakker et al. (2014), Bocken et al. (2016), Van den Berg and Bakker (2015), Gerritsen (2015), Geldermans (2016) and Poppelaars (2014), and includes the following parameters: (1) *systemics* of the technical design. Referring to the unravelling of the building into elements (from building to component, to part to material). (2) The *material* used. (3) The *energy* used. (4) The *applied circular design strategy*. (5) The *expected lifespan* of a system element. (6) The *amount* of a system element measured in [pieces or m^3].

Parameters which influence circularity of the industrial model were identified in the business model canvas (Osterwalder and Pigneur 2010) and were complemented with industrial parameters used in LCA assessments (e.g., see Scheepens et al. 2016). The following parameters are included in the matrix: (1) The *key partners* in the supply chain. (2) *Activities* carried out by the partners, including their (re-) production processes. (3) *Key resources* needed in the supply chain. This includes the facility in which the activity takes place and the system elements (e.g., components, parts) which move through the supply chain. (4) Mode of *transport* in the supply chain. (5) Type of *process energy* used.

Parameters which influence circularity of the business model were found in the business model canvas (Osterwalder and Pigneur 2010), and the circular business model frameworks by Mentink (2014), Lewandowski (2016) and Nußholz (2017). The following parameters are included in the matrix: (1) *Key partners* in the business model. (2) *Customer segments* in the business model (3) The *supply chain relations* between partners. (3) The *cost structure* per partner. (4) The *revenue streams* per product or service offered, including the financial arrangement (lease, sale, etc.). (5) The *value proposition* which specifies: the product or service proposition offered to the customer, the value delivery and capturing per partner. Value delivery clarifies how the product brings value to customers and value capturing how the business model brings value to a partner. Both are needed to align incentives within the supply chain, and it is this alignment that is crucial for the feasibility of the business model. (6) The *channels* used to reach the customer. (7) The *take-back systems*

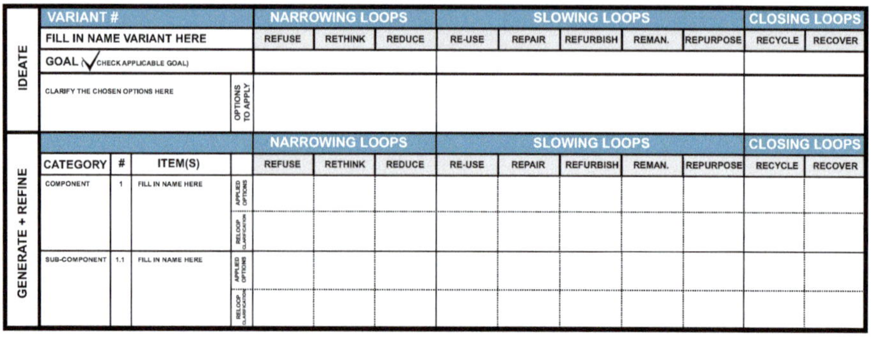

				NARROWING LOOPS			SLOWING LOOPS				CLOSING LOOPS	
IDEATE	**VARIANT #**				REFUSE	RETHINK	REDUCE	RE-USE	REPAIR	REFURBISH	REMAN.	REPURPOSE	RECYCLE	RECOVER
	FILL IN NAME VARIANT HERE													
	GOAL (✓ CHECK APPLICABLE GOAL)													
	CLARIFY THE CHOSEN OPTIONS HERE			OPTIONS TO APPLY										
					NARROWING LOOPS			SLOWING LOOPS				CLOSING LOOPS		
GENERATE + REFINE	CATEGORY	#	ITEM(S)		REFUSE	RETHINK	REDUCE	RE-USE	REPAIR	REFURBISH	REMAN.	REPURPOSE	RECYCLE	RECOVER
	COMPONENT	1	FILL IN NAME HERE	APPLIED OPTIONS										
				RE-LOOP CLARIFICATION										
	SUB-COMPONENT	1.1	FILL IN NAME HERE	APPLIED OPTIONS										
				RE-LOOP CLARIFICATION										

Fig. 39.1 Design template for the technical model generator

applied to ensure the return of key *resources* for re-looping (Lewandowski 2016). (8) The *adoption factors* which determine how the business model can be implemented within the organisation of a partner, regulations, and society (Lewandowski 2016).

39.3.6 The Design Templates

The matrices are complemented with a design template (see Fig. 39.1). How elaborate these templates need to be filled in is prescribed by the operational pathways 'ideate', 'generate' and 'refine'. The template consists of a table, whose use is twofold. First, it forms the frame in which options are systematically combined – applying them as building blocks – to form logical combinations for a design. Second, applying the concept of the 'circular building matrix' as developed by Geldermans (2016), in this table the circular potential of the combined options can be explored and refined.

For these purposes, the horizontal axis of this table lists several categories in which the selected options can be organised, according to how they contribute to achieving circularity. The categories apply the taxonomy of the circular design framework developed by Bocken et al. (2016): 'narrowing, slowing or closing resource loops'. Designing to 'narrow resource loops' aims to reduce resource use, or achieve resource efficiency. Designing to 'slow resource loops' aims to slow down the flow of resources through extension or intensification of the utilization period of the designed artefact. When a design is made to 'close resource loops', it is designed so all used materials are recycled at the end of life. This categorisation is further nuanced with the 9R model. (0) Refuse, (1) Rethink, (2) Reduce, (3) Re-use, (4) Repair, (5) Refurbish, (6) Remanufacture, (7) Repurpose, (8) Recycle and (9) Recover – as developed by Potting et al. (2017) and Van Buren et al. (2016).

The vertical axis of the design table is used to list the technical, industrial or business model design, depending on the operational pathway, from its entirety to more and more specified per parameters.

39.3.7   Testing the CBC-Generator: The Development of the Circular Kitchen

The CBC-Generator has been tested through application in the development of an exemplary building component: The Circular Kitchen (CIK), following the modi operandi 'ideate' (39.3.8), 'generate' (39.3.9) and 'refine' (39.3.10).

39.3.8   From a Blank Page to the First Ideas: Ideating a Circular Building Component

Applying the CBC-generator's operational pathway 'ideate' several ideas for circular kitchen design variants were developed. To illustrate how we have used the CBC-generator, we elaborate on the development of one of these ideas: 'The plug-and-play kitchen'.

The ideation process started by conceiving an inspirational direction (e.g., requirement, guiding theme, example) for the design variant. In this case, we started from the idea to make a kitchen which has a long life, can be recycled and, subsequently, saves resources. The parameter-option matrices were consulted by systematically looking at each parameter. Design options which helped to achieve the inspirational direction were selected. The technical model matrix was consulted first. Various design strategies to prolong the lifespan through re-use, repair, refurbishment, remanufacturing and recycling were selected. Subsequently, we turned to the accompanying business model, which needed to make the long-life design interesting to the manufacturer. From the business model matrix, the options: 'the manufacturer as owner' and 'revenue stream generated through service and updates' were selected. Then, for the industrial model, options were selected for the various 're-loop' activities, initiated by the manufacturer. The options were organised in the design template, creating a cohesive set of technical, industrial and business model options (see Fig. 39.2).

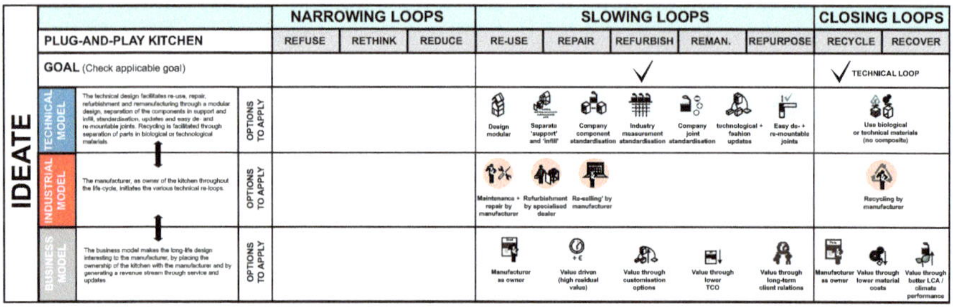

Fig. 39.2 Ideation design template for the technical, industrial and business model

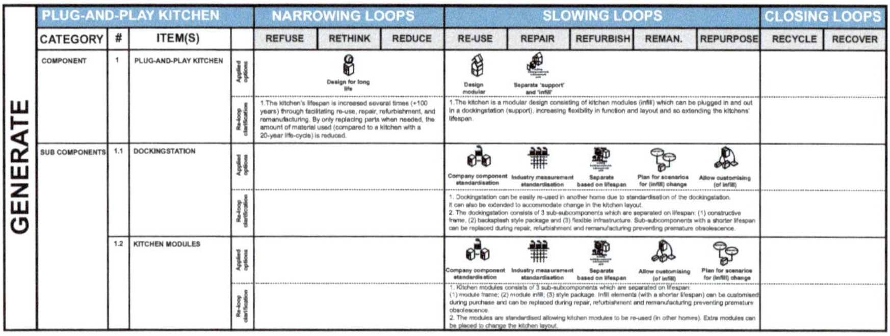

Fig. 39.3 Part of the technical model design template as filled in during 'generation'

39.3.9 Generating a Concept Design for the CIK

The combination of options for 'the plug-and-play kitchen' – as selected during ideation – were the starting point to 'generate' the CIK concept design. The selected options were applied as building blocks to develop an initial design for the CIK. Then, all the *system elements* (the component, sub-components, parts, and materials) of this initial design were placed on the vertical axis of the technical model design template, and the *key partners* of the initial design on the industrial and business model design templates. Per *system element* and *key partner*, the matrices were consulted. The options for each parameter were reviewed and additional design options were selected. The selected options were placed in the design templates, exploring how they could 'narrow, slow or close the loops' for each of the *system elements*, *key partners* and for the design as a whole (see Fig. 39.3). The 'generate' process was highly iterative: the exploration of circular potential would feed the design and, ultimately, help create a cohesive and comprehensive concept design for the CIK.

The synthesised CIK design facilitates various re-loops by separating parts based on *lifespan*. The kitchen consists of a docking station in which modules can be easily plugged in and out, allowing for future changes in lay-out. The kitchen modules themselves are also divided in a long-life frame to which 'module infill' (e.g., appliances) and 'style packages' (e.g., front, countertop, handles) can be easily attached using click-on connections. The high level of modularity and customisability of this design allowed for additional opportunities in the business model, such as: diversification of *revenue streams* and enlargement of the targeted *customer segments*. The business model parameter-option matrix was reviewed and additional options were selected and applied in the design. In the business model, the kitchen manufacturer sells the docking station and base modules directly to the housing associations, with a take-back guarantee, maintenance subscription and circular KPI's. This arrangement offers a clear incentive for the manufacturer to realise a kitchen which is easy to repair and to give a second life, or more. The extra modules and style packages are made available to users through financial arrangement such as lease and sale-with-deposit, that motivate returning the product at the end of their use cycle. The industrial model was aligned with the technical and business model. As the repair, re-use, refurbishment, and remanufacturing possibilities increased, the *mode of transport* and/or *location* of these re-loop processes became increasingly important to define. As the selected *transport* option relies on fossil fuels, options were selected from the matrix which reduce the distance between the user and the location where frequent re-loop activities take place. A local 'Return-Street' is introduced in which collected products are sorted to be traded, resold, lightly refurbished or sent back to the kitchen manufacturer. Products that come back to the manufacturer are sorted in their national 'Return-Factory' to be refurbished, remanufactured or recycled.

39.3.10 Refining the CIK

The concept design of the plug-and-play kitchen was refined to a full design proposal. The design templates were completed and further detailed (see Fig. 39.4). In this process, the parameter- option matrices were reviewed to select

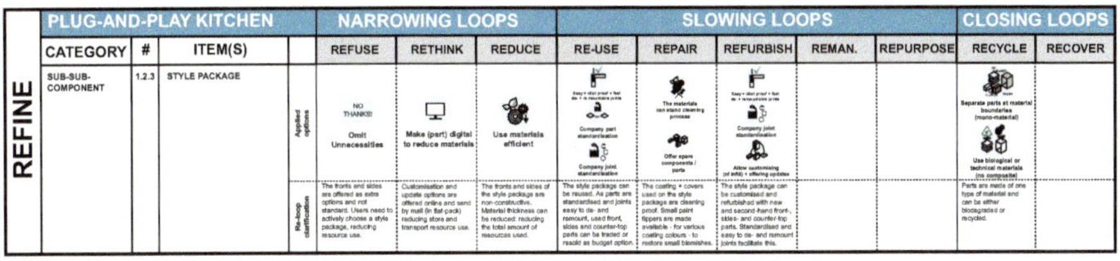

Fig. 39.4 Part of the technical model design template filled in during 'refinement'

additional options to complete parts of the templates which were previously left underdeveloped. Options, which were dismissed by the group, were reviewed with the parameter-option matrix and alternative options were selected. For example, to increase longevity, *the material* of the kitchen module frames was initially metal. For reasons of feasibility and poor environmental performance this material was dismissed. Alternative options were reviewed in the matrix and a (technical looped) wood was selected. For an image of the refined CIK, see attachment A.

39.4   Discussion

The example of the CIK shows that the CBC-generator can support integral synthesis of circular building components in different stages of the design. It supports designers as follows: (1) it provides designers all the design parameters which should be considered when making a circular design; (2) it gives designers an extensive list of circular design options for each parameter; (3) through the design templates, in which selected design options can be systematically mixed and matched, the CBC-generator supports the synthesis of a cohesive and comprehensive circular design. Yet, several limitations should be noted. Firstly, the framework analysis focused on frameworks explicitly related to the circular economy. The literature of circular economy precedents' design frameworks – such as eco design tools – were considered to a far lesser extent. Secondly, no systemic literature review was performed to identify a complete list of all possible design options. Therefore, the options included in the matrices could benefit from further specification. Furthermore, the CBC-generator does not show any causal link yet between different options nor between the technical, industrial, and business models. For example, if for the parameter *transport energy*, the option fossil fuel is selected, then the parameter *distance* should not offer any long(er) distance options such as global, continental, and national. The long transport with fossil fuels would likely have such a negative impact on the environmental performance that the process had better be performed locally, or not at all. The lack of advice on what makes 'logical combinations' of options makes the CBC-generator less suitable for use without assistance of a CE expert. The frameworks as developed by Bakker et al. (2014) and Moreno et al. (2016) did manage to develop such a correlation by linking technical model strategies to business model archetypes, but the 'design freedom' to investigate alternative options is strongly reduced. This does raise the question to what extent the CBC-generator enables a skilled and unskilled design team to develop a more circular design. Additional testing of the CBC-generator, through a comparative analysis, could create a better understanding of how to improve the tool for both skilled and unskilled designers. Additionally, the design template allows to systematically mix design options, but does not support the translation of options to a design. The canvasses of Mentink (2014) and Scheepens et al. (2016) do support translation and could be integrated in further development of the CBC-generator. Finally, the CBC-generator only provides support in the synthesis and not in the assessment of the most circular design. For example, when it is more 'circular' to facilitate upgrading of a product or when to choose for recycling does not become evident in the framework. Scheepens et al. (2016), propose that the environmental assessment of circularity should include quantitative assessment of material consumption, environmental impact and the value of the designed artefact. Bradley et al. (2018) suggest that the financial assessment of circularity could consist of an analysis of the 'Total Cost of Ownership (TCO)'. Further research can contribute to develop complementary circular assessment tool(s).

39.5   Conclusion

This article has sought to develop an integral design framework to support the design synthesis of circular building components. Through analysis of previous developed circular design frameworks, gaps and employable elements were identified. By combining, adapting and specifying the employable elements of the design frameworks, an integral design framework for circular building components: The Circular Building Components (CBC) Generator has been proposed. The CBC-Generator is a three-tiered design tool, consisting of a technical, industrial and business model generator. The generators are 'parameter based', consisting of a design template and parameter-option matrix. By filling the templates through systematically 'mixing and matching' the options, applying them as building blocks, design variants for circular building components can be synthesised. The application of the CBC-generator in the development of the Circular Kitchen shows that it can successfully be applied in the integral development of circular building components, and makes an important step towards supporting the building industry in developing and implementing such components in the built environment. Further development can contribute to adding design options, supporting translation of options to a circular design proposal, establishing causal links between design options and identification of 'the most circular variant', improving user-friendliness for industry.

Appendices

Appendix A: Example Case: Design Proposal for the CIK

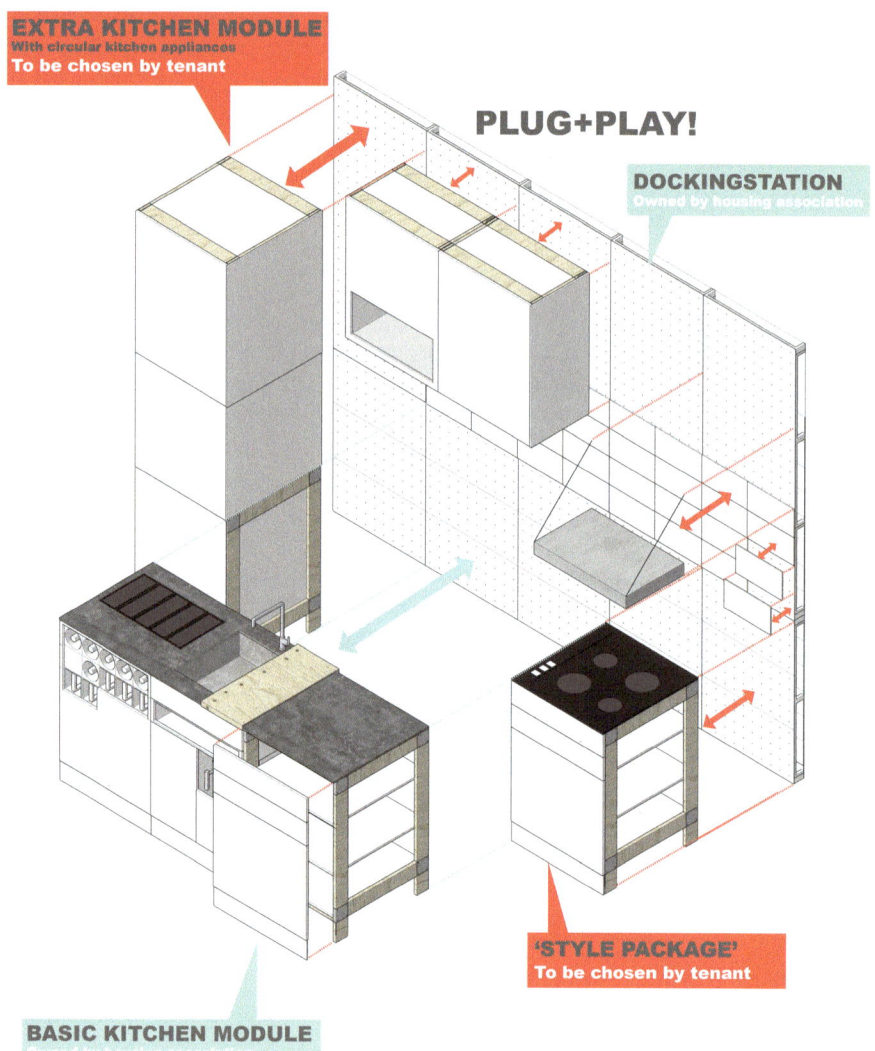

Appendix B: Parameter-Option Matrix of the Technical Model Generator

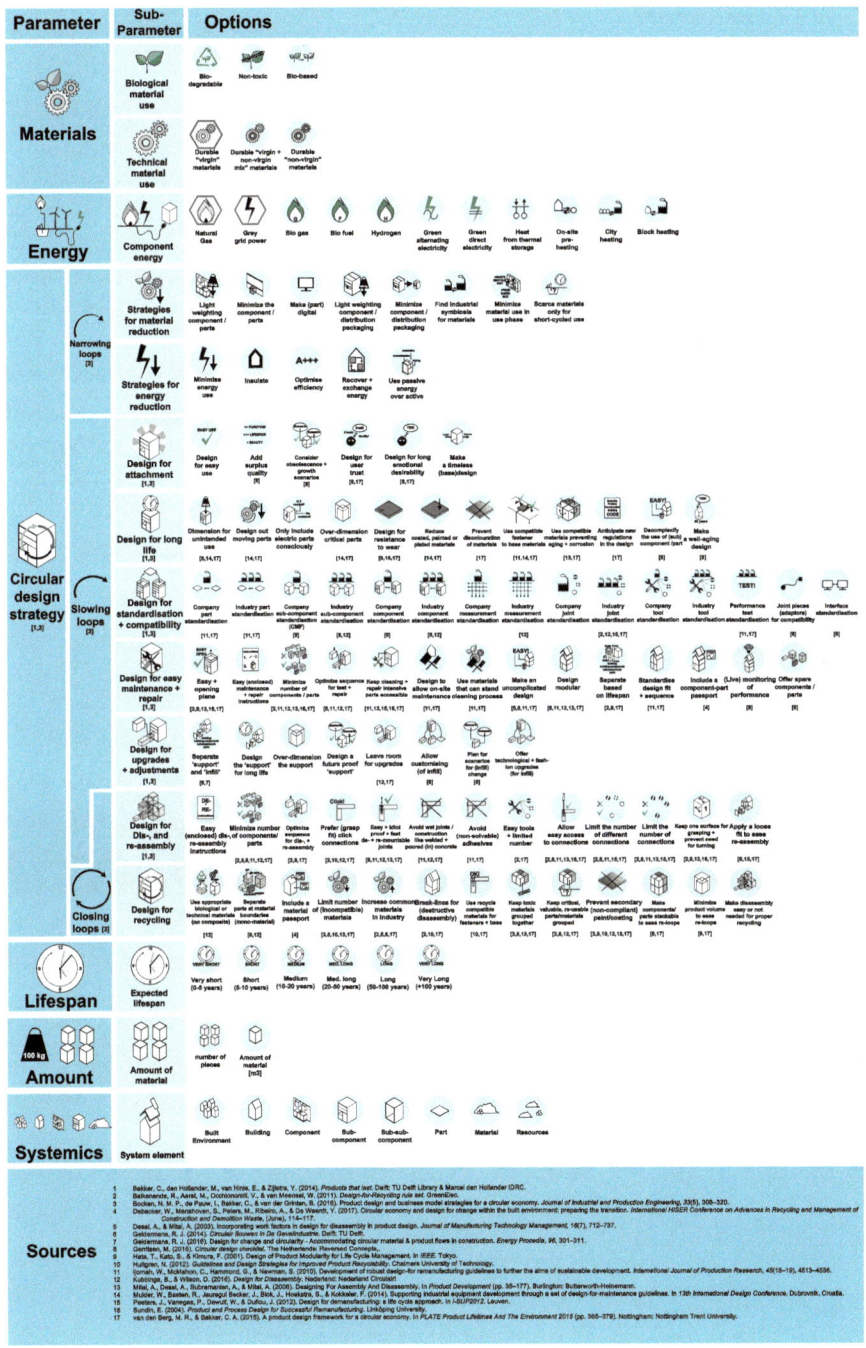

Appendix C: Parameter-Option Matrix of the Industrial Model Generator

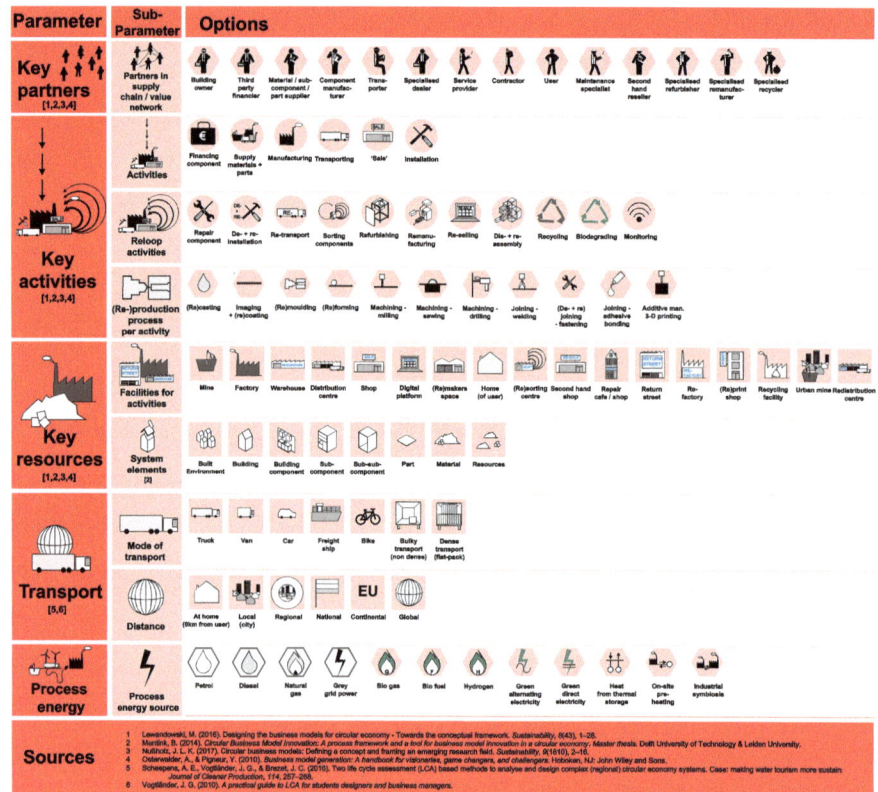

Appendix D: Parameter-Option Matrix of the Business Model Generator

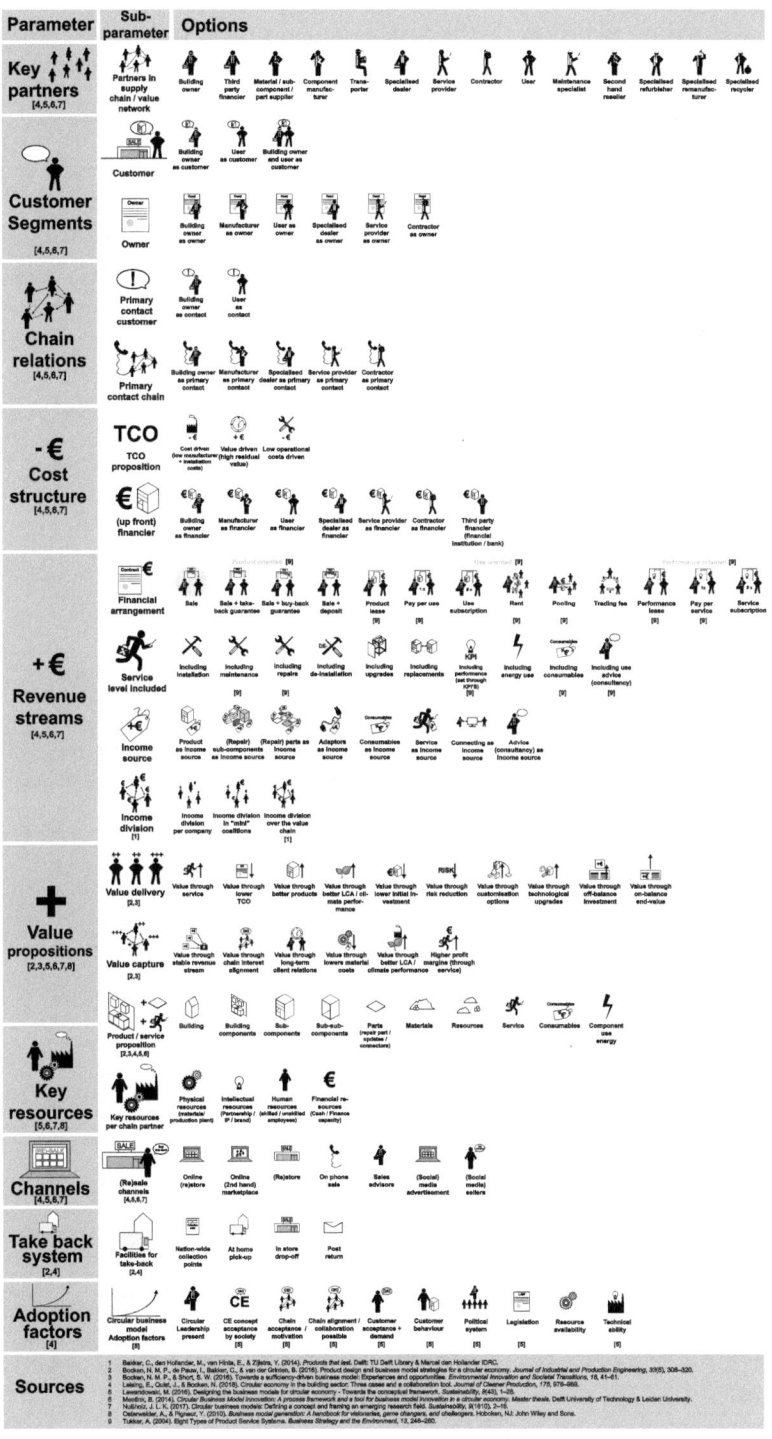

References

Achterberg E, Hinfelaar J, Bocken N (2016) Master circular business with the value hill. Sustainable Finance Lab, Circle Economy, Het Groene Brein, Nuovalente & Delft University of Technology, Amsterdam

Antikainen M, Valkokari K (2016) A framework for sustainable circular business model innovation. Technol Innov Manag Rev 6(7):5–12

Bakker C, Den Hollander M, Van Hinte E, Zijlstra Y (2014) Products that last: Product design for circular business models. Delft: TU Delft Library

Bocken NMP, De Pauw I, Bakker C, Van der Grinten B (2016) Product design and business model strategies for a circular economy. J Ind Prod Eng 33(5):308–320

Bocken NMP, Farracho M, Bosworth R, Kemp R (2014) The front-end of eco-innovation for eco-innovative small and medium sized companies. J Eng Technol Manag 31(1):43–57

Bovea MD, Pérez-Belis V (2012) A taxonomy of ecodesign tools for integrating environmental requirements into the product design process. J Clean Prod 20(1):61–71

Bradley R, Jawahir IS, Badurdeen F, Rouch K (2018) A total life cycle cost model (TLCCM) for the circular economy and its application to post-recovery resource allocation. Resour Conserv Recycl 135:141–149

De Koeijer B, Wever R, Henseler J (2017) Realizing product-packaging combinations in circular systems: shaping the research agenda. Packag Technol Sci 30:443–460

Ellen MacArthur Foundation (2013a) Towards the circular economy: economic and business rationale for an accelerated transition. Ellen MacArthur Foundation, Cowes

Ellen MacArthur Foundation (2013b) Towards the circular economy: opportunities for the consumer goods sector. Ellen MacArthur Foundation, Cowes

Ellen MacArthur Foundation (2015) Growth within: a circular economy vision for a competitive Europe. Ellen MacArthur Foundation, Cowes

Ellen MacArthur Foundation and IDEO (2017) The circular design guide. Retrieved August 27, 2018, from https://www.circulardesignguide.com/

Evans JL, Bocken NMP (2014) A tool for manufacturers to find opportunity in the circular economy: https://www.circulareconomytoolkit.org. In: Setchi R, Howlett RJ, Naim M, Seinz H (Eds) KES transactions on sustainable design and manufacturing I. Future Technology Press, Cardiff, Wales, pp 303–320

Fischer A, Achterberg E (2016) Create a financeable circular business in 10 steps. Circle Economy & Sustainable Finance Lab, Amsterdam

Fitzgerald DP, Herrmann JW, Sandborn PA, Schmidt LC (2005) Beyond tools: a design for environment process. International Journal of Performability Engineering 1(2):105–120

Forum for the Future and Novelis (n.d.) Design for demand. Retrieved August 28, 2018, from http://designfordemand.forumforthefuture.org/

Forum for the Future and Unilever (n.d.) Circular business models. Retrieved August 28, 2018, from https://www.forumforthefuture.org/the-circular-economy-business-model-toolkit

Geldermans RJ (2016) Design for change and circularity – accommodating circular material & product flows in construction. Energy Procedia 96:301–311

Geng Y, Fu J, Sarkis J, Xue B (2012) Towards a national circular economy indicator system in China: an evaluation and critical analysis. J Clean Prod 23(1):216–224

Gerritsen M (2015) Circular design checklist. Reversed Concepts, Amsterdam

Gispen (n.d.) Design framework. Retrieved August 24, 2018, from https://www.gispen.com/nl/circulaire-economie/het-ontwerpproces-circulaire-economie

Goldsworthy K (2017) The Speedcycle: a design-led framework for fast and slow circular fashion lifecycles. Des J 20(sup1):S1960–S1970

Heyes G, Sharmina M, Mendoza JMF, Gallego-Schmid A, Azapagic A (2018) Developing and implementing circular economy business models in service-oriented technology companies. J Clean Prod 177:621–632

Joustra, D.J., de Jong, E. and Engelaer, F. (2013) Guided choices towards a circular business model. Eindhoven: C2C BIZZ & Samenwerkingsverband Regio Eindhoven

Leising E, Quist J, Bocken N (2018) Circular economy in the building sector: three cases and a collaboration tool. J Clean Prod 176:976–989

Lewandowski M (2016) Designing the business models for circular economy - towards the conceptual framework. Sustainability 8(43):1–28

Mendoza JMF, Sharmina M, Gallego-Schmid A, Heyes G, Azapagic A (2017) Integrating backcasting and eco-design for the circular economy: the BECE framework. J Ind Ecol 21(3):526–544

Mentink B (2014) Circular business model innovation: a process framework and a tool for business model innovation in a circular economy. Master thesis. Delft University of Technology & Leiden University

Mestre A, Cooper T (2017) Circular product design. A multiple loops life cycle design approach for the circular economy. Des J 20(sup1):S1620–S1635

Moreno M, De los Rios C, Rowe Z, Charnley F (2016) A conceptual framework for circular design. Sustainability 8(937):1–15

Ness DA, Xing K (2017) Toward a resource-efficient built environment: a literature review and conceptual model. J Ind Ecol 21(3):572–592

Nußholz JLK (2017) Circular business models: defining a concept and framing an emerging research field. Sustainability 9(1810):2–16

Osterwalder A, Pigneur Y (2010) Business model generation: a handbook for visionaries, game changers, and challengers. Wiley, Hoboken

Pomponi F, Moncaster A (2017) Circular economy for the built environment: a research framework. J Clean Prod 143:710–718

Poppelaars F (2014) Designing for a circular economy: the conceptual design of a circular mobile device. Master thesis. Delft University of Technology

Potting J, Hekkert M, Worrell E, Hanemaaijer A (2017) Circular economy: measuring innovation in the product chain. PBL Netherlands Environmental Assessment Agency, The Hague

Saidani M, Yannou B, Leroy Y, Cluzel F (2017) How to assess product performance in the circular economy? Proposed requirements for the design of a circularity measurement framework. Recycling 2(6):1–18

Scheepens AE, Vogtländer JG, Brezet JC (2016) Two life cycle assessment (LCA) based methods to analyse and design complex (regional) circular economy systems. Case: making water tourism more sustainable. J Clean Prod 114:257–268

Sempels C (2014) Implementing a circular and performance economy through business model innovation. In: Ellen MacArthur Foundation (ed) A new dynamics: effective business in a circular economy. Ellen MacArthur Foundation, Cowes

Telenko C, Seepersad CC, Webber ME (2008) A compilation of design for environment principles and guidelines. DETC/CIE 2008, ASME 2008 International Design Engineering Technical Conferences & Computers and Information in Engineering Conference, (June 2014), pp 1–13

The Great Recovery and RSA (2013) Investigating the role of design in the circular economy. The Great Recovery and RSA, London

Ullman DG (2010) The mechanical design process. McGraw-Hill, New York

Van Buren N, Demmers M, Van der Heijden R, Witlox F (2016) Towards a circular economy: the role of Dutch logistics industries and governments. Sustainability 8(647):1–17

Van Dam SS, Bakker CA, De Pauw I, Van Der Grinten B (2017) The circular pathfinder: development and evaluation of a practice-based tool for selecting circular design strategies. In: PLATE product lifetimes and the environment 2017. IOS Press, Amsterdam, pp 102–107

Van den Berg MR, Bakker CA (2015) A product design framework for a circular economy. In: PLATE product lifetimes and the environment 2015. Nottingham Trent University, Nottingham, pp 365–379

Van Dooren E, Boshuizen E, Van Merriënboer J, Asselbergs T, Van Dorst M (2014) Making explicit in design education: generic elements in the design process. Int J Technol Des Educ 24(1):53–71

WIITHAA (n.d.) Tools - Circulab board, the business game to activate your circular business model. Retrieved August 27, 2018, from https://circulab.eu/tools/#board

Chapter 40
Cradle to Cradle Building Components Via the Cloud: A Case Study

David Ness, Ki Kim, John Swift, Adam Jenkins, Ke Xing, and Nick Roach

Abstract This paper expands upon 'cradle to cradle carpets and cities' presented by Ness and Field (Cradle to cradle carpets and cities. In Proceedings of SASBE 03, Brisbane, 2003) at SASBE 03, where the notion of providing modular carpets as a service was introduced, and a paper at SASBE 06, where the theme of providing C2C products as a service was further developed by Ness and Pullen (Decoupling resource consumption from growth: new business model towards a sustainable built environment in China. In: Proceedings of SASBE 06, Shanghai, 2006). It reports on the outcomes of an ARUP Global Research Challenge Project 2017, undertaken by University of South Australia, ARUP, Prismatic Architectural Research and other partners, under the theme of adapting the circular economy to the built environment. The project addresses the challenge of reusing building components, so they deliver more value over their extended life-cycle, with consequent reductions in resource consumption, greenhouse gas emissions, pollution and waste, coupled with creation of new enterprises and jobs. A universally accessible 'Cloud-based building information management platform' is being developed, which enables components to be identified, reclaimed reused and exchanged multiple times over their lifecycle, within the same or different facilities. A cyber-physical information exchange system was established between physical building components and their virtual counterparts, known as Building Information Models, so that their life cycle information including history of ownership, condition, maintenance history, technical specifications and physical performance could be tracked, monitored and managed. In addition, designers could identify reused components via the cloud platform, and assess their suitability for incorporation in building projects when compared with

D. Ness (✉) · K. Kim · A. Jenkins · K. Xing
University of South Australia, Adelaide, Australia
e-mail: david.ness@unisa.edu.au

J. Swift
Prismatic Architectural Research, Clarence Park, Australia

N. Roach
ARUP, Adelaide, Australia

© Springer Nature Switzerland AG 2020
R. Roggema, A. Roggema (eds.), *Smart and Sustainable Cities and Buildings*,
https://doi.org/10.1007/978-3-030-37635-2_40

new products. This research was complemented by an innovative business model, whereby components and products can be provided as a service, with producers retaining responsibility for their repair, remanufacturing and/or reuse over their life cycle. The methodology involved establishing a Cyber-Physical System by connecting a series of existing technologies, including Radio Frequency Identification (RFID) and Building Information Modelling (BIM). Using a case study of a section of a major new hospital for 'proof of concept', information on the history, location, properties and performance of physical components could be exchanged in real-time from RFID tags to a local BIM system and thence to the cloud platform. Complemented by interviews within Australia and Europe with key stakeholders including designers, project managers, manufacturers, owners, investors and facility managers, the research led to the development of a 'products as service' business model and associated business case for the new paradigm. In short, a self-populating relational database that can execute predefined multiple/ conditional ownership exchange via a graphical user interface and/or web site front end. The findings are expected to drive increased reuse, adaptation and life-cycle stewardship within the building industry, whereby more value can be derived from built resources, new business opportunities created in the service sector, and adverse environmental impacts reduced. This is consistent with the pursuit of a 'circular economy', where the construction and management of the built environment can exert a major influence.

Keywords Reuse · Adaptability · RFID · BIM · Products as service · Business case

40.1   Introduction

The paper expands upon themes presented to SASBE 03 (Ness and Field 2003), where the notion of providing modular carpets as a service was introduced, and a subsequent paper to SASBE 06 where the theme of providing "cradle to cradle" (C2C) products as a service was further developed (Ness and Pullen 2006).

It outlines research associated with the ARUP Global Research Challenge (GRC) 2017, under the theme of adapting the built environment to the concept of a circular economy (CE). While the CE is often viewed in terms of recycling, more value can be obtained from buildings and physical components by their reuse, whereby their life cycle is extended and they are kept in circulation, aided by their stewardship and remanufacture to ensure they retain their highest possible performance capability.

The submission for ARUP GRC 2017 was based upon the notion that reuse of building components could be enabled by a): cyber-physical connectivity whereby the location, condition, performance and other properties of components could be identified and monitored remotely via a cloud system, and b): use of a product-service system (PSS) business model that enabled producers to retain ownership responsibilities for products over their life cycle and to provide them to various users as a service. It was postulated that the symbiosis created by integrating these two mechanisms within a 'Cloud BIM data service platform' could facilitate a substantial increase in reuse.

Hence, the paper first describes the nature and impacts of present practices, resulting in around 40 per cent of building materials being deposited to landfill, over-extraction of scarce resources, and excessive use of energy and production of GHG emissions. Moves to overcome this challenge are then outlined, within the context of CE principles and practices. Gaps in the literature in regard to state of knowledge and practice are then identified, leading to the expression of the research aims and methodology. The research is then explained, the findings presented, conclusions drawn, and further opportunities and limitations outlined.

40.2   Background

Whilst global attention has centered on reducing operational GHG emissions, the importance of embodied CO_2 emissions or "capital carbon" has recently gained increased attention (HM Treasury 2013). This has been aided by the adoption of CE principles and policies in many jurisdictions, including the EU and China, and the promulgation of the UN Sustainable Development Goals (SDGs) that include Goal 12: Sustainable Production and Consumption. The construction industry, though, has been slow in the up-take of such approaches, seemingly due to a widespread belief that resources are limitless, and that increased energy efficiency will suffice.

40.2.1   Literature and Research

The notion that parts of buildings may be changeable and/or adaptable is not new. Alex Gordon coined the 3Ls concept – long life, low energy, loose fit – in the early 1970s, N.J. Habraken introduced the notion of 'open building' in the early 1990s, while Stewart Brand (1965) developed the concept of 'layers of change'; these ranged from less changeable, longer-life elements such as site and structure, to more adaptable 'space' (fitout), 'services' (e.g. lighting), 'skin' (e.g. façade) and 'stuff' (loose fittings).

ARUP (2016), ARUP and BAM (2017), and the Ellen MacArthur Foundation (EMF 2015) have published guidelines related to a circular built environment, including the role of circular business models. These highlight the opportunity for a platform to be used to track and advertise products and components that are currently 'locked in' buildings. Thus, when a building approaches its end of life, these resources could be procured by another development, thus enabling their reuse and circularity.

Baker-Brown (2017, 180) envisaged an online market that tapped into all industries and quantified new material flows as soon as they became available: "Enlighted designers and manufacturers would borrow or lease the material before returning it to the 'Material Flow Market'".

EU researchers are active in exploring circular material and product flows in buildings and the role of alternative business models (Geldermans 2016). The BauHow5 group of universities conducted a seminar on *Circular Building Construction*, examining business models involving "material ownership retention by suppliers and delivery of buildings as living, dynamic and adaptable platforms" (Klein 2017). Simultaneously, the Technical University Delft embarked upon *The Façade Leasing Project* in 2016.

One of the most transformative areas of research, which brings together the notions of "open building" and PSS, is the proposition that the changeable, moveable parts of a facility may be provided as a service, which is separate to the base building or real estate. This was first raised by Yashiro (2000, 2004), who sought to disentangle the relationship between skeleton and infill, so that infill could be viewed separately as 'moveable property' and provided as an 'Infill service'. More recently, Zuidema (2015) has sought to elevate thinking on this topic in Europe, recognising that such approaches assisted by "material tracking parameters in BIM" could greatly facilitate open, circular building.

40.2.2 Practices

ARUP introduced what was claimed to be the world's first 'circular building' at the London Exhibition in December 2016. This prototype pavilion incorporated modular and adaptable components with reused content, and employed scanning technologies to identify components and their leasing to enable take-back and reuse. Further prototypes have included the ABN AMRO *Circle Pavilion* in Amsterdam.

With countries such as the Netherlands setting targets for a fully CE by 2050 (Government of the Netherlands 2016), including the construction industry, there has been rapid increase in the introduction of "platforms" such as *Madaster* and *BAMB* (Buildings as Material Banks) to identify and facilitate the reuse of components from existing buildings. Both these initiatives aim to increase and sustain the value of building materials via the "material passport" concept, and pursue the idea that every building is a "material depot".

40.2.3 The Gaps in Research and Practice

While the above EU initiatives involve platforms whereby building owners may upload digital data on existing buildings as 'material depots', they do not enable the real-time identification of reusable components by designers and others, and online interrogation of their location, condition, performance and availability. Although BIM has a capability to capture and maintain the essential knowledge among relevant stakeholders, and enable them to examine the feasibility of reusing building components for projects by importing component models into new designs, research to explore the potential of BIM for improving circularity has rarely been studied (Ness et al. 2015).

The levels of sophistication in the use of RFID, BIM and the Cloud have been uneven (Swift et al. 2015). While some tracking of items by bar or QR codes is practised as inventory management, this stops when the items are in-situ and does not extend to their life cycle management.

While the *Madaster* database can be accessed by 'service providers', a mechanism does not exist for these providers to remotely monitor and manage their components. In addition, although PSS business models have been applied in other fields, there have been few attempts to introduce these to the construction industry where buildings are normally viewed as long life, immovable entities (as in the French word '*immeuble*'. Doubts have also been expressed about the efficacy of providing building and other products as a service.

While this has appeared promising in theory, the reality has thrown up a number of barriers. A survey of businesses, government officials and others revealed that low virgin material costs and high upfront investment represented major impediments to adoption of circular business models (Deloitte and Utrecht University 2017). However, a report on the circular phone (Circle Economy 2018) describes legal, operational and financial solutions to unlock the potential of the *Fairphone-as-a-Service* business model. Such approaches have the potential to be translated to the building sector, whereby reusable components are offered as a service (Ness and Xing 2017).

40.3   Research Aims and Method

40.3.1   Aims

To fill the above gaps in research and practice, the research aims were therefore expressed as: a) Develop seamless interchange of data between physical components using attached RFID tags, BIM and Cloud storage and functionality; b) test the efficacy of a PSS business model for reuse of building components, and c) develop and test a prototype Cloud-BIM data service platform.

40.3.2   Case Study

A University of South Australia (UniSA) led research team previously established an integrated agenda for reuse of components enabled by envisaging the connection of a series of well-known digital technologies (RFID, BIM and Cloud-based technology), supported by a PSS business model (Ness et al. 2015). This was followed by a paper to the 'Open Building' session of UIA 2017, which demonstrated that a seamless connection could be made between RFID and BIM (Swift et al. 2017).

This led to the approach for the ARUP GRC Project, which firstly identified – via a scoping study and according to Brand's (1995) 'shearing layers of change' – the types of building components with most potential for disassembly, exchange and reuse. These included: framed glazed systems, ceiling systems ('space'); lighting and air handling systems ('services'); and façade systems ('skin'). The *Uniclass* classification system, based on the framework set out in ISO 12006-2: 2015 and able to be assigned to BIM objects, was also used as a typology – with a particular focus on the 'systems' level. Part of the new Royal Adelaide Hospital (nRAH), which incorporated the various system elements, was then selected as a case study. The nRAH provided the most appropriate test site for this project for several reasons: (a) It was at the time one of the most extensively documented major building in Australia, hence offered a lot of scope from which to choose a bounded small test site; (b) it was accessible to majority of researchers involved in the study; and (c) The hospital utilised BIM not only for construction, but also for life-cycle facility management.

Access to the model was negotiated via a representative of BuildingSMART Australasia, also employed by Spotless Facility Management in the ongoing operation of the nRAH project.

Due to the political and commercial implications of making the BIM universally accessible, only a limited number of components were considered prudent to use.

This research on connecting RFID-BIM-Cloud was accompanied by investigations on the viability of providing and reusing these component systems as a service (PSS). PSS enables the performance of a component and its system, such as a framed glazed system, to be managed and optimised by the supplier/service provider over its extended life, such as by regular maintenance, repair or remanufacture as necessary. On the other hand, if a component is continually bought and sold over its life, with no single company able to exercise overall "extended product responsibility", its performance is likely to gradually deteriorate.

The PSS aspect of the research was conducted in collaboration with the Collaborating Centre for Sustainable Consumption and Production (CSCP), Wuppertal, Germany. This involved a series of interviews, within both Australia and Europe, with key stakeholders including designers, project managers, owners, investors, manufacturers and, in particular, a supplier of both internal glazed windows and doors and more complex glazed façade systems.

The supplier was helpful in comparing the economic viability of the PSS approach with business as usual (BAU), using the nRAH for simulating the two scenarios. This comparison was extended to both internal and external glazed systems, with the expectation that a modular, "unitized" and "active" façade system that delivered higher and changeable performance requirements would yield a greater income stream for the service provider.

Finally, the 'back-end' RFID-BIM data was integrated with the cloud-based 'front-end' user interface that enabled users to identify reusable components, to determine their suitability for alternative application(s), and to procure these either as a service or by purchase.

40.4   Results

40.4.1   RFID-BIM-Cloud Data Transfer

Using the case study, the research was successful in demonstrating the Cloud-BIM information flow process, as is illustrated in Fig. 40.1.

1. A building owner or service provider with an existing building about to be dismantled (or refurbished) could generate (if not already available) a BIM of that building.
2. Once the BIM was completed, it could be converted into IFC format (a universal non- proprietary language) that could be interoperable, regardless of authoring BIM tools such as Revit, ArchiCAD and Tekla (Kim and Park 2018).
3. A RFID tag could be attached (if not already integrated during manufacture) to a physical reusable building component, containing the basic information of the tagged objects and systems: this included the "Uniclass" classification code according to ISO 1206-2: 2015 (ISO 2015), description, material types, dimensions, ownership, life expectancy and geographic location.
4. When changes were made to the tagged object, the updated information could then be recorded within the IFC data and, simultaneously, the information could be updated to the RFID tag. Thus, Step 3 and 4 involved an ongoing periodic

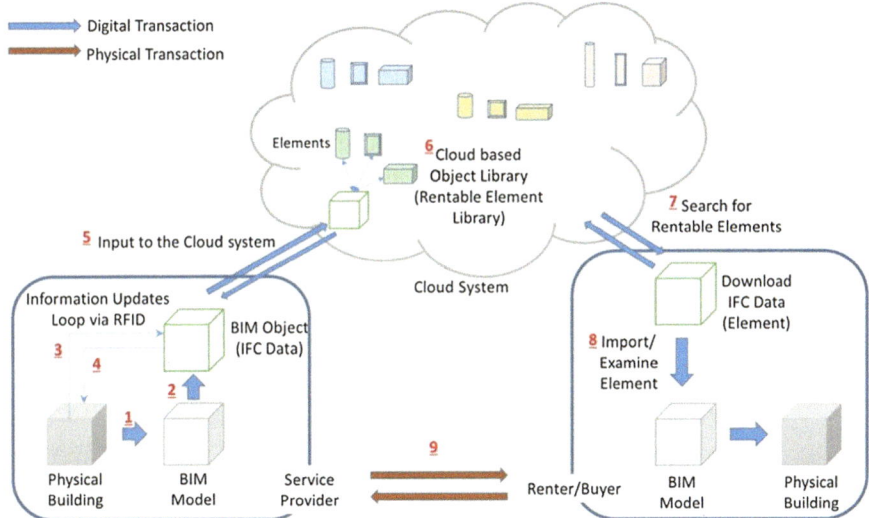

Fig. 40.1 Cloud-BIM information flow

update. When the feedback loop was established between the tagged objects and IFC, the IFC data could be uploaded onto the Cloud System in parallel with the full BIM.

5. At this point, the feedback loop was extended to the Cloud System and, finally, seamless information exchange and updates were established between the physical building component and the Cloud System. Consequently, the information feedback loop became a Cloud-based BIM/IFC data system (i.e. Cyber Physical System) underpinned by the AutoID system.

6. Once the information of the reusable elements/components became available via the Cloud system and multiple service providers uploaded their products, a Cloud-based reusable/ rentable object library) could be populated.

7. Via the Cloud-based library, it was shown that potential renters/buyers could search via a dialogue based and/or graphical interface for reusable (singular/ multiple) objects (products and components).

When a suitable object(s) was identified, a query could be initiated which revealed the dataset associated with that particular component. Due to the limitations of electronic storage space, this full dataset could not be accommodated on the RFID tag, hence the Cloud-based data that includes more visual information such as what it is, its dimensions, materials and ownership to be interrogated, along with other more detailed performance information. For example, exposure to weather could affect the quality and suitability of the object for another application. The query could involve examination of the original BIM model via the IFC data, Cloud-based viewing applications, or using the website searchable interface to conduct multiple parameter- based queries for available building components offered by multiple vendors, or via our software that supplies individual elements. The distance of the reusable component from the destination project could also be ascertained by GPS capability within the search parameters, as this could affect transport and carbon costs.

When a potential user confirmed the suitability of an element, they could download the IFC version of that element via their proprietary modelling package into their model and examine whether it is appropriate for that particular application (8). Therefore, the process from 2 to 8 could be termed as "Digital Inter-Party Transaction". Once the digital transaction was completed with a confirmation of use of the reusable element, the final step, which is the "Physical Transaction", could then take place (9).

At this time, due to the physical range associated with writing/reading the RFID tags selected, a hand-held bespoke device would be required to update the data in situ. This process would need to be carried out concurrently with any change of ownership or other variable.

40.4.2   The Product-Service Business Model

Concerns at the quality and performance of reused components had been raised by stakeholders from various disciplines. The ability to identify, track, monitor and manage reusable components via digital means opens up a number of possibilities to address such challenges. It may not only enable their procurement and reuse as part of traditional purchase-based procurement processes, but also facilitate the addition of various types of services to provide quality and performance assurance.

Multiple buy/sell or conditional transactions could be incorporated into the offerings of available services. These could range from basic services attached to the purchase, such as delivery, warranty and maintenance services, to more advanced services akin to 'performance guarantee' coupled with rental/lease options, where the supplier retains ownership responsibilities over the component's extended life-cycle and licensing its use by customers in appropriate locations and circumstances. The latter arrangement, although a bigger step for suppliers and service providers, had important benefits. Firstly, it could enable the service provider to defray the initial cost of manufacturing the component, the longer it was used. Secondly, it was more likely to ensure that the performance and value of the component could always be retained (i.e. guaranteed) at its optimum level, as specified and contracted in a 'Service Level Agreement', by regular maintenance and upgrading (Circle Economy 2018). Thus, customers for advanced services could be more confident of the quality and performance of the reused component. Such innovative business models can potentially shift from one-off transactions to a continuing, longer- term partnership. For manufacturers, this creates opportunities to provide an increased suite of services to their customers, with increased profit pools. Meanwhile, customers may benefit from ongoing maintenance and performance support accompanying the products over the term of the contract. In addition to facilitating product recovery and reuse, opportunities also arise for new offerings and profit centres to provide innovative information management services. Installing RFID tags and managing RFID-enabled BIM and life-cycle data can be performed by manufacturers or by third party stakeholders in the supply chain.

With this in mind, the research team explored the various digitally-assisted options with a supplier of aluminium-framed glazing panels. The supplier saw advantages in being able to offer follow-up services attached to sale, such as cleaning and maintenance. However, the greatest opportunity for providing advanced services was with higher-value, modular façades – which could theoretically be disassembled, taken back and reused a number of times. Financial, insurance and other barriers to such a scheme would first need to be overcome, especially the need for additional investment to offset the initial acquisition of the glazed panels and the negative cash-flow at the beginning of the service contract. In this regard, the type of contract and financial arrangement devised for the Circular Phone (Circle Economy 2018) may offer a solution.

40.4.3 Discussion

By means of the case study, it has been demonstrated that a well-integrated system comprised of robust individual sub-systems such as Cloud, BIM, RFID and data security systems can guarantee the full functionality of the Cloud BIM based Cyber Physical System by establishing a seamless information and physical circularity of building components. The RFID-BIM-Cloud platform can serve as an information repository and exchange platform, across disciplines and stakeholders, to share and identify the information of reusable building components via utilizing IFC data format, regardless of proprietary BIM software or other system employed by end-users.

The mechanism is aimed initially at new buildings, although it could be extended to existing structures in future. New building systems could be tagged by the manufacturer before delivery to site, with basic data related to the physical dimensions and other information. Enhanced lifetime data could then be added as needed or when re-allocation of ownership occurs. The original tag or "material birth certificate" (akin to the "material passport" concept), if installed by the manufacturer, would improve familiarity and ensure consistency of tag placement, as it would form part of the systematic approach employed as part of the fabrication process.

As the essence of the Cloud-based information exchange system is the 'Universal Accessibility', the security of data related to building components needs to be carefully planned and managed. More importantly, the integrated cyber physical system can only be established properly with a team of multi-disciplinary subject matter experts based on both technical and intellectual aspects, until the system is effectively self-sustaining.

Whilst the cyber-physical exchange system was an important breakthrough towards reuse of building components, its integration with PSS enabled its full capabilities to be delivered. Reusable components could be managed at their optimum performance over their extended life, keeping them in closed loops for a much longer period.

40.5 Conclusions and Implications

Whilst the research has demonstrated "proof of concept" via a prototype case study and "mock- up" Cloud platform, further research is required to implement the platform and ensure the findings are translated into practice. This is expected to eventuate in a widely adopted, universal and "self- populating" form of eBay, where construction resources are automatically and potentially listed for possible reuse, coupled with their provision as a service and life cycle stewardship.

This phase of the research, however, has confirmed that the Cloud-based Cyber-Physical System, coupled with PSS, is capable of enhancing the circularity of building components, by serving as a platform for capturing, restoring, updating and managing their lifecycle information within the Cloud system. This finding is believed to an important transformational innovation that has the potential, after further development, to greatly increase reuse and life-cycle stewardship within the building industry.

Using the system, it is expected that designers will be able to assess the condition, reusability, performance and carbon implications of components when considering their suitability for reuse. With the inevitable advent of carbon pricing for new building materials and components, the reuse of existing resources – which makes use of "sunk carbon" – is likely to be more financially viable. Reuse will only generate carbon emissions associated with transport, rather than via full resource extraction, manufacturing and production processes – as was shown by a previous case study of structural steel components (Ness et al. 2015).

The successful implementation of the new paradigm can be facilitated by greater adoption of modularization, prefabrication and design for disassembly (Kieran and Timberlake 2004), coupled with increased recognition of the value of building resources, and supportive government policies.

Further investigation is required on the legal responsibilities associated with reliability of data and components procured via the Cloud platform. However, it is expected that responsibility should rest with those who upload data and manage components over their life, whilst the providers of the platform should exercise responsibility for oversight and processes.

Thus, this research is expected to serve as a stepping stone in the era of Industry 4.0 and illuminate a more sophisticated way to manage built assets based on circular economy principles.

Acknowledgements The assistance of nRAH and Spotless Facility Management is greatly acknowledged, in particular Mr Chris Penn (Deputy Chair of buildingSMART Australasia).

Mr Marc Kovacic of Construction Glazing, Australia, kindly assisted with advice on providing framed glazed systems as a service, and is testing procurement scenarios.

The advice and assistance of Prof Walter R. Stahel, recognised as the founder of the CE concept, is much appreciated.The research was funded under the ARUP Global Research Challenge 2017.

References

ARUP (2016) Circular economy in the built environment. ARUP, London. September
ARUP, BAM and CE100 (2017) Circular business models for the built environment, ARUP and Royal BAM Group
Baker-Brown D (2017) The re-use atlas. RIBA Enterprises Ltd, London
Brand S (1995) How buildings learn: what happens after they're built? Penguin
Circle Economy (2018) The circular phone. Circle Economy, Netherlands, 8 Jan.
Deloitte and Utrecht University (2017) Breaking the barriers to a circular economy
EMF (2015) Delivering the circular economy: a toolkit for policymakers, Ellen MacArthur Foundation, June

Geldermans RJ (2016) Design for change and circularity – accommodating circular material and product flows in construction. Energy Procedia 96:301–311

Gordon A (1972) Designing for survival: the president introduces his long life/loose fit/low energy study. R Inst Br Archit J 79(9):374–376

Government of the Netherlands (2016) A circular economy in the Netherlands by 2050, Ministry of Infrastructure and the Environment and Ministry of Economic Affairs, 14 Sept

Habraken NJ (2003) Open building as a condition for industrial construction, Automation and Robotics in Construction

HM Treasury (2013) Infrastructure carbon review. London, November

ISO (2015) Building construction – organisation of information about construction works – part 2. Framework for classification, ISO 12006-2: 2015

Kieran S, Timberlake J (2004) Refabricating architecture: how manufacturing methodologies are poised to transform building construction. McGraw-Hill, New York

Kim KP, Park KS (2018) Housing information modelling for BIM-embedded housing refurbishment. J Facil Manag 16(3):299–314

Klein T (2017) Towards circular building construction. In: Positions on circularity in the built environment (BauHow5), Department of Architecture, Technical University of Munich, 13–15 March

Ness D, Field M (2003) Cradle to cradle carpets and cities. In: Proceedings of SASBE 03, Brisbane

Ness D, Pullen S (2006) Decoupling resource consumption from growth: new business model towards a sustainable built environment in China. In: Proceedings of SASBE 06, Shanghai

Ness D, Xing K (2017) Toward a resource efficient built environment: a literature review and conceptual model. J Ind Ecol 21:572–592

Ness D, Swift J, Ranasinghe D, Xing K, Soebarto V (2015) Smart steel: new paradigms for the reuse of steel enabled by digital tracking and modelling. J Clean Prod 98:292–303

Swift J, Ness D, Chileshe N, Xing K, Gelder J (2015) Enabling the reuse of building components: a dialogue between the virtual and physical worlds. In: Proceedings of unmaking waste 2015 conference, University of South Australia, 22–24 May, pp 252–260

Swift J, Ness D, Kim K, Gelder J, Jenkins A, Xing K (2017) Towards adaptable and reusable building elements: harnessing the versatility of the construction database through seamless RFID and BIM data transfer. In: UIA 2017 Seoul World Architects Conference, 3–10 September

Yashiro T (2000) Construction sectors as 'service providers': alternative business model for sustainable construction. In: Open building Tokyo 2000: continuous customisation in housing. Architectural Institute of Japan, Tokyo

Yashiro T (2004) Regeneration of small/medium size building stock by flexible re-fitting system: case illustration of 'conversion' in Japan. In OECD/IEA Workshop on sustainable buildings: towards sustainable use of building stock, 15–16 January, Tokyo

Zuidema R (2015) Open building as the basis for circular economy buildings. ETH Zurich, Zurich

Chapter 41
Producing Work-Ready Graduate for the Construction Industry

Sadegh Aliakbarlou, Suzanne Wilkinson, Seosamh B. Costello, Hyounseung Jang, and Hamid Aliakbarlou

Abstract While educating building professionals is an essential component of the construction education sector, producing work-ready graduates to meet the construction industry requires greater effort. The factors that influence graduate students' successful transition into the competitive construction environment need to be further explored. This study investigates these factors through a literature review and interviews six experts in academia who have industry experience. The findings of this study show that factors such as 'team-working skills' and 'communication skills' are the two top factors for producing work ready graduates for the construction industry. The findings can help academia to understand the important factors for producing work-ready graduates for the construction industry.

Keywords Construction · Employability · Work-ready graduates

S. Aliakbarlou (✉)
Departement of Building and Construction Services, Unitec, Auckland, New Zealand
e-mail: saliakbarlou@unitec.ac.nz

S. Wilkinson · S. B. Costello
Department of Civil and Environmental Engineering, University of Auckland, Auckland, New Zealand
e-mail: s.wilkinson@auckland.ac.nz; s.costello@auckland.ac.nz

H. Jang
School of Architecture, Seoul National University of Science & Technology, Seoul, South Korea
e-mail: jang@seoultech.ac.kr

H. Aliakbarlou
Department of Civil Engineering, Amirkabir University of Technology, Tehran, Iran

© Springer Nature Switzerland AG 2020
R. Roggema, A. Roggema (eds.), *Smart and Sustainable Cities and Buildings*,
https://doi.org/10.1007/978-3-030-37635-2_41

41.1 Introduction

Graduate employability depends on the quality of education. As defined by Brynteson (2012), employability is about skills that can help graduates to find proper employment, in accordance with their experiences and knowledge. Carroll (2011) stated that employability of graduates can be considered as a remarkable benchmark to measure the quality of higher education.

Achieving excellent graduate results depends on producing employable graduates. This is due to the fact that individuals select their tertiary education in order to achieve the skills they need for a rewarding work life. They select what to study because of "what they are capable of and what will get them started on a career" (New Zealand Ministry of Education 2013). Hence, educational service providers are required by the governments to improve the employability agenda across their organisations, and many of them have attempted to meet this challenge, with varying degrees of success (Culkin and Mallick 2011).

The response by the educational service providers is not sufficient, as there is a need for a more radical approach that can help in empowering new graduates with the knowledge and skills required for setting up a business rather than focusing on simply 'getting a job' (Culkin and Mallick 2011). Similarly, Herrmann et al. (2008) stated that graduates, in addition to achieving work skills, are required to develop skills that help them to take risks and opportunities, work flexibly, handle complexity, and achieve team-work skills and awareness.

This paper is structured as follows: first, the significance of this study is presented by reviewing the available literature. Next, key success factors for producing work ready graduates, as the basis for conducting a qualitative study, are identified. Following this, the qualitative findings of this study are presented and they are also compared with the literature findings. Finally, the study's limitations and suggestions for future research are presented.

41.2 Significance of the Study

A report by Lambert (2003) provides the UK employer's perspective on the skills market that educational service providers need to consider in their education curricula. This view was revised and supported by the Confederation of British Industry (2009), which developed Figs. 41.1 and 41.2 to show the role and responsibility that universities need to consider in providing employability skills. As can be seen from the both figures, educational services providers have a significant responsibility for helping students to achieve employability skills.

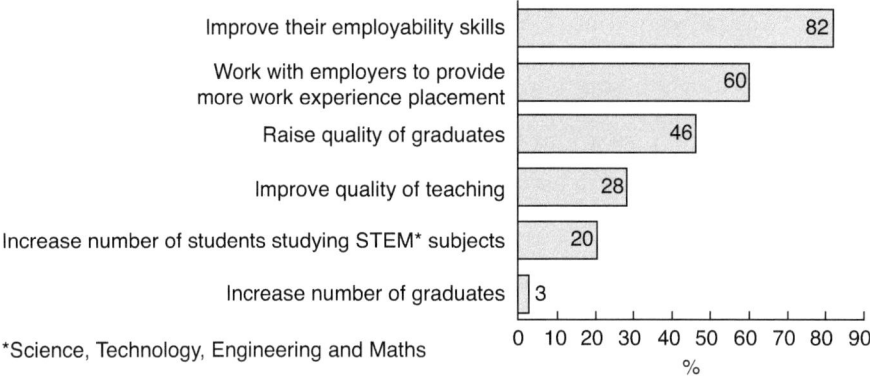

Fig. 41.1 Universities priorities in terms of undergraduates. (Source: Confederation of British Industry 2009)

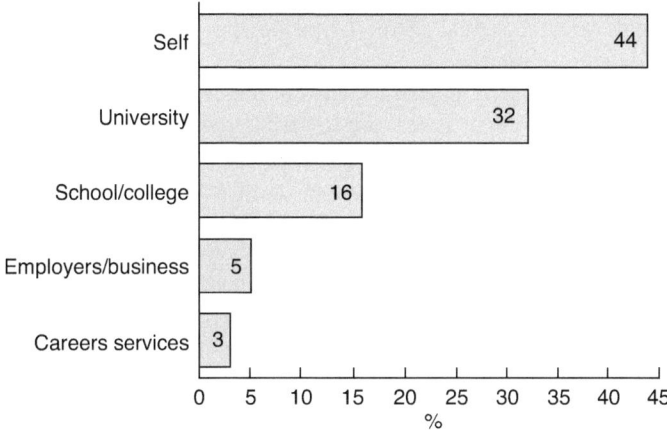

Fig. 41.2 Responsibility for helping students to achieve employability skills. (Source: Confederation of British Industry 2009)

41.3   The Economic Aspects of Graduates' Employability Skills

Universities in New Zealand are large scale institutions which form an important backbone to the country's economy. For example, they employ around 20,000 full-time staff, have a turnover of more than a $3.8 billion annually, and they spend about $960 million on research. They teach around 174,000 at any time and graduate around 43,000 students per year (Universities New Zealand 2018).

New Zealand universities are not only a growth driver for the New Zealand economy, they also play key roles within the construction industry in the country. The New Zealand construction industry is a small but competitive construction market and very important for the overall economy of the country (Construction Strategy XE "Strategies" Group 2015). For example, it contributed 8% to GDP and employed 10% of the workforce in 2015 (PWC 2016). Despite the importance of the construction sector for New Zealand, its performance is still low in comparison with some other comparable sectors of the New Zealand economy (PWC 2016), and other similar economies and countries (Constructing Excellence 2009). According to Building and Economics Research Limited (2003), a 10% improvement in New Zealand construction performance would increase the country's GDP by about $2 billion. To help achieve this, producing work ready graduates can be considered as an effective approach. Hence, the responsibility for the employability of graduates has been imposed on the government's respective higher education systems. This is aligned with the Theory of Human Capital introduced by Becker (1975), which states that governments are responsible for improving the stock of human capital, as an essential factor for achieving success and performance of knowledge-based economies.

Exploring the earnings of graduates helps in understanding the contribution that the tertiary education system is making towards producing work ready graduates. The key findings of a report by the New Zealand Ministry of Education (2013) entitled "looking at the employment outcomes of tertiary education" shows that: "earnings increase with the level of qualification completed; employment rates increase with the level of qualification gained; very few young people who complete a qualification at diploma level or above are on a benefit in the first five years after study; earnings vary considerably by field of study; some qualification types and some fields are associated with high rates of further study; and graduate certificate and diploma graduates have very high employment rates".

41.4 Key Success Factors for Producing Work Ready Graduates

A study entitled 'employers' perceptions of graduate employability' by Eurobarometer (2010) provides insights into the needs of graduate recruiters by assessing the perceptions of employers. In conducting the study, 11 factors such as team-working skills, sector-specific skills, communication skills, computer skills, ability to adapt to and act in new situations, good reading/weriting skills, analytical and problem-solving skills, planning and organisational skills, decision- making skills, good with numbers, and foreign language skills were considered as key success factors for producing work ready graduates. These factors can be determined as instrumental factors which can be considered as a means to an end (Aliakbarlou et al. 2017b), where the end is about achieving a high level of graduate employability. Based on these 11 factors, this paper reports a pilot study which applies the expert interviews method to prioritise the key success factors for producing work ready graduates in New Zealand.

41.5 Research Methodology

This study takes a social constructionist stance where a reality is constructed by individuals' activities (Aliakbarlou et al. 2017a). A qualitative research methodology, as used in this study, suggests an in-depth approach to exploring experts' insights about a research question (Aliakbarlou et al. 2018; Mbachu and Nkado 2004). In the first step, a literature review was conducted to define the main factors affecting students' employability, as well as to develop a basis to prioritise the key success factors from the identified factors based on New Zealand experts' perspectives. In the second step, by focussing on educational organisations as detailed in Table 41.1, it was decided to conduct expert interviews with individuals who are experts in the particular field that is being researched (Sugar and Schwen 1995). Hence, selection of the research participants from the educational organisations (10 interviewees)' was based on their experience in a university, institute of technology and polytechnic as well as their construction industry experience. Initially, ten organisations were contacted using the researcher's own connections. However, due to, individual availability within the research time frame, six interviewees from five organisations participated in interviews. Selection of the interviewees was based on their knowledge and experience in construction management in New Zealand, Australia and Iran. Table 41.2 represents the interview participants' details. As Table 41.2 shows, the participants were senior within their organisations. This helped them to understand what value meant to them from an organisational perspective (Aliakbarlou et al. 2017c). In addition, they had extensive experience and clear ideas about construction employers' recruiting preferences as they had played various roles in construction projects.

Table 41.1 Organisations' description

Code	Educational organisation type	Country
P1	Public – Institute of technology and polytechnics	New Zealand
P2	Private – Institute of technology and polytechnics	New Zealand
P3	Private – University	New Zealand
P4-P5	Private – University	Australia
P6	Public – University	Iran

Table 41.2 Interview participants' profiles

Field of expertise	Code	Years of experience
Construction industry experience	P1	30
	P2	10
	P3	10
	P4	8
	P5	20
	P6	5
Educational experience	P1	10
	P2	20
	P3	20
	P4	15
	P5	10
	P6	10

Table 41.3 Factors influencing graduates' successful transition into construction market

Factors	Rank (this study)	Rank (Europe)
Team-working skills	1	1
Sector-specific skills	6	2
Communication skills	2	3
Computer skills	5	4
Ability to adapt to and act in new situations	4	5
Good reading/writing skills	3	6
Analytical and problem-solving skills	7	7
Planning and organisational skills	11	8
Decision-making skills	10	9
Good with numbers	9	10
Foreign language skills	8	11

41.5.1 Types of Questions and Structure to Address the Objectives of the Research

The participants were briefed on the research objectives. In so doing, the list of 11 identified factors (as outlined in Table 41.3) were explained to the participants. They were then asked to prioritise the 11 factors by assigning a range of scores from 1 to 11 to each factor. This method was selected because the data was collected by conducting interviews that allowed appropriate time for the participants to review the factors and to rank them. The interviews helped to achieve insights into participants' experiences (Aliakbarlou et al. 2017d). By conducting the interviews in this study some underlying contextual information was gained regarding graduate employability. Although the number of interviewees was limited, the interviewees were expert in education. Also, the sample of interviewees represents a range of experts with considerable expertise in construction projects. Conducting interviews with experts in their particular field of expertise is recognised as an appropriate method to justify a research finding (Egbelakin et al. 2015). After obtaining the scores for each factor, by calculating the average for the scores given to each factor, the ranks, as shown in Table 41.3, have been developed based on the average score for each factor.

In addition, the participants were asked to provide explanations as to how the employers perceive value from each of the 11 factors. However, this paper only represents the identified ranks for the factors.

41.6 Findings

A study by Ray et al. (2012), indicated that there is a significant skills gap between graduates and entry-level requirements. "There is an issue with education systems that fail to produce future workers with the kinds of skills required by today's organisations – let alone those of tomorrow" (Ray et al. 2012). In what follows, the key success factors for producing work ready graduates are addressed.

41.6.1 Prioritising Key Success Factors for Producing Work Ready Graduates

Table 41.3 lists the key success factors in rank order, as determined by this study. Reviewing the interview results (with 3 experts from New Zealand), factors such as 'team-working skills', 'communication skills', and 'good reading/writing skills' were shown to be the 3 most important factors for producing work ready graduates for the New Zealand construction industry. Similarly, these 3 factors were recognised as the 3 most important factors by the Australian experts in this study.

The participant from Iran selected 'team-working skills', 'communication skills' and 'computer skills' as the 3 most important factors.

41.6.2 Comparison Between Studies

Table 41.3 shows how the perception of graduate employability in Europe differs from the participants of this study.

41.6.2.1 Similarities Between Studies

Team-working skills and communication skills were identified among the most important factors by both studies. 'Team-working skills' was ranked first by both studies. 'Communication skills' was ranked third and second by the European and this study's participants, respectively. In addition, 'computer skills' and 'ability to adapt to and act in new situations' were regarded as important factors by both sets of respondents, since they were ranked (fourth and fifth) and (fifth and fourth) by European and this study's participants, respectively. 'Analytical and problem- solving skills' was ranked seventh by both sets of respondents. Finally, 'decision-making skills' and 'good with numbers' were ranked (ninth and tenth) and (tenth and ninth) by European and this study's participants, respectively.

41.6.2.2 Differences Between Studies

'Sector-specific skills' was ranked as the second most important factor by the European participants while it was ranked sixth in this study. Also, 'planning and organisational skills' was ranked as the eighth most important factor by the European participants while it was ranked as the eleventh most important factor in this study.

Finally, while communication skills are ranked highly important for both sets of respondents, there is not strong agreement among them regarding 'good reading/ writing skills' as well as "foreign language skills'. 'Good reading/writing skills' and 'foreign language skills' were ranked (sixth and eleventh) and (third and eighth) by European and this study's participants, respectively.

41.6.3 Implications of the Findings

The study's findings help in better understanding the factors that influence graduates' successful transistion into the construction industry. While the findings highlighted factors necessary to achieve successful results for the students and educational services providers, they can also help employers to recruit new graduates who have the skills and knowledge that their organisations need to succeed (Finch et al. 2013). This is also critical for the growth of the constructing industry which is a knowledge-based economy.

Developing an educational strategy that recognises, prioritises and satisfactorily delivers these factors is essential to ensure that the highest levels of student employability are achieved. For example, as 'team-working skills' and 'communication skills' are the two key factors for producing work ready graduates, promoting collaborative learning cultures gains importance. However, individual assignments and activities are preferred to team assignments and activities by students (Raidal and Volet 2009), due to, for example, poorly designed assignments and activities with lack of support from their supervisor, unequal same contributions from team members working on the same assignment, and lack of having an appropriate space (classroom) to conduct a meeting and group activities (Scager et al. 2016). To help this improve, using a collaborative space (instead of a traditional classroom) and a team project task that lends itself to both collaboration and distinct contributions of students is required. This helps students to develop the capabilities and attitudes that are required for their employers. For example, in addition to the two aforementioned factors, achieving 'higher-level reasoning skills', 'accurate and creative problem-solving skills', 'willingness to take on difficult tasks', 'intrinsic motivation', and 'ability to transfer learning from one situation to another' are some other key benefits of collaborative learning for students (Johnson et al. 2014), which highlights the significance of collaborative learning for the employability of students (Hayward and Horvath 2000).

41.7   Conclusion and Further Research

This study considers employers' views of the factors that are important for employ-ability of graduates. Based on the qualitative findings of this study, 11 employability factors were prioritised based on employers' perspectives. The findings contribute to the existing literature on employability of new graduates. In addition, the study examined how employers' perspectives on significance of the identified factors differ between European's employers and this study's participants.

The findings of this study show that factors such as 'team-working skills' and 'communication skills' are the two most important factors for producing work ready graduates for the New Zealand construction industry. The findings can be used to improve the employability of new graduates, and can be considered by educational services providers when reflecting on employers' recruiting preferences.

By understanding the significance that employers place on key factors for achiev-ing graduate employability, educational service providers can develop curricula based on the development and improvement of important skills that employers look for. Also, it helps graduates to position themselves in the marketplace by highlighting the required factors when they are applying for rewarding work.

An important limitation is related to the scope of the study. The identification of the key success factors for producing work ready graduates in this study is limited, based on the findings of literature and expert interviews. Although the literature review drew on a recognised international source, further interviews and surveys may find that other factors should be added to the list of the factors developed in this study. In addition, this study's findings are subject to the limitations of interview based research. Also, the number of interviewees is small and may not be represen-tative of the views of the wider industry.

Conducting a similar study in other countries is recommended, to determine the adaptability and replicability of the research results in other contexts. Future studies may also identify measurements and benchmarks for the factors in this study. In so doing, the identified factors (and their measurements) can be categorised under 'absolute measures' and 'relative measures' or 'objective measures' and 'subjective measures'. This will help in practical application of this study's findings and bring the graduates closer to the construction industry needs and requirements.

References

Aliakbarlou S, Wilkinson S, Costello SB (2017a) Exploring construction client values and qualities: are these two distinct concepts in construction studies? Built Environ Project Asset Manag J 7 (3):234–252

Aliakbarlou S, Wilkinson S, Costello SB, Jang H (2017b) Achieving Postdisaster reconstruction success based on satisfactory delivery of client values within contractors' services. J Manag Eng 34(2):04017058

Aliakbarlou S, Wilkinson S, Costello SB, Jang H (2017c) Client values within post disaster reconstruction contracting services. Disaster Prev Manag 26(3):348–360

Aliakbarlou S, Wilkinson S, Costello SB, Jang H (2017d) Conceptual client value index for post disaster reconstruction contracting services. KSCE J Civ Eng 22(4):1067–1076. https://doi.org/10.1007/s12205-017-0432-1

Aliakbarlou S, Wilkinson S, Costello Seosamh B (2018) Rethinking client value within construction contracting services. Int J Manag Projects Bus 11:1007–1025. https://doi.org/10.1108/IJMPB-07-2017-0076

Becker G (1975) Human Capital. Chicago University Press, Chicago

Brynteson R (2012) Innovation at work: 55 activities to spark your team's creativity. Amacom Books, New York

Building and Economics Research Limited (2003) Assessment of the economic impact of efficiency improvements in building and construction. Retrieved 1 Sep 2016, from http://www.branz.co.nz/cms_show_download.php?id=9355bd8b85c79bb5337bcfbe26c9952ae9f5e95b

Carroll C (2011) Accessing the graduate labour market: assessing the employability of Irish non-traditional graduates of trinity college, Dublin. Widening Participation Lifelong Learn 13 (2):86–104

Confederation of British Industry (2009) Future fit. Preparing graduates for the world of work, CBI London

Constructing Excellence (2009) DBH building & construction taskforce: procurement working group. Retrieved 1 Sep 2016, from http://www.constructing.co.nz/files/file/114/Report%20for%20DBH%20Procurement%20Working%20Group%20220109%20Issue%201.0.pdf

Construction Strategy Group (2015) The NZ industry. Retrieved 1 March 2015, from http://www.constructionstrategygroup.org.nz/industry.php

Culkin N, Mallick S (2011) Producing work-ready graduates: the role of the entrepreneurial university. Int J Mark Res 53(3):347–368

Egbelakin T, Wilkinson S, Ingham J (2015) Integrated framework for enhancing earthquake risk mitigation decisions. Int J Constr Supply Chain Manag 5(2):34–51

Eurobarometer F (2010) Employers' perception of graduate employability. Available online: fl_304

Finch DJ, Hamilton LK, Baldwin R, Zehner M (2013) An exploratory study of factors affecting undergraduate employability. Educ Train 55(7):681–704

Hayward G, Horvath P (2000) The effect of cooperative education on occupational beliefs. J Cooperative Educ 35(1):7

Herrmann K, Hannon P, Cox J, Ternouth P, Crowley T (2008) Developing entrepreneurial graduates: putting entrepreneurship at the centre of higher education. NESTA London, London

Johnson DW, Johnson RT, Smith KA (2014) Cooperative learning: improving university instruction by basing practice on validated theory. J Excell Univ Teach 25(4):1–26

Lambert R (2003) Lambert review of business-university collaboration

Mbachu J, Nkado R (2004) Reducing building construction costs; the views of consultants and contractors. In: Paper presented at the proceedings of the international construction research conference of the royal institution of chartered surveyors, Leeds Metropolitan University

New Zealand Ministry of Education (2013) Looking at the employment outcomes of tertiary education. Retrieved 1 June 2018, from https://www.educationcounts.govt.nz/__data/assets/pdf_file/0020/143561/Looking-at-the-employment-outcomes-of-tertiary-education-ii.pdf

PWC (2016) Valuing the role of construction in the New Zealand economy. Retrieved 1 Sep 2016, from http://infrastructure.org.nz/resources/Documents/Reports/CSG%20PwC%20Value%20of%20Construction%20Sector_final%20report_2016_10_16.pdf

Raidal SL, Volet SE (2009) Preclinical students' predispositions towards social forms of instruction and self-directed learning: a challenge for the development of autonomous and collaborative learners. High Educ 57(5):577–596

Ray R, Mitchell C, Abel AL (2012) The state of human capital 2012: false summit: Why the human capital function still has far to go, research report no. R-1501-12-RR, McKinsey and Company

Scager K, Boonstra J, Peeters T, Vulperhorst J, Wiegant F (2016) Collaborative learning in higher education: evoking positive interdependence. CBE—Life Sci Educ 15(4):ar69

Sugar W, Schwen T (1995) Glossary of technical terms. In: Kelly LE (ed) ASTD technical and skills training handbook. McGraw-Hill, New York

Universities New Zealand (2018) Growing New Zealand's economy. Retrieved 1 June 2018, from https://www.universitiesnz.ac.nz/sector-research/growing-new-zealand-economy

Part XII
Performance

How do buildings perform? Their energy use, and efficiency, the light performance in high rise buildings are main discussion points in this theme.

Chapter 42
Tower Blocks in Different Configurations – Aspects of Daylight and View

Bengt Å. Sundborg, Barbara Szybinska Matusiak, and Shabnam Arbab

Abstract Groupings of buildings can be made in different regular geometric patterns or in more irregular arrangements. In the PLEA-study "Urban Form, Density and Solar Potential", (Cheng et al. Urban form, density and solar potential. In: PLEA2006 – the 23rd conference on passive and low energy architecture, Geneva, Switzerland, 2006) tested some alternatives with uniform/random heights and patterns. The research team concluded with pointing out the advantages of randomly positioned buildings compared to repetitive patterns with respect to daylight access and solar potential. In our opinion, the conclusions about random layouts should be interpreted in terms of specific variations. This suggests that there may be strategies for patterns and heights – not simply random arrangements – which this new study clearly confirms. Although the authors of the 2006 study underlined the need for further studies, as far as we know, no research of building groupings has been done.

In this research, a series of geometrical patterns of tower blocks was developed to examine daylight conditions. Some are already used in practice while others seemed to be very promising. The choice of evaluation criteria was based on the discourse in the scientific community on daylighting and on practical experience in urban planning. The view was also included as in the new EU standard. The study is carried out for an assumed FAR (Floor Area Ratio) of 1,12 with buildings of seven floors. Advanced computerbased daylighting simulations and calculations of view parameters have been done for seven different designs of building groupings of equal density. All seven groupings have good daylight conditions with Vertical Sky Components over 40%. The six alternatives to the quadratic reference model have

B. Å. Sundborg (✉)
Research Institutes of Sweden, RISE, Stockholm, Sweden

Faculty of Architecture and Design, Department of Architecture and Technology, Norwegian University of Science and Technology (NTNU), Trondheim, Norway
e-mail: bengt@dtark.se

B. S. Matusiak · S. Arbab
Faculty of Architecture and Design, Department of Architecture and Technology, Norwegian University of Science and Technology (NTNU), Trondheim, Norway
e-mail: barbara.matusiak@ntnu.no; shabnam.arbab@ntnu.no

© Springer Nature Switzerland AG 2020   619
R. Roggema, A. Roggema (eds.), *Smart and Sustainable Cities and Buildings*,
https://doi.org/10.1007/978-3-030-37635-2_42

higher sunlight radiation on façades, especially on lower floors, due to their less perpendicular orientation to the surrounding blocks. The same alternatives have sightlines up to 3–7 times longer than in the reference model. These advantages depend on the oblique, triangular and scattered configurations as well as the different shapes of the ground floor area. The quadratic group is the most common pattern for tower blocks. Unfortunately, it also has the worst possibilities for view with a perpendicular view of 30 m compared to 50.7–93.3 m for the alternatives. Local conditions as well as technical requirements must – as always – influence layouts. However, the six alternatives can still produce tangible consequences thanks to considerations of daylight and view.

Keywords Daylight · Views · Urban Morphology · Tower Blocks

42.1   Introduction

Urban geometry can be created in many different ways. Mark DeKay worked on the evaluation of different forms of a city quarter to find out which has the greatest potential for daylight autonomy by asking: " What would the form of the city be like if we were to take seriously the provision of daylight to all buildings?" (2010). He developed different geometries of quarters and verified daylight availability. Following this, Peter Andreas Sattrup and Jakob Strømann-Andersen made energy simulations to calculate the total energy consumption and the daylight autonomy for six different building patterns (2013). Simulations examined how density ratio in urban blocks affects solar gains and daylight autonomy. Of all studied patterns the results were best for free standing tower blocks.

A previous study by Vicky Cheng et al. (2006) focused on differences between uniform groups of tower blocks (equal height and repetitive parallel patterns) and groups with variations in height and position. The research team pointed out the advantages of randomly positioned buildings of different height compared to uniform groups with respect to solar potential, but concluded there was a need for further research. The present study goes deeper into this topic and broadens the investigation to include views.

This paper presents a study of seven different groups of tower blocks which evaluates the solar radiation on façades and plots, the daylighting accessibility (vertical and standard sky factors) and the potential views. In this paper only a part of the study is presented, i.e. alternatives where all blocks are of equal height. The evaluations have been done with computer simulations. We will continue the tower block research and also comment on the present study in detail, especially with respect to strategies and "random" solutions. This ongoing project will also include other settlement configurations e.g. perimeter blocks (Fig. 42.1).

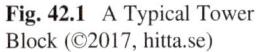
Fig. 42.1 A Typical Tower Block (©2017, hitta.se)

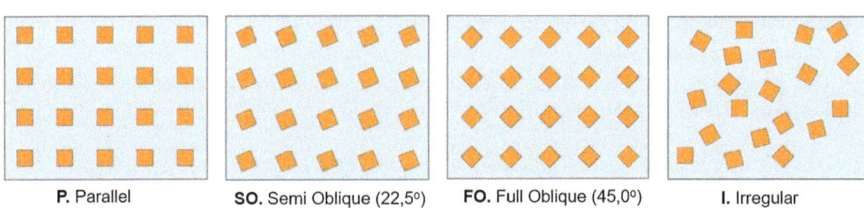

P. Parallel   SO. Semi Oblique (22,5º)   FO. Full Oblique (45,0º)   I. Irregular

Fig. 42.2 Four groupings of tower blocks

The basic geometries in the modelled alternatives are simplifications of common modern examples together with some new patterns. The patterns of similar buildings in existing urban settlements are usually grouped in one of two typical patterns, parallel, P or oblique, FO, see Fig. 42.2. Even in smaller groups such as 4–5 buildings along a street the same strategies are common. Regarding the interior and the exterior qualities for daylight and view based on practical observations this study confirms that the oblique pattern is the best. A third type, an irregular grouping, is used mainly in some low-density developments in order to adapt to the terrain.

The oblique positions can be fully rotated 45o, Full Oblique, from the parallel as well as to other angles as in the illustration below with half that rotation, SO. In a current report focusing on urban structures and energy, that pattern is also mentioned by Philipp Rode, et al. (2014). Here we classify the pattern as Semi Oblique. It is sometimes used on a small scale.

We regard six important parameters in the urban morphology of tower blocks. This ongoing study deals with all of them:

1. Block heights.
2. The shape of a single block, especially the plan form of the built ground area – the urban foot print, and the proportion between the height and the foot print dimensions.
3. Orientation of the single block in different compass directions (In Trio, alternative 4 in Sect. 42.2 below, every building in each group is orientated in a different direction).
4. The shape of the area for the whole group.
5. The patterns of the groups of the blocks.
6. Orientation of the groups of blocks in different compass directions.

42.2   The Alternatives

The developed alternatives represent both existing settlements and new proposals (4, 5, 6 and 7 are developed within this project). It is possible to mirror alternatives 2 and 5 to improve their adaptation to local conditions. The alternatives have been developed in Sketch Up- drawing, see Fig. 42.3. The intentions for each of the seven alternatives are stated below.

The blocks in the three *"pattern"* alternatives have variations according to the following:

1. QUADRATIC blocks are grouped in a quadratic pattern. This simple and very typical alternative is used as a reference.
2. SEMI-OBLIQUE. The Quadratic alternative with every block obliquely positioned – rotated 22,5°.
3. FULL-OBLIQUE. The Quadratic alternative with every block obliquely positioned – rotated 45,0°.

Both groups in the two *"courtyard"* alternatives are orientated around a courtyard:

4. TRIO. Blocks of three are grouped in a triangular pattern.
5. OBLIQUE-FOUR. Blocks of four are grouped in an "oblique pattern". This alternative is similar to TRIO but the buildings are orientated according to a rectangular grid, so it is easier to adapt them to a quadratic street grid.

The tower blocks in the two alternatives with different *"ground forms"* are arranged to develop better light and sight conditions than in the usual rectilinear layout:

6. HEXAGON. Hexagon-shaped blocks in a pattern of hexagonal plots. It is important that the blocks are rotated clockwise 30,0° in order to improve the sight and the light perpendicular to the windows.

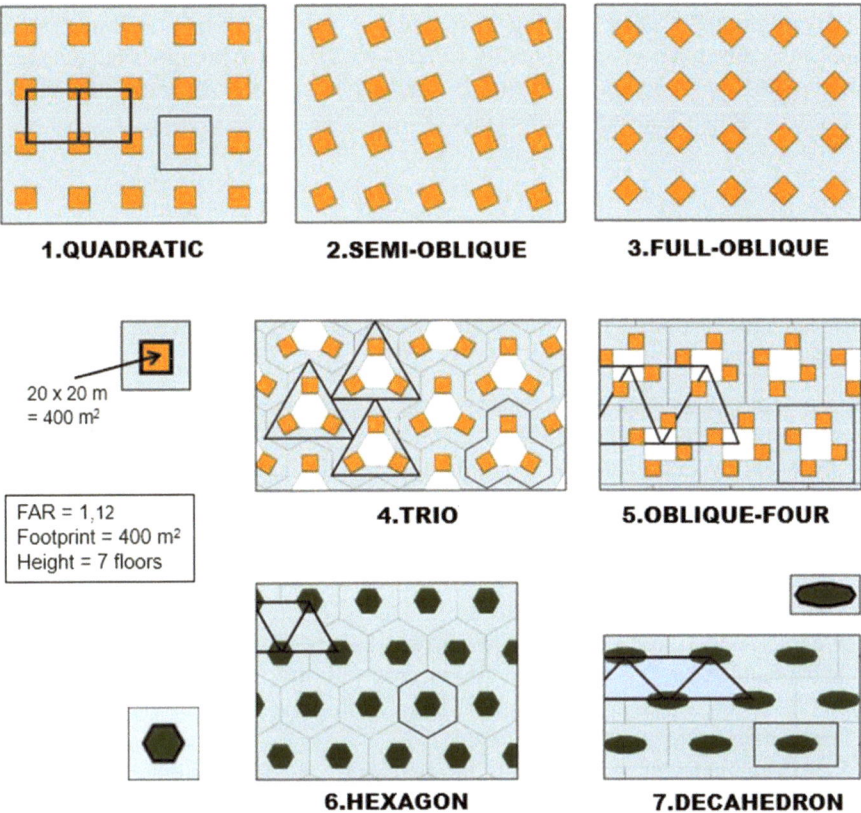

Fig. 42.3 The seven alternatives for groups of tower blocks

7. DECAHEDRON. Polygon-shaped blocks (ten façades) in a pattern of rectangular plots. All façades are orientated in order to improve the sightlines and the light perpendicular to the windows.

Modern tower blocks are often 20 m wide with four apartments (one in each corner) which are wider than previous blocks of around 18 m. This is because of new building regulations for disabled people and economic considerations. The following dimensions have been used in all alternatives (except the footprints in the alternative 6 and 7):

w = width = 20 m A_{fl} = floor area = w x w = 400 m^2 A_{bg} = building area (built area on the ground, foot print) = w x w = 400 m^2 n = number of floors = 7 hfl = 3 m (height of each floor) H = 21 m (total height) A_{pl} = plot area 50 x 50 m = 2500 m^2 FAR = floor area ratio = total floor area/plot area = n x A_{fl}/A_{pl} = 7 x 400/2500 = 1,12 *FAR = FSR (floor space ratio), FSI (floor space index), site ratio and plot ratio.*

Even groups of taller buildings (skyscrapers) and smaller ones (villas) demonstrate the same relative advantages for patterns with the same grouping and the same foot print. However, in this study, it was important to investigate the specific conditions for each group of tower blocks.

42.3   The Daylight in the Urban Areas

The daylight conditions in the seven alternatives, see Fig. 42.4, have been simulated in three ways:

- Sunlight radiation on façades and on plots, average value, during the 1st of May from sunrise to sunset (kWh/m^2).
- Vertical Sky Component on the façades (VSC), average of all façades;
  50% is the maximum value.
- Sky Component on the plot area (SC), average across the plot; 100% is the maximum value including visual access to the whole hemisphere.

All the simulations have been done with DIVA for Rhino, a well-recognized tool for climate based and static daylighting calculations. The DIVA (Design, Iterate, Validate and Adapt), a plug-in for Rhinoceros software, enables effective calculations of daylight metrics, e.g. daylight factor, using the Radiance/Daysim engine. Climate data for Stockholm was used.

By keeping the reflection factor of the block surfaces and the ground close to zero (0.01%) the daylight factor script in DIVA was used to calculate SC and VSC. All simulations have been executed by Postdoctoral Fellow Shabnam Arbab at the NTNU, Trondheim, Norway.

All seven alternatives have good daylight accessibility, e.g. the average VSC, Vertical Sky Components, is over 40% (27% is recognized as good enough and 50% is the maximum). The average SC, Sky Components, on plots is also high, see Fig. 42.4. HEXAGON has highest values both on façades (VSC) and on the plot (SC).

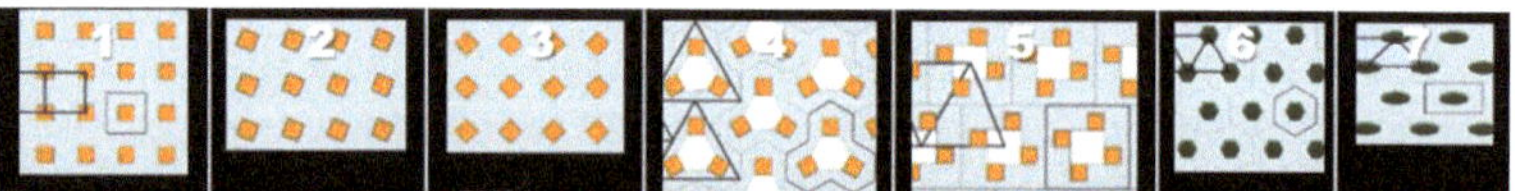

DAY-LIGHT	QUADRATIC	SEMI-OBLIQUE	FULL-OBLIQUE	TRIO	OBLIQUE-FOUR	HEXA-GON	DECA-HEDRON
Sunlight Radiation 1ST of May: Facades (kWh/m²)	2.94	2.96	3.04	3.04	3.00	3.06	2.95
Plots (kWh/m²)	3.88	3.86	3.84	3.83	3.81	3.95	3.85
Vertical Sky Component: Facades (%)	43.78	43.82	43.87	44.07	43.53	44.34	43.28
Sky Component: Plots (%)	76.72	76.61	77.53	75.12	76.61	78,19	76.52

Fig. 42.4 Daylight conditions in the alternatives

All the six alternatives to the QUADRATIC have slightly greater sunlight radiation on the façades due to the less perpendicular positions of the surrounding blocks. Regarding sunlight radiation on the plot areas, only HEXAGON has a higher value compared to the QUADRATIC. It is different to optimize for the buildings compared to optimizing for the plot. The exception, HEXAGON, has more daylight and an even better view depending on its form and a small rotation of the building (30.0°) from the original position on the plot.

The differences between average solar radiations on façades in the alternatives are not strong as very high values of top and middle floors dominate calculations of average values. However, deeper consideration of the details of the simulation results confirms that solar radiation on façades at the level of 1st floor (worst places on façades) on all alternatives is significantly higher than in QUADRATIC.

42.4   The Views in the Urban Areas

People's views in urban settlements consist of their visual experience of the outdoor environment. Qualitative details of the views of beautiful streets with well-designed outdoor furniture and decorative façades are important to that perception. The aspects of people's experiences in views from windows are many; see Matusiak & Klöckner (2015).

However, the crucial aspects to consider in the early stages of urban planning are the geometrical structures which limit both quantitative and qualitative opportunities of the fields of vision. Here we only consider simplified building alternatives.

Wide angle views of long distances give better opportunities for attractive views than short and narrow views even in detailed urban design. Large open spaces and strategically positioned openings between buildings are examples of how settlements can create such views. The measurements for the views described below, 1–4, are part of the criteria for the comparison of alternatives.

The views are multifaceted and include important aspects such as privacy and safety. It is also important to remember that there is a correlation between the assessment of daylight and the view quality as well as indoor privacy. The presence of people in the views is also appreciated in most cases even if exceptions exist such as noisy outdoor events.

Aesthetic values depend greatly on individual preferences which differ from person to person. However, general aspects of the aesthetics are important in urban design and are correlated with complexity, maintenance, age, composition of the view, etc. In a current EU-standard the diverse aspects of views are described as follows:

> View windows provide contact with the surrounding, information about orientation, weather changes and time over the day. A composition of a view, which includes layers of sky, city or landscape, and ground, could counteract fatiguing monotony and contribute to relief from the feeling of being closed in. All occupants should have the opportunity for the refreshment and relaxation afforded by a change of scene and focus. A natural view is preferred over a view towards man-made environment and a wide and distant view is appreciated more than a

narrow and near view. A diverse and dynamic view is more interesting than a monotonous view. View to the nature may have positive influence on people's sense of wellbeing, on job satisfaction, and recovery of surgical patients. *see* CEN/TC, Annex C (2018).

The view described as the length of the sightline in different directions is of primary interest. Some settlements have good possibilities for views and some do not despite the same density measured as floor area ratio. This is of importance especially in the planning of high-density settlements.

42.5   The Geometry of the Views Impacts the Privacy

Increasing urbanization including the replacement of low buildings with taller ones and infill with new buildings leads not only to darker indoor and outdoor environments. Privacy will also be reduced.

A good view consists of many attractive elements. In contrast to the visual connection between the parents and their children, the sight of a stranger often reduces privacy. However, privacy is difficult to achieve in dense urban settlements and requires intelligent geometry such as zigzag façades, star block houses and chamfered. The six alternative groupings of tower blocks give more privacy than the QUADRATIC group depending on longer sightlines and less central views to the surrounding tower blocks in the alternatives.

42.5.1   Obstructions in Views from the Windows

42.5.2   Two Different Views from the Windows (Fig. 42.5)

With the assumed measures of the window in the simulations it is possible to see 138° using different viewing positions. On both sides along the façade, up to 21°, there are absolute obstructions.

42.6   Viewing Conditions for the Alternatives

The following four ways describe the views in urban alternatives. All views are from one window in the building (calculated as an average of different positions along the façade):

1. $D_{average-138}$ Average distance to the surrounding buildings within 138° excluding 42° of the 180° of the sight which are obstructed by the frame of the window, see Fig. 42.6. $D_{average-54}$ Average distance to the surrounding buildings within 54° (m) – angle width for a central field view, (from CEN/TC 169).

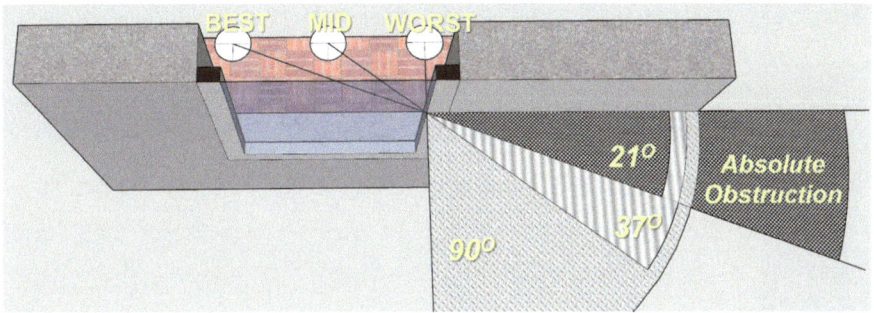

Fig. 42.5 Three different viewing positions have different possibilities for width of view

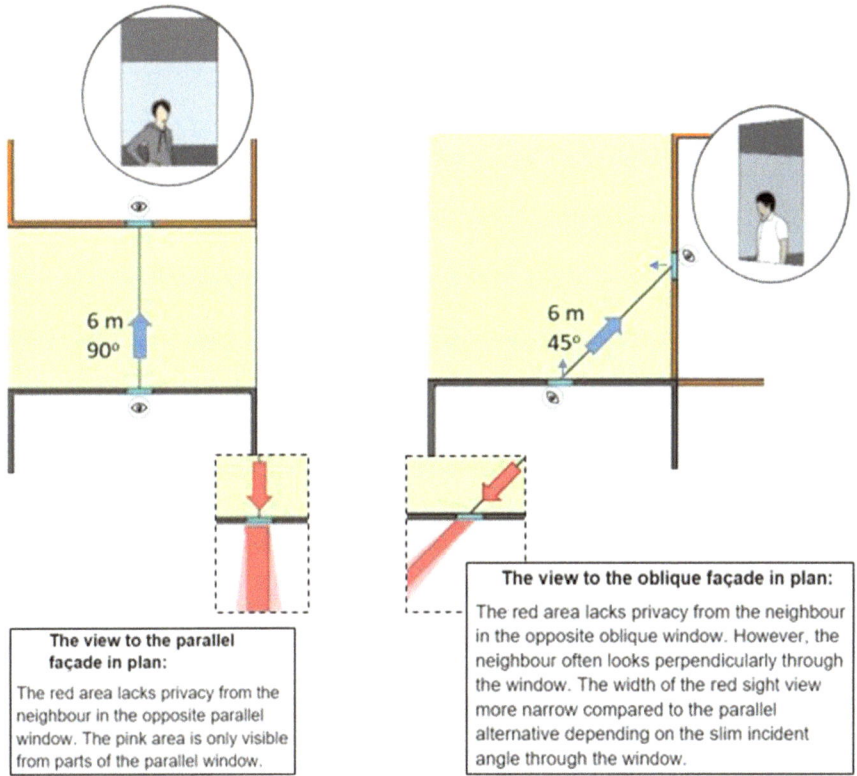

The view to the parallel façade in plan:

The red area lacks privacy from the neighbour in the opposite parallel window. The pink area is only visible from parts of the parallel window.

The view to the oblique façade in plan:

The red area lacks privacy from the neighbour in the opposite oblique window. However, the neighbour often looks perpendicularly through the window. The width of the red sight view more narrow compared to the parallel alternative depending on the slim incident angle through the window.

Fig. 42.6 The view to a parallel and an oblique façade

2. $D_{perpendicular}$ Perpendicular distance to a building (m) – measured along the perpendicular sight line to the window. Broad views out consist of that sightline.
3. D_{max} Maximal distance to a building within 138° (m). A long maximal distance is good for the view and for daylight distribution. **C centrality** angle for the maximal distance (°). The best is the perpendicular view from the window, 90°

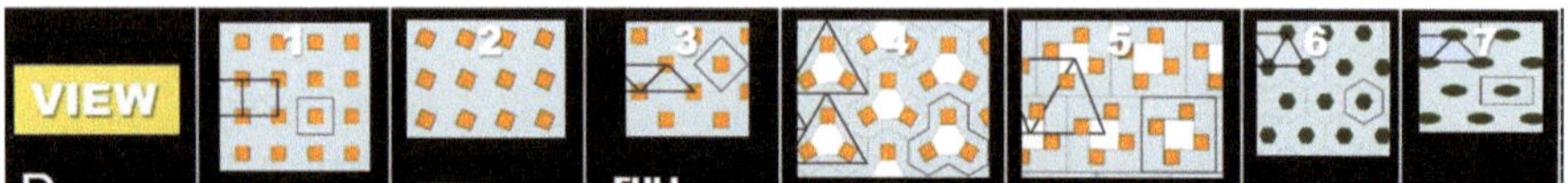

VIEW	QUADRATIC	SEMI-OBLIQUE	FULL-OBLIQUE	TRIO	OBLIQUE-FOUR	HEXA-GON	DECA-HEDRON
D$_{average}$ 138°	78,5	103,9	80,4	80,3	95,0	91,2	-
54°	75,7	88,5	87,6	96,8	90,2	119,7	-
D$_{perpendicular}$	30,0	90,8	50,7	73,0	90,0	71,5	93,3
D$_{max}$	166,3	299,6	192,8	172,6	610,7	228,0	274,2
C, Centrality 21-90° =Perpendicularity	51,5°	27,9°	54,4°	72,0°	33,7°	68,0°	83,2°
D$_{min}$	30,0	27,7	29,6	32,5	22,6	32,4	27,5
P, Periphery 0-69° =Obliqueness	0°	19,2°	58,8°	45,9°	26,6°	39,5°	11,5°

Fig. 42.7 View values in the alternatives

and worst is 21° which is the lower limit depending of the obstructing beam of the window.

4. **D**$_{min}$ Minimal distance to a building within 138° (m). For good privacy the distance should be further than 20 m according to practical guidelines. **Pperiphery**, The peripheral angle is the deviation in the minimal distance from the perpendicular normal (°). Oblique distances with very peripheral angles preserve privacy better and are limited to 69° depending on the obstructing frame of the window.

The views are calculated and analysed for the QUADRATIC group and the six alternatives, see Fig. 42.7. Summing up the view analysis:

- In alternative groups of tower blocks, the sightlines can be up to 3–7 times longer than the QUADRATIC and the interior privacy can be improved and secured.
- The QUADRATIC group has the worst possibilities for views with a perpendicular view of only 30 m compared to distances 50.7–93.3 m for the alternatives.
- In all the distance parameters, the results point in the same way except for some minimum distances which are up to 7.4 m lower than the quadratic value, 30 m. Those exceptions consist of more peripheral sightlines than the perpendicular view for the QUADRATIC so the difference in perception of privacy is not so big.
- The maximal distances are longer in all the alternatives to the QUADRATIC group – up to 610.7 m compared to 166.3 m.
- Each alternative has its own specific profile of view conditions. Depending on those differences in the alternatives, local needs and local conditions can often be satisfied.

- The secret to the advantages of the alternative groups is the different geometries of the oblique and triangular groups as well as the different configuration of the ground area of the blocks.

42.7 Conclusions

All seven models have good daylight conditions with average Vertical Sky Components on façades well above 40%. The average Sky Components on plots are also high in all alternatives, over 75%. HEXAGON scores highest values, both on façades and on the plot. The alternatives to the QUADRATIC have higher sunlight radiation, which is the result of less perpendicular positions of the surrounding blocks in the alternatives. This is especially significant at first floor level. The alternatives to the QUADRATIC group of tower blocks have up to 3–7 times longer sightlines. The special geometries give advantages to the alternative oblique, triangular and scattered groups as well as the different shapes of the ground area of the blocks. The QUADRATIC group is the most used pattern for tower blocks in town planning. Unfortunately, it also has the worst possibilities for offering views with a perpendicular view of 30 m compared to distances 50.7–93.3 m for the other alternatives. Local conditions as well as technical requirements must – as always – influence layouts. However, the alternative tower blocks can still produce tangible consequences thanks to considerations of daylight and view. After some practical applications, we will know better how big these impacts can be.

References

CEN/TC 169 17037 (2018) Daylight in buildings

Cheng V, Steemers K, Montavon M, Compagnon R (2006) Urban form, density and solar potential. In: PLEA2006 – The 23rd conference on passive and low energy architecture, Geneva, Switzerland

DeKay M (2010) Daylighting and urban form: an urban fabric of light. J Archit Plann Res 27:1. Locke Science Publishing Company, Inc. Chicago, IL, USA

Matusiak B, Klöckner C (2015) How we evaluate the view out through the window. Archit Sci Rev 59:203–211

Rode P, Keim C, Robazza G, Viejo P, Schofield J (2014) Cities and energy: urban morphology and heat energy demand. Environ Plann B Plann Des 41:138–162. LSE Cities and EIFER at Karlsruhe Institute of Technology

Sattrup PA, Strømann-Andersen J (2013) Building typologies in Northern European cities: daylight, solar access and building energy use. J Archit Plann Res 30(1):56

Chapter 43
Assessing the Lighting Performance of an Innovative Core Sunlighting System

Liliana O. Beltrán

Abstract This paper presents an efficient daylighting technology to improve the lighting conditions in deep interior spaces of multi-story buildings without the penalties of increased solar heat gains. A passive horizontal solar light pipe is proposed to efficiently redirect sunlight to distances between 5 m and 10 m from the building façade. Photometric measurements throughout a year show that the system can consistently provide more than 300 lx for 9 h (9:00 to 18:00) under clear and partly cloudy skies, and during 4 h under overcast sky conditions (exterior global horizontal illuminance, EXGH, over 18–20 klx). In addition, the illuminance values, over 1000 lx are achieved consistently between 10:30 and 16:30 under clear sky conditions. At the back of the space (7.6 m under the light pipe), the Useful Daylight Illuminance autonomous (UDI-a, 300–3000 lx), UDI autonomous for aging eyes (UDI-a AE, 600–3000 lx) and UDI-a Bright Daylight (BD, 1000–3000 lx) ranged between 60–88%, 44–64%, and 28–45% respectively. The light pipe introduces consistently illuminance levels ranging between 300 lx and 2500 lx throughout the year, saving energy during peak load electricity demand, and providing the daily bright light doses necessary to regulate and entrain building occupant's circadian rhythms.

This passive solar system proves to be an energy efficient sustainable technology that utilizes direct solar energy, and provides high illuminance levels of full-spectrum light without the negative effects of glare and solar heat gains that are found in buildings with large expanses of glass.

Keywords Daylighting · Sustainability · Energy-efficiency · Solar light pipes · Circadian light

L. O. Beltrán (✉)
Department of Architecture, Texas A&M University, College Station, TX, USA
e-mail: lbeltran@arch.tamu.edu

© Springer Nature Switzerland AG 2020   631
R. Roggema, A. Roggema (eds.), *Smart and Sustainable Cities and Buildings*,
https://doi.org/10.1007/978-3-030-37635-2_43

43.1 Introduction

It has been demonstrated that the use of daylighting in commercial office buildings is an effective strategy to offset electrical lighting, reduce cooling, and heating loads; as well as to increase human comfort and productivity (Heschong and Mahone 2003). This paper intends to demonstrate that this passive horizontal light pipe provides adequate light levels in building cores without introducing additional solar heat gains.

We live and work in buildings that are often isolated from natural light and where electric light is often around 200 lx and seldom exceeds 400–500 lx (Foster 2011). In recent years light pipes have been explored because of their potential to introduce daylight further into the building core. One of the first developments of a passive horizontal light pipe suitable for deep plan office buildings was developed by LBNL (Beltrán et al. 1994, 1997). The characteristics of the light pipe presented in this paper are based on the preliminary design concepts developed by the author at LBNL. Other researchers adapted the passive horizontal light pipe to locations at low latitudes (3°N, 14°N) (Chirarattananon et al. 2000; Garcia and Edmonds 2003), where the light pipes were oriented to face the sun towards the East or West limiting the light pipes' daylight performance. An anidolic (non-imaging) ceiling was developed to collect light rays from the sky and redirect the emitted light i n a6 m room. This system is suitable for locations with predominantly overcast skies (Courret et al. 1998). Recent developments include active light guiding systems that integrate electric lighting, as backup lighting, along with heliostats and tracking mirrors to redirect sunlight (Rosemann et al. 2007; Mossman et al. 2018).

43.2 Description

The light pipe system is designed to be placed within the ceiling plenum or can be exposed hanging from the underside of a floor slab above. The light pipe collector extends 0.25 m from the building facade plane, so that it could be used with flush and articulated facades. This system can be used in new buildings, and in existing buildings. The current light pipe was designed to be used in combination with a lower, sidelight window, and was tested in a room 6 m wide by 9.1 m deep. It provides supplementary illumination at distances between 4.6 m and 10 m.

Our proposed light pipe system was designed to introduce daylight passively in any floor of deep-plan multistory buildings. The light pipe uses a small inlet glazing area, 0.3 m by 1.6 m, to efficiently redirect sunlight at distances up to 10 m from the window wall (Fig. 43.1). The window wall ratio (WWR) and window floor ratio (WFR) of the light pipe are 4% and 1.6%, while the sidelight window has a WWR of 41% and a WFR of 16%. The glass area of the light pipe is less than one tenth of the sidelight window. The challenge of the design stems from the large variation in solar position and daylight availability throughout the day and year. Several reflectors are used to collimate incoming sunlight to minimize inter-reflections within the transport section of the light pipe, and to maximize the efficiency of the system (Beltrán et al.

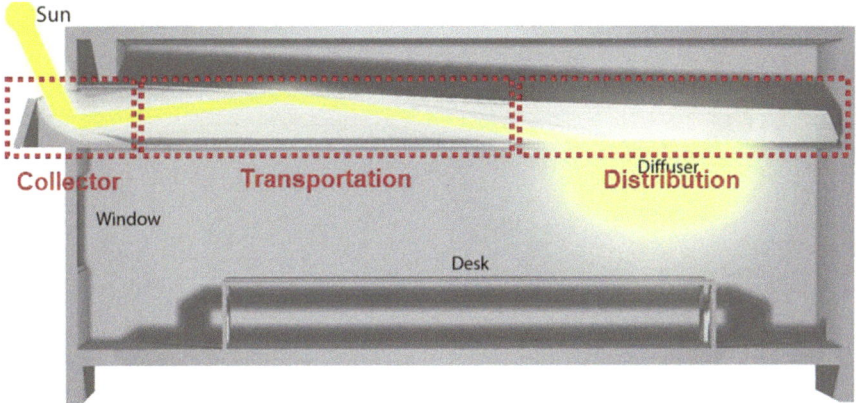

Fig. 43.1 Light pipe longitudinal section

1997). The current light pipe design has been redesigned to improve performance and adapted to our current location. The pipe is coated with a 99.3% specular reflective film. The distribution element at the back end of the light pipe consists of a 4.6 m long diffusing radial film located at the ceiling plane with an 87% visible transmittance (Tvis) and an area of 4.2 m^2.

The light pipe design was optimized by computer-assisted ray-tracing calculations that determined the optimum geometry of the various light-redirecting optical elements (Beltrán and Uppadhyaya 2008). Hourly sun rays were traced to verify that rays are directed along the light pipe shaft (Fig. 43.2). Efforts were focused on determining the optimum aperture size, reflectors size and shape, taking advantage of the optical properties of the films, and accommodating the sun path viewed by the window for a specific orientation and building latitude.

43.3   Methodology

43.3.1   Experimental Set-Up

The experimental facility consists of a 360° rotating room that represents a section of a deep open plan office space of 3.0 m high, 6 m wide and 9.1 m long (Fig. 43.3). The facility is located in College Station at Texas A&M University Riverside campus (latitude 30.6°N) in an open area with no obstructions around it. This paper presents the photometric measurements of the light pipe facing south. The space includes two sidelight windows of 2.74 m wide by 1.52 m high (4.16 m^2 each) with Tvis of 51%. The WWR of the sidelight windows is 45% and WFR of 7.5%. The windows have external moveable blinds with a reflectance of 0.8. The interior surface reflectances are 0.81 for the ceiling, 0.88 for the walls, and 0.15 for the floor. The light pipe glass area is 5.5% of the sidelight windows.

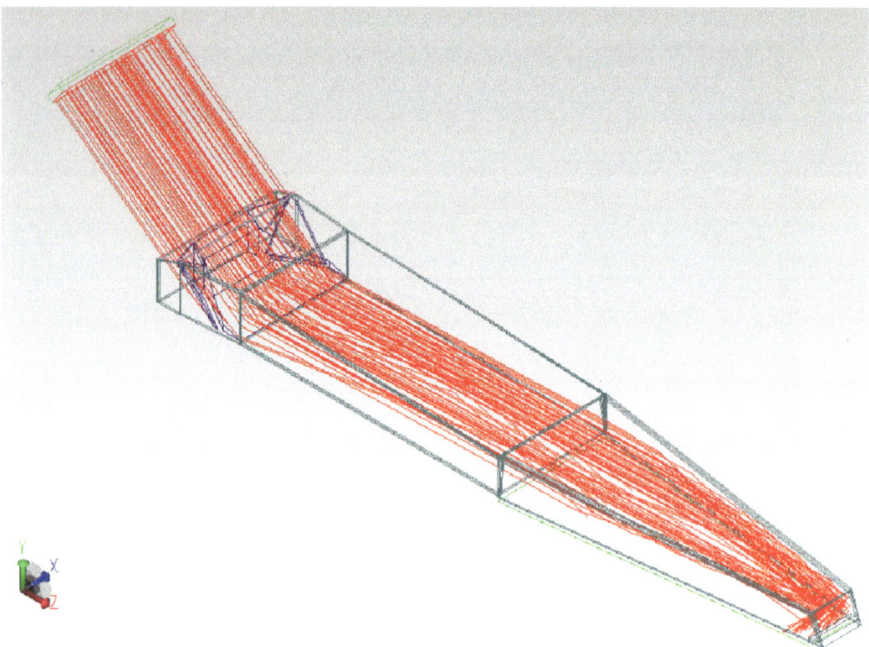

Fig. 43.2 Ray-tracing analysis of optimized light pipe for December 21, 12:00

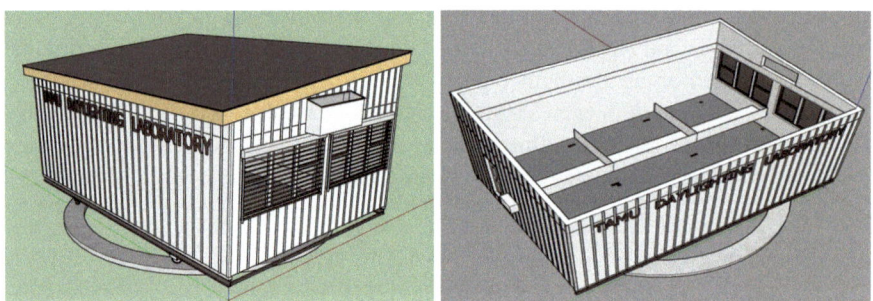

Fig. 43.3 Experimental facility, exterior and interior views

43.3.2 Measurements

Photometric interior illuminance measurements were taken at 25 reference points at work plane height, 0.76 m. Twenty five cosine- and color-corrected LI-COR photometric sensors (LI-210SA) were placed over the work plane at equal distances, 1.5–7.6 m from the window wall, at centerline (Fig. 43.4). Outside the test room, two sensors were placed on the roof and façade to take global horizontal illuminance (GH) and global vertical illuminance (GV). Data was collected every 30 seconds for

Fig. 43.4 Location of the
25 photometric work plane
sensors (#14 and #15 are
located at 6 m and 7.6 m
from the window wall)

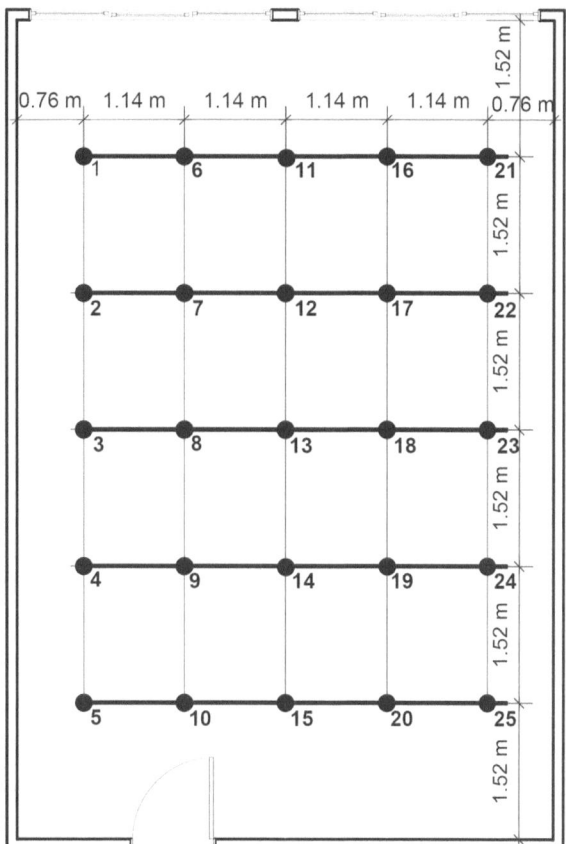

an entire year. The analysis of illuminance levels is based on 10 h, 8:00 to 18:00 true local time (TLT), which is typical office building schedule.

High Dynamic Range (HDR) images were created using the programs Photo-sphere to assess visual qualities in the testing room. HDR images were created from 11 bracketed exposures to cover from 1–20,000 cd/m². A Nikon Coolpix 5400 camera and a FC-E9 Nikon fisheye lens were used to capture a wide view of the space as well as the external conditions. False-color images were created from the HDR images to visualize the spatial luminance distribution, and measure the lumi-nance variability across the space.

43.3.3   Performance Metrics

Daylight autonomy (DA) and Useful Daylight Illuminance (UDI) are two metrics to assess the annual occurrence of illuminance across the workplane maintained by daylight alone and that are within a range considered "sufficient" and "useful" by occupants. Four thresholds were used to report the daylight levels achieved at the rear of the room (7.6 m) modified from the UDI paradigm of Nabil and Mardaljevic (2005): (1) UDI autonomous, UDI-a (300–3000 lx); (2) UDI supplementary, UDI-s (100–300 lx); (3) UDI autonomous for aging eye, UD-a AE (600–3000 lx); and (4) UDI autonomous and bright daylight, UDI-a BD (1000–3000 lx);

43.4   Results and Discussion

43.4.1   Work Plane Illuminance

The light pipe and sidelight window provide natural light evenly distributed over the workplace and throughout the space. The space shows an overall uniform daylight distribution (Fig. 43.5), the sidelight window illuminates the front of the room and the light pipe illuminates the back. Illuminance values at 4.5 m–8.5 m from the window wall are higher than at 3.6 m. The high illuminance levels introduced by the light pipe at the back of the space demonstrates the efficiency of the light pipe design, which with an opening of 1/18th of the sidelight window area provides 5–6 times higher illuminance levels than those provided by the sidelight window at the back of the space. Figure 43.5 compares the illuminance distribution throughout the space with and without the light pipe on April 23 at around solar noon. It is noticeable how daylight is uniformly distributed throughout the space. It is also worth mentioning that a single light pipe is able to introduce adequate illuminance levels across a 6 m wide space. Long-term illuminance measurements confirmed that

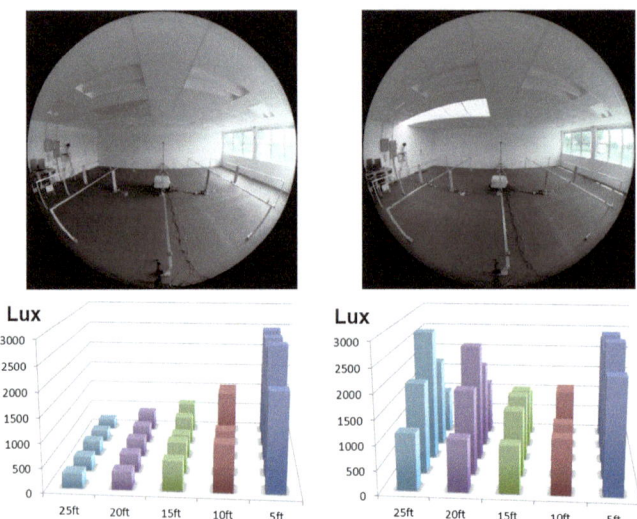

Fig. 43.5 Daylight distribution in the room without (left) and with (right) the light pipe

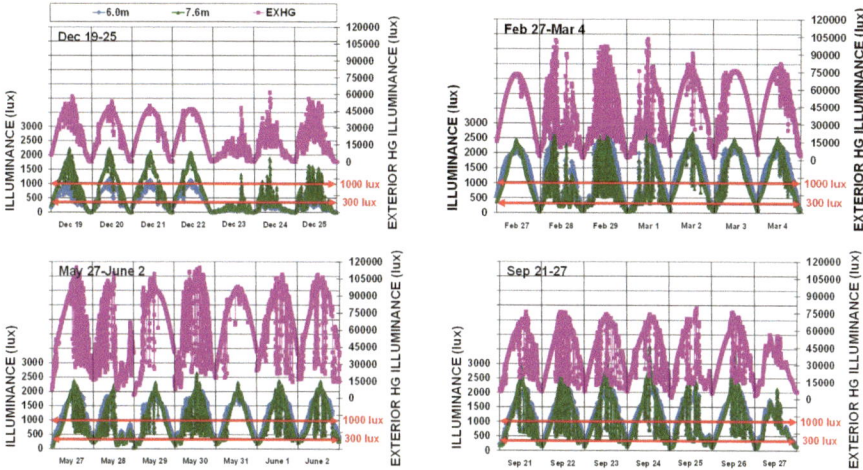

Fig. 43.6 Illuminance at rear work plane sensor #15, solstices and equinoxes

the light pipe provides similar lighting levels at the back of the room as in areas adjacent to the windows (Fig. 43.5); for example, at 1.5 m from the window, light levels reach over 2500 lx while at the back of the space (beyond 6 m) light levels reach over 2000 lx. Daylight delivered by the light pipe at the back of the space provides high illuminance levels of full-spectrum daylight in interior office cubicles or basement throughout the day.

Figure 43.6 depicts the illuminance levels at the back of the room (6 m–7.6 m, Sensors #14 and #15) under different sky conditions throughout a week around the solstices and equinoxes. On clear and partly cloudy days, the light pipe provides between 300 and 2500 lx for about 9 h at the back of the space (7.6 m) throughout the year. Under partly cloudy sky conditions the light pipe introduces higher illuminance values at the back of the space than under clear skies, as observed during the vernal and autumnal equinox weeks in Fig. 43.6. The highest illuminance values (over 1000 lx) are achieved consistently between 10:30 and 16:30 under clear sky conditions. Under overcast sky conditions the light pipe introduces at the back of the space more than 300 lx when EXGH was over 18–20 klx. When the EXGH falls below 18 klx, electric lighting will be used to supplement the 150–200 lx needed throughout the space, front and back.

Figure 43.7 depicts plots of the EXHG and workplane illuminance at 7.6 m (Sensor #15) from the window wall during 1 week around the solstices and equinoxes. It is noticeable the large number of hours when illuminance levels are above the threshold of 300 and 600 lx. The highest illuminance levels (>2500 lx) are achieved when EXHG is above 70–75 klx.

Table 43.1 presents a summary of the UDI values of Sensor #15 at the back of the testing room (see Fig. 43.4). The light pipe helps to achieve UDI-a ranges (300–3000 lx) from 60% to 88% (average UDI 80%) during the weeks around the solstices and equinoxes. The high annual UDI-a means that a building with the light

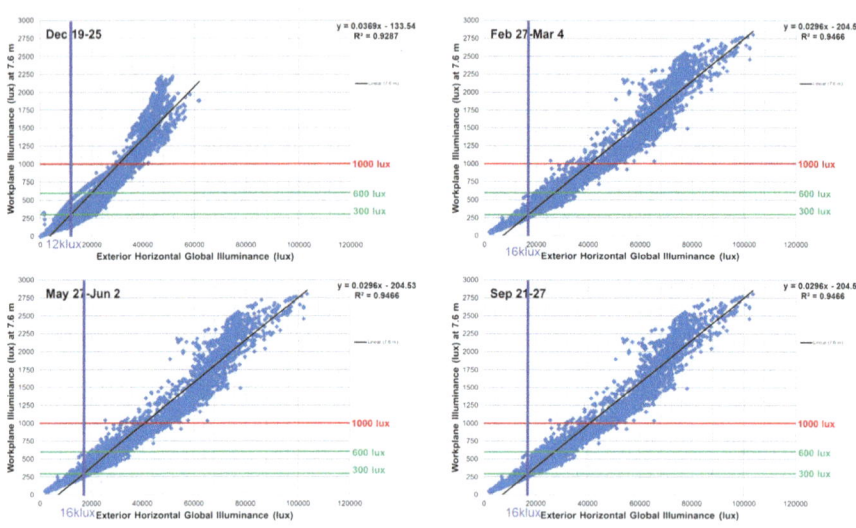

Fig. 43.7 Work plane illuminance (#15) vs. EXGH solstices and equinoxes, 8:00–18:00

Table 43.1 Summary of UDI around the weeks of solstices and equinoxes (8:00–18:00)

	Useful daylight Illuminance (UDI)	Dec 6–12	Feb 27-Mar 4	Jun 21–27	Sep 27-Oct 3
Illuminance at rear work plane sensor	Number of overcast days	3	1	1	1
< 100 lx	UDI 'fell-short' (or UDI-f)	19%	3%	0%	0%
100–300 lx	UDI supplementary (or UDI-s)	21%	13%	12%	13%
300–3000 lx	UDI autonomous (or UDI-a)	60%	84%	88%	87%
600–3000 lx	UDI 'for elderly' (or UDI-ae)	44%	64%	61%	62%
1000–3000 lx	UDI 'bright light' (or UDI-a BD)	28%	45%	42%	43%

pipe can be illuminated by daylight alone most hours of a typical 10-hour working schedule year-round (including weekends), and may need to use supplementary electric lighting for less than 1 h per day throughout the year mostly between 8:00–9:00 and 17:00–18:00 during the winter months and under overcast conditions. The annual UDI-a BD (1000–3000 lx) at the back of the space during solstices and equinoxes ranges from 28% to 45% (average 39.5%). The fact that the light pipe system introduces more than 1000 lx of full spectrum light for about 3–4 h most days of the year is extremely beneficial for building occupants, especially to those that cannot be outdoors, e.g. nursing homes, hospitals, etc. The illuminance levels introduced by the light pipe (>600 lux, with annual average UDI-a AE of 58%)

exceed current recommendations for reading and writing in spaces where at least half of the occupants are over 65 years old (DiLaura et al. 2012). This is particularly important due to an increased aging population (65 years or older) with needs for higher illuminance levels due to the fact that less light reaches the retina of an aging eye than it does in a younger eye.

In recent field studies, researchers have demonstrated the benefits of bright light in building occupant's well-being. Subjects exposed to bright light showed reduced sleepiness, shortened reaction times on psychomotor vigilance task, increased alertness and vitality (Iskra-Golec and Smith 2008; Smolders et al. 2012; Phipps-Nelson et al. 2003). The benefits of having bright light in building cores become extremely important for occupants that spend most of their time indoors.

43.4.2   Luminous Flux Distribution

Figure 43.8 depicts a time lapse of the daylight distribution in the room on May 3. Luminance levels across the space are uniformly distributed throughout the space, as can be observed over the floor and ceiling between 9:00 and 17:00. Around noon (12:00–15:00) hours when the sun is perpendicular to the façade wall, the back-wall luminance is significantly greater than on the side walls, indicating that the luminous flux would have been distributed much deeper than the physical limits of the testing room (9.1 m).

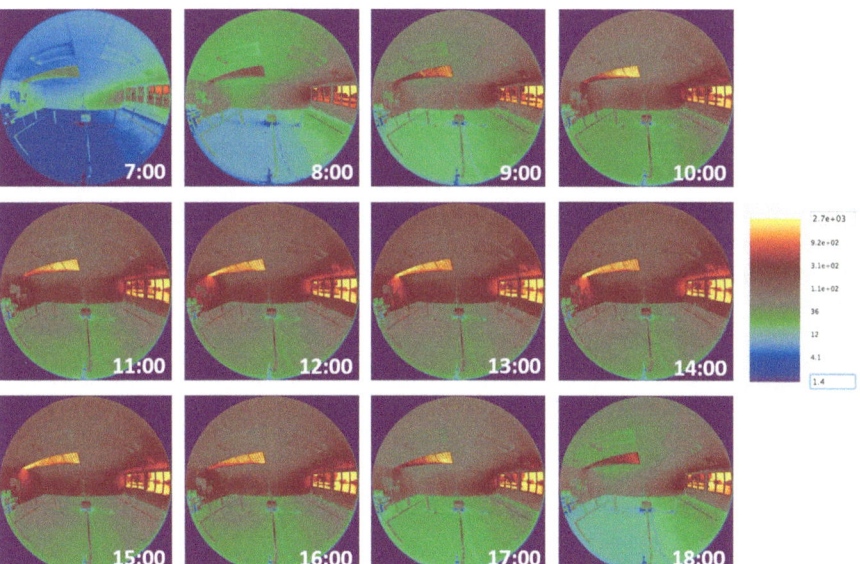

Fig. 43.8 Time lapse of high dynamic range luminance maps (cd/m2) distribution, clear sky conditions, May 3. DGP 0.21, DGI 14

43.5 Conclusions

The passive solar horizontal light pipe presented in this paper is an effective system that can provide healthy lighting in South-facing deep floor plan spaces for more than 9 h under clear and partly cloudy sky conditions. The light pipe introduces consistently throughout the year illuminance levels between 300–2500 lx at 9 m from the window wall. Exposing building occupants to bright light (>1000 lx) can help them regulate the timing of their circadian rhythms, which also has a direct effect on sleep patterns, alertness and performance (Foster 2011). The light pipe is a building component designed to meet not only visual needs, but to increase the circadian light exposures in deep-floor plan buildings.

The lighting levels provided by the light pipe at the back of the space are similar to the ones provided by the sidelight window at the front of the space, even though the light pipe's glass area is only 5.5% of the sidelight window area. The cooling loads generated by the light pipe glazing will be insignificant compared to the ones generated by the sidelight window, and to the cooling loads generated by the electric lighting it offsets. In addition, the light levels distributed uniformly throughout the space creates a visually comfortable space for occupants of deep floor plan buildings.

The light pipe is a sustainable technology that can change the way buildings will be designed in the future. It may not be necessary to have large expanses of glass to introduce more daylight to the core of buildings and deal with the effects of increased cooling loads. Several building types (e.g. offices, schools, nursing homes, hospitals, housing for the elderly and visual impaired people) can benefit from this technology, which utilizes direct solar energy with no operational costs, provides high illuminance levels of full-spectrum light, and regulates occupants' circadian rhythms.

References

Beltrán LO, Uppadhyaya K (2008) Displacing electric lighting with optical daylighting systems. In: Proceedings PLEA 2007 conference, Dublin
Beltrán LO, Lee ES, Papamichael K, Selkowitz SE (1994) The design and evaluation of three advanced daylighting systems: light shelves, light pipes, and skylights. In: National passive solar conference, pp 229–234
Beltrán LO, Lee ES, Selkowitz SE (1997) Advanced optical daylighting systems: light shelves and light pipes. J Illum Eng Soc 26(2):91–106
Chirarattananon S, Chedsiri S, Renshen L (2000) Daylighting through light pipes in the tropics. J Solar Energ 69(4):331–341
Courret G, Scartezzini J, Franzioli D, Meyer J (1998) Design and assessment of an anidolic light-duct. Energ Buildings 28(1):79–99
DiLaura DL, Houser KW, Mistrick RG, Steffy GR (2012) Illuminating Engineering Society the lighting handbook, 10th edn. Illuminating Engineering Society, New York
Foster RG (2011) Body clocks, light, sleep and health, Daylight & Architecture Magazine by Velux, Spring 15, pp 7–12

Garcia V, Edmonds I (2003) Natural illumination of deep-plan office buildings: light pipe strategies. In: ISES Solar World Congress, Goteborg, Sweden

Heschong L, Mahone D (2003) Windows and offices: a study of office worker performance and the indoor environment. California Energy Commission Report, San Francisco

Iskra-Golec I, Smith L (2008) Daytime intermittent bright light effects on processing of laterally exposed stimuli mood, and light perception. Chronobiol Int 25(2–3):471–479

Mossman M, Whitehead L, Beltrán L, Friedel P, Krahe F, Mead D, Papamichael K, Samla C, Suvagau C, Tanteri M, Walsch J (2018) Design guide on active core sunlighting for buildings (DG-31-18). Illuminating Engineering Society, New York

Nabil A, Mardaljevic J (2005) Useful daylight illuminance: a new paradigm to access daylight in buildings. Lighting Res Technol 37(1):41–59

Phipps-Nelson J, Redman JR, Dijk DJ, Rajaratman SMW (2003) Daytime exposure to bright light, as compared to dim light, decreases *sleep*iness and improves psychomotor vigilance performance. Sleep 26:695–700

Rosemann A, Cox G, Upward A, Friedel P, Mossman M, Whitehead L (2007) Efficient dual-function solar/electric light guide to enable cost-effective core daylighting. Leukos 3 (4):259–276

Smolders KCHJ, De Kort YAW, Cluitmans PJM (2012) A higher illuminance induces alertness even during office hours: findings on subjective measures, task performance and heart rate measures. Physiol Behav 107(1):7–16

Chapter 44
Vertical Light Pipe Potentiality for Buildings in Surabaya, Indonesia

Hanny Chandra Pratama and Yingsawad Chaiyakul

Abstract In Surabaya, Indonesia, retail shops and offices are a popular trend of building typologies. Those buildings usually have a thick plan to maximize their land area usage. This affects daylight contribution in the deeper space of the buildings. Since Surabaya is located at $-7.25°$ S Latitude and $112.75°$ E Longitude, sky conditions in Surabaya are predominantly in intermediate sky condition (Rahim R, Mulyadi R. Preliminary study of horizontal illuminance in Indonesia. Department of Architecture, Hasanuddin University, Makasar. Retrieved Makassar, from http://repository.unhas.ac.id/handle/123456789/2566, 2000). Those conditions support the use of light pipes where the sun is at around zenith most of the time. The potential of daylight strategy under the circumstance is to bring daylight from the top.

This study demonstrates the potential of using vertical light pipe in Surabaya as one strategy to obtain daylight for deep space building. This research was conducted by model simulation through DIALux 4.13. A room of $16 \times 16 \times 3$ m is created in DIALux using virtual model. Wide space of the room was created to avoid any reflectance effect from walls, which fall into the calculation surfaces. The height was set at three meters, as this is common in Indonesian buildings. Furthermore, the light tube was set in the middle of the virtual model which has various tube diameters from 0.6 m, 0.8 m and 1 m. Several factors were set to get the actual result, such as location of Surabaya and sky conditions, and were calculated on four critical dates. The result proclaimed that a vertical light pipe is proper to extend daylight availability since the light pipe provides illumination starting from approximately 40 lux – 250 lux. The final part of this study suggests a method for predicting illuminance by length and diameter of the tube.

Keywords Lighting · Low energy · Light tube · Daylighting guideline

H. C. Pratama (✉) · Y. Chaiyakul ·
Faculty of Architecture, Khon Kaen University, Khon Kaen, Thailand
e-mail: hannychandrapratama@kkumail.com; cyings@kku.ac.th

© Springer Nature Switzerland AG 2020
R. Roggema, A. Roggema (eds.), *Smart and Sustainable Cities and Buildings*,
https://doi.org/10.1007/978-3-030-37635-2_44

44.1   Introduction

In the last decades, the developing construction has been increasing in developing countries. This causes an increase in the city's energy consumption. Indonesia, as a country which is still developing infrastructure, has built many buildings and its cities continue to expand. Energy consumption is mainly caused by cooling, lighting, lift and escalators, among other causes (The Department of Climate Change and Energy Efficiency 2012). Lighting was determined as the second largest factor causing energy consumption in buildings. Moreover, Indonesia has huge potential to expand daylight availability, since the sun will shine 12 h a day throughout the whole year. Indonesia is located in Southeast Asia, and lies between the Pacific and Indian Ocean and between the continents of Asia and Australia. Geographically, Indonesia lies between 6 N – 11 S latitudes and 95 E – 141 E longitudes.

Surabaya was chosen as the setting for this study since Surabaya is a metropolitan city and the second largest city in Indonesia. It is located at 7 S latitude and 112 E longitude. Several buildings in Surabaya have a wider plan, and this has an impact on daylight penetration with not enough daylight entering the building to illuminate all areas. Consequently, artificial lightings are used during daytime. As shown in Figs. 44.1 and 44.2, Retail and department stores that have wide plans need to turn on lighting during daytime when Surabaya has the potential to optimize daylighting as the sun travels around zenith during the whole year, as shown in Fig. 44.3. A vertical light pipe guideline is created from a simulation model, which analyzed the illuminance result of light pipes. Vertical light pipes are described as daylight

Fig. 44.1 Wider plan of retail store

Fig. 44.2 Luminaires configuration in department store

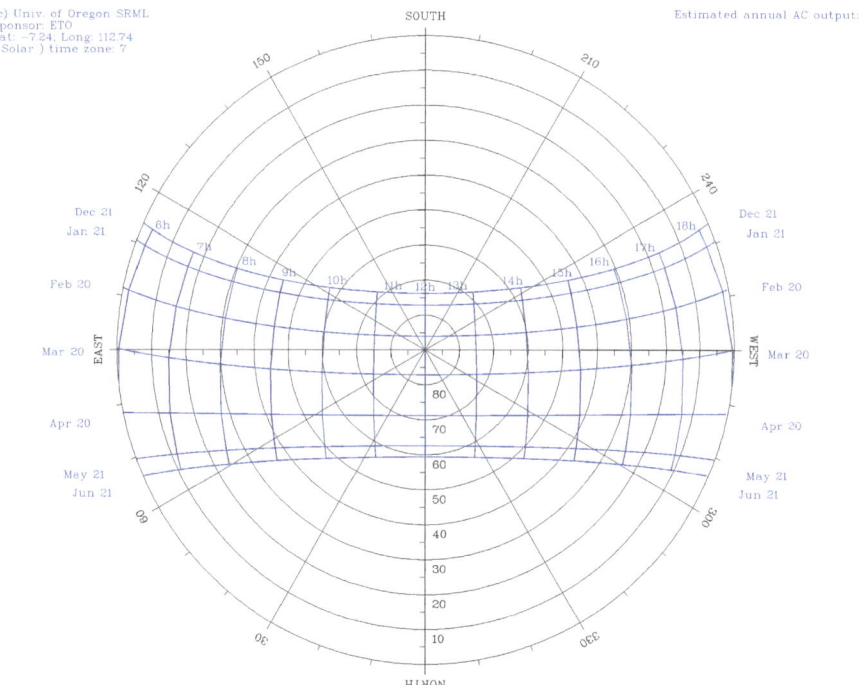

Fig. 44.3 Correlation of Surabaya's sun path towards light pipe reflected

entering from the protruding glasses into the duct wall, reflected in every side of the duct wall, exiting from the other end of the tube and lighting up the room (Wu and Ma 2005). Through a vertical light pipe, internal daylight penetration can be maximized, to decrease building energy consumption. During the equinox, the sunlight reflected into the pipe is merely one reflectance. Thus, sun illuminance will not reduce much from input to the output luminaire. However, during the solstice, the highest reflection number in June is up to six times, and in December the reflection number is only four times. Solar altitude between June and December is different where June is at $59°$ and December at $73°$ (University of Oregon 2012).

To achieve appropriate lighting in buildings, Indonesia has developed the National Standard of Indonesia as a reference. The level of awareness from design practitioners should be enhanced in considering these documents. The National Standard of Indonesia provides the guidelines for lighting design as in SNI no. 03-6197-2000 of the Energy Conservation in Lighting System (Badan Standarisasi Nasional 2000). There are also a number of other documents related to lighting configurations to support SNI documents, for instance from CIBSE and IES (CIE Standard 2002). SNI set several minimum illumination standards for specific tasks as the recommendation for determining lighting design. The illumination started from the lowest at 60-100 lux for garages, terraces and corridors. For common task areas, for instance offices, schools, shopping complexes, the lux level is approximately 100–500 lux. Other specific tasks such as industrial work started from 1000–2000 lux. If the standard has been set, this provides the requirement which an architect should consider while designing especially for daylighting.

However, a previous study about implication of SNI document in Surabaya, Indonesia (Pratama and Chaiyakul 2018) found that most architects and design practitioners did not consider the lighting issues mentioned in the SNI document. It causes that hard to achieve SNI standards. Thus, this study proposes an option to solve the difficulty in assessing SNI when using the vertical light pipe guideline.

44.2   Method

The aim of the simulation is to determine the average illuminance of each light pipe considered. A simulation in DIALux 4.13 program (DIAlux 2011) has been conducted with several controlled aspects (described in Fig. 44.4). The steps to conduct the simulation are:

1. The vertical tube has created in the DIALux with 3 different diameters Ø 0.6 m tube, Ø 0.8 m tube, and Ø 1.0 m tube with length 1–6 m in 1 m interval in each tube.
2. A 16 m × 16 m × 3 m simulation room was created with the tube placed at the center of the room. To avoid any obstruction illuminance from walls, the room was created larger than usual.
3. Reflectance value was set as ceiling: wall: floor = 70: 10: 20.

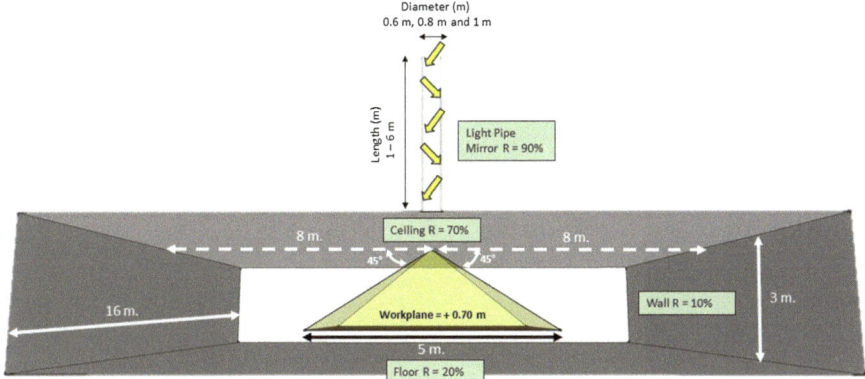

Fig. 44.4 Virtual room simulation on DIALux

4. Sky option simulation was set to intermediate sky as this is the sky condition most common in Indonesia.
5. Time simulation was set at 12.00 as the most potential sunlight in a day.
6. Material glass reflectance was set to the highest option of DIAlux at 0.90.
7. All these data arrangements were simulated for equinox and solstice periods which are on March 21st, June 21st, September 23rd, and December 22nd.
8. Work plane was set at 0.75 m above the floor, area of work plane 5 × 5 m based on the rule of thumb of light that a light source has spread angle of 45o. Area of work plane derived from straight line of the 45o angle that was intended from architectural drawing.

44.3   Discussion and Results

From intermediate sky results, it showed that the three various diameter pipes and six various pipe lengths provide an illuminance level range of 40 lux up to 250 lux. Interestingly, differences in the illuminance level provided by the pipe is affected by the diameter. If the diameter is narrower, the rising illuminance value for the different lengths do not significantly increase. However, if the diameter is larger, each enlarged length will provide much difference in value. As illustrated in Fig. 44.5 (a–d), it was found that a 0.6 m diameter pipe provides illuminance starting from 40–90 lux. This meet the SNI minimum requirement of 60 lux for terraces and garages, which means that several pipes can be used for completing terrace and garage tasks. However, to reach 60 lux, it is suggested to use at least Ø 0.6 m or higher diameter and 3 m pipe length or shorter length.

Figures 44.5a–d show the result from the topic above as graph of illuminance level on 4 critical days.

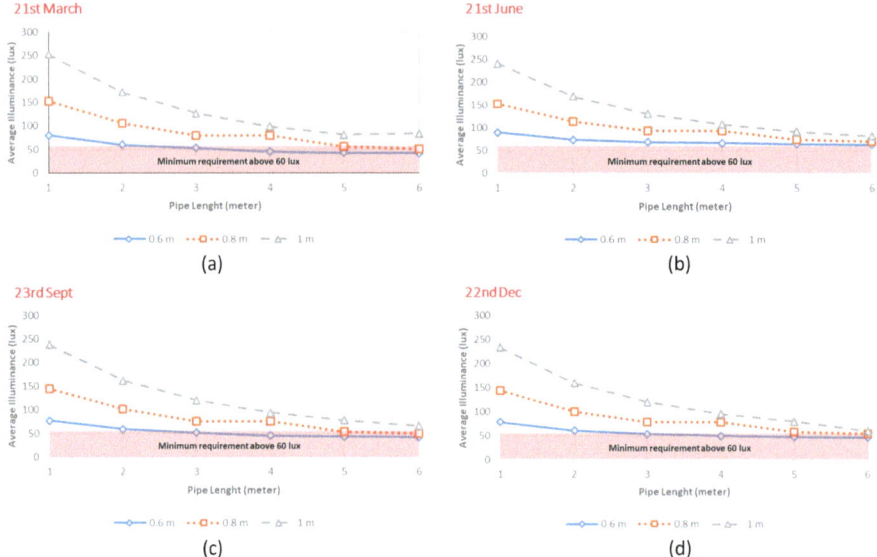

Fig. 44.5 (**a–d**) Average Illuminance on the work plane at 12.00 during intermediate sky

On the other hand, vertical light pipes can be used also to light up corridors that have a requirement of 100 lux, by using at least an Ø 0.8 m pipe or a 3m length pipe. In case of common tasks, we referred to SNI document, which requires levels between 150–300 lux, and Ø 0.6 m and Ø 0.8 m pipes were unable to provide proper illuminance. Ø 1.0 m of pipe and 3meter length can be possibly used for common task.

Secondly, the study focused on differences of illuminance level on equinox and solstice days. Generally, the results from each critical day are quite close to each other, because during intermediate sky, direct sun was covered by 30–70 % cloud (Boyce and Raynham 2009) which means the sun might not directly shine into the pipe. The illuminance values of selected dates are slightly different under the intermediate sky. Still, the number of reflections in June is more than the other months. This may be the effect of the light from other sky and clouds under this sky.

Factors of light pipe configuration are shown in Fig. 44.6. It is shown that shorter pipe length will result in illuminance deviation. In the 1 m pipe length, Ø 0.6 m pipe has an approximate deviation of 50 lux in comparison to Ø 0.8 m pipe. Furthermore, Ø 0.8 m pipe has an approximate deviation of 80 lux in comparison to Ø 1 m pipe. Three equations were derived from the dates for three pipe diameters as shown in Table 44.1.

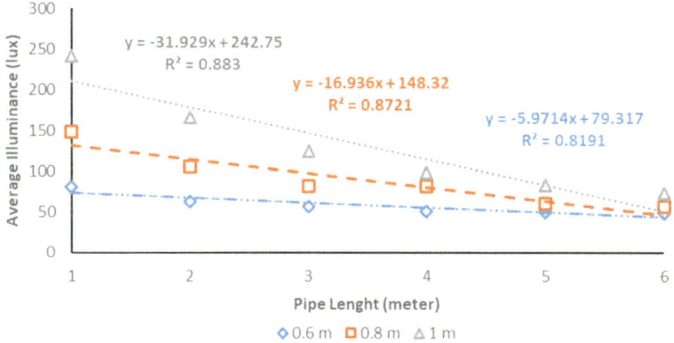

Fig. 44.6 Average illuminance of pipe the whole year in the equation

Table 44.1 Average illuminance in the equation

No.	Pipe Diameter (m)	Model	R2
1.	0.6	Eav = −5.9714ln(L) + 78.489	0.8191
2.	0.8	Eav = −16.936ln(L) + 148.32	0.8721
3.	1.0	Eav = −31.929ln(L) + 242.75	0.883

44.4 Conclusions

Daylight availability is important to decrease building energy consumption (Asian Institute of Technology 1998). Besides, it is better than artificial light for human sight requirements. This study aims to obtain one solution for architects who want to design buildings using optimized daylight availability. Moreover, there are a number of vertical light pipe products on sale in the market place (Chatron-Ltd 2010). Thus, the architect can predict the dimensions of the appropriate vertical light pipe before considering the actual product.

To conclude, light pipes can be applied as a daylighting strategy to pass task requirements determined by the SNI document (Badan Standarisasi Nasional 2000), since light pipes provides proper light itself. Even though the lux levels provided are not enough to pass the requirements, vertical light pipes can be adjusted by increasing the number of pipes, enlarging the diameter, or decreasing pipe length. There are two points that must be noted. Firstly, the illuminance level can be greatly increased by enlarging the pipe diameter. However, this may have impact on the building structure. Secondly, vertical light pipes may not be practical for multi-story buildings that have a story height of more than 6 m, as the levels of illuminance will drop rapidly.

Acknowledgement The authors would like to acknowledge the Khon Kaen University, Thailand which has supported scholarship for this study

References

Asian Institute of Technology (1998) Daylighting for buildings in the tropic I: daylighting availability and heat gain into buildings. Thailand

Badan Standarisasi Nasional (2000) Konservasi energi pada sistem pencahayaan. Badan Standarisasi Nasional, Jakarta

Boyce P, Raynham P (2009) The SLL lighting handbook. The Society of Light and Lighting, London

Chatron-Ltd. (2010) In: Industrial TDZ (ed) Solar light tube 100% natural lighting. Travessa da Zona Industrial, Vale de Cambra, pp 1–4

CIEStandard (2002) Lighting of Indoor Work Places: International Organization for Standarization

DIAlux (2011) DIAlux 4.13. from http://www.dial.dc

Pratama HC, Chaiyakul Y (2018) Implication of the National Standard Indonesia (SNI) on Lighting Conservation as a Basis of Architectural Design. J Build Energy Environ 1:51–57

Rahim R, Mulyadi R (2000) Preliminary study of horizontal illuminance in Indonesia. Department of Architecture, Hasanuddin University, Makasar. Retrieved Makassar, from http://repository.unhas.ac.id/handle/123456789/2566

The Department of Climate Change and Energy Efficiency (2012) Baseline energy consumption and greenhouse gas emissions in comercial building in Australia. Australia: attribution 3.0 Australia Licence

University of Oregon (2012) Sun Path Chart Program 2012. Retrieved December 20, 2012, from http://solardat.oregon.edu/SunChartProgram.html

Wu Y, Ma C (2005) Daylight performance of top lighting light pipes and side lighting light pipes under sunny conditions in Beijing. The 2005 World Sustainable Building Conference (September 2005), 6

Chapter 45
Energy Efficiency of a High-Rise Office Building in the Mediterranean Climate with the Use of Different Envelope Scenarios

Soultana (Tanya) Saroglou, Isaac A. Meir, and Theodoros Theodosiou

Abstract This paper investigates different strategies towards advancing the energy efficiency of a high-rise office building by focusing on the building envelope. More specifically, the focus is on the thermal properties of the building envelope and the effect of altitude on energy performance. The studies are focused on Tel Aviv, Israel, a city with a vibrant high-rise building activity. Studying this typology in this Mediterranean climate will be of relevance for other cities with similar climate (e.g. in Middle East, S. Europe, N. Africa) that undergo similar processes of high-rise development. The study is based on thermal simulations of an office reference model at different heights: 8 m (ground level), 82 m, 168 m, 254 m, and 340 m. Alternative façade scenarios were implemented for gradually upgrading the building envelope and studying its relationship with the changing microclimate with altitude (wind speed increase and dry bulb temperature drop) between ground and top level. The envelope scenarios range from clear single glazing, to LowE double-glazing, triple glazing, the addition of external shading devices, and a double skin façade (DSF). The popularity of DSFs has grown over the last decades, due to their potential of improving a building's energy performance, when compared to a single-glazing curtain wall. The high levels of solar radiation common in latitudes and climates like the one in discussion, result in high cooling loads, especially in relation to the design of glass façades. As a result, studies on the energy performance of DSFs in comparison to a single-glazing envelope become very important for improving energy efficiency. Moreover, published research on DSFs is currently mainly on cold and moderate climates. Energy consumption between ground-to-top floors

S. (T). Saroglou (✉)
Kreitman School for Advanced Graduate Studies, Ben-Gurion University of the Negev (BGU), Be'er Sheva, Israel
e-mail: saroglou@post.bgu.ac.il

I. A. Meir
Department of Structural Engineering, Faculty of Engineering Sciences, Ben-Gurion University of the Negev (BGU), Be'er Sheva, Israel

T. Theodosiou
Department of Civil Engineering, Aristotle University of Thessaloniki, Thessaloniki, Greece

© Springer Nature Switzerland AG 2020   651
R. Roggema, A. Roggema (eds.), *Smart and Sustainable Cities and Buildings*,
https://doi.org/10.1007/978-3-030-37635-2_45

alters with the changing microclimate in relation to height: heating increases, while cooling drops. Moreover, for an office building in Tel Aviv the energy loads for cooling are much higher compared with those for heating. Results show that single clear glazing performs the worst, while using LowE double glazing in scenario B reduces energy consumption for cooling by 25% from scenario A. The addition of external shading devices in scenario C reduces cooling loads by a further 50% from scenario B. In scenario D triple LowE glazing performs by 1% worse in cooling loads from the double-glazing option, however, comparing scenario E of triple glazing with external shading with scenario C of double glazing with external shading, scenario E has 20% higher cooling loads. The comparison between scenario F of the DSFs with scenario C shows that scenario C performs 8% better in cooling loads. However, the shading devices shade a large portion of the facades, while in the DSF option transparency and visibility are maintained, as well as improving the energy performance of the building. The results prompt for further studies on the energy efficiency of DSF in warm climates.

Keywords Energy efficiency · EnergyPlus · Building envelope · High-rise office building · Mediterranean climate

45.1 Introduction

Urban development sustainability is defined as the process of integration and co-evolution of economic, social, physical and environmental subsystems. The carrying capacity of a city is another aspect of sustainability within the city (Wei et al. 2016). The aim of sustainable development is to enhance population wellbeing without compromising development possibilities. In terms of city infrastructure and building technology, these need to display technological progress and environmental design principles in order to successfully sustain the growing population.

Population growth, intense urbanization, and expanding industrialization have promoted the typology of the skyscraper as an important solution for the future of high-density urban centers around the world. This is achieved by increasing city-space vertically, as opposed to a continuous expansion outwards that devours natural and agricultural landscapes/resources. However, the skyscraper has to also comply with current strict regulations on building energy efficiency. As a result, the sky-scraper as a positive addition within the urban fabric calls for further research and experimentation.

Today, there is a global movement of 'green' high-rise buildings that utilize sustainable technologies on a whole new scale, and possibly celebrate a green certification award, e.g. LEED certification. Nevertheless, the towers are still linked to high-energy demand, environmental and social imbalances (Girardet 2006). An example is 'The Bank of America' in New York, completed in 2010. The tower received 'Core and Shell' LEED Platinum certification, which is the highest level of LEED certification, and was awarded the 2010 'Best Tall Building Award –

Americas' by the Council of Tall Buildings and Urban Habitat. However, in 2013 a New York City public report on the building operations revealed that in 2012 the tower's site energy use was higher than any comparably sized office building in Manhattan (Donnolo et al. 2014). This leads to the conclusion that there is a potential gap between green certification and operational energy, and that a highly-certified building can still be consuming high amounts of energy and producing high amounts of greenhouse gasses.

The above is a general observation of twentieth century's architecture where the availability of cheap energy, and use of HVAC systems, had a profound impact on the design and operation of buildings. Curtain wall facades replaced heavyweight-building envelopes, creating the glass-box style of architecture. However, the building envelope is important to the regulation of a building's thermal behavior relative to the climate and microclimate around it (Trefil 2003). Its special physical properties, i.e., the thermal resistance and heat capacity of the materials used, must interact appropriately with the ambient climatic conditions to reach healthy, comfortable indoors (Givoni 1969).

The desired transparency of the building envelope and lightness of the structure initiated in the mid-twentieth century resulted in high-energy demands in buildings, both for cooling and heating. Large-scale buildings, like skyscrapers, have even higher energy demands on the urban fabric. This paper, which is part of a wider research on skyscraper energy efficiency (Saroglou et al. 2017a, b), considers this building typology as an urban phenomenon closely related to contemporary and future city living, and investigates design strategies towards improving energy efficiency with a focus of the building envelope.

45.2 Methodology

The present study is based on thermal simulations of an office high-rise reference model up to 400 m high located in Tel Aviv, Israel. The greater Tel Aviv metropolitan area is already growing upwards and is expected to have a significant number of new skyscrapers in the near future, while the city's Municipality Planning and Construction Committee issued the 2025 city master plan that supports new sky-rise development. At the moment, there are 21 skyscrapers over 150 m, 29 between 120–150 m, 35 between 100–120 m, and 18 under construction. The average height of these high-rise buildings is 150 m, already much higher than the existing skyline. The study of Tel Aviv high-rise building typology will be relevant for other cities with a similar climate (e.g. in Middle East, S. Europe, N. Africa) that undergo comparable processes of high-rise development (Meir et al. 2012). This paper focuses on an office high-rise, not least because of the high internal heat gains and the increased windows ratio of this building typology, in comparison to a residential option, and their effect on cooling needs in this Mediterranean climate.

Simulations are performed for different single-skin envelope scenarios and a double skin façade (DSF). The single-skin curtain walls are simulated with and without external shading devices, and comparisons are made between the different scenarios and the DSF option. The incorporation of DSF in buildings has increased rapidly over the last decades, as an option of energy conservation and sustainable development, while retaining flexibility and transparency in architecture. DSF studies around the world have shown that the double layer façade envelope can significantly improve the thermal performance of the structure, by mitigating the impact of the exterior environmental conditions on the interior of the building (Chan et al. 2009; Joe et al. 2013). However, contrary to the number of DSF studies, there are still no official specifications on the simulation of DSFs and their energy savings (Joe et al. 2014; Ahmed et al. 2015). In this research paper, the DSF is simulated as a naturally ventilated DSF, and its energy efficiency is compared in relation to height with the other envelope scenarios.

45.2.1   Building Simulation Data

Simulations are conducted at heights of 8 m (ground level), 82 m, 168 m, 254 m, and 340 m (CTBUH 2015). EnergyPlus 8.8 was used as the energy simulation engine that can model wind acceleration with height according to ASHRAE (2009), and air temperature drop by elevation. The proposed location for the tower is within an urban environment, yet no other structures were included in its proximity during the simulations presented here. Energy consumption is calculated in relation to indoor thermal comfort standards (Givoni 1981), with the temperature range according to the Predicted Mean Vote (PMV) model by Fanger at 20 °C during winter, and at 26 °C during summer (Fanger 1970). Envelope characteristics (U-values of walls and window- to-floor ratio – WFR) are designed in accordance with Israel's Green Building Standards, SI 5282 (SII 2011). Internal heat gains are calculated for 10 m² of office space per person. Table 45.1 shows the envelope characteristics used in the simulations. In regard to the WFR, the model was designed with the minimum values stated in the standards, noted in bold. Comparisons towards energy efficiency are made between six different curtain wall envelope scenarios, and energy consumption is compared, in relation to height for the climatic conditions of Tel Aviv. The material properties of the glazing are shown in Table 45.2.

The scenarios as follows:

- **Scenario A:** Clear Single-Glazing
- **Scenario B:** LowE Double-Glazing
- **Scenario C:** LowE Double-Glazing and Shading
- **Scenario D:** LowE Triple-Glazing
- **Scenario E:** LowE Triple-Glazing and Shading
- **Scenario F:** Double Skin Façade DSF (single double LowE)

Table 45.1 Thermal properties of building envelope and WFR according to Green Building Standards, SI 5282: Energy Rating of Buildings

Mass Wall GBS (SOUTH) U-value 1.02 [W/m^2°K]	Mass Wall GBS (N/ E / W) U-value 0.54 [W/m^2°K]	Window Sgl Clr 6 mm U-value 5.778 [W/m^2°K]	Window double Low-E Spec Selection 6 mm/13 mm Air U-value 1.626 [W/m^2°K]	Window triple Low-E Film 6 mm/ 13 mm Air U-value 1.22 [W/m^2°K]
19 mm gypsum board	19 mm gypsum board	Clear 6 mm	LowE spec Sel TINT 6 mm	Clear 6 mm
Extruded poly-styrene – 15 mm	Extruded polystyrene – 40 mm		Air 13 mm	Air 13 mm
				LowE Film 6 mm
300 mm heavy-weight concrete	300 mm heavy-weight concrete		Clear 6 mm	Air 13 mm
				Clear 6 mm
F07 15 mm stucco	F07 15 mm stucco	SHGC: 0.819	SHGC: 0.421	SHGC: 0.361
		V.T.: 0.881	V.T.: 0.682	V.T.: 0.535

Note: windows G3 and/or G4 Low-E Glazing (W/m^2°K) specifications: G3: Uglass = 1.8/ Uframe = 3.5/SHGC = 0.6/Daylight trans. = 0.6; G4: Uglass = 1.8/Uframe = 3.5/SHGC = 0.5/ Daylight trans. = 0.5. *Visual Transmittance (V.T.)*

Table 45.2 Thermal properties of mass wall and glazing in accordance with Israel's Green Building Standards GBS

Window-to-floor ratio (WFR) %	Windows thermal properties per orientation	U-value of Walls (W/m^2 K)
S = **23**–35, N = **23**–27 E = **23**–32, W = **18**–27	N/E = G3 S/W = G3 or G4	N/E = 0.6 S = 1.2 W = 0.6

45.2.2   *Design of Reference Models*

The typical floor layout is an open plan layout, positioned on a north-south axis with elevations on all secondary orientations: SE, SW, NE and NW (Fig. 45.1 left). The design of the office reference model took into consideration the requirement for natural day-lighting in SI 5282 Energy Rating of Buildings of 5 m depth of windows to usable floor space, and measures just 460 m^2 per floor. The hatched area indicates circulation area within the offices, while the central rectangle marked with diagonals as X is the core of the building. Real life office buildings have a usable floor are 3–4

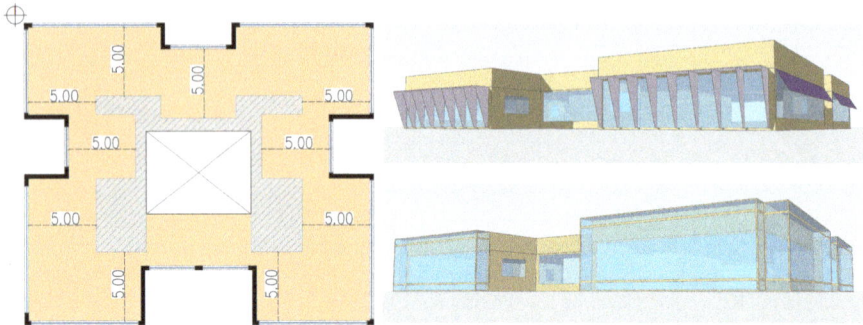

Fig. 45.1 Left office typical floor plan layout with the requirement set in SI 5282 Energy Rating of Buildings for 5 m depth of windows to usable floor space. Right top: shading devices on south and west elevations. Right bottom: view of typical floor plan's façades with DSFs on all orientations

times bigger and incorporate into the design ceiling reflectors, light wells, central atriums etc. Such multi-layered designs were not part of the simulations conducted at the moment.

The optimum design of the shading devices was configured with the Ecotect 'Shading Design Wizard'. Shading devices were positioned on the south, east and west elevations (Fig. 45.1 right top). The south elevation uses a horizontal shade with vertical fins every 1 m. For the east and west elevations, a 45° angle shading blocks 50% of the window, due to the low angle sun on these orientations. The shading devices were designed for optimum energy efficiency and did not take into consideration issues like 'view out' and natural light.

In regard to the DSF design scenario, in total there are eight DSFs on every floor level: SE-**E**, SE-**S**, SW-**S**, SW-**W**, NE-**E**, NE-**N**, NW-**N**, NW-**W**. The height of the DSF corresponds to the height of each floor level that is 3.9 m (Fig. 45.1 right bottom). This design of the DSF, as opposed to a multi-story DSF, allows for similar temperature gradients within the cavity, while each story has more or less the same temperature (Hensen et al. 2002). In addition, this type of DSF behaves better in relation to acoustical, fire and ventilation issues (Poirazis 2004). The simulated DSF is a naturally ventilated corridor façade with inlet (bottom) and outlet (top) openings for each DSF, at each floor level. During the cold season (November 1 – March 31) the DSF is airtight, while during the hot season (April 1 – October 31), the inlet and outlet openings are open (outdoor air curtain type DSF). The composition of the DSF under study is: LowE double-glazing (interior layer), an air cavity of 1 m deep in the middle, and single glazing (exterior layer). The 1 m-cavity is considered a good choice in hot climates due to the increasing air volume that allows for significant pre-cooling of the air during summer (Balocco 2002; Hamza 2008; Papadaki et al. 2014) (Fig. 45.2).

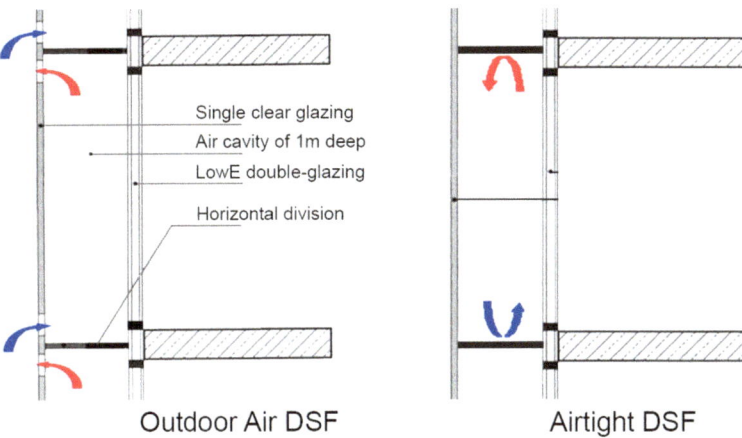

Outdoor Air DSF   Airtight DSF

Fig. 45.2 DSF spatial configurations for hot season (outdoor air DSF) and cold season (airtight DSF)

Table 45.3 Heating (H) and cooling (C) loads of six envelope scenarios. **A**: Single Clear and no shading (Sgl. Cl. Gl. n.Sh.), **B**: DoubleLowE and no shading (Dbl. LowE n.Sh.), **C**: DoubleLowE and shading (Dbl. LowE Sh.), **D**: TripleLowE and no shading (Trp. LowE n.Sh), **E**: TripleLowE and shading (Trp. LowE Sh.), and **F**: double skin facade (DSF) (single_doubleLowE). U values of wall and windows ratio according to Israel's GBS

Meters high/kWh/m²/year		8 m	82 m	168 m	254 m	340
Scenario A Sgl.Cl.Gl.n.Sh	H	1.79	2.82	3.86	5.03	6.47
	C	133.6	116.3	108.3	101.3	94.3
Scenario B Dbl. LowE n.Sh.	H	0.24	0.3	0.4	0.6	0.8
	C	99.5	87.5	82.3	77.6	73.1
Scenario C Dbl. LowE Sh.	H	0.9	1.38	1.9	2.52	3.06
	C	**49.3**	**43.8**	**39.7**	**36.1**	**35.8**
Scenario D Trp. LowE n.Sh	H	0.15	0.2	0.3	0.4	0.6
	C	99.5	88.3	83.4	78.9	74.6
Scenario E Trp. LowE Sh.	H	0.37	0.6	0.8	1.13	1.34
	C	**60.8**	**56.2**	**52.1**	**48.3**	**48.0**
Scenario F DSF	H	0.2	0.4	0.7	1.03	1.37
	C	**53.2**	**46.5**	**41.6**	**38.3**	**36.0**

45.3   Simulation Results

Table 45.3 presents the cumulative changes in energy efficiency between the six different envelope scenarios of the office reference model from ground level to top floor. The improvements in the building envelope are as follows:

Scenario A: Energy consumption of reinforced concrete (RC) structure with clear glazing (6 mm), infiltration 0.9 ACH. This scenario reflects the design of high-rise buildings in the beginning of twentieth century and is included for comparison purposes with the following scenarios.

Scenario B: Replacement of windows with low-emissivity, spectrally selective, tinted double- glazing (LoE Spec Sel Tint 6 mm/13 mm air/clear glass 6 mm), infiltration 0.6 ACH. The reduction in ACH from 0.9 to 0.6 is based on an assumption that changing window systems from clear-glass, to LowE double-glazing, including improved sealants and frames, will reduce infiltration (El-darwish and Gomaa 2017). High cooling loads decreased by approximately 25% for all subsequent heights, while heating decreased by approximately 90%, however very low to begin with (e.g. 1.79 kWh/m^2/year > 0.24 kWh/m^2/year).

Scenario C: Incorporation of shading devices (Fig. 45.1 right top) on the south, east and west elevation. Cooling energy decreased considerably by a further 50% (e.g. 99.5 kWh/m^2/year > 49.3 kWh/m^2/year), while heating increased by almost 4 times, however, is still quite low (e.g. 0.24 kWh/m^2/year > 0.9 kWh/m^2/year).

Scenario D: Replacement of windows with low-emissivity, spectrally selective, tinted triple- glazing (clear glass 6 mm/13 mm air/LoE Film 6 mm/13 mm air/clear glass 6 mm). This scenario is like scenario B, but instead of double LowE glazing it uses triple LowE glazing. Comparing scenario B with scenario D we see that cooling loads start by being identical for the ground floor, while with height cooling loads for scenario D perform worst by approximately 1% (e.g. @82 m high 87.5 kWh/m^2/year < 88.3 kWh/m^2/year). Heating loads in scenario D are lower by 80–90%, but still very low (e.g. 0.9 kWh/m^2/year > 0.15 kWh/m^2/year).

Scenario E: Incorporation of external shading devices in scenario D with triple glazing. Cooling loads decreased by approximately 40% from scenario D, however comparing scenario C of double-glazing with external shading devices with scenario E of triple glazing with external shading devices, we see that the cooling loads of scenario E are by 20% higher (e.g. 49.3 kWh/ m^2/year < 60.8 kWh/m^2/year).

Scenario F: Incorporation of eight DSFs for all orientations: SE-**E**, SE-**S**, SW-**S**, SW-**W**, NE-**E**, NE- **N**, NW-**N**, NW-**W**. Comparing scenario C, with scenario E, and scenario F, we see that scenario B of double-glazing and external shading performs the best with the lowest cooling loads by 20% from scenario E, and 8% from scenario F at ground level. However, when studying more closely scenario B with scenario F, we see that the differences in cooling diminish with height, where for example at 340 m become almost identical (e.g. 35.8 kWh/m^2/year < 36 kWh/m^2/year). In regard to heating loads, scenario F performs better by approximately 70%. However, while heating loads are substantially lower in comparison with the cooling loads at lower levels, at 340 m high heating loads for scenario B are 3.06 kWh/m^2/year, and for scenario F 1.37 kWh/m^2/year.

45.4 Discussion

For all the scenarios studied above, cooling energy decreased considerably from ground floor to top. In scenario A 30%, in scenario B 26%, in scenario C 27%, in scenario D 25%, in scenario E 21%, and in scenario F 32%. Comparing scenario B of double LowE glazing with scenario D triple LowE glazing, we see that the differences are not great, but scenario B performs slightly better in regard to cooling loads. However, with the addition of external shading their differences escalate. Heating loads for scenario C are about 2.5 times higher from scenario E, but still considerably lower than cooling, for an office building in the Mediterranean climate of Tel Aviv, making the double LowE glazing with external shading a more favorable scenario from the triple glazing with external shading one. Scenario F of DSFs is more closely compared with scenario C of double LowE glazing with external shading. The average cooling loads for scenario C is 40.9 kWh/m^2/year, while for scenario F is 43.12 kWh/m^2/year. On the other hand, the average heating for scenario C is 1.95 W/m^{2}K and for scenario F is 0.74 W/m^{2}K. While the differences in energy loads between the two scenarios are not great, the possibilities of implementing external shading on a high-rise building envelope are much lower in comparison with the option of a double skin façade.

45.5 Conclusions

This paper discussed the changes in energy loads for heating and cooling between different envelope scenarios of an office high-rise building reference model located in Tel Aviv. Comparisons were made in relation to heating vs. cooling loads between five subsequent heights: 8 m (ground level), 82 m, 168 m, 254 m, and 340 m. The significant differences in heating vs. cooling are due to the high internal heat gains of the office use, in combination with the climatic conditions of Tel Aviv.

In consideration of the increased windows ratio especially noticeable in office buildings today, the study focused on increasing the energy efficiency of a glass curtain wall construction. The improvements of the building envelope, e.g. thermal properties of glazing and addition of external shading devices, resulted in considerable reductions in cooling loads. It was noticeable that the double LowE with external shading option performed considerably better from the triple LowE with external shading option, with 20% less cooling loads in the Mediterranean climate of Tel Aviv.

In the DSF scenario, cooling energy loads were by 8% higher from the envelope option with double LowE glazing and external shading, however, given the popularity of DFSs, and the difficulties of implementing external shading devices on high-rise buildings, further studies on their energy efficiency is required. This is especially valid in hot climates, where the combination of high solar gains and a glass-building envelope affect greatly cooling loads. Furthermore, regarding skyscraper energy efficiency, actual monitoring of climatic conditions over and around tall buildings in a real urban environment with all its wind and temperatures complexities, becomes vital in order to verify the accuracy of the simulations.

Acknowledgements This research is partly supported by the Tsin Mid Way Scholarship for outstanding Ph.D. students, Kreitman School of advanced Graduate Studies, BGU, and the Rieger Foundation-Jewish National Fund in Environmental Studies.

References

Ahmed MMS, Abel-Rahman AK, Ali AHH, Suzuki M (2015) Double skin Façade: the state of art on building energy efficiency. J Clean Energy Technol 4(1):84–89. https://doi.org/10.7763/JOCET.2016.V4.258

Balocco C (2002) A simple model to study ventilated facades energy performance. Energ Buildings 34(5):469–475. https://doi.org/10.1016/S0378-7788(01)00130-X

Chan ALS, Chow TT, Fong KF, Lin Z (2009) Investigation on energy performance of double skin facade in Hong Kong. Energ Buildings 41(11):1135–1142. https://doi.org/10.1016/j.enbuild.2009.05.012

CTBUH (2015) Calculating the height of a tall building where only the number of stories is known. www.Ctbuh.Org 9:1–3

Donnolo M, Galatro V, Janes L (2014) Bank of America Tower at One Bryant Park: New York City, NY: Ventilation in Wonderland. High Performing Buildings, pp 51–58

El-darwish I, Gomaa M (2017) Retrofitting strategy for building envelopes to achieve energy efficiency. Alexandria Eng J 56(4):579–589. https://doi.org/10.1016/j.aej.2017.05.011

Fanger PO (1970) Thermal comfort: analysis and applications in environmental engineering. McGraw-Hill, Malabar

Girardet H (2006) Which way China? | Herbert Girardet – China dialogue, Arup and SIIC

Givoni B (1969) Man Climate Architecture. Van Nostrand Reinhold, New York

Givoni B (1981) Man, climate and architecture. Van Nostrand Reinhold, New York

Hamza N (2008) Double versus single skin facades in hot arid areas. Energ Buildings 40:240–248. https://doi.org/10.1016/j.enbuild.2007.02.025

Hensen J, Bartak M, Kal F (2002) Modeling and simulation of a double-skin Façade system. *Am Soc Heating Refrig Air-Cond Eng* 108(Part 2):1251–1259

Joe J, Choi W, Kwak Y, Huh JH (2014) Optimal design of a multi-story double skin facade. *Energ Buildings* 76:143–150. https://doi.org/10.1016/j.enbuild.2014.03.002

Joe J, Choi W, Kwon H, Huh JH (2013) Load characteristics and operation strategies of building integrated with multi-story double skin facade. *Energ Buildings* 60:185–198. https://doi.org/10.1016/j.enbuild.2013.01.015

Meir IA, Peeters A, Pearlmutter D, Halasah S, Garb Y, Davis J-M (2012) An assessment of regional constraints, needs and trends. Adv Building Energ Res 6(2):173–211. https://doi.org/10.1080/17512549.2012.740209

Papadaki N, Papantoniou S, Kolokotsa D (2014) A parametric study of the energy performance of double-skin façades in climatic conditions of Crete, Greece. Int J Low-Carbon Technol 9 (4):296–304. https://doi.org/10.1093/ijlct/cts078

Poirazis H (2004) Double skin Façades for office buildings literature review. Division of Energy and Building Design, Lund Institute of Technology, Lund University, Lund

Saroglou T, Meir IA, Theodosiou T (2017a) Quantifying energy consumption in skyscrapers of various heights. SBE16 international conference on sustainable synergies from buildings to the urban scale, Oct. 17–19, Aristotle University, Thessaloniki, Greece. In: Procedia – environmental sciences. In print

Saroglou T, Meir IA, Theodosiou T, Givoni B (2017b) Towards energy efficient skyscrapers. Energ Buildings 149:1–13. https://doi.org/10.1016/j.enbuild.2017.05.057

SII (2011) SI -5281 part 3, sustainable building (green-building), office building guidelines, Standards Institute of Israel. Technical guide, Standards Institute of Israel, Tel Aviv

Trefil JS (2003) The nature of science: an A–Z guide to the laws and principles governing our universe. Houghton Mifflin Harcourt, New York

Wei Y, Huang C, Li J, Xie L (2016) An evaluation model for urban carrying capacity: a case study of China's mega-cities. Habitat International 53:87–96. https://doi.org/10.1016/j.habitatint.2015.10.025

Correction to: Smart and Sustainable Cities and Buildings

Rob Roggema and Anouk Roggema

Correction to:
R. Roggema, A. Roggema (eds.), *Smart and Sustainable*
Cities and Buildings,
https://doi.org/10.1007/978-3-030-37635-2

This book was inadvertently published with an incorrect version of Chapter 9, and which was incorrectly named as Chapter 11.

The correct version of Chapter 9 "Deep Renovation in Sustainable Cities: Zero Energy, Zero Urban Sprawl at Zero Costs in the Abracadabra Strategy" by Annarita Ferrante, Anastasia Fotopoulou, Lorna Dragonetti, and Giovanni Semprini is now included in the book.

The correct version of Chapter 11 "The Role of Smart City Initiatives in Driving Partnerships: A Case Study of the Smart Social Spaces Project, Sydney Australia" by Homa Rahmat, Nancy Marshall, Christine Steinmetz, Miles Park, Christian Tietz, Kate Bishop, Susan Thompson, and Linda Corkery, is now included in the book.

The book has been repaginated due to these changes.

The updated online version of the book can be found at
https://doi.org/10.1007/978-3-030-37635-2

Index

© Springer Nature Switzerland AG 2020   663
R. Roggema, A. Roggema (eds.), *Smart and Sustainable Cities and Buildings*,
https://doi.org/10.1007/978-3-030-37635-2